Lecture Notes in Artificial Intelli

Edited by R. Goebel, J. Siekmann, and W. Wa

Subseries of Lecture Notes in Computer Science

John Darzentas   George A. Vouros
Spyros Vosinakis   Argyris Arnellos (Eds.)

Artificial Intelligence: Theories, Models and Applications

5th Hellenic Conference on AI, SETN 2008
Syros, Greece, October 2-4, 2008
Proceedings

 Springer

Series Editors

Randy Goebel, University of Alberta, Edmonton, Canada
Jörg Siekmann, University of Saarland, Saarbrücken, Germany
Wolfgang Wahlster, DFKI and University of Saarland, Saarbrücken, Germany

Volume Editors

John Darzentas
University of the Aegean, Department of Product and Systems Design Engineering
84100 Syros, Greece
E-mail: idarz@aegean.gr

George A. Vouros
University of the Aegean, School of Sciences
Department of Information and Communication Systems Engineering
83200 Samos, Greece
E-mail: georgev@aegean.gr

Spyros Vosinakis
University of the Aegean, Department of Product and Systems Design Engineering
84100 Syros, Greece
E-mail: spyrosv@aegean.gr

Argyris Arnellos
University of the Aegean, Department of Product and Systems Design Engineering
84100 Syros, Greece
E-mail: arar@aegean.gr

Library of Congress Control Number: 2008935555

CR Subject Classification (1998): I.2, I.5.1, D.3.2, D.3.3, I.4

LNCS Sublibrary: SL 7 – Artificial Intelligence

ISSN 0302-9743
ISBN 978-3-540-87880-3 Springer Berlin Heidelberg New York

Springer is a part of Springer Science+Business Media

springer.com

© Springer-Verlag Berlin Heidelberg 2008

Typesetting: Camera-ready by author, data conversion by Scientific Publishing Services, Chennai, India
Printed on acid-free paper   SPIN: 12530045   06/3180   5 4 3 2 1 0

Preface

Artificial intelligence (AI) is a dynamic field that is constantly expanding into new application areas, discovering new research challenges and facilitating the development of innovative products. Today's information overload and rapid technological advancement raise needs for effective management of the complexity and heterogeneity of knowledge, for intelligent and adaptable man–machine interfaces and for products and applications that can learn and take decisions by themselves. Although the mystery of human-level intelligence has just started to be uncovered in various interdisciplinary fields, AI is inspired by the respective scientific areas to explore certain theories and models that will provide the methods and techniques to design and develop human-centered applications that address the above-mentioned needs.

This volume contains papers selected for presentation at the 5th Hellenic Conference on Artificial Intelligence (SETN 2008), the official meeting of the Hellenic Society for Artificial Intelligence (EETN). Previous conferences were held at the University of Piraeus (1996), at the Aristotle University of Thessaloniki (2002), at the University of the Aegean (2004) and at the Institute of Computer Science at FORTH (Foundation for Research and Technology - Hellas) and the University of Crete (2006).

SETN conferences play an important role in the dissemination of innovative and high-quality scientific results in AI which are being produced mainly by Greek scientists in institutes around the world. However, the most important aspect of SETN conferences is that they provide the context in which AI researchers meet, get to know each other's work, as well as a very good opportunity for students to get closer to the results of innovative AI research.

SETN 2008 was organized by the Hellenic Society of Artificial Intelligence and the Department of Product and Systems Design Engineering of the University of the Aegean. The conference took place in Hermoupolis, on the island of Syros, during October 2–4, 2008. We wish to express our thanks to the sponsors of the conference, the University of the Aegean, the Department of Product and Systems Design Engineering, the Prefecture of the Cyclades and the Holy Metropolis of Syros.

The program of the conference included presentations of research results and applications from distinguished Greek and foreign scientists in institutes all over the world; 38% of the submitted papers were authored or co-authored by non-Greek researchers. Each submitted paper was evaluated by two independent reviewers of the Program Committee, with the invaluable help of additional reviewers, on the basis of their relevance to AI, originality, significance, technical soundness and presentation. The selection process was hard, with only 27 papers out of the 76 submitted being accepted as full papers. In addition, another 17 submissions were accepted as short papers. We must emphasize the high quality of the majority of the submissions and we would also like to thank all those who submitted papers for review and for publication in the proceedings.

This proceedings volume also includes two prestigious invited papers presented at SETN 2008 by the two distinguished keynote speakers:

- "Grounding Concrete Motion Concepts with a Linguistic Framework" by Yiannis Aloimonos of the University of Maryland, USA
- "Emotion in Cognitive Systems Architectures" by Tom Ziemke of the University of Skövde, Sweden

Finally, SETN 2008 included in its program an invited session on "Applications of AI Research in Engineering Design," chaired by Argiris Dentsoras.

The editors would like to thank all those who contributed to the success of SETN 2008. In particular, they would like to express their gratitude to the Organizing Committee, for implementing the conference schedule in a timely and flawless manner, the Steering Committee, for its assistance and support, the Program Committee and the additional reviewers, for the time and effort spent in reviewing the papers, and the two invited speakers for their kind participation. Special thanks go to Alfred Hofmann and Ursula Barth and the Springer team for their continuous help and support.

October 2008

John Darzentas
George Vouros
Spyros Vosinakis
Argyris Arnellos

Organization

SETN 2008 was organized by the Department of Product and Systems Design Engineering of the University of the Aegean and EETN (Hellenic Association of Artificial Intelligence).

Conference Chair

John Darzentas   University of the Aegean, Greece

Conference Co-chair

George Vouros   University of the Aegean, Greece

Organizing Committee

Argyris Arnellos   University of the Aegean, Greece
Toula Dafna   University of the Aegean, Greece
Ariadne Filippopoulou   University of the Aegean, Greece
Anastasia-Evaggelia Papadopoulou   University of the Aegean, Greece
Thomas Spyrou   University of the Aegean, Greece
Spyros Vosinakis   University of the Aegean, Greece
Ioannis Xenakis   University of the Aegean, Greece

Steering Committee

Grigoris Antoniou   University of Crete and FORTH, Greece
Ioannis Hatzilygeroudis   University of Patras, Greece
Themistoklis Panayiotopoulos   University of Piraeus, Greece
Ioannis Vlahavas   Aristotle University, Greece

Program Committee

Christos-Nikolaos Anagnostopoulos   University of the Aegean, Greece
Ioannis Anagnostopoulos   University of the Aegean, Greece
Ion Androutsopoulos   Athens University of Economics and
                                            Bussiness, Greece
Dimitris Apostolou   University of Piraeus, Greece
Argyris Arnellos   University of the Aegean, Greece
Alexander Artikis   NCSR "Demokritos", Greece

Nikolaos Avouris	University of Patras, Greece
Grigorios Beligiannis	University of Ioannina, Greece
Nick Bassiliades	Aristotle University of Thessaloniki, Greece
Basilis Boutsinas	University of Patras, Greece
Yannis Dimopoulos	University of Cyprus, Cyprus
Christos Douligeris	University of Piraeus, Greece
Theodoros Evgeniou	INSEAD, France
Nikos Fakotakis	University of Patras, Greece
Eleni Galiotou	Technological Educational Institution of Athens, Greece
Efstratios Georgopoulos	TEI Kalamata, Greece
Manolis Gergatsoulis	Ionian University, Greece
Katerina Kabassi	University of Piraeus, Greece
Dimitris Kalles	CTI, Greece
Nikos Karacapilidis	University of Patras, Greece
Vangelis Karkaletsis	NCSR "Demokritos", Greece
Dimitrios Karras	Chalkis Institute of Technology, Greece
Sokratis Katsikas	University of Piraeus, Greece
Petros Kefalas	City Liberal Studies, UK
Yiannis Kompatsaris	CERTH, Greece
Stasinos Konstantopoulos	NCSR "Demokritos", Greece
Constantine Kotropoulos	Aristotle University of Thessaloniki, Greece
Manolis Koubarakis	Technical University of Crete, Greece
Panayiotis Koutsambasis	University of the Aegean, Greece
Michail Lagoudakis	Technical University of Crete, Greece
Aristidis Likas	University of Ioannina, Greece
Spyridon Likothanassis	University of Patras, Greece
Ilias Maglogiannis	University of the Aegean, Greece
George Magoulas	Birkbeck College, University of London, UK
Filia Makedon	Dartmouth College, USA
Bill Manaris	College of Charleston, South Carolina, USA
Vassilis Moustakis	FORTH, Greece
George Paliouras	NCSR "Demokritos", Greece
Ioannis Papadakis	Ionian University, Greece
Christos Papatheodorou	Ionian University, Greece
Dimitris Plexousakis	FORTH, Greece
Aristodemos Pnevmatikakis	AIT, Greece
George Potamias	FORTH, Greece
Ioannis Pratikakis	NCSR "Demokritos", Greece
Ioannis Refanidis	University of Macedonia, Greece
Timos Sellis	Technical University of Athens, Greece
Kyriakos Sgarbas	University of Patras, Greece
John Soldatos	AIT, Greece
Andreas Stafylopatis	Technical University of Athens, Greece
Efstathios Stamatatos	University of the Aegean, Greece
Panayiotis Stamatopoulos	University of Athens, Greece
Kostas Stergiou	University of the Aegean, Greece

Ioannis Tsamardinos	FORTH, Greece
Giorgos Tsekouras	University of the Aegean, Greece
Nikos Vasilas	TEI Athens, Greece
Michalis Vazirgiannis	Athens University of Economics and Bussiness, Greece
Maria Virvou	University of Piraeus, Greece
Spyros Vosinakis	University of the Aegean, Greece

Additional Reviewers

George Anastassakis
Nick Avradinis
Yiorgos Chrysanthou
Evangelos Dermatas
Peter Dietz
Alex Duffy
Katerina Fountoulaki
Dimitris Gavrilis
Nikos Papayannakos
Georgios Papadakis
Duc Pham
Emmanouil Psarakis
Modestos Stavrakis
Tetsuo Tomiyama
Dimitrios Vogiatzis
Elias Zavitsanos

Table of Contents

Invited Talks

Full Papers

Short Papers

Grounding Concrete Motion Concepts with a Linguistic Framework

Gutemberg Guerra-Filho[1] and Yiannis Aloimonos[2]

[1] Department of Computer Science and Engineering,
University of Texas at Arlington,
Nedderman Hall, Arlington, TX, USA 76019
guerra@cse.uta.edu
[2] Department of Computer Science,
University of Maryland,
A.V. Williams Bldg, College Park, MD, USA 20742
yiannis@cfar.umd.edu

Abstract. We have empirically discovered that the space of human actions has a linguistic framework. This is a sensorimotor space consisting of the evolution of the joint angles of the human body in movement. The space of human activity has its own phonemes, morphemes, and sentences formed by syntax. This has implications for the grounding of concrete motion concepts. We present a Human Activity Language (HAL) for symbolic non-arbitrary representation of visual and motor information. In phonology, we define basic atomic segments that are used to compose human activity. We introduce the concept of a kinetological system and propose basic properties for such a system: compactness, view-invariance, reproducibility, and reconstructivity. In morphology, we extend sequential language learning to incorporate associative learning with our parallel learning approach. Parallel learning solves the problem of overgeneralization and is effective in identifying the kinetemes and active joints in a particular action. In syntax, we suggest four lexical categories for our Human Activity Language (noun, verb, adjective, adverb). These categories are combined into sentences through specific syntax for human movement.

Keywords: concrete concept grounding, linguistic framework, sensorimotor intelligence.

1 Introduction

For the cognitive systems of the future to be effective, they need to be able to share with humans a conceptual system. Concepts are the elementary units of reason and linguistic meaning. A commonly held philosophical position is that all concepts are symbolic and abstract and therefore should be implemented outside the sensorimotor system.

This way, meaning for a concept amounts to the content of a symbolic expression, a definition of the concept in a logical calculus. This is the viewpoint that elevated AI to the mature scientific and engineering discipline it is today.

There exists however an alternative which states that concepts are grounded in sensorimotor representations. This sensorimotor intelligence considers sensors and motors in the shaping of the cognitive hidden mechanisms and knowledge incorporation.

J. Darzentas et al. (Eds.): SETN 2008, LNAI 5138, pp. 1–12, 2008.

There exists a variety of studies in many disciplines (neurophysiology, psychophysics, cognitive linguistics) suggesting that indeed the human sensorimotor system is deeply involved in concept representations. In this paper, we investigate this involvement in concept description and the structure in the space of human actions.

In the sensorimotor intelligence domain, our scope is at the representation level for human activity. We are not concerned with visual perception, nor with actual motor generation. This way, we leave the issues of motion capture from images and computation of torque at joints to Computer Vision and Robotics, respectively.

We discovered that human action has the structure of a formal language, with its own phonology, morphology and syntax. In this paper, we show how we could obtain this language using empirical data. The availability of a language characterizing human action has implications with regard to the grounding problem, to the origin of human language, and its acquisition/learning process.

The paper follows with a brief review of representative related work. We introduce the concept of kinetology with its basic properties and the symbolization process. The morphology is described through sequential and parallel language learning. Syntax concerns lexical categories for a Human Activity Language (HAL) and syntactic rules constraining sentence formation. In the conclusion, we summarize our main results, speculate about a new way of achieving AI, and indicate future research.

2 Related Work

The functionality of Broca's region in the brain [8] and the mirror neurons theory [3] suggests that perception and action share the same symbolic structure as a knowledge that provides common ground for sensorimotor tasks (*e.g.*, recognition and motor planning) and higher-level activities.

Furthermore, spoken language and visible movement use a similar cognitive substrate based on the embodiment of grammatical processing. There is evidence that language is grounded on the motor system [4], what implies the possibility of a linguistic framework for a grounded representation.

The sensorimotor projection of primitive words leads to language grounding. Language grounding for verbs has been addressed by Siskind [9] and Bailey *at al.* [1] from the perspective of perception and action, respectively.

In our linguistic framework, we aim initially to find movement primitives as basic atoms. Fod *at al.* [2] find primitives by k-means clustering the projection of high-dimensional segment vectors onto a reduced subspace. Kahol *at al.* [5] use the local minimum in total body force to detect segment boundaries. In Nakazawa *at al.* [6], similarities of motion segments are measured according to a dynamic programming distance and clustered with a nearest-neighbor algorithm. Wang *at al.* [11] segment gestures with the local minima of velocity and local maxima of change in direction. The segments are hierarchically clustered into classes by using Hidden Markov Models to compute a metric. A lexicon is inferred from the resulting discrete symbol sequence through a language learning approach.

Language learning consists in grammar induction and structure generalization. Current approaches [7, 10, 12] accounts only for sequential learning by combining statistics and rules. In this paper, we extend language learning to consider parallel learning which is related to associative learning.

3 Kinetology: The Phonology of Human Movement

A first process in our linguistic framework is to find structure in human movement through basic units akin to phonemes in spoken language. These atomic units are the building blocks of a phonological system for human movement which we refer as kinetological system. We propose this concept of kinetology, where a kinetological system consists in a representation of 3D movement, a specification of atomic states (segmentation and symbolization), and satisfies some basic principles. We introduce principles on which such a system should be based: compactness, view-invariance, reproducibility, and reconstructivity. Without loss of generality, we use human walking gait to illustrate and evaluate one proposed kinetological system. Actual movement data of human walking gait is analyzed.

3.1 Compactness and View-Invariance

The compactness principle is related to describing a human activity with the least possible number of atoms in order to decrease complexity, improve efficiency, and allow compression.

We define a state according to the sign of derivatives of the original 3D motion (joint angle function). The derivatives used in our segmentation are velocity and acceleration. This way, four states correspond to all possible sign combinations.

A view-invariant representation provides the same 2D projected description of an intrinsically 3D action even when captured from different viewpoints. View-invariance is desired to allow visual perception and motor generation under any geometric configuration in the environment space.

A compactness/view-invariance (CVI) graph for a DOF shows the states associated with the movement according to two dimensions: time and viewpoint (see Fig. 1). In order to evaluate the compactness and view-invariance, a circular surrounding configuration of viewpoints is used. For each time instant (horizontal axis) and for each viewpoint in the configuration of viewpoints (vertical axis), the movement state is associated with a representative color.

A compactness measurement consists in the number of segments when the movement varies with time. For each viewpoint, the compactness measurement is plotted on the left side of the CVI graph. The view-invariance measurement concerns the fraction of the most frequent state among all states for all viewpoints at a single instant in time. For each time instant, the view-invariance measure is computed and plotted on the top of the CVI graph.

Note that the view-invariance measure is affected by some uncertainty at the borders of the segments and at degenerate viewpoints (see Fig. 1). The border effect shows that movement segments are not completely stable during the temporal transition between segments. The degenerate viewpoints are special cases of frontal views where the sides of a joint angle tend to be aligned.

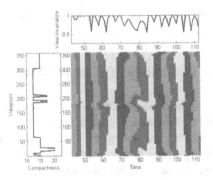

Fig. 1. Compactness/View-Invariance Graph

3.2 Reproducibility

An important requirement for a kinetological system is the ability to represent actions exactly in the same way even facing inter-personal or intra-personal variability. A kinetological system is reproducible when the same symbolic representation is associated with the same action performed by different subjects.

Each segment corresponds to an atom α, where α is a symbol associated with the segment's state. The atomic symbols (**R, Y, B, G**), called kinetemes, are the phonemes of our kinetological system. The symbol **R** is assigned to negative velocity and negative acceleration segments; the symbol **Y** is assigned to negative velocity and positive acceleration segments; the symbol **B** is assigned to positive velocity and positive acceleration segments; and the symbol **G** is assigned to positive velocity and negative acceleration segments (see Fig. 2).

A reproducibility measure is computed for each joint angle considering a gait database. The reproducibility measure of a joint angle is the fraction of the most frequent symbolic description among all descriptions for the database files. A very high reproducibility measure means that symbolic descriptions match among different gait performances and the kinetological system is reproducible.

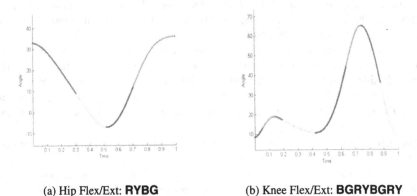

(a) Hip Flex/Ext: **RYBG**   (b) Knee Flex/Ext: **BGRYBGRY**

Fig. 2. Symbolic representation of joint angle functions

The reproducibility measure is very high for the joint angles which play a primary role in the walking action (see Fig. 3). Using our kinetological system, six joint angles obtained very high reproducibility: pelvic obliquity, hip flexion-extension, hip abduction-adduction, knee flexion-extension, foot rotation, and foot progression. These variables seem to be the most related to the movement of walking forward. Other joint angles obtained only a high reproducibility measure which is interpreted as a secondary role in the action: pelvic tilt and ankle dorsi-plantar flexion. The remaining joint angles had a poor reproducibility rate and seem not to be correlated to the action: pelvic rotation, hip rotation, knee valgus-varus, and knee rotation.

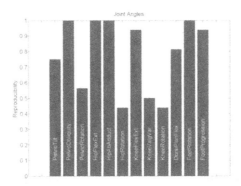

Fig. 3. Reproducibility measure for 12 DOFs during gait

Our kinetological system performance on the reproducibility measure for all the joint angles shows that the system is reproducible for the DOFs intrinsically related to the action. Further, the system is useful in the identification (unsupervised learning) of the variables playing primary roles in the activity. The identification of the intrinsic variables of an action is a byproduct of the reproducibility requirement of a kinetological system. That will be further explored in the morphology analysis.

3.3 Reconstructivity

Reconstructivity is associated with the ability to approximately reconstruct the original movement signal from a compact representation. The reconstructivity property provides visuo-motor ambivalence which denotes the capacity of performing both visual perception and motor generation tasks by using the same representation. Once the movement signal is segmented and converted into a non-arbitrary symbolic representation, this compact description is only useful if we are able to recover the original joint angle function or an approximation.

In order to use a sequence of kinetemes for reconstruction, we consider one segment at a time and concentrate on the state transitions between two consecutive segments. Based on transitions, we determine constraints about the derivatives at border points of the segment. Derivatives will have zero value (equation) or a known sign (inequality) at these points.

Let's assume that the signs of velocity and acceleration don't change simultaneously. This way, each segment can have only two possible states for a next neighbor segment. However, the transition **B → Y (R → G)** is impossible, since velocity cannot become negative (positive) with a positive (negative) acceleration. Therefore, each of the four segment states has only two possible state configurations for previous and next segments and, consequently, there are eight possible state sequences for three consecutive segments. Each possible sequence corresponds to two equations and two inequality constraints associated with first and second derivatives at border points. Other two inequalities come from the derivatives at interior points of the segment.

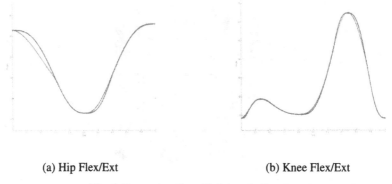

(a) Hip Flex/Ext   (b) Knee Flex/Ext

Fig. 4. Reconstruction of joint angle functions

A simple model for the joint angle function during a segment is a polynomial. However, low degree polynomials don't satisfy the constraints originated from the possible sequences of kinetemes. For example, a cubic function has a linear second derivative which is impossible for sequences where the second derivative assumes zero value at the borders and non-zero values at interior points (e.g. GBG). The least degree polynomial satisfying the constraints imposed by all possible sequences of kinetemes is a fourth degree polynomial. This way, the reconstruction process consists in finding the five parameters defining this polynomial with the two associated equations for the particular sequence of kinetemes.

However, this involves an under-constrained linear system and, consequently, additional constraints are required to reconstruct the polynomial modeling the joint angle function in a segment. We introduce two more equations using the joint angle value at the two border points.

With four equations, a linear system is solved up to one variable. This last free variable is constrained by four inequalities and it can be determined using some criteria such as jerk (third derivative) minimization (see Fig. 4).

3.4 Symbolization

The kinetological segmentation process results into atoms with two parameters: angular displacement and time length. A kineteme is represented as $\alpha(d, t)$, where α denotes the atomic state, d corresponds to the quantized angular displacement, and t is the discrete time length.

The atoms observe some natural variability on the quantitative parameters (d, t). Our goal is to identify the same kineteme amidst this variability. A classification process associates each kineteme with a class/cluster which is denoted with a literal. Due to a low-dimensional parametric space, we use clustering instead of HMMs (high-dimensional) or PCAs (dimensional reductor).

Different from model-based probabilistic clustering, we use a generalized probabilistic clustering algorithm. Our algorithm partitions the 2D parameter domain into regions of any shape. The sample data for our clustering algorithm is several performances of a particular action. Our goal is to learn the kinetemes in this action. All kinetemes are obtained by executing the same process for a vocabulary of human actions.

Each joint angle is considered independently and the movement corresponding to a specific DOF is segmented into a sequence S of n atoms $\alpha_i(d_i, t_i)$. The input for our algorithm is this sequence of atoms. A class label is assigned to each atom such that the "same" kinetemes will be assigned to the same cluster.

Initially, we compute probability distributions over the 2D parameter space. We find one distribution P_α for each of the four possible states by considering only the atoms where $\alpha_i = \alpha$. Each atom contributes with the probability modeled as a Gaussian filter $h(n_1, n_2)$ centered at (d_i, t_i) with size $W_D \times W_T$ and standard deviation σ. This way, the probability distribution is defined as

$$P_\alpha(d,t) = \frac{1}{n} \sum_{\substack{i=1 \\ \alpha_i=\alpha}}^{n} \sum_{\substack{n_1=-W_D \\ d_i+n_1=d}}^{W_D} \sum_{\substack{n_2=-W_T \\ t_i+n_2=t}}^{W_T} h(n_1,n_2) \; . \tag{1}$$

Once the probability distribution is computed, each local maximum is associated with a class. This way, the number of clusters is selected automatically.

The partitioning of the parameter space is performed by selecting a connected region for each cluster (peak). For a cluster c with center peak p_c, we find the minimum value v_c such that the region r_c in the parameter space satisfying $P_\alpha(d, t) > v_c$ contains only the peak p_c and no other.

Each sample atom $\alpha_i(d_i, t_i)$ is assigned to the cluster c which maximizes the expected probability

$$e_c(d_i,t_i) = \sum_{n_1=-W_D}^{W_D} \sum_{n_2=-W_T}^{W_T} h(n_1,n_2) \bullet R_c(d_i + n_1, t_i + n_2) \; , \tag{2}$$

where R_c is a binary matrix specifying the connected region r_c corresponding to the cluster c. This way, an atom $\alpha_i(d_i, t_i)$ becomes $\alpha_i L_i$, where L_i is the literal associated with the assigned cluster. This simple probabilistic clustering algorithm is faster and uses a more general model than standard probabilistic clustering techniques.

4 Morphology

Morphology is concerned with the structure of words, the constituting parts, and how these parts are aggregated. In the context of human activity, morphology involves the structure of each action and the organization of a praxicon in terms of common subparts. A praxicon is a lexicon of human actions. This way, our methodology consists in determining the morphology of each action in a praxicon and then in finding the organization of the praxicon.

The morphology of an action includes the identification of its kinetemes, the selection of which joints are active during the action, and the ordering relation between kinetemes in different active joints.

4.1 Sequential Grammar Learning

In order to analyze the morphology of a particular action, we are given the symbolic representation for the movement associated with the concatenation of several performances of this action. For each joint angle, we apply a grammar learning algorithm to this representation, which ultimately consists in a single string of symbols instantiating the language to be learned. Each symbol in the string is associated with a time period.

Fig. 5. Grammar tree for the hip joint during "walk forward"

The learning algorithm induces a context-free grammar (CFG) corresponding to the structure of the string representing the movement. The algorithm is based on the frequency of digrams in the sequence. At each step i, a new grammar rule $N_i := AB$ is created for the most frequent digram AB of consecutive symbols and every occurrence of this digram is replaced by the non-terminal N_i in the sequence of symbols. This step is repeated until the most frequent digram in the current sequence occurs only once. Each non-terminal N_i is associated with the time period corresponding to the union of the periods of both symbols A and B. The CFG induced by this algorithm corresponds to a binary grammar tree which represents the structure of the movement (see Fig. 5).

4.2 Parallel Grammar Learning

The problem with this sequential learning is the overgeneralization that takes place when two unrelated non-terminals are combined in a rule. This happens mostly in higher-levels of the grammar tree, where the digram frequencies are low. In order to overcome this problem, we rely on parallel grammar learning which considers all joint angles.

A parallel grammar consists in a set of simultaneous CFGs related by parallel rules. A parallel rule between two non-terminals of different CFGs constrains these non-terminals to have an intersecting time period in the different strings generated by their respective CFGs. This grammar models a system with different strings occurring at the same time. In human activity, each string corresponds to the representation of the movement for one joint angle.

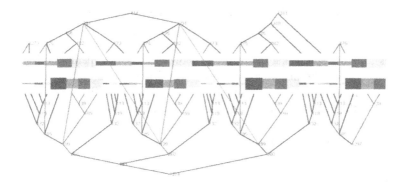

Fig. 6. Parallel grammar learning

The execution of a human action involves the achievement of some goal and, therefore, requires consistency in a single string (sequential grammar) and coordination among different strings (parallel grammar). This way, sequential and parallel grammar learning are combined to infer the morphology of a human action.

The parallel learning algorithm executes sequential learning considering all the joint angles simultaneously. The digram frequency is still computed within the string corresponding to each joint angle independently. When a new rule is created, the new non-terminal is checked for possible parallel rules with non-terminals in the CFGs of other joint angles.

A parallel rule relating two non-terminals in different CFGs is issued if there is a one-to-one mapping of their occurrences in the associated strings. Furthermore, any two mapped occurrences must intersect the corresponding time periods (see Fig. 6 and Fig. 7 for an example).

Initially, parallel rules are difficult to be issued for low-level non-terminals (closer to the leaves of the grammar tree) because they have a high frequency and some atom occurrences are spurious. However, high-level non-terminals are more robust and parallel rules are reliably created for them. This way, when a parallel rule is issued for

```
S1 :=  A B C D B C A D A B C D B C A D A B C D B C A D
S2 :=  B C A D A B C D B C A D A B C D B C A D A B C D
S3 :=  C D A B C D A B C D A B C D A B C D A B C D A B
S4 :=  A D A D C A B C A D B B D B C A C D C B B A A D
S5 :=  A B C D B C A D A B C D B C A D A B C D B C A D

5 :=  B C  [6 occurrences in S2]

S1 :=  A B C D B C A D A B C D B C A D A B C D B C A D
S2 :=  -5- A D A -5- D -5- A D A -5- D -5- A D A -5- D
S3 :=  C D A B C D A B C D A B C D A B C D A B C D A B
S4 :=  A D A D C A B C A D B B D B C A C D C B B A A D
S5 :=  A B C D B C A D A B C D B C A D A B C D B C A D

6 :=  C D  [6 occurrences in S3]
6 || 5   [Parallel Rule]

S1 :=  A B C D B C A D A B C D B C A D A B C D B C A D
S2 :=  -5- A D A -5- D -5- A D A -5- D -5- A D A -5- D
S3 :=  -6- A B -6- A B -6- A B -6- A B -6- A B -6- A B
S4 :=  A D A D C A B C A D B B D B C A C D C B B A A D
S5 :=  A B C D B C A D A B C D B C A D A B C D B C A D

7 :=  6 A  [6 occurrences in S3]
7 || 5   [Parallel Rule]

S1 :=  A B C D B C A D A B C D B C A D A B C D B C A D
S2 :=  -5- A D A -5- D -5- A D A -5- D -5- A D A -5- D
S3 :=  --7-- B --7-- B --7-- B --7-- B --7-- B --7-- B
S4 :=  A D A D C A B C A D B B D B C A C D C B B A A D
S5 :=  A B C D B C A D A B C D B C A D A B C D B C A D
```

Fig. 7. Parallel learning example

a pair of non-terminals A and B, their descendents in the respective grammar trees are re-checked for parallel rules. This time, considering only instances of these descendent non-terminals which are concurrent with A and B, respectively.

Besides formally specifying the relations between CFGs, the parallel rules are effective in identifying the maximum level of generalization for an action. Further, the set of joints related by parallel rules corresponds to the active joint angles concerned intrinsically with the action. The basic idea is to eliminate nodes of the grammar trees with no associated parallel rules and the resulting trees represent the morphological structure of the action being learned.

5 Syntax

In a sentence, a noun represents the subjects performing an activity or objects receiving an activity. A noun in a Human Activity Language (HAL) sentence corresponds to the body parts active during the execution of a human activity and to the possible objects involved passively in the action. The initial posture for a HAL sentence is analogous to an adjective which further describes (modifies) the active joints (noun) in the sentence. The sentence verb represents the changes each active joint experiences during the action execution. A HAL adverb is a string of multiplicative constants modeling the variation in the execution time of each segment.

The Subject-Verb-Object (SVO) pattern of syntax is a reflection of the patterns of cause and effect. An action is represented by a word that has the structure of a sentence: the agent or subject is a set of active body parts; the action or verb is the motion of those parts.

A HAL sentence $S := NP\ VP$ consists of noun phrase (noun + adjective) and verbal phrase (verb + adverb), where $NP := N\ Adj$ and $VP := V\ Adv$ (see Fig. 8). The organization of human movement is simultaneous and sequential. This way, the basic HAL syntax expands to parallel syntax and sequential syntax. The parallel syntax concerns simultaneous activities represented by parallel sentences $S_{t,j}$ and $S_{t,j+1}$ and constrains the respective nouns to be different: $N_{t,j} \neq N_{t,j+1}$. This constraint states that simultaneous movement must be performed by different body parts.

The temporal sequential combination of action sentences $S_{t,j}\ S_{t+1,j}$ must obey the cause and effect rule. The HAL noun phrase must experience the verb cause and the joint configuration effect must lead to a posture corresponding to the noun phrase of the next sentence. Considering noun phrases as points and verb phrases as vectors in the same space, the cause and effect rule becomes $NP_{t,j} + VP_{t,j} = NP_{t+1,j}$ (see Fig. 8). The cause and effect rule is physically consistent and embeds the ordering concept of syntax.

The lexical units are arranged into sequences to form sentences. A sentence is a sequence of actions that achieve some purpose. In written language, sentences are delimited by punctuation. Analogously, the action language delimits sentences using motionless actions. In general, a conjunctive action is performed between two actions, where a conjunctive action is any preparatory movement that leads to an initial position required by the next sentence.

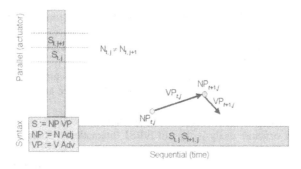

Fig. 8. Parallel and sequential syntax

6 Conclusion

In this paper, we presented a linguistic framework used for the grounding of AI through human activity. We introduced the concept of a kinetological system and proposed four basic properties for such a system: compactness, view-invariance, reproducibility, and reconstructivity. We also designed a new probabilistic clustering algorithm to perform the symbolization of atomic segments.

In morphology, we extended sequential language learning to incorporate associative learning with our parallel learning approach. Parallel learning solves the problem of overgeneralization and is effective in identifying the kinetemes and active joints in a particular action. Our parallel learning algorithm is also a result in natural language acquisition if used in this domain with agreement, a sort of parallel feature in speech, to resolve overgeneralization.

In syntax, we suggested four lexical categories for our Human Activity Language (noun, verb, adjective, adverb). These categories are combined into sentences through basic syntax and other specific syntax for human movement.

From a methodological viewpoint, our paper introduced a new way of achieving Artificial Intelligence through the study of human action, or to be more precise, through the study of the sensorimotor system. We believe this study represented initial steps towards the grounding of AI. The closure of this semantic gap will lead to the foundation of concepts into a non-arbitrary meaningful symbolic representation based on sensorimotor intelligence. This representation will serve to the interests of AI in higher-level tasks and open the way to more effective techniques with powerful applications.

For future work, we intend to construct a praxicon (vocabulary) and explore its morphological organization towards the discovery of more structure in the human activity language. We also expect more development concerning human movement syntax from the empirical study of this praxicon.

References

1. Bailey, D., Chang, N., Feldman, J., Narayanan, S.: Extending Embodied lexical development. In: 20th Annual Meeting of the Cognitive Science Society (1997)
2. Fod, A., Mataric, M., Jenkins, O.: Automated derivation of primitives for movement classification. Autonomous Robots 12(1), 39–54 (2002)

3. Gallese, V., Fadiga, L., Fogassi, L., Rizzolatti, G.: Action recognition in the premotor cortex. Brain 119(2), 593–609 (1996)
4. Glenberg, A., Kaschak, M.: Grounding language in action. Psychonomic Bulletin & Review 9(3), 558–565 (2002)
5. Kahol, K., Tripathi, P., Panchanathan, S.: Automated gesture segmentation from dance sequences. In: IEEE International Conference on Automatic Face and Gesture Recognition, pp. 883–888 (2004)
6. Nakazawa, A., Nakaoka, S., Ikeuchi, K., Yokoi, K.: Imitating human dance motions through motion structure analysis. In: IEEE/RSJ International Conference on Intelligent Robots and Systems, pp. 2539–2544 (2002)
7. Nevill-Manning, C., Witten, I.: Identifying hierarchical structure in sequences: A linear-time algorithm. Journal of Artificial Intelligence Research 7, 67–82 (1997)
8. Nishitani, N., Schurmann, M., Amunts, K., Hari, R.: Broca's region: from action to language. Physiology 20, 60–69 (2005)
9. Siskind, J.: Grounding the lexical semantics of verbs in visual perception using force dynamics and event logic. Journal of Artificial Intelligence Research 15, 31–90 (2001)
10. Solan, Z., Horn, D., Ruppin, E., Edelman, S.: Unsupervised learning of natural languages. Proceedings of National Academy of Sciences 102(33), 11629–11634 (2005)
11. Wang, T.-S., Shum, H.-Y., Xu, Y.-Q., Zheng, N.-N.: Unsupervised analysis of human gestures. In: IEEE Pacific Rim Conference on Multimedia, pp. 174–181 (2001)
12. Wolff, J.: Learning syntax and meanings through optimization and distributional analysis. In: Levy, Y., Schlesinger, I., Braine, M. (eds.) Categories and processes in language acquisition, pp. 179–215. Lawrence Erlbaum Associates, Inc., Hillsdale (1988)

Emotion in Cognitive Systems Architectures

Tom Ziemke

University of Skövde
School of Humanities & Informatics
PO Box 408, 54128 Skövde, Sweden
tom.ziemke@his.se

Abstract. Inspired by theories of situated and embodied cognition, much AI and cognitive systems research today, autonomous robotics in particular, is biologically inspired, although at very different levels of abstraction. This ranges from explicitly biomimetic approaches, concerned with biological plausibility and scientific modeling, to the use of abstractly biologically inspired techniques such as evolutionary and neural computing in engineering. However, it is quite clear that even today the state of the art in cognitive systems research hardly lives up to the original (e.g. Brooksian) ambition to build *complete* robotic autonomous agents. The 'incompleteness' is particularly evident in the somewhat superficial treatment of the issue of *embodiment* (and its reduction to the physical body's sensorimotor interaction with the environment) as well the widespread neglect of *emotion* (or more broadly, *affect*) and its grounding in homeostatic/bioregulatory processes. This talk discusses work in the European project *ICEA – Integrating Cognition, Emotion and Autonomy* (cf. www.iceaproject.eu). The twofold hypothesis behind the project is (1) that the emotional and bioregulatory mechanisms that come with the organismic embodiment of living cognitive systems also play a crucial role in the constitution of their "high-level" cognitive processes, and (2) that models of these mechanisms can be usefully integrated in artificial cognitive systems architectures, which will constitute a significant step towards more autonomous robotic cognitive systems capable of dealing with issues such as energy management, self-monitoring, self-repair, etc. A crucial question addressed in the ICEA project is how in hierarchically organized, layered embodied cognitive architectures lower-level (e.g. bodily, homeostatic) mechanisms can modulate processing at higher levels through emotional/affective mechanisms.

J. Darzentas et al. (Eds.): SETN 2008, LNAI 5138, p. 13, 2008.

Application of Naturalistic Decision Making to Emergency Evacuation Simulations

Fatemeh Alavizadeh, Behzad Moshiri, and Caro Lucas

Control and Intelligent Processing Center of Excellence
School of ECE, University Of Tehran, Tehran, Iran
f.alavi@ece.ut.ac.ir,
moshiri@ut.ac.ir,
lucas@ut.ac.ir

Abstract. Evacuation simulations are among essential analysis to assess the level of life safety in buildings. Currently, a number of evacuation modeling tools are available, each with unique characteristics. But, complexity of human behaviors made most of these tools ignore behavioral data in their models; and among those models which account for human behaviors, rule based modeling and rational decision making approaches are used for decision process. However, to achieve more realistic evacuation calculations, human-like decision making should be integrated to evacuation simulations. In this research we present a novel approach to build multi-agent evacuation simulations that exhibit more realistic simulation behaviors. We propose a model that integrates individual differences, such as personality, age, and gender, with naturalistic decision making, where human decision making in natural environments is investigated. The resulted decision process is applied to BDI agent architecture within a building evacuation system.

1 Introduction

Understanding human social behavior in emergency contexts and applying it to evacuation models is a crucial issue concerning human lives. Many injuries and consequences of a disaster are resulted from human social behaviors such as overcrowding, pushing, rushing, and herding during emergency situations. However, instability in the theoretical issues and complexity of humans decision process made most of current simulation tools ignore behavioral data which made them unreliable or unrealistic.

On the other hand, most agent-based human models use rational choice for decision making, whereby all possible choices are considered and ranked, and the highest ranked plan is chosen. But, evidence suggests that people rarely use this approach in their decisions. Specifically in emergency contexts panic and stress can affect decision making process, and rational decision approaches may lead to unreliable simulations. Therefore, the field of study called "naturalistic decision making" that offers several descriptive models of human decision making in different sorts of situations, should be investigated for agent models of humans [24].

J. Darzentas et al. (Eds.): SETN 2008, LNAI 5138, pp. 14–25, 2008.

Although NDM is a great step toward realistic human models, people are different in their decision process. A great part of this difference is due to individual differences such as personality, gender, and age. While studies [16] validated the predictive value of personality on decision outcomes, consideration of personality is missed from current simulation tools. Therefore, in this work, we propose a variation of current naturalistic decision making strategies that accounts for individual differences. In our model, mentioned individual differences are applied to NDM and the resulted decision model is integrated to BDI agent architecture.

Personality traits have been studied for a long time. Recently, some works are done on geographic distribution of personality traits [17,21]. Therefore, based on known personality distribution of a society, the emergent social behavior would be predictable which can effectively enhance related systems. For instance, based on the specific trends in a culture, disaster response management systems can predict behavior of people and train forces accordingly. Also, by different simulations based on the dominant personality of a culture, buildings can be constructed with more effective exit plans.

2   Related Works

Before exploring our model we should briefly discuss related topics. First we present a brief review of available projects on emergency evacuation; then we will discuss theoretical studies on NDM and its application to human decision modeling. Finally, we will present a brief review on personality traits and studies which are concerned with the impact of individual differences on decision process.

2.1   Emergency Response and Evacuation Analysis

Evacuation modeling is an open research topic in many fields of knowledge such as crowd management, fire safety engineering, and operation research. Different evacuation models are proposed in past decades like Steps [11], Evacnet4 [12], Simulex [8], GridFlow [10], buildingExodus [7], Crisp [13], DrillSim [3], Legion [27], and MassEgress [6]. Three main architectures which are used in these systems are flow-based modeling, cellular automata, and agent-oriented modeling [5]. In flow based model, a network of nodes is considered where each node represents a physical structure and certain nodes are designated as destination nodes; graph theories will be applied to this model. In cellular automata node density in shown in individual floor cells. The evacuees are modeled as individuals on a grid, and individuals move from one cell to another in series of time steps. Agent-oriented models usually account for individuality. The key characteristic of these models is the fact that occupants are modeled as independent modules with different levels of awareness and intelligence.

Most of these models are focused on movement and physical aspects of simulation rather than human behavior. Among those which accounts for behavior, Legion, Simulex and MassEgress are more similar to this work. In Legion [27], occupants are intelligent individuals and social, physical, and behavioral characteristics are assigned probabilistically. The social characteristics include gender, age, culture, and pedestrian type. The behavioral characteristics include memory, willingness to adapt, and preferences for unimpeded walking speeds, personal space, and acceleration. Simulex [8] is another example which assigns a certain set of attributes to each person, such as physical motions and gestures, the proximity of other evacuees, and the influence of gender and age. The program assumes the presence of a rational agent able to assess the optimal escape route. Finally MassEgress [6] is also similar to this work in accounting for individual differences such as gender and age. Decision making process of MassEgress is a rule based process which is not concerned with context or personality.

The main differences between our model and discussed systems are consideration of personality as well as application of NDM rather that rational choice which is more feasible in human modelings. Moreover, in our model, while some occupants are familiar with exit plans, most of them have bounded knowledge and are unaware of exit directions which is a common situation in public buildings such as libraries or museums. Therefore, cooperation between individuals with behaviors like answering other people or announcing the exit paths, has great effect on evacuation which will be captured during simulation. These interactions are also modeled based on individual differences.

2.2   Naturalistic Decision Making

Naturalistic decision making established as a methodological and theoretical perspective since the first naturalistic decision-making conference in September 1989. In general, researchers observed that people often did not make optimal decisions so they began to investigate how and why deviations occurred. Therefore, rather than focusing on how people ought to make decisions in complicated situations, NDM investigates how people actually make decisions in such situations [23]. Different conditions in which naturalistic decision making is typically employed are discussed in work of Orasanu and Connolly [9]. They suggested that NDM is well suited in ill-structured problems where uncertain conditions and vague goals exists. Moreover, time pressure and high stakes as well as team and organizational constraints are among other situations leading to NDM. Clearly, emergency can be a typical example of these conditions.

Perhaps the best-known naturalistic decision-making theory is Kleins recognition-primed decision model [24]. The model describes how experts use their experiences to arrive at decisions quickly and without the computational analysis of traditional normative decision-making approaches. RPD employs situation assessment to generate a likely course of action and then uses mental simulation to envision and evaluate the course of action.

Since Naturalistic Decision Making process and Belief-Desire-Intention architecture are both inspired by human behavior process, they can be integrated naturally. Where agents perceives the surroundings; then they select between available options based on individual differences and mental assessment of selection, and acts on the environment. Finally, the agent evaluate his selection over time and may change it in case of failure or difficulties. In [24] authors discussed three different approaches to integrate RPD into BDI architecture. In the first approach, they consider an expert agent able to recognize all possible situations. In the second approach, agents have a preference weighting on plans. And finally, in the last approach a form of learning would be used to adjust plan context. In our work we proposed a variation that integrates preference model and learning model with consideration of individual differences in both phases.

2.3 Personality Traits

Personality has been conceptualized from a variety of theoretical perspectives, and at various levels of abstraction or breadth. One of available general taxonomies of personality traits is the "Big Five" personality dimensions [15,14] which were derived from analysis of the natural-language terms people use to describe themselves over decades. While the first public mention of this model was by Thurstone in 1933; In 1981, available personality tests of the day were reviewed by four prominent researchers; they concluded that the tests which held the most promise measured a subset of five common factors. This event was followed by widespread acceptance of the five factor model among personality researchers. These factors are often called *Openness, Conscientiousness, Extraversion, Agreeableness,* and *Neuroticism.*

Openness to Experience describes a dimension of personality that distinguishes imaginative and creative people from conventional people. Open people are intellectually curious and appreciative of art. They therefore tend to hold unconventional and individualistic beliefs. Conscientiousness concerns the way in which people control, regulate, and direct their impulses. Highly conscientiousness people have strong control of their impulses. They achieve high levels of success through purposeful planning and persistence. Extraversion is marked by engagement with the external world. Extraverts enjoy being with people; they often experience positive emotions. Introverts tend to be quiet, and less dependent on the social world. Agreeableness concerns with cooperation and social harmony. Agreeable individuals are optimistic, considerate, friendly, helpful, and willing to compromise with others. Disagreeable individuals place self-interest above getting along with others. Neuroticism refers to the tendency to experience negative emotions. People high in Neuroticism are emotionally reactive. They respond emotionally to events that would not affect most people, and their reactions tend to be more intense than normal. At the other end of the scale, individuals who score low in Neuroticism tend to be calm, emotionally stable, and free from persistent negative feelings [14,15].

Studies on big five personality traits show that even though these dimensions are extracted from English dictionaries they are valid in many other languages such as German, Dutch, Japanese, and so on [15]. Also there are studies on geographical distribution of these traits among more than 50 nations [17,21]. In addition, many studies have confirmed the predictive value of the Big Five across a wide range of behaviors. For instance, in the area of job performance, in [20] Barrick and Mount found that conscientiousness showed consistent relations with all performance criteria for all occupational groups.

Although no theoretical research was found on the impact of big five model on emergent behaviors in disaster, considering the defined individual level actions: "Moving to nearest exit door", "following the crowd", and "moving to the less crowded door" risk taking behavior has a strong impact in action selection. In [18] it is argued that openness and extraversion have the most impact on risk taking behavior.

3   Proposed Model

The proposed model is a multi-agent framework for simulation of human and social behavior during emergency evacuations. What distinguish this model from existing systems are application of naturalistic decision making based on individual differences and adaptation of this decision model to BDI agent architecture. A multi-agent simulation framework is a computational methodology that allows building an artificial environment populated with autonomous agents, which are capable of interacting with each other. Such a framework is particularly suitable for simulating individual cognitive processes for exploring emergent phenomena such as social or collective behaviors [22]. BDI architecture has been used in this model. A BDI agent is a particular type of bounded rational software agent, with particular mental attitudes such as Beliefs, Desires and Intentions.

On the other hand, as discussed earlier, NDM considers 4 stages:

- Situation awareness: In this phase agent senses the environment. Theoretically, in this step the raw data from physical models is transformed into more complex symbolic descriptors.
- Situation assessment: After sensing the environment, agent tries to recognize the situation based on the output of the previous step.
- Tactics selection: Agent selects a tactic based on the current set of goals and the situation.
- Operating procedures: After sensing the environment, recognizing the situation, and deciding on the action, agent performing the action and evaluate it.

This model can be naturally integrated to BDI architecture[24]. The difference between common BDI architecture and NDM is in tactic selection phase. In BDI, the agent must reason about beliefs, goals, and intentions. In theory, agent uses bounded rationality to select a plan. However, in NDM model [24], agent has a preference weighting on plans. Basically, all plans would be equally weighted. When an agent was faced with a choice between several applicable plans, it would pick the one with the highest preference weighting or randomly choose from those with the highest preference weighting. If the plan succeeded, the preference weighting would increase, and if it failed it would decrease [24].

However, considering equal preferences on the plans also seems unrealistic. People seldom think of different options equally. This is the reason why people act differently from each other in a same situation over and over. Therefore, in our model, agents have different preference weighting on plans. These preferences are calculated based on personality, age, and gender. Thus, when an agent faces with a choice between several applicable plans, it would pick the one that better matches his personality and characteristics. While the agent is following his selected plan, he will monitor the results; the preference weighting of the selected plan will increase in case he gets in a better situation otherwise the preference will decrease. When the weight of the plan becomes lower than a threshold the agent would switch to the second highest rated plan. Yet, we should consider that even pace of this learning process, the actual value of the switching threshold, should be calculated based on individual characteristics.

In this model, at micro level, each individual has the choice between possible behaviors such as "Moving to nearest exit door", "moving to the less crowded door", "following the crowd", and "moving with relatives". At Macro level, from individuals' choices emerges social behaviors such as "Queuing", "Herding", and "Pushing". For behavior selection process, a weight will be assigned to each dimension which makes a personality vector on facets of the big five factors: Neuroticism, Extroversion, Agreeableness, Conscientiousness, and Openness. This merged with gender and age would be a base for individual's heightening schema. Defined behaviors and their descriptions are as follows:

Finding Exit door by following the signs: This behavior encourages searching for exit doors by looking for exit signs or following known exit paths.

Moving to less crowded exit doors: In case the agent face with more than one exit path another decision should be made about moving to least crowded path or the most crowded one.

Moving with the crowd: This behavior, which inspires moving in the direction where most people are going, is known as flocking or herding, and it is a phenomenon that is highly probable in emergency contexts where agents affected by stress are reluctant to be alone. Herding usually leads to over-utilization of some exit doors while others are not utilized [6].

Moving with relatives: This behavior suggests that an agent may give up a plan since his related neighbor has decided to do the other alternative. for instance a child which would have the intention to follow exit sign probably gives up his plan and follows his father.

For each behavior a selection likelihood vector is defined on the big five dimensions. Gender and age are two additional dimensions of this vector. An individual would select the behavior that has the least weighted distance from his own personality vector. Behavior vectors and dimension weights are defined based on [19,18] which discussed big five and risk taking behavior. For instance, finding your own way compared to moving with the crowd has a significant level of risk taking which means higher levels of openness and extraversion. Yet, we should consider that in emergency context neuroticism plays a major rule since it makes people panic and leads to irrational behaviors. So a common sense guides us to consider this dimension as well. Moreover, as finding exit path is a kind of planning conscientiousness may also play a part. But emergency context makes the impact of conscientiousness on decision making less than that of a normal context, so we assigned a lower weigh to conscientiousness. At the doors individuals decide whether to push others in a competitive manner or to queue for exit. Queuing is an effective behavior that can enhance evacuation significantly; thus these individuals' choices at social level can affect evacuation process. The second behavior is cooperation with others. Based on personality some people tend to help others with behaviors such as announcing the exit directions. Yet, others affected by personality or stress context may even not answer questions about the directions.

4   Implementation

The implemented tool is a grid world representation of a building written in Java where walls, doors, and exit signs are read from a text file containing the building plan. Population can be generated based on gender, age, mean, and standard deviation of a personality trait in a specific culture, which are provided in [17,21], or randomly. Each agent has a specific body type ranged from thin to fat. Also based on this body type and the age a movement speed will be assigned to each agent. Each Cell has specific capacity and agents can enter the cell in case it has room for their body type. In each simulation iteration agents decide on their next move, thus they can change their plan in each time tick. The simulation ends whenever all people exit the building.

Each agent has a specific sight power which is assigned in a randomly in a range based on age. Moreover, similar to cellular automata approaches, the agent has interaction with neighbors and gets information from these agents. The agent scans his visible area to see whether he sees a door or not; if not, he will decide whether to move with the crowd or to look for the exit signs. In first case, he will ask his neighbors for their direction. Based on personality these neighbors may answer or not. Then the agent will follow the direction that most of neighbors are following. If the individual decides to follow signs, he will look for a sign in his visible area. However, based on relation of the agent and his neighbors, the agent may give up his own decision and follow relatives. Age and closeness of relations are two important factors for this action. But, moving with crowd or following the sign may not be possible. For example, the surrounding cells may be full of people or there may be no knowledge of paths, or no sign or crowd. In these cases, the individual will walk randomly.

Since each cell has a specific capacity individuals would know whether their front cell is crowded or not. So they can decide whether to enter the cell or not. This affects their approach toward the doors where they have the chance to queue or push. Door's performance is related to the amount of people in surrounding cells. If the cell at a door is full or over-crowded, only one agent can exit in a time tick. But if it's moderately crowded with half of its capacity then all people in that cell can exit. This can simulate the negative impact of pushing at the doors. Figure 1 shows a screenshot of the system.

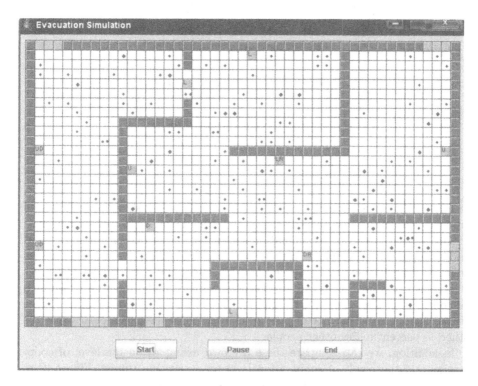

Fig. 1. Agents are shown based on their body type. Doors are shown in light gray, and exit signs are shown with the direction letters: U, D, L, R.

5   Results

Here we present results for three different personality distributions which are shown in table 1. Based on these personality distributions, and random assignment of age and gender following results are gathered.

Table 1. Average in each dimension on 0 to 99 scale

	Extraversion	Agreeableness	Conscientiousness	Neuroticism	Openness	SD
crowd1	46	47	52	53	41	10
crowd2	55	51	48	46	57	10
crowd3	50	50	50	50	50	10

As it can be seen in table 1, in first crowd, people are scored higher on neuroticism and score lower in extraversion and openness. These dimensions make us expect that these people tend to stay with the crowd during the evacuation since they get nervous easily and they are not open to risk taking. The average values of 10 simulation executions also shows that only 15 percent of people tend

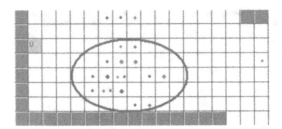

Fig. 2. Emergence of group movement

to follow their own paths by searching for sign. Figure 2 shows an example of group movement as a sample of emergent herding behavior in this simulation.

The average values of 10 simulation executions for second crowd shows that about 75 percent of people tend to follow their own paths by searching for sign which is expected due to higher levels of openness and extraversion and lower level of neuroticism. The third crowd scores average in all dimension which help us consider it as a base for our comparisons. Simulations with average values show that about 60 percent of people tend to follow their own path while the other 40 percent follow other people.

In addition, we can compare these 3 crowd examples on the level of cooperation and the result of cooperation level on evacuation timing. Figure 3 investigates the effect of cooperation on evacuation among the third crowd. This diagram compares the evacuation timing for different cooperation percentages. For instance, if no one help other people by announcing exit paths or answering others it would take about 500 iteration to evacuate the building. However, in same simulation conditions when all people cooperate this value reaches to 140.

Another important issue that should be considered is the evacuation timing for different population types. Figure 4 compares evacuation timing for mentioned crowds, based on different initial simulation populations. This diagram shows that the best evacuation timing, regardless of initial population, happens

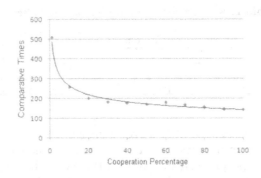

Fig. 3. Effect of cooperation

when about 70 percent of people move in their own direction. We can also conclude that herding is not an unwanted behavior in all conditions, it can lead to unwanted situations in highly crowded buildings but when the building is not highly populated and people are mostly unfamiliar with exit paths, which is the case in public buildings evacuation such as museums and libraries, herding may help people exit faster since it represents a type of cooperation that let people share their knowledge when the knowledge is limited. Another important factor is placement of exit signs and the way they guide people through exits. Different placements of signs should be applied to different social behaviors. For instance, when people tend to follow the crowd, signs should be placed in the way that it divides population equally to different exit paths. However, when people tend to follow the sign, better solution is to place the exit signs in the way that they guide people to nearest door.

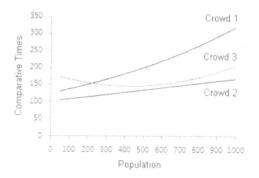

Fig. 4. Evacuation Timing

6   Conclusions and Future Works

Here a system for realistic models of human decision process based on individual differences such as personality, age, and gender is proposed. We have applied personality-based naturalistic decision making to BDI agent architecture in emergency evacuation case study; a simulation environment has been developed in which each human perceives the surroundings, decides how to behave according to his individuality and acts on the environment. We present evacuation simulation for three sample crowds. Also, simulations can be done for specific nations based on values of [17,21]

As a part of future work, individuality should be considered in other stages of NDM. For instance, people in certain age may scan their environment and evaluate their decisions more precisely than others. Also in most evacuation cases there are some people who are familiar with exit paths. Their decision process is different from those who are unfamiliar with the path as discussed in this work.

Therefore, a good extension to current system would be addition of this type of decision process. Group formation is another part of our future work that better reflects the herding behavior.

References

1. Gilbert, N.: Agent-based social simulation: dealing with complexity. University of Surrey (2004)
2. Jain, S., McLean, C.: An architecture for modeling and simulation of emergency response. In: Proceedings of the IIE conference (2004)
3. Balasubramanian, V., Massaguer, D., Mehrotra, S., Venkatasubramanian, N.: Drill-Sim: A Simulation Framework for Emergency Response Drills. ISI (2006)
4. Kuligowski, E.D., Peacock, R.D.: Review of Building Evacuation Models, Rep. No. NIST TN 1471 (2005)
5. Santos, G., Aguirre, B.E.: Critical Review of Emergency Evacuation Simulation Models. In: Workshop on Building Occupant Movement During Fire Emergencies, pp. 27–52. National Institute of Standards and Technology, Gaithersburg (2005)
6. Pan, X., Han, C.S., Law, K.H.: A Multi-agent Based Simulation Framework for the Study of Human and Social Behavior in Egress Analysis. In: The International Conference on Computing in Civil Engineering, Cancun (2005)
7. Gwynne, S., Galea, E.R., Lawrence, P., Filippidis, L.: Modelling Occupant Interaction with Fire Conditions Using the building EXODUS Evacuation Model, University of Greenwich (2000)
8. Thompson, P.A., Wu, J., Marchant, E.W.: Modelling Evacuation in Multi-storey Buildings with Simulex. Fire Engineering 56, 7–11 (1996)
9. Orasanu, J., Connolly, T.: The reinvention of decision making, Decision Making in Action: Models and Methods, pp. 3–20. Ablex Publishing Corporation, Greenwich (1993)
10. Bensilum, M., Purser, D.A.: Gridflow: an object-oriented building evacuation model combining pre-movement and movement behaviours for performance-based design. In: 7th International Symposium on Fire Safety Science Worcester, Worcester Polytechnic Institute, MA (2002)
11. MacDonald, M.: STEPS Simulation of Transient Evacuation and Pedestrian Movements User Manual, unpublished Work (2003)
12. Kisko, T.M., Francis, R.L., Nobel, C.R.: EVACNET4 User's Guide, University of Florida (1998)
13. Boyce, K., Fraser-Mitchell, J., Shields, J.: Survey Analysis and Modelling of Office Evacuation Using the CRISP Model. In: Shields, T.J. (ed.) Proceedings of the 1st International Symposium on Human Behaviour in Fire, pp. 691–702 (1998)
14. Lim, M.Y., Aylett, R., Jones, C.M.: Emergent Affective and Personality Model, vol. IVA, pp. 371–380 (2005)
15. John, O.P., Srivastava, S.: The Big-Five trait taxonomy: History, measurement, and theoretical perspectives. In: Handbook of personality: Theory and research, vol. 2, pp. 102–138. Guilford Press, New York (1999)
16. Roberts, B., Kuncel, N., Shiner, R., Caspi, A., Goldberg, L.: The Power of Personality: The Comparative Validity of Personality Traits, Socioeconomic Status, and Cognitive Ability for Predicting Important Life Outcomes. Perspectives on Psychological Science 2(4), 313–345 (2007)

17. Reips, S., et al.: Geographic distribution of Big Five personality traits: Patterns and profiles of human self-description across 56 nations. Journal of Cross-Cultural Psychology 38(2), 173–212 (2007)
18. Maryann, G.: High Rolling Leaders: The Big Five Model of Personality and Risk Taking During War. Annual meeting of the International Studies Association, San Diego, California, USA (2006)
19. Lee, J., Deck, C., Reyes, J., Rosen, C.: Measuring Risk Attitudes Controlling for Personality Traits. Working Papers, Florida International University, Department of Economics (2008)
20. Barrick, M.R., Mount, M.K.: The big five personality dimensions and job performance: A meta-analysis. Personnel Psychology 44, 1–26 (1991)
21. McCrae, R.R., Terracciano, A.: Personality profiles of cultures: Aggregate personality traits. Journal of Personality and Social Psychology 89(3), 407
22. Robinson, S.: Modelling Human Decision-Making. In: Proceedings of the 17th European Simulation Multiconference, pp. 448–455 (2003)
23. Shattuck, L.G., Miller, N.L.: Naturalistic Decision Making in Complex Systems: A Dynamic Model of Situated Cognition Combining Technological and Human Agents. Organizational Behavior: Special Issue on Naturalistic Decision Making in Organizations (in press, 2005)
24. Norling, E., Sonenberg, L., Ronnquist, R.: Enhancing multi-agent based simulation with human-like decision-making strategies. In: Moss, S., Davidsson, P. (eds.) MABS 2000. LNCS (LNAI), vol. 1979, pp. 214–228. Springer, Heidelberg (2001)
25. Collier, N.: RePast: An extensible framework for agent simulation (2003)
26. Minar, N., Burkhart, R., Langton, C., Askenazi, M.: The Swarm Simulation System, A Toolkit for Building Multi-Agent Simulations (1996)
27. Legion International L, http://www.legion.biz/system/research.cfm

A Multi-agent Environment for Serving Proof Explanations in the Semantic Web

Grigoris Antoniou[1], Antonis Bikakis[1], Polyvios Damianakis[1], Mixalhs Foukarakis[1],
Giorgos Iacovidis[1], Marianna Karmazi[1], Haridimos Kondylakis[1], Antreas Makridakis[1],
Giorgos Nikiforos[1], Grigoris Papadourakis[1], Manolis Papoutsakis[1],
Aggeliki Psyharaki[1], Giorgos Stratakis[1], Panagiotis Tourlakis[1], Petros Tsialiamanis[1],
Giorgos Vasileiadis[1], Gerd Wagner[2] and Dimitris Velegrakis[1]

[1] Computer Science Department, University of Crete,
Institute of Computer Science, FORTH-ICS
kondylak@ics.forth.gr
[2] Cottbus University of Technology

Abstract. In this work we present the design and implementation of a multi-agent environment for serving proof explanations in the Semantic Web. The system allows users or agents to issue queries, on a given RDF& rules knowledge base and automatically produces proof explanations for answers produced by a popular programming system (JENA), by interpreting the output from the proof's trace and converting it into a meaningful representation. It also supports an XML representation (a R2ML language extension) for agent communication, which is a common scenario in the Semantic Web. The system in essence implements a proof layer for rules on the Semantic Web empowering trust between agents and users.

Keywords: Multi-agent Systems, Semantic Web, Proof Explanation.

1 Introduction

The evolvement of World Wide Web is changing the way people communicate with each other and the way business is conducted [1]. At the heart of this revolution is the Semantic Web. The Semantic Web is an evolving extension of the World Wide Web in which the meaning of information and services on the web is defined, making it possible for the web to understand and satisfy the requests of people and machines to use web content [5].

The development of the Semantic Web proceeds in steps, each step building a layer on top of another. The pragmatic justification for this approach is that it is easier to achieve consensus on small steps, while it is much harder to get everyone on board if too much is attempted [1].

The next step in the development of the Semantic Web will be the logic and proof layers. The implementation of these two layers will allow the user to state any logical principles, and permit the computer to infer new knowledge by applying these principles on the existing data. Rule systems appear to lie in the mainstream of such activities.

J. Darzentas et al. (Eds.): SETN 2008, LNAI 5138, pp. 26–37, 2008.

Besides rule systems, the upper levels of the Semantic Web have not been researched enough and contain critical issues, such as accessibility, trust and credibility.

Specifically for the proof layer, little has been written and done. In this work we describe the implementation of a multi-agent environment for serving query explanations. The main difference between a query posed to a traditional database system and a semantic web system is that the answer in the first case is returned from a given collection of data, while for the semantic web system the answer is the result of a reasoning process.

While in some cases the answer speaks for itself, in other cases the user will not be confident in the answer unless s/he can trust the reasons why such an answer has been produced. In addition, it is envisioned that the semantic web is a distributed system with disparate sources of information. Thus a semantic web answering system, to gain the trust of a user must be able, if required, to provide an explanation or justification for an answer. Since the answer is the result of a reasoning process, the justification can be given as a derivation of the conclusion with the sources of information for the various steps. Our system is capable of presenting such explanations to users for the answers it computes in response to their queries.

More specifically, the main contributions of this paper are that:

- We implemented a proof layer in the Semantic Web.
- We designed and implemented a multi-agent environment for serving proof explanations.
- We extended R2ML in order to support proof explanations. This extension is also used for the communication between agents.
- The system produces proof explanations for answers produced by a popular programming system (JENA), by interpreting the output from the proof's trace and converting it into a meaningful representation.

The rest of the paper is organized as follows: In Section 2 we discuss related work while in Section 3 we present the whole system architecture. Then, Section 4 describes the extension of R2ML used for representing proofs. The entire workflow of a query in our system is presented in Section 5. Finally Section 6 concludes the paper and gives directions for further research.

2 Related Work

Many recent studies have focused on the integration of rules and ontologies, including DLP [9], [14] and [17], TRIPLE [20] and SWRL [11]. Ongoing standardisation work includes the RuleML Markup Initiative [18], the REWERSE Rule Markup language [23] and the Rule Interchange Format (RIF) W3C Working Group. Apart from classical rules that lead to monotonic logical systems, recently researchers started to study systems capable of handling conflicts among rules and reasoning with partial information. Recently developed non-monotonic rule systems for the Semantic Web include DR-Prolog [2], DR-DEVICE [4], SweetJess [10] and dlvhex [6]. Rule-based expert systems have been very successful in applications of AI, and from the beginning, their designers and users have noted the need for explanations in their recommendations.

However, besides teaching logic [3], a little effort has been devoted on providing explanation in reasoning systems so far. In expert systems like [19] and Explainable Expert System [22], a simple trace of the program execution / rule firing appears to provide a sufficient basis on which to build an explanation facility and they generate explanations in a language understandable to its users. Work has also been done in explaining the reasoning in description logics [15]. The authors present a logical infrastructure for separating pieces of logical proofs and automatically generating follow-up queries based on the logical format.

The most prominent work on proofs in the Semantic Web context is Inference Web [16]. The Inference Web (IW) is a Semantic Web based knowledge provenance infrastructure that supports interoperable explanations of sources, assumptions, learned information, and answers as an enabler for trust. It supports provenance, by providing proof metadata about sources, and explanation, by providing manipulation trace information. It also supports trust, by rating the sources about their trustworthiness. What is still missing in Inference Web is support for more types of explanations, e.g. explanations for *why not*, or *how to* queries. Moreover the current IW infrastructure cannot support explanations in negative answers about predicates.

3 System Architecture

The architecture of our system is shown in Figure 1 and consists of four main modules. The GUI, the Agens, the Jena to R2ML translator and the reasoning engine which is the Jena 2 Reasoning Engine.

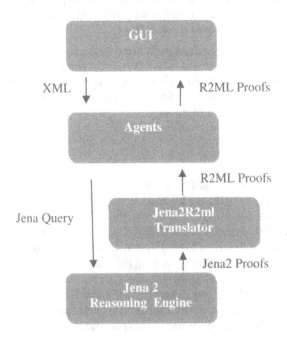

Fig. 1. System Architecture

3.1 GUI

In the top layer of our system stands the graphical user interface (GUI). The GUI provides a simple and intuitive interface to issue queries and to receive and visualize the corresponding answers.

Users can specify:

1. the name of the agent that they want to query,
2. the query to be answered and
3. Whether they want a simple true or false answer or a proof explanation as well.

The user issues queries of the form: *Predicate (atom1, atom2)*. The *"Predicate"* is a valid Jena predicate followed by one, or two, atomic expressions. Those atomic expressions are referring to objects inside our knowledge base.

```
<?xml version="1.0" encoding="UTF-8"?>
<message>
        <queryType>
                        Proof
        </queryType>
        <queryBody>
                        cous(markos, pakos)
        </queryBody>
</message>
```

Fig. 2. The XML format of a simple Query

The GUI transforms the queries into XML [7] format (a simple example is depicted in Figure 2) and sends them to the agent that the user has selected.

The user can request for a simple true/false answer (*"queryType"* is *"Answer"*), or a proof explanation (*"queryType"* is *"Proof"*). In case the user has requested a proof explanation, the XML tree representing the derivation process followed by the reasoning engine is returned. In case of failure the appropriate error messages are presented to the user.

An image showing a typical use case of the graphical inreface is shown in Figure 3. The user has issued the query *"cous (markos, pakos)"* wishing to know whether *markos* and *pakos* are *cousins*. The answer computed by the reasoning engine is *true* and the proof for this answer is presented to the user. The answer is *true* since *bernardos* and *manwlas* are *brothers*, *manwlas* is the *parent* of *macros* and *bernardos* is the *parent* of *pakos*.

The proof is rendered as a tree structure, in which each node represents a single predicate. A tree node may have child nodes that represent the simpler, lower level, predicates that are triggered by the evaluation of the parent predicate. Thus, the leaf nodes represent the lowest level predicates of the proof system, which correspond to the basic atoms of a theory (facts and derivation rules) which can not be further explained.

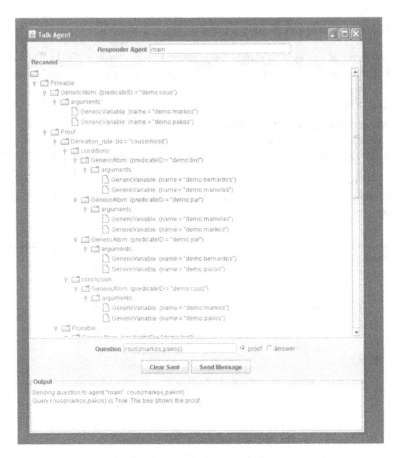

Fig. 3. The Graphical User Interface

3.2 Agents

The next layer of the system architecture comprises the agents which interface with the reasoning system and are responsible for providing responses to the queries issued through the GUI. Agents are based on Java Agent DEvelopment Framework (JADE version 3.4) [12]. JADE simplifies the implementation of multi-agent systems through a middle-ware that complies with the Foundation for Intelligent Physical Agents (FIPA) [8] specifications. The agent platform can be distributed across machines and the configuration can be controlled via a remote GUI. The configuration can be even changed at run-time by moving agents from one machine to another one, as and when required.

The agents are the endpoint for user queries, and the client which invokes the reasoning engine which will answer them. They extract the predicate from the query and send them to the reasoning engine with the right parameters in order to be executed. In order to provide a meaningful facility for exchanging messages between agents and support potential extensions in user queries, we use the simple XML format already presented in Figure 2. Each agent uses the Java Runtime Environment to invoke:

a) the reasoning engine which will accept the queries and compute the answers to them,

b) the tool which will transform the answer returned by the reasoning engine to a meaningful XML document. This XML document is either dispatched to another agent in order to be further processed or returned to the GUI in order to be visualized.

3.3 Jena2R2ML Translator

The main purpose of this component is to translate Jena2 proofs to R2ML proof representations that will be used by the Agents. The tool is implemented in Java 6 and takes as input a text file which contains the Jena2 proofs and translates it in an equivalent R2ML proof. A detailed description about proof representation in R2ML is available in Section 4.

A simple Jena proof is shown in Figure 4, which is the response to the query depicted in Figure 2. This is trace from the Jena2 reasoning process. A proof output form Jena2 contains only the conclusions of the rules used to explain an answer. But in order to justify each rule we have to be able to check both the conditions and the conclusions. Moreover the R2ML proof representation requires both of them. One solution would be to process the theory file(s) given as input to the Jena2 tool.

```
Statement is
     [http://jena.hpl.hp.com/demo#pakos,
      http://jena.hpl.hp.com/demo#cous,
      http://jena.hpl.hp.com/demo#markos] ------> TRUE

Rule cousinhood concluded
        (demo:pakos demo:cous demo:markos) <-
--- Rule symmetricRule concluded
    (demo:manwlas demo:bro demo:bernardos) <-
--- --- Rule symmetricRule concluded
        (demo:bernardos demo:bro demo:manwlas) <-
--- --- --- Fact (demo:manwlas demo:bro demo:bernardos)
            already shown
--- Fact (demo:bernardos demo:par demo:pakos)
--- Rule fatherhood concluded
    (demo:manwlas demo:par demo:markos) <-
--- --- Rule symmetricRule concluded
        (demo:manos demo:bro demo:markos) <-
--- --- --- Rule symmetricRule concluded
            (demo:markos demo:bro demo:manos) <-
--- --- --- --- Fact (demo:manos demo:bro demo:markos)
                already shown
--- --- Fact (demo:manwlas demo:par demo:manos)
```

Fig. 4. Jena Proof

Alternatively, we could extract the rule conditions from Jena2 Proof graphs. We adopted the second solution since it is more elegant. Using the first one would involve exchange of files between agents apart from the R2ML answers.

To implement this solution, we build the graph of the proof in memory an then we extract the conditions of a rule using the graph structure. We can find the rule conditions since they are available in the layer above the conclusion of the rule. In the previous example, the rule with conclusion *"Rule cousinhood concluded (demo:pakos demo:cous demo:markos)"* has as conditions two rules *"Rule symmetricRule concluded (demo:manwlas demo:bro demo:bernardos)"*, *"Rule fatherhood concluded (demo:manwlas demo:par demo:markos)"* and one Fact *"Fact (demo:bernardos demo:par demo:pakos)"* (Figure 4).

3.4 Reasoning Engine

Our choice for the reasoning engine was Jena2 [13]. Jena is a Java framework for building Semantic Web applications. It provides a programmatic environment for RDF, RDFS and OWL, SPARQL and includes a rule-based inference engine.

It has a key RDF package the API of which has been defined in terms of interfaces so that application code can work with different implementations without change. This package contains interfaces for representing models, resources, properties, literals, statements and all the other key concepts of RDF, and a ModelFactory for creating models.

The Jena2 inference subsystem is designed to allow a range of inference engines or reasoners to be plugged into Jena. Such engines are used to derive additional RDF assertions, which are entailed from some base RDF, together with any optional ontology information and the axioms and rules associated with the reasoner. The primary use of this mechanism is to support the use of languages, such as RDFS and OWL, which allow additional facts to be inferred from instance data and class descriptions. However, the machinery is designed to be quite general and, in particular, it includes a generic rule engine that can be used for many RDF processing or transformation tasks.

There are many different kinds of reasoners some of which are:

- Transitive reasoner
- RDFS rule reasoner
- OWL, OWL Mini, OWL Micro Reasoners
- DAML micro reasoner
- Generic rule reasoner.

In our environment we used the generic rule reasoner. A rule processed by the rule-based reasoner is defined by a Java Rule object with a list of body terms (premises), a list of head terms (conclusions) and an optional name and optional direction. The difference between the forward and backward rule syntax is only relevant for the execution strategy.

The rules are written in the form of:

[rule: (?A pre:property ?B), (?B pre: property ?C) -> (?A pre: property ?C)].

"pre:" defines a prefix which can be used in the rules. The prefix is local to the rule file .The rules are written in a rule file (shown in Figure 5) which is imported in the code. For example the rule (1) denotes that if A is brother of B and B of C then A is borther of C.

```
(1)[transitiveRule:(?A  demo:bro  ?B),(?B  demo:bro  ?C)->(?A
demo:bro ?C)]
(2)[fatherhood:(?A demo:bro ?B),(?C demo:par ?A)->(?C
demo:par ?B)]
(3)[unclehood:(?A demo:bro ?B),(?B demo:par ?C)->(?A
demo:uncl ?C)]
(4)[cousinhood:(?A demo:bro ?B),(?B demo:par ?C), (?A
demo:par ?D)->(?C  demo:cous ?D)]
```

Fig. 5. Jena Proof

The facts (data) are as well written in a rdf file. An example file presenting the facts used in our example is shown in Figure 5.

4 R2ML Extension

R2ML stands for The REWERSE II Rule Markup Language [23]. It is a comprehensive and user-friendly XML format that allows interchanging rules between different systems and rules.We describe R2ML as comprehensive because it integrates:

- the *Object Constraint Language (OCL)* - a standard used in information systems engineering and software engineering,
- the *Semantic Web Rule Language (SWRL)* - a proposal to extend the Semantic Web ontology language OWL by adding implication axioms,
- the *Rule Markup Language (RuleML)*- a proposal based on Datalog/Prolog,

Moreover, it includes four rule categories: derivation rules, production rules, integrity rules and ECA/reaction rules. It is a *usable* language in the sense that it allows structure-preserving markup and does not force users to translate their rule expressions into a different language paradigm such as having to transform a derivation rule into a FOL axiom, an ECA rule into a production rule, a function into a predicate, or a typed atom into an untyped atom.

We used R2ML for proof representation as it follows the goal of providing a general rule markup language to make the deployment, execution, publishing, and communication of rules on the Web possible. But R2ML can only represent rules and not proofs. We extended R2ML in order to support proofs and those extensions are depicted in Figure 5.

A "Fact" comprises certain knowledge. It is a complex type which consists of a sequence of "*GenericAtoms*", as defined in R2ML. The element "*Definitely_provable*" is also a complex type consists of a sequence of "*GenericAtoms*", as described in

```
<rdf:RDF xmlns:rdf="&rdf;" xmlns:rdfs="&rdfs;"
         xmlns:demo="&demo;" xmlns="&demo;">

<demo:brother rdf:about="&demo;bro" /> // property bother
<demo:parent rdf:about="&demo;par" />  // property parent
<demo:uncle rdf:about="&demo;uncl" />  // property uncle
<demo:cousin rdf:about="&demo;cous" /> // property cousin

 <rdf:Description rdf:about="&demo;manos">
    <bro rdf:resource="&demo;nikiforos" />
 </rdf:Description>      // manos brother of nikiforos

 <rdf:Description rdf:about="&demo;maria">
   <par rdf:resource="&demo;manos" />
 </rdf:Description>     // maria parent of manos

  <rdf:Description rdf:about="&demo;manwlas">
     <par rdf:resource="&demo;manos" />
  </rdf:Description>  // manwlas parent of manos

  <rdf:Description rdf:about="&demo;manos">
     <bro rdf:resource="&demo;markos" />
  </rdf:Description>  // manos brother of markos

  <rdf:Description rdf:about="&demo;manwlas">
     <bro rdf:resource="&demo;bernardos" />
  </rdf:Description>  // manwlas brother of bernardos

  <rdf:Description rdf:about="&demo;manwlas">
     <par rdf:resource="&demo;nikiforos" />
  </rdf:Description>  // manwlas parent of nikiforos

  <rdf:Description rdf:about="&demo;bernardos">
     <par rdf:resource="&demo;pakos" />
  </rdf:Description>  // bernardos parent of pakos

</rdf:RDF>
```

Fig. 6. Jena Proof

R2ML and *"Proofs"*, as described in the same XML schema. Finally, a *"Definite Proof"* is also a complex type. In order something to be a *"Definite_Proof"* it must be consisted of a sequence of *"DerivationRules"* and *"Definitely_provable"* elements or to be a *"Fact"*.

Moreover, we extended R2ML in order to support defeasible proofs. The defeasible proofs were not finally used and are not presented here since Jena2 produces only definite rules. Having the extension we were able to build the translator that could translate proofs output from Jena2 to the extended R2ML.

```
<!-- Fact -->
<xs:element name="Fact">
    <xs:complexType>
        <xs:sequence>
            <xs:element ref="r2ml:GenericAtom" />
        </xs:sequence>
    </xs:complexType>
</xs:element>

<!-- Definitely_provable -->
<xs:element name="Provable">
    <xs:complexType>
        <xs:sequence>
            <xs:element ref="r2ml:GenericAtom"/>
            <xs:element ref="r2mlext:Proof"/>
        </xs:sequence>
    </xs:complexType>
</xs:element>

<!-- Definite_Proof -->
<xs:element name="Proof">
    <xs:complexType>
        <xs:choice>
            <xs:sequence>
                <xs:element
                    ref="r2ml:DerivationRule"/>
                <xs:element ref="r2mlext:Provable"
                    minOccurs="0"
                    maxOccurs="unbounded" />
            </xs:sequence>
            <xs:element ref="r2mlext:Fact"/>
        </xs:choice>
    </xs:complexType>
</xs:element>
```

Fig. 7. R2ML extension to support proofs

We used the previous version of R2ML 0.4 as the current 0.5 is still a beta version and the new version of Jena2. For this reason we created an XML schema that represents the interrelationship between the attributes and elements of our R2ML objects.

5 Query Workflow

Having presented the main components, in this section we show a whole workflow in detail in order to return a response to a given query.

1. A user issues a query using the GUI to be executed by an agent.
2. GUI transforms the query into XML and sends it to the agent.
3. The agent sends the query to the Jena2 system. The query is extracted from the XML message and passed onto the Jena2 engine for execution.

4. After the predicate is received, the Jena2 system executes it against the knowledge base and provides an answer in a text format. Note that the answer is pruned to only provide the tree of rules which led to the query being true.

5. After an answer is obtained from the Jena2 engine, a translator tool is employed to convert the answer into the extended R2ML format, to be further used in our agent environment. An XML structure which describes the proof process is then supplied in the form of a file.

6. After the XML result is obtained, it is sent to the GUI agent which made the original request together with a value that defines whether the answer to the query is True, False, or an Error in case some error occurred during the process. Types of errors may include, but are not limited to:

   - The predicate not being syntactically correct
   - The query predicate is not defined at our knowledge base
   - The type of response not being defined (something other than Answer or Proof)
   - Some exception that occurs while computing the result at the Jena2 engine, and has no relation to the reasoning process (system failures and crashes, etc.)

7. The GUI receives the answer and visualizes it.

6 Conclusion

This work presents a proof system that aims to increase the trust of the users in the Semantic Web applications. The system automatically generates an explanation for every answer it computes in response to user's queries, in a formal and useful representation. This can be used by individual users who want to get a more detailed explanation from a reasoning system in the Semantic Web, in a more human readable way. Also, an explanation could be fed into a proof checker to verify the validity of a conclusion; this is important in a multi-agent setting. Our reasoning system is based on monotonic rules as it is implemented on top of the Jena reasoning engine. We developed a tool that reads the trace from Jena and produces a R2ML graph in order to formulate a sensible proof. Furthermore, the system can be used by agents, a common feature of many applications in the Semantic Web. Another contribution of our work is a R2ML extension for a formal representation of an explanation using definite logic. Additionally, we provide a web style representation for the facts that is an optional reference to a URL. We expect that our system can be used by multiple applications, mainly in E-commerce and agent-based applications.

There are interesting ways of extending this work. The explanation can be improved to become more intuitive and human-friendly, to suit users unfamiliar with logic. The XML Schema should be made fully compatible with the latest version of R2ML. Finally, integration with the Inference Web infrastructure will be explored. It is an interesting and open issue how our implemented proof system could be registered in the Inference Web, so as to produce PML proofs (an OWL-based specification for documents representing both proofs and proof meta-information). This would possibly require the registration of our inference engine.

References

1. Antoniou, G., van Harmelen, F.: A Semantic Web Primer. MIT Press, Cambridge
2. Antoniou, G., Bikakis, A.: DR-Prolog: A System for Defeasible Reasoning with Rules and Ontologies on the Semantic Web. IEEE Transactions on Knowledge and Data Engineering (accepted for publication)
3. Barwise, J., Etchemendy, J.: The language of first-order logic. Center for the study of Language and Information (1993)
4. Bassiliades, N., Antoniou, G., Vlahavas, I.P.: Dr-device: A defeasible logic system for the semantic web. In: Ohlbach, H.J., Schaffert, S. (eds.) PPSWR 2004. LNCS, vol. 3208, pp. 134–148. Springer, Heidelberg (2004)
5. Berners-Lee, T., Hendler, J., Lassila, O.: The Semantic Web, May 17, 2001. Scientific American Magazine (2001)
6. Eiter, T., Ianni, G., Schindlauer, R., Tompits, H.: dlvhex: A System for Integrating Multiple Semantics in an Answer-Set Programming Framework. In: WLP, pp. 206–210 (2006)
7. Extensible Markup Language, http://www.w3.org/XML/
8. Foundation for Intelligent Physical Agents, http://www.fipa.org/index.html
9. Grosof, B.N., Horrocks, I., Volz, R., Decker, S.: Description logic programs: combining logic programs with description logic. WWW 48, 57 (2003)
10. Grosof, B.N., Gandhe, M.D., Finin, T.W.: SweetJess: Translating DAMLRuleML to JESS. In: RuleML (2002)
11. Horrocks, I., Patel-Schneider, P.F.: A proposal for an OWL Rules Language. In: WWW 2004: Proceedings of the 13th international conference on World Wide Web, pp. 723–731. ACM Press, New York (2004)
12. Java Agent DEvelopment Framework, http://jade.tilab.com/
13. Jena – A Semantic Web Framework for Java, http://jena.sourceforge.net/
14. Levy, A.Y., Rousset, M.C.: Combining Horn rules and description logics in CARIN. Artificial Intelligence 104(1-2) 165, 209 (1998)
15. McGuinness, D.L., Borgida, A.: Explaining subsumption in description logics. In: IJCAI, vol. (1), pp. 816–821 (1995)
16. McGuinness, D.L., da Silva, P.P.: Explaining answers from the semantic web: the inference web approach. J. Web Sem. 1(4), 397–413 (2004)
17. Rosati, R.: On the decidability and complexity of integrating ontologies and rules. WSJ 3(1), 41–60 (2005)
18. RuleML: The RuleML Initiative website (2006), http://www.ruleml.org/
19. Shortliffe, E.: Computer-based medical consultations: MYCIN. American Elsevier, New York (1976)
20. Sintek, M., Decker, S.: TRIPLE - A Query, Inference, and Transformation Language for the Semantic Web. In: International Semantic Web Conference, pp. 364–378 (2002)
21. Sotiriou, C., Piccart, M.J.: Taking gene-expression profiling to the clinic: when will molecular signatures become relevant to patient care? Nature Reviews 7, 545–553 (2007)
22. Swartout, W., Paris, C., Moore, J.: Explanations in knowledge systems: Design for explain able expert systems. IEEE Expert: Intelligent Systems and Their Applications 06(3), 58–64 (1991)
23. The REWERSE I1 Rule Markup Language, http://oxygen.informatik.tu-cottbus.de/rewerse-i1/?q=R2ML

A Study of SAT-Based Branching Heuristics for the CSP

Yannis Argyropoulos and Kostas Stergiou

Department of Information and Communication Systems Engineering
University of the Aegean, Samos, Greece

Abstract. Constraint Satisfaction Problems (CSPs) and Propositional Satisfiability (SAT) are two closely related frameworks used for solving hard combinatorial problems. Despite their similarities regarding the problem formulation and the basic backtracking search algorithms they use, several advanced techniques have been developed and standardized in one framework but have been rather ignored in the other. One class of such techniques includes branching heuristics for variable and value ordering. Typically, SAT heuristics are highly sophisticated while CSP ones tend to be simplistic. In this paper we study some well known SAT heuristics with the aim of transferring them to the CSP framework. Through this attempt, new CSP branching techniques are developed; exploiting information not used by most of the standard CSP heuristics. For instance such information can be the arity of the constraints and the supports of values on the constraints. Preliminary empirical results on random problems show that this unexploited information can be used to design new efficient CSP heuristics or to enhance the performance of existing ones, like dom/wdeg.

Keywords: SAT, CSP, Branching Heuristics.

1 Introduction

CSPs and SAT are two closely related frameworks that are widely used to model and solve hard combinatorial problems from areas such as planning and scheduling, resource allocation, bioinformatics, configuration, software and hardware verification etc. Although they have fundamental similarities regarding the problem formulation and the search strategies they employ, there exist numerous advanced techniques that have been developed and widely used in one framework but have been rather ignored by the other one. For example, the binary (Boolean) domain size and the clausal form, under which constraints are stated in SAT, lead to a low-level and non-intuitive problem modeling compared to the CSP equivalent which is closer to human perception. Hence, modeling techniques have received much attention in CSPs but have been rather ignored by the SAT community. On the other hand, SAT solvers employ advanced methods for no-good learning and intelligent backtracking, while corresponding methods are under-developed and rarely used by the CSP community.

This difference in the evolution of specific methods in the two research communities applies for branching heuristics too. In both CSPs and SAT, branching heuristics aim at minimizing the size of the explored search tree by making intelligent decisions when called to select the next variable and value assignment [2]. But while most of

J. Darzentas et al. (Eds.): SETN 2008, LNAI 5138, pp. 38–50, 2008.

the well known CSP heuristics tend to be simple implementation of the "fail-first" principle [10], most SAT heuristics are highly sophisticated focusing on either enforcing the greatest possible unit propagation or exploiting as much as possible the learned clauses. SAT heuristics of the first type are typically embedded into look-ahead solvers while the latter are embedded into clause-learning solvers.

In this paper, we make a study of some well known SAT branching heuristics with the aim of transferring them to CSPs, in the spirit of [12] and [14]. We are particularly interested in heuristics used by look-ahead solvers because such heuristics bear similarities with the well-known CSP ones. We show that a direct translation of common SAT heuristics, like Jeroslow-Wang, MOM, and Bohm, to CSP counterparts results in inefficient and expensive heuristics. However, through a careful examination of these heuristics, we are able to adapt them; deriving new CSP branching techniques which exploit information not used by most of the standard CSP heuristics. Such information is the arity of the constraints and the supports of values on the constraints. Although these new heuristics are not always effective, or optimized in the best possible way for the CSP, preliminary empirical results demonstrate that the new techniques can potentially contribute to the development of effective and competitive branching methods. This is evident in the case of dom/wdeg, one of the most successful general purpose CSP heuristics, which can be significantly enhanced by incorporating information about the arity of the constraints.

The rest of the paper is structured as follows. In Section 2 we give some necessary background on CSPs and SAT and briefly review some standard heuristics for the two frameworks. Section 3 describes ways to transfer SAT heuristics to the CSP culminating in the development of efficient new heuristics. In Section 4, preliminary experimental results from random problems are presented. Finally, in Section 5 we conclude and briefly discuss future work.

2 Background

A CSP is defined as a triple (X,D,C), where $X=\{x_1,x_2,..x_n\}$ is a finite set of n variables, $D = \{D(x_1), D(x_2)...., D(x_n)\}$ is the set of their respective (finite) domains, and C is a set of constraints. For any constraint c, on m variables $x_1,x_2,..x_m$, $vars(c)$ denotes the variables involved in it, and $rel(c) \subseteq D(x_1) \times D(x_2) \times .. \times D(x_m)$ is the set of combinations (or *tuples*) of assignments for the variables $x_1,x_2,..x_m$ that satisfy the constraint. A tuple $T \in rel(c)$ is called *valid* when all the values assigned to the respective variables $x_1,x_2,..x_m \in vars(c)$ are available in the corresponding domains. For any variable x, $|dom(x)|$ denotes the cardinality of the variable's current domain and *forward degree* (*fwdeg*) denoted the number of constraints with unassigned variables where x is involved in. The *arity* of a constraint c is the number of variables involved in c. The *current arity* of c is the number of unassigned variables in c.

A binary constraint c between variables x_i and x_j is *arc consistent* (AC) iff for any value a in $D(x_i)$ there exists at least one value b in $D(x_j)$ s.t. a and b satisfy c, and vice versa [13]. In this case b is called a support of a on c. This definition has been extended to non-binary constraints, giving *generalized arc consistency* (GAC). For simplicity, in the following we use the term AC to refer to both the binary and the non-binary cases.

A common way to solve CSPs is using a backtracking-based search method. The most widely used such algorithm is MAC (*Maintaining Arc Consistency*) [1]. At each step MAC makes an assignment of a value to a selected variable x and then applies AC. The application of AC results in the deletion of any unsupported value from the corresponding domain. If some domain becomes empty, a new value for x is tried. If all available values of x fail, in which case we have reached a *dead-end*, then MAC backtracks to the previously assigned variable and tries a new value for it. The choice of which variable to select next and which value to assign it is done though the use of variable and value ordering heuristics, collectively called *branching* heuristics.

SAT is the problem of determining satisfiability for a set φ of propositional logic clauses expressed in CNF (conjunctive normal form). That is, the clauses are of the form $l_1 \vee l_2 \vee \ldots \vee l_n$, where each $l_i, i \in [1,n]$ is a *literal* i.e. the positive or negative appearance of a Boolean variable. As in CSPs, a common way to solve SAT problems is using backtracking search, and namely variations of the DPLL algorithm [5]. At each step DPLL assigns a variable to one of its two possible values and applies a restricted form of resolution called *Unit Propagation (UP)*. UP automatically sets to true the only literal appearing in a unary clause. As a result, any clause containing this literal is removed from the problem (as it is definitely true) while the size of any clause containing the opposite of this literal is reduced by one (as the literal is definitely false). This process is repeated until no more unary clauses appear or the empty clause is derived. In the latter case, the other value of the currently selected variable must be tried. If this causes an empty clause too (dead-end) then the algorithm backtracks to the previously assigned variable. Through UP, the SAT problem is reduced to fewer and smaller clauses, while dead ends are discovered earlier. Due to DP, clauses contain literals of unassigned variables only. The *length* of a clause C is the number of such literals in C.

2.1 Heuristics for CSPs and SAT

Variable and value ordering heuristics play an important role in both CSPs and SAT as different branching decisions (especially about the variables) may lead to vastly different search trees being explored.

Early SAT heuristics such as MOM [7,17], Bohm [3] and Jeroslow Wang (JW) [11] relied on information about the occurrence of variables in clauses and the clauses' lengths. Intuitively, variables appearing in many clauses may cause more propagation and clauses of small length are more likely to trigger UP. MOM selects the variable that has the highest number of occurrences in minimum length clauses, ignoring clauses of bigger lengths. In case of ties, it selects the variable having the positive/negative literal appearances ratio as close as possible to 1. Bohm works in a similar way, but breaking ties by considering the number of occurrences in clauses of the next smallest clause length. JW follows another strategy, counting clauses of all lengths, but using a weight based on each clause's length. In this way, the chosen variable will either have a large degree, or will be present in many minimum length clauses, or both. The GRASP system [15] introduced DLCS and DLIS which were cheaper and less complex, but yet effective. The main concept was the selection of the variable with the greatest number of occurrences in clauses of any arity.

Heuristics like the ones described above are typically used by look-ahead SAT solvers like GRASP. Heuristics used by clause-learning solvers are quite different as they focus search on variables involved in recently learned clauses. A standard heuristic of this type is VSIDS, introduced in the clause-learning solver CHAFF [16].

Common CSP variable ordering heuristics mainly follow the "fail-first" principle which states that in order to succeed you have to fail as soon as possible. In other words, they try to discover dead-ends as soon as possible. Such heuristics are *mindom* [10] which selects the variable with the smallest current domain size, and max forward degree (*fwdeg*) [6] selecting the variable involved in the most constraints with unassigned variables. An effective combination of the above is *dom/fwdeg* [1], which selects the variable with minimum ratio of current domain size and forward degree. In analogy to SAT heuristics like VSIDS, the recently proposed weighted degree heuristics *wdeg* and *dom/wdeg* [4] base their choices on information learned from conflicts discovered during search. These heuristics are nowadays considered as among the most efficient general-purpose CSP heuristics.

A common heuristic for value ordering is Geelen's promise [8]. For each value of the selected variable, this heuristics computes the number of supports that the value has in the domains of future variables and takes the product of these counts. The value having maximum product is then selected.

Adapting SAT heuristics to the CSP has received some (limited) attention in the past. Most notably, [12] derived CSP value ordering heuristics based on the JW SAT heuristic by studying encodings of CSP into SAT. In [14] ideas from SAT heuristics, like MOM, and deduction techniques, like detecting prime implicants, were used to devise new CSP heuristics.

3 Transferring SAT Heuristics to the CSP

SAT problems can be considered as a special subclass of CSPs, under the following limitations: 1) all variables have Boolean (0-1) domains, and 2) constraints are stated in clausal form. The method under which we transfer SAT heuristics to the CSP framework is by studying how they would behave if these two limitations did not stand. In most cases, the behaviour of the derived heuristics is not efficient, so additional analysis is required. In what follows we first discuss the various stages of adaptation required to derive an efficient CSP heuristic from the JW SAT heuristic. We then discuss the corresponding heuristics derived from MOM and Bohm. Due to space limitation, we omit the intermediate stages for these two heuristics and we only give the final adapted CSP heuristics. Finally, we demonstrate how lessons learned from this process can be used to enhance the performance of weighed degree heuristics.

3.1 Jeroslow Wang

The JW heuristic [11] first computes $J(l) = \sum_{l \in \omega, \omega \in \phi} a^{-|\omega|}$, where a is a constant determined heuristically and usually set to 2, for every literal l. The so called two-sided JW then chooses the variable x that maximizes $J(x) + J(\neg x)$. JW assigns the variable to true if $J(x) \geq J(\neg x)$ and false otherwise. We can note that JW prefers variables appearing (as either literal) in many clauses, giving priority to ones appearing in short clauses. JW achieves both variable and value ordering. In order to compute the score of a variable, JW adds the scores of the variable's two values. Taking this approach literally, a first attempt to transfer JW to CSPs would be:

JW-CSP1:

- Ωx,i – The set of constraints Cm for which : $i \in$ dom(x) and $\exists$ valid T$\in$ rel(Cm) with T[x]=i and x$\in$ vars(Cm) and $\exists$y: y$\in$ vars(Cm), y an unassigned variable, y$\neq$x

- |Cm| – The number of unassigned variables y, y$\in$ vars(Cm)

- Select the variable x be that maximizes $\displaystyle\sum_{i \in dom(x)} \sum_{Cm \in \Omega x,i} a^{-|Cm|}$, $\alpha \in$ R+

- Assign to x the value i $\in$ dom(x) that maximizes $\displaystyle\sum_{Cm \in \Omega x,i} a^{-|Cm|}$

Following JW for SAT, JW-CSP1 sums the scores of individual values for each variable. The score of a value is computed by counting the constraints in which it has at least one support, taking in mind the current arity of the constraints. Unfortunately, the heuristic suffers from important flaws. First of all, it is not effective for value ordering when MAC is used because, after AC filtering is applied, all values will have at least one support on every constraint (otherwise they would have been deleted). As a result, the value ordering part would always lead to a tie between all the available values of any selected variable. In order to resolve this issue, we can count the valid tuples per constraint supporting a value and select the one with most supports.

A second problem with the heuristic arises from the non-binary domain sizes. In contrast to the fail-first principle, the variable ordering part favors variables with large domains. To overcome this we can normalize by dividing each variable's score by its current domain size. After these necessary changes, the new version of the heuristic would be as follows:

JW-CSP2:
- Tx,i,cm – The Set of tuples T for which : T$\in$ rel(Cm), T[x]=i, Cm$\in \Omega$x,i

- Select the variable x that maximizes $\displaystyle\sum_{i \in dom(x)} \sum_{Cm \in \Omega x,i} \sum_{T \in Tx,i,cm} a^{-|Cm|}$ /|dom(x)|

- Assign to x the value i $\in$ dom(x) that maximizes $\displaystyle\sum_{Cm \in \Omega x,i} \sum_{T \in Tx,i,cm} a^{-|Cm|}$

JW-CSP2 would prefer variables involved in many constraints, like dom/fwdeg. But it will also give preference to those involved in constraints with small current arity. Still, some improvements are necessary to make the heuristic applicable in practice. The problem is that in order to compute the score of a variable, JW-CSP2 counts all valid tuples that support its values. This can be very expensive and should be avoided or replaced with equivalent information. Moreover, the heuristic gives ambiguous information for the variable ordering: A variable can have a high score because it has many supports for its values in a few constraints, rather than being involved in many constraints.

To overcome these problems we can avoid counting supporting tuples when computing the score of a variable, and instead, only take into account the forward degree and current arity information of each variable. The division by the current domain size would be still useful, as the heuristic would give preference to variables with small domain sizes, therefore resulting in fewer branches in the search sub-tree. Our final version of JW for CSPs is as follows:

JW-CSP3:

- Ωx – The Set of Constraints Cm for which $x \in \text{vars}(Cm)$ and $\exists y$: $y \in \text{vars}(Cm)$, y an unassigned variable, $y \neq x$
- $|Cm|$ – The number of unassigned variables y, $y \in \text{vars}(Cm)$
- Select the variable x that maximizes $\sum_{Cm \in \Omega x} a^{-|Cm|} \Big/ |dom(x)|$, $a \in R+$

- Assign to x the value $i \in dom(x)$ that maximizes $\sum_{Cm \in \Omega x, i} \sum_{T \in Tx, i, cm} a^{-|Cm|}$

Compared to standard CSP variable ordering heuristics such as dom/fwdeg, we can note that JW-CSP3 offers the current arity of the constraints as an extra parameter when computing the heuristic value of the variables. In problems consisting of binary constraints only, JW-CSP3 and dom/fwdeg are essentially equivalent[1]. Regarding value ordering, JW-CSP3 again diverts from standard heuristics such as Geelen's promise by taking into account the current arity of the constraints.

3.2 MOM

The MOM heuristic for SAT [7,17] works as follows. For any variable x let $f^*(x)$ be the number of clauses of minimum length that contain x and at least another unassigned variable. MOM chooses the variable x that maximizes $((f^*(x) + f^*(\neg x)) * 2^k + f^*(x) * f^*(\neg x)$. k is a constant whose value is chosen heuristically. MOM prefers variables whose literals appear in many clauses of minimum length, and among them is breaks ties by the one having the most "balanced" occurrences of its literals. Transferring MOM to CSPs gives rise to similar problems as with JW. Hence, after some adaptations the final version of the heuristic is:

MOM-CSP:

- $|Cm|$ – The number of unassigned variables y, $y \in \text{vars}(Cm)$
- μ – The minimum $|Cm|$ between all the constraints Cm in the problem

---

[1] JW-CSP3 maximizes the ratio of forward degree to current domain size which is equivalent to minimizing the inverse ratio like dom/fwdeg does.

- $\Omega\mu x$ – The Set of Constraints Cm for which : $|Cm|=\mu$ and $x \in vars(Cm)$ and $\exists$ y: y $\in$ vars(Cm), y an unassigned variable, y≠x
- $|\Omega\mu x|$ – # of constraints $\in \Omega\mu x$.
- g(Cm,x,i) – # of valid tuples T$\in$rel(Cm), with T[x]=i, Cm$\in$ C
- Let X' be the set of variables that maximize $|\Omega\mu x|/|dom(x)|$.
- If X' is singleton, select the only variable in X'. Otherwise, select the vari-

  able $x \in$ X' that minimizes $\sum\limits_{i \in dom(x)} \sum\limits_{Cm \in C} g(Cm,x,i)^{-|Cm|} \Big/ |dom(x)|$

MOM-CSP first finds the variables involved in the maximum number of constraints with minimum current arity. If there is only one such variable, it selects it. Otherwise, it breaks ties by considering the supports that the values of the variables have in the constraints. Following the fail-first principle, the heuristic chooses the variable having minimum average number of supports for its values in constraints of small arity. Note that MOM has been adapted to CSPs before, albeit in a different way [14]. The CSP heuristic derived from MOM in [14] computes for each variable x the ratio of the sum of lengths of constraints where x appears to the number of these constraints. The variable having smallest ratio is selected for instantiation. This heuristic tries to capture the intuition of MOM rather than directly transferring MOM to the CSP as we do, and has the drawback that it does not consider domain sizes which is a very important piece of information for CSP heuristics.

3.3 Bohm

Bohm's heuristic [3] selects the variable x that maximizes the vector : $(H_1(x), H_2(x),\ldots,H_n(x))$ under the lexicographical order, where $H_i(x) = a$ max $(h_i(x), h_i(-x)) + b$ min$(h_i(x), h_i(-x))$ and h_i = #of clauses with i unassigned literals, containing literal x. a and b are constants which are usually set to 1 and 2 respectively. This heuristic again favors variables appearing in many clauses of minimum length. But in contrast to MOM, it breaks ties by considering the occurrences of variables in clauses of second shortest length. Any ties remaining are broken by considering clauses of third shortest length, and so on. Transferring Bohm to CSPs again gives rise to similar problems as with JW and MOM. Our final version of Bohm is similar to MOM-CSP with the only difference being the tie breaking mechanism. To break ties Bohm-CSP selects the variable with maximum number of occurrences in constraints of second smallest current arity. If there is more than one such variable, the heuristic considers occurrences in constraints of third smallest arity, and so on.

3.4 Weighted Degree Heuristics

The *conflict-driven* weighted degree heuristics of Boussemart et al. [4] are designed to enhance variable selection in CSPs by incorporating knowledge gained during search, in particular knowledge derived from failures. These heuristics work as follows. All constraints are given an initial weight of 1. During search the weight of a constraint is incremented by 1 every time the constraint removes the last value from a domain, i.e. it causes a domain wipe-out (DWO). The *weighted degree* of a variable is then the sum of the weights of the constraints that include this variable and at least another unassigned variable. The weights are continuously updated during search by using information learnt from previous failures. The basic *wdeg* heuristic selects the variable having the largest weighted degree. Combining weighted degree and domain size yields heuristic *dom/wdeg* that selects the variable with the smallest ratio of current domain size to current weighted degree. The advantage that these heuristics offer is that they use previous search states as guidance, while older standard heuristics either use the initial state or the current state only.

The two weighted degree heuristics, and especially dom/wdeg, have displayed very good performance on a variety of problems. However, it has been noted that these heuristics do not make very informed choices at the first few steps of search where there is not enough information about constraint weights (because they have not been updated much) [9]. So a natural question is whether we can further improve the heuristics.

Using the lessons learned from the transfer of SAT heuristics to CSPs, we propose an extension to the weighted degree heuristics that takes into account information based on 1) the current arity of the constraints, and 2) on the current domain sizes of the unassigned variables in the constraints. For each constraint, these two pieces of information are combined with the number of times the constraint has caused a DWO (i.e. the weight as defined in [4]) to compute the constraint's weight. The new heuristic is simply called *extended wdeg* (*ewdeg*). To explain its behavior we first need the following definitions. For any constraint Cn in the problem:

- W1Cn is the number of times that Cn caused the DWO of a variable during search.
- W2Cn = $2^{-|Cn|}$
- XCn is the set of unassigned variables $x \in$ vars(Cn)
- W3Cn = $\prod\limits_{x \in XCn} |dom'(x)|/|dom(x)|$, where |dom'(x)| is the initial domain size of variable x

Note that the product W3Cn gives an approximation for the constraint tightness of Cn, as the valid tuples of that constraint cannot be more than this number. Now the weight of a constraint Cn as computed by heuristic ewdeg is WCn = W1Cn * W2Cn * W3Cn. Having computed WCn for each constraint, ewdeg selects the variable x that

maximizes $\sum\limits_{Cn \in vars(x)} WCn \Big/ |dom(x)|$. At the start of search, where W1Cn and W3Cn are

1 for all constraints, heuristic ewdeg gives preference to variables involved in many constraints of small arity. As search progresses W1Cn, mainly, and W3Cn, to a lesser extent, become increasingly more important.

4 Experimental Results

Preliminary experimental results to evaluate the performance of the new heuristics were carried out on randomly generated CSPs. A class of random problems is defined according to extended model B by a tuple <k, n, d, p, q>, where *n* is the number of variables, all with uniform domain size d, k is the uniform initial constraint arity, p is the proportion of constraints present in the problem among the $n!/k!(n-k)!$ total possible constraints, and q is the uniform proportion of allowed tuples in each constraint among the d^k total tuples. The CSP solver used implements MAC.

The results presented include average CPU time in seconds, and average number of nodes (i.e. variable assignments made). In addition, we sometimes report these measures by differentiating between soluble and insoluble instances. The time-out limit and the number of time-outs for any heuristic is given as "*T.O.*", when necessary. For each class, a set of instances was generated randomly and all heuristics were run on the same set. For an objective comparison of variable ordering heuristics alone, the value ordering was *lexicographical*.

We first compare heuristics JW-CSP3 and MOM-CSP, referred to as JW and MOM for simplicity, to dom/fwdeg on various classes of problems. Then we compare ewdeg to dom/fwdeg and dom/wdeg. The following SAT heuristics were transferred to the CSP and tested, but results are omitted:

- DLIS and the 1-sided version of JW are SAT heuristics that directly select a variable-value pair instead of differentiating between the variable and value ordering parts. Their transfer to CSPs resulted in counter-intuitive heuristics that do not follow the fail first principle.
- DLCS is omitted because its CSP version is equivalent to using dom/fwdeg for variable ordering together with Geelen's promise for value ordering.
- Bohm-CSP is omitted because its performance is between JW-CSP3 and MOM-CSP.

4.1   JW and MOM

The first class of problems we tested the heuristics on involved binary constraints. Experiments from Class 1 show that, as expected, JW makes the same variable ordering choices as dom/fwdeg when all constraints are binary, but it is slightly more expensive. MOM would also make the same choices as JW and dom/fwdeg, if it had no tie-breaking mechanism. As we can see, the tie breaking of MOM has a negative effect on the performance of the heuristic as the number of node visits and the CPU time increase.

Class 1: < 2, 35, 6, 0.842, 0.889 >			
Heuristic \ Output	JW	dom/fwdeg	MOM
CPU	6,49	5,83	8,01
CPU Unsat	8,40	7,53	10,28
CPU SAT	3,11	2,80	3,98
Nodes	9963,32	9963,32	10360,90
Nodes unsat	12645,09	12645,09	13168,63
Nodes sat	5195,72	5195,72	5369,39
sat instances	18/50 (36%)		

Results from classes 2 and 3, which consist of 3-ary and 4-ary constraints respectively, show that JW and MOM can outperform dom/fwdeg once the arity of the constraints starts to grow. Therefore, it seems that preferring variables constrained by minimum arity constraints can be an effective approach. JW is better than MOM which is again hampered by the tie breaking method used.

Class 2: < 3, 30, 6, 0.01847, 0.5 >			
Heuristic Output	JW	dom/fwdeg	MOM
CPU	9,00	11,62	12,16
CPU Unsat	10,85	13,75	18,09
CPU SAT	8,20	10,71	9,63
Nodes	9479,40	12352,30	11610,45
Nodes unsat	12528,50	15569,33	16552,67
Nodes sat	8172,64	10973,57	9492,36
sat instances	14/20 (70%)		

Class 3 < 4, 14, 5, 0.05, 0.63 >			
Heuristic Output	JW	dom/fwdeg	MOM
CPU	22,18	27,83	24,43
CPU Unsat	29,63	37,71	30,25
CPU SAT	14,74	17,95	18,60
Nodes	1556,80	1810,90	1697,90
Nodes unsat	2171,80	2539,60	2284,60
Nodes sat	941,80	1082,20	1111,20
sat instances	10/20 (50%)		

In Class 4 there is a wider gap between the performance of dom/fwdeg compared to JW and MOM. In this experiments the tie breaking of MOM was switched off (i.e. tie breaking was done lexicographically). Results confirm that the proposed tie breaking is ineffective as switching it off boosts the performance of MOM to the extent that it is more efficient than JW and almost twice better than dom/fwdeg.

Class 4 < 3, 20, 10, 0.05, 0.45 >			
Heuristic Output	JW	dom/fwdeg	MOM*
CPU	70,19	97,27	55,18
CPU Unsat	16,96	24,89	10,77
CPU SAT	67,47	88,59	62,16
Nodes	21449,85	28247,00	18519,10
Nodes unsat	5498,95	7469,41	3876,21
Nodes sat	19503,45	24875,73	19928,18
sat instances	11/20 (55%)		

*MOM without the tie breaking function

The preliminary experiments presented in this section demonstrate that the extra information that JW and MOM use has the potential to improve the performance of standard CSP heuristics, especially on non-binary problems. However, more experiments are required to verify this. Also, the tie breaking method of MOM needs to be further investigated. Note that the CSP version of Bohm that employs a different tie-breaking method usually outperforms MOM but is less robust than JW.

4.2 Weighted Degree Heuristics

We now examine the potential of the new weighted degree heuristic ewdeg. Classes 5 and 6 are example classes where dom/fwdeg outperforms or is very close to dom/wdeg. This is not surprising given that dom/wdeg is by its nature better suited to structured rather than random problems. However, in hard random CSPs requiring the exploration of more than 100.000 nodes, dom/wdeg usually has better performance than dom/fwdeg. In any case, the early variable ordering selections made by dom/wdeg are not very informed because at that stage of search only a few (if any) DWOs have been encountered, and hence not much has been learned about the weights of the constraints. It is notable that in both Classes 5 and 6 ewdeg displays the best performance among the three heuristic and significantly outperforms dom/wdeg. Also, it is important to note that ewdeg times out in fewer instances than dom/fwdeg and dom/wdeg in both classes. Heuristics JW and MOM are not competitive in these problems.

Class 5: < 2, 100, 12, 0.4, 0.236 >			
Heuristic Output	ewdeg	dom/fwdeg	dom/wdeg
CPU	11,46	13,17	23,56
Nodes	12521,15	17305,48	28256,25
T.O.(60 sec)	7	10	13
sat instances	31/50 (62%)		

Class 6: < 3, 300, 5, 0.00022446, 0.1>			
Heuristic Output	ewdeg	dom/fwdeg	dom/wdeg
CPU	18,89	37,70	37,56
Nodes	6848,49	14640,07	14565,73
T.O.(600 sec)	4	9	9
sat instances	76/150 (50%)		

As these results demonstrate, the extra information that ewdeg utilizes helps it make better variable ordering decisions at the early stages of search which has a considerable impact on the overall search efficiency. Note that after a certain amount of nodes, the weight factor based on DWOs (W1Cn) becomes the dominant factor for the decisions of ewdeg. However, more experiments, mainly with structured problems, are required to get a more accurate picture about the benefits of ewdeg. A general lesson we have learned is that the successful weighted degree heuristics can be further improved by augmenting them with additional information about the features of the problem.

5 Conclusions

We studied some well known SAT heuristics with the aim of transferring them to the CSP framework. From this study, new CSP branching methods we developed, exploiting information not used by most of the standard CSP heuristics. Through an analysis of the transfer process for the Jeroslow-Wang heuristic, we showed that the direct transfer of a SAT heuristic to the CSP can result in an inefficient and counterintuitive heuristic. However, with additional study and some necessary adjustments we can extract the core information offered by SAT heuristics and exploit it to devise new, potentially competitive, CSP counterparts.

As future work we aim to continue this study by exploring ways to transfer clause learning SAT heuristics like VSIDS to the CSP. On a more general note, the CSP framework can greatly benefit from ideas concerning branching heuristics or other techniques that are better developed in SAT. Of course, the converse is also true.

References

1. Bessiere, C., Regin, J.: Mac and Combined Heuristics: two Reasons to Forsake FC (and CBJ?) on Hard Problems, CP (1996)
2. Bordeaux, L., Hamadi, Y., Zhang, L.: Propositional Satisfiability and Constraint Programming: A Comparative Survey. ACM Computing Surveys 38(4) (2006)
3. Buro, M., Kleine-Buning, H.: Report of a SAT competition, Technical Report, University of Paderborn (1992)
4. Boussemart, F., Hemery, F., Lecoutre, C., Lakhdar, S.: Boosting systematic search by weighting contraints. In: ECAI (2004)
5. Davis, M., Logemann, G., Loveland, D.: A machine program for theorem-proving. Communications of the ACM 5(7), 393–397 (1962)
6. Dechter, R., Meiri, I.: Experimental evaluation of preprocessing algorithms for constraint satisfaction problems. Artificial Intelligence 68, 211–241 (1994)
7. Freeman, J.W.: Improvements to propositional satisfiability search algorithms, Ph.D. thesis, Department of Computer and Information Science, University of Pennsylvania (1995)
8. Geelen, P.A.: Dual viewpoint heuristics for binary constraint satisfaction problems. In: ECAI 1992, Vienna, Austria, pp. 31–35 (1992)
9. Grimes, D., Wallace, R.J.: Sampling Strategies and Variable Selection in Weighted Degree Heuristics, CP, pp. 831–838 (2007)
10. Haralick, R.M., Elliot, G.L.: Increasing tree search efficiency for constraint satisfaction problems. Artificial Intelligence 14, 263–313 (1980)
11. Jeroslow, R.J., Wang, J.: Solving propositional satisfiability problems. Annals of Mathematics and Artificial Intelligence 1, 167–188 (1990)
12. Lecoutre, C., Sais, L., Vion, J.: Using SAT Encodings to Derive CSP Value Ordering Heuristics. Journal of Satisfiability, Boolean Modeling and Computation 1, 169–186 (2007)
13. Mackworth, A.: Consistency in networks of relations. Artificial Intelligence 8, 99–118 (1977)

14. Mali, A.: Puthan Veettil, Sandhya: Adapting SAT Heuristics to Solve Constraint Satisfaction Problems, Technical Report, University of Wisconsin (2001)
15. Marques– Silva, J.P., Sakallah, K.A.: GRASP - A new search algorithm for satisfiability. In: International Conference on Computer Aided Design (ICCAD), pp. 220–227 (1996)
16. Moskewicz, M.W., Madigan, C.F., Zhao, Y., Zhang, L., Malik, S.: Chaff: Engineering an efficient SAT solver. In: International Design Automation Conference, pp. 530–535 (2001)
17. Zabih, R., McAllester, D.A.: A rearrangement search strategy for determining propositional satisfiability, pp. 155–160. AAAI, Menlo Park (1988)

Autonomy in Virtual Agents: Integrating Perception and Action on Functionally Grounded Representations

Argyris Arnellos[1], Spyros Vosinakis[1], George Anastasakis[2], and John Darzentas[1]

[1] Department of Product and Systems Design Engineering,
University of the Aegean, Hermoupolis, Syros, Greece
{arar,spyrosv,idarz}@aegean.gr
[2] Department of Informatics,
University of Piraeus, Greece
anastas@unipi.gr

Abstract. Autonomy is a fundamental property for an intelligent virtual agent. The problem in the design of an autonomous IVA is that the respective models approach the interactive, environmental and representational aspects of the agent as separate to each other, while the situation in biological agents is quite different. A theoretical framework indicating the fundamental properties and characteristics of an autonomous biological agent is briefly presented and the interactivist model of representations combined with the concept of a semiotic process are used as a way to provide a detailed architecture of an autonomous agent and its fundamental characteristics. A part of the architecture is implemented as a case study and the results are critically discussed showing that such architecture may provide grounded representational structures, while issues of scaling are more difficult to be tackled.

Keywords: Autonomy, representation, interaction, functional grounding, intelligent virtual agent, anticipation, belief formation.

1 Introduction

Intelligent Virtual Agents (IVAs) today play an increasingly important role in both fields of Artificial Intelligence (AI) and Virtual Reality (VR), for different reasons in each case [1]. The need for flexible and dynamic interactions between embodied intelligent entities and their environment as well as among themselves has risen quite early as the field of AI matured enough to set higher targets towards believable simulations of complex, life-like situations. Similarly, the added element of complexity, emerging as a natural consequence of realistic behaviour exhibited by embodied actors, either as units or in groups, and enhancing the feeling of user presence at an impressive rate, was soon recognized as the inevitable next goal of Virtual Reality. In effect, the two fields have converged towards a common aim as they concurrently progressed driven by their respective and, admittedly, quite different scientific concerns and individual goals. This has created new, fascinating possibilities, as well as a range of design and implementation issues [2].

Contemporary research in Virtual Environments has marked the need for autonomy in virtual agents. Autonomy has many interpretations in terms of the field it is being

J. Darzentas et al. (Eds.): SETN 2008, LNAI 5138, pp. 51–63, 2008.

used and analysed, but the majority of the researchers in IVEs are arguing in favour of a strong and life-like notion of autonomy, which should first of all replace omniscience in virtual worlds. As such, even from a practical perspective, autonomy is not a needless overhead. Since believability is considered as a crucial factor, virtual agents should appear to have limitations in their interaction with the environments, just as agents in the real world have. In this case, virtual agents should be able to interact with other agents and users in unexpected events and circumstances under fallible anticipations, to have limited perception capabilities and plausible action, to create and communicate new meanings about their environments and to exhibit novel interactions. Such agents could be used in dynamic and open-ended scenarios, where adaptability is needed.

As such, the notion of autonomy seems to play a very crucial role in the design of IVAs. Design and implementation benefits seem equally probable and significant. The IVA will be re-usable across a variety of different instances of a particular class of virtual worlds, not requiring re-engineering and additional implementation in case the virtual world model has to change. In general, behavioural flexibility in dynamic environments is an inherently-desired feature of any IVA design [3], [4]. Even more importantly, the process of generating the IVA's behaviour-control modules and modelling a particular IVA personality shall be disentangled at a great degree from the respective process of designing the virtual world and unconstrained by specific semantics – potentially unsuitable for cognitive processes – imposed by the virtual world's design, as predefined environment knowledge required would be reduced to a minimum [5]. However, for the time being, these issues remain theoretical to a large extent, reflecting the contemporary immaturity of the field and the diversity of the problems to be tackled. As it will be explained in the following sections, the main problem in the design of an autonomous IVA is that the respective models approach its interactive, environmental and representational issues as separate to each other, while the situation in biological agents, where one can find genuinely autonomous agents, is quite different.

In this paper an attempt towards the designing of autonomous IVAs is presented, at the theoretical, architectural and implementation level. Specifically, in section 2 a theoretical framework indicating the fundamental properties and characteristics of an autonomous biological agent is presented. Section 3 suggests the interactivist model of representations combined with the concept of a semiotic process as a way to model an autonomous agent and its fundamental characteristics and lays out a detailed architecture in order to implement such an autonomous IVA. A partial implementation of the proposed architecture is presented as a case study in Section 4. The conclusions of this work are mentioned in Section 5.

2   Designing Autonomous Artificial Agents

In almost any typical architecture of an artificial agent there are several components responsible for critical cognitive capacities, such as perceiving, reasoning, decision-making, learning, planning etc, regardless of whether the agent is situated in a virtual or in a physical environment (i.e. a robot). Interaction between IVAs and their environment is two-fold. IVAs sense their environment and generate knowledge about it according to perception mechanisms, and they act upon their environment by applying actions generated by behaviour-control methodologies. Sensing and acting are achieved based on a processor of symbols, which in turn is connected to distinct modules of sensors and actuators/effectors. The processor relates together encoded symbols to form abstract representations of the environment. Each encoding results in a representational content which relates the cognitive system with the environment. Therefore, the respective environmental knowledge, as well as the generated actions is usually encoded in conceptual, symbolic forms, capable of expressing high-level, abstract semantics and compatible with the cognition processes supporting them. On the contrary, virtual environments are typically modelled according to low-level, platform-dependent symbols, which are quite efficient in both expressing detailed geometrical and appearance-related data, as well as effectively hiding deducible higher-level semantics [1], [6]. These approaches in modelling an IVA have been successful on tasks requiring high-level cognition, but they have been quite unsuccessful on everyday tasks that humans find extremely easy to manage. Additionally, such approaches cannot connect low-level cognition to higher-level behaviours, which is the key in the evolution of the autonomy of real-life biological agents. The problem is that in the respective architectures, the syntactic and semantic aspects of an IVA are separated, making the creation and enhancement of inherent meaning structures almost impossible. Particularly, reasoning and behaviour control involve relationships among generic, non-grounded representations of environment-related concepts, while sensory data retrieved and actions requested have to be explicit and grounded in order to be meaningful regarding a given instance of the underlying virtual environment. This is widely known as the symbol-grounding problem [7], which is a global problem in the design of artificial agents, and as such, it is also the most crucial problem of the merging between AI and VR inherent in IVAs, creating substantial implications at all levels of their design and implementation.

On top of that, the frame problem comes as a natural consequence. Specifically, since agent's functionality is based on predetermined and non-inherent representations, it will neither have the capacity to generalise its meanings in order to act on new contexts presenting similar relations and conditions, nor the ability to develop new representations and hence to function adaptively whenever is needed [8]. These problems have posed great difficulty in creating autonomous IVAs capable of successful interaction in complex and ill-defined environments. Agent's autonomy is compromised and belongs solely to its designer. In the realm of virtual agents, some attempts have been made to ground representations in the sensorimotor interaction with the environment [4], [9] but a cognitivist grounding theory should also explain the interdependence of each subsystem participating in the acquisition of the signal (i.e. the transducing system) with its environment and the central computational system, in order to be complete [10]. Everything that an IVA does, if it is to be an autonomous system, should, first of all, be intrinsically meaningful to itself.

In order to disentangle the designer from providing ad-hoc solutions to the interactive capabilities of an IVA, (through the introduction of pre-determined and ad-hoc semantics) one should try to see what the biological approach to autonomy –where symbol-grounding is not a problem and the frame problem is much more loose – may provide to the design of autonomous IVAs. The last two decades, there has been a growing interest in several theories of the development of autonomous biological agents. A thorough analysis of these theoretical frameworks is out of the scope of this paper, but the reader may see [11] for a critical review, as well as for an analysis of an integrated framework of autonomous agents. What should be kept in mind is that autonomy and agency have many definitions with respect to the domain they are being used. Furthermore, autonomous systems are acting in the world for their self-maintenance, which is their primary goal.

For the purposes of this paper, agency is defined in a way that the suggested definition is more susceptible to an analysis of its functional characteristics. Therefore it is being suggested that a strong notion of agency calls for: *interactivity*, that is, the ability of an agent/cognitive system to perceive and act upon its environment by taking the initiative; *intentionality*, the ability of an agent to effect a goal-oriented interaction by attributing purposes, beliefs and desires to its actions; and *autonomy*, which can be characterized as the ability of an agent to function/operate intentionally and interactively based on its own resources. These three fundamental capacities/properties should be exhibited in a somewhat nested way regarding their existence and their evolutionary development and this makes them quite interdependent, especially when one attempts to understand if it is possible for each one of them to increase qualitatively while the others remain at the same level.

As such, it should be mentioned that there appears to be an interesting interdependence between the three fundamental properties, in the form of a circular connection between them. Specifically, Collier [12] suggests that there is no *function* without *autonomy*, no *intentionality* without *function* and no *meaning* without *intentionality*. The circle closes by considering meaning as a prerequisite for the maintenance of system's autonomy during its interaction. Indeed, this circle is functionally plausible due to the property of self-reference of an autonomous system. This is not just a conceptual interdependence. As it is analysed in [13], this is also a theoretical interdependence with a functional grounding, and as such, it sets some interesting constraints in the capacities that contribute to agency and it brings about some requirements in terms of the properties that an agent should exhibit independently of its agential level or in other words, of its level of autonomy. In short, these properties and their interdependence are considered as emergent in the functional organisation of the autonomous agent. The term 'functional' is used here to denote the processes of the network of components that contribute to the autonomy of the agent and particularly, to the maintenance of the autonomous system as a whole (see e.g. [14]). On the other hand, meaning should be linked with the functional structures of the agent. Hence, meaning should guide the constructive and interactive processes of the functional components of the autonomous agent in such a way that these processes maintain and enhance its autonomy. In this perspective, the enhancement of autonomy places certain goals by the autonomous system itself and hence, the intentionality of the system is functionally guiding its behaviour through meaning.

It should now be clear that the interactive, the environmental and the representational aspects of an autonomous agent cannot be separated to each other. This emergent nature of agency does not allow for the partitioning of agency in 'simpler problems' or/and the study of isolated cases of cognitive activity (e.g. perceiving, reasoning, planning, etc.). Nevertheless, these phenomena are quite typical in the research of autonomous artificial agents. However, the notion of a 'simpler problem' should always be interpreted with respect to the theoretical framework upon which the design of the artificial agent relies.

As such, the primary aim of an attempt to design an autonomous virtual agent is not to design an agent that will mimic in a great detail the activities of a human. On the contrary, the aims of such research attempts should be the design of a complete artificial agent, that is, a design which will support, up to a certain satisfying level, the set of the abovementioned fundamental and characteristic properties of autonomy, by maintaining its systemic and emergent nature in different types of dynamically changing environments. The theoretical basis for such architecture, as well as the architecture itself is presented in the next section.

3 Virtual Agent Model

The design of an IVA following the principles sketched in Section 2 is based on the interactivist model of representation [15], [16], which favours more intentional and anticipatory aspects of the agent's representations. Due to space limitations a brief presentation of the model follows. In the interactivist model two properties are required for a system (and its functional subsystems) to be adaptable towards a dynamic environment. The system should have a way of differentiating instances of the environment and a switching mechanism in order to choose among the appropriate internal processes. In the system, several differentiating options should be available. These differentiations are implicitly and interactively defined by the final state that a subsystem would reach after the system's interaction with a certain instance of environment. One should be aware that although such differentiations create an epistemic contact with the environment, they do not carry any representational content, thus, they are not representations and they do not carry any meaning for the agent. What they do is that they indicate the interactive capability of system's internal process.

Such differentiations can occur in any interaction and the course of the interaction depends on the organization of the participating subsystem and of the environment. A differentiated indication constitutes emergent representation, the content of which consists of the conditions under which an interactive strategy will succeed in the differentiated instances of the environment. These conditions play the role of "dynamic presuppositions" and the respective representational content emerges in the anticipations of the system regarding its interactive capabilities. In other words, the interactive capabilities of the agent are constituted as anticipations and it is these anticipations that could be inappropriate and this is detectable by the system itself, since such anticipations are embedded in the context of a goal-directed system. It should be noticed that the possibility of internally detectable error on a functional basis, is the prerequisite for learning. Error guides learning in an autonomous system, where its capacity for directed interaction (towards certain goals) in a dynamic environment results in the anticipation of the necessity to acquire new representations [16].

In the proposed model, autonomy and the resulting intelligence is not considered as an extra module, but as an asset emerging from the agent's functionality for interaction. Specifically, the use of the proposed architecture aims at the unification of the modality of interaction, perception and action with the smallest possible number of representational primitives. What is missing is the way that the representations of the anticipated interactions will be constructed in the system. In order to do this we use the form of the semiotic processes suggested by Peirce [17]. This semiotic framework is totally compatible with the theory of autonomous agents sketched above at the theoretical level [18]. At the design level of the suggested architecture the idea is that there is a *sign* (or *representamen*), which is the form which a signal takes (and it not necessarily be material) and which designates an *object* (its *referent*) through its relation to the meaning (or *interpretant*) that it is created in the system while the latter interacts with the sign.

In this respect and for the purposes of the suggested architecture, the sign is everything that is comprised in the agent's perception of the environment (at any given moment of its interaction with it), the object are the set of the state of affairs present in the environment at the respective time (i.e. the agent's context) and the meaning is the functional connection between the sign and the object, which is being constructed by the agent during its interaction with the object. It should be mentioned that according to Peirce, meaning arise in the interpretation of the sign. Therefore, in our case, meaning depends on how and with what functional aspects (agent's internal processes) the form of the perceived sign is related to its object.

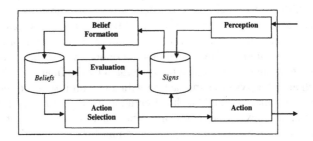

Fig. 1. The proposed agent architecture

Specifically, the resulting representations acquire a functional substrate, and as such, what really becomes represented in their contents is the outcome of the anticipated interactions internal to the agent. Since the agent interacts with the environment in order to achieve its goal and there is a way that the agent will understand whether its actions are towards a certain goal or not, then, the resulting representations necessarily, and by default, contain a model of the environment and of the way the agent may interact with it. Hence, the features of any object in the environment (material or not) are related to its functional properties for the agent. As a result, the agent perceives signs and constructs more developed signs as anticipations of successful interactions, with a functional substrate grounded on its organisation.

The agent architecture is presented in fig. 1. The design of the system is based on the assumption that the agent is equipped with a number of sensors and effectors that receive / send signals from / to the environment and with at least one internal status variable that describes the agent's degree of satisfaction. In the context of the proposed architecture, perception is defined as the process that reads values originating from the sensors and generates signs and action as the process of transmitting a value to one or more effectors at one time instance. Each time an action is performed, a respective sign is generated as well. The belief formation process transforms stored signs into beliefs, i.e. the anticipated interactions of the agent, and the evaluation process continuously compares feedback from the actual sensors (incoming sings) to the anticipated interactions in order to detect errors and restructure these representations accordingly. Finally, the action selection process is the deliberative part of the agent that plans ahead and selects the action that is expected to bring the environment into a state that improves the agent's status. The individual processes of the proposed architecture will be explained in more detail in the next paragraphs.

3.1 Sign Production and Functional Relation

Signs are stored in the sign repository and contain all the information that the agent has collected during its interaction with the environment: the entities, properties, and events it has observed, the status changes it has noticed and the actions it has performed. The purpose of the sign collection is to process this data in order to generate beliefs about the environment and its dynamics. The detection of functional relations between signs and the formation of more complex signs, the generalization of similar signs into rules, and the formation and evaluation of hypotheses, will drive the meaningful interaction of an agent inside a dynamic environment.

We will present the notion of signs and the functionality of the architecture components that are utilizing them, through a simple example of a virtual environment. Let us think of an artificial environment, which contains agents and passive objects of various types. The environment is supported by a simple physically based modeling process: accelerated motion, collision detection and response, and friction. Agents can only perform the actions *move forward, rotate left* and *rotate right*, and therefore their only ability is to move around and to hit objects. Finally, a function defined by the designer updates the agent's status based on the state of the environment, e.g. the distance of a given object from a given region.

Let us define level-0 signs as any distinct property that the agent can observe. Each individual level-0 sign has a type (the property to which it refers) and a value. In our example, the level-0 signs could be:

- *time*: an integer value that increases by one after every interaction loop of the agent
- *object*: the unique id of an object
- *position*: an object's position in the 2D Cartesian space
- *orientation*: an object's orientation defined as the angle between its front vector and the x-axis.
- *action*: the type of action that the agent has performed
- *event*: any event of the environment that the agent can detect through its senses, e.g. a collision between two objects
- *status*: the internal status value of the agent.

Signs of level-1 are functionally grounded level-0 signs and are the ones generated by the agent's perception and action processes. In the above-mentioned example, four level-1 signs can be produced: The first is the *object-perception* sign that relates a time value t, an object o, a position (x,y) and an orientation r. The functional relation (semantics) of this sign is that object o at time t was located in (x,y) and had the orientation r. Object-perception signs are generated every time an agent observes an object through its sensors, e.g. whenever the object is in its field of view. A special case of this sign is the one generated about the agent itself, which describes its own property values at every timeframe. The second sign is the *event-perception* sign that associates a time value t with an event e, stating that the event e was perceived at time t. The third sign is the *status* sign, which associates a status value s with a time value t stating that the value of the agent's internal status at time t was s. Finally, the fourth sign is the *action* sign, which associates an action a with a time value t to denote that the agent performed a at time t.

Multiple level-1 signs may contain the same level-0 sign. E.g. two object-perception signs that refer to the same object, or an event-perception and an object-perception that took place at the same time. In these cases, the level-1 signs can be functionally related based on the fact that they share one or more common properties, and form level-2 signs. Equivalently, two or more level-2 signs can be related if they contain one or more common level-1 or level-2 signs and form signs of level-3, etc.

3.2 Belief Formation, Evaluation and Action Selection

Apparently, from any given set of level-1 signs a huge number of functional relations leading to higher level signs can be produced. Some of these actually contain data that may be a useful basis for the agent's formation of meaning, provided that they are properly processed. The challenge for the agent's cognitive abilities is to be able to generate the appropriate signs, to detect similarities and to formulate hypotheses, in order to produce empirical laws that describe the environment's dynamics. However, the generation and comparison of all possible signs will face the problem of combinatorial explosion after very few interaction cycles. A possible solution to this problem is to have the designer insert a number of heuristic rules that will drive the production of higher level signs towards more meaningful representations. E.g. one heuristic rule could be to assign higher priority to the combination of two object-perception signs that refer to the same object and to successive time frames. The application of this rule will lead to higher level signs that describe *fluents*, i.e. object property values that change in time. E.g. such a rule could create level-2 signs that associate two successive positions of an object. This sign actually represents the velocity of the object at that time frame. Two such successive signs could be associated using the same heuristic rule and form a level-3 sign which will represent the acceleration of the same object.

One possible set of heuristic rules that can be employed in order to drive the belief formation process into representations that may assist the agent's deliberative behaviour is the following;

1. *fluents*: Two signs that include the same object sign and successive time signs are joined to form a fluent. The semantics of the fluent sign are that an objects property values $v_1, v_2, ..., v_k$ at the time point t have changed to $v'_1, v'_2, ..., v'_k$ at $t+1$.
2. *action effects*: A sign that includes the agent's property values at time t, an action sign at time t and a third sign with the agent's property values at t+1 can be joined to form a new sign that describes the effects of the agent's action at time t.

3. *event effects*: Similarly to actions effects, a sign that describes an event at time t can be connected to signs that include other objects' property values at times t and t+1.
4. *desired states*: A status sign at time t is connected to other signs that include the time point t. The aim of these signs is to lead to hypotheses about states of the environment that affect the agent's internal status value regarding its goals.

Even if a number of 'interesting' higher level signs can be generated using some heuristic rules, the information about past events and property changes of specific objects is of not much practical use to the agent. These signs will be useful if they can be generalized and lead to the formation of laws, which the agent will use to predict future states of the environment. In the proposed architecture, this sort of abduction is carried out by the belief formation process, which compares signs of the same type and attempts to generalize them. The process examines a series of signs of the same type and tries to detect patterns: which values remain unchanged, if the rate of change of a property value is constant or linear, etc. The generalization attempts to replace the level-0 signs into variables and assumes that the same belief holds for all possible values of each variable, e.g. for all time points t, for all objects o, etc. Additionally, in the cases of fluents, action effects and event effects, it is assumed that for each variable v'_i that changed value at $t+1$ there is a function f_i for which holds: $v'_i - v_i = f_i(v_1, v_2, ..., v_{i-1}, v_{i+1}, ..., v_k)$, where $v_1, v_2, ..., v_k$ are the variable values at time t. If the belief holds and this function can be properly approximated based on previous observations, the agent will be able to predict the next state of (part of) the environment based on its current state. A simple approximation of a function for a given set of input variables is to detect the nearest neighbors of the input variable set and to perform an interpolation on the output variable based on the Euclidean distance from the neighbors.

Let us present a simple example of generating beliefs. A generated action-effect sign could associate the agent's own property values (object-perception sign) at time t, the action it performed at time t, and the agent's property values at time $t+1$. Let p, r be the agent's position and rotation angle at time t respectively, a be the action it performed, and p', r' be the agent's position and rotation angle at time $t+1$. If a belief is formed that the agent's position and angle are affected by each one of its actions, then the belief formation algorithm will have to approximate the functions f_1 and f_2, where $p'-p = f_1(r, a)$ and $r'-r=f_2(p, a)$. The constant inspection of action-effect signs will provide more samples of f_1 and f_2 and will lead to the better approximation of the functions. Assuming that the agent is rotating at a constant rate and that it is moving forward by a constant distance, the comparison of a multitude of samples will result to the following values: if a is an action *rotate left* or *rotate right*, f_1 will be always equal to zero and f_2 will be equal to a constant number, whilst if a is a *move forward* action, f_1 is not constant and will have to be approximated and f_2 will be equal to zero.

The beliefs generated by the belief formation process are subject to failure, as they are the product of abduction and may not be true in all contexts. E.g. a belief about the implications of a collision on the motion of two objects may be true for objects of similar mass, but will fail if one of the objects is significantly heavier, or even inanimate. The aim of the belief evaluation process is to test the validity of the agent's beliefs by examining the existing and the incoming signs and to determine whether they support the beliefs or not. If a sign is detected that does not support a belief, the sign is treated as an exception, and the belief is being restructured in an attempt to exclude the exception. In that case, one of the variables of the belief is selected, and the respective value of the exceptional sign is excluded from it. However, since this variable is randomly selected, the result of the restructuring may still be subject to failures. Belief generation and evaluation is a continuous iterative process that drives the agent's representations towards more meaningful functions.

The distinction between cases that a belief applies and others that it does not, leads to the introduction and formation of categories concerning the values of perceivable properties. E.g. an action-effect belief that is until now proven to be applicable only to a subset of perceivable objects, categorizes objects. If a category proves to be persistent in time, it may describe the value of an essential property of the environment, which is perceived indirectly by the agent, e.g. mass, friction, velocity, etc. In such a case, the agent generates a symbol that describes the category. The formation of new symbols and signs will ultimately to lead to a higher layer of abstraction.

The action selection process decides about the next agent action to perform on the basis of maximizing the agent's status value, i.e. the agents 'satisfaction' as defined by the designer. Assuming that the agent has formulated a number of beliefs concerning fluents and events of the environment, states that increase its status value, and effects of its actions on its own property values, the agent can generate plans with a simple forward chaining algorithm in order to select the action that is expected to lead to the achievement of the its goal. The algorithm will estimate the effects of applying every possible action in the current time frame using the functions obtained in the respective beliefs and will estimate all the possible states of the environment in the next time frame. Using this process repeatedly up to a maximum number of states, a plan of action can be generated. It is, however, possible that the plan will fail because the beliefs that determine the effects of actions and the evolution of the environment are based solely on the agent's observations and approximations and will never grasp the actual 'physical' laws of the environment in their entirety. If the agent does not possess enough beliefs to construct plans, it may select actions that are expected to increase its knowledge.

4 Case Study

A partial implementation of the proposed architecture in the context of a simple virtual environment with agents, passive objects, and a collision detection and response mechanism is presented as a case study (fig. 2.a). Initially, the agents are unaware of the function of the actions they can perform. Consequently, they observe the effects of their own actions in an indirect fashion. In addition, agents are capable of building action-effect signs as their sign repository grows over time, and to transform them into beliefs. Their internal status value (satisfaction) increases as they minimize their distance from a specific target in space. In the experiments performed, targets could be set to be static points or actual objects in the environment. The latter lead to constant re-locating of targets as the agent hits objects when it collides with them. The agent is equipped with a field-of-view sensor, thanks to which it can receive signals describing locations of objects in the world that it stores as signs in its repository. During its interaction loop the agent tries to select the most effective action towards its target based on the beliefs it has generated concerning the effects of its actions.

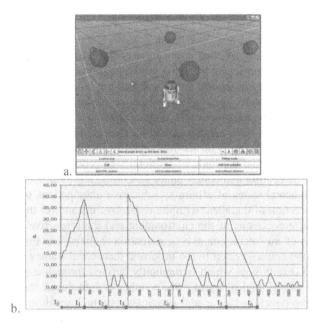

Fig. 2. a. Virtual world application **b.** agent's distance to target over time

Experiments were focused on an analysis of the agent's behavior in terms of deliberate modification of its own location over time by means of the available actions (move forward, rotate left and rotate right), with respect to target locations. The diagram in fig.2b above illustrates a sampling of 561 readings of the agent's Euclidean distance to a specific target on the x-z plane over time. As shown on the diagram, the agent's initial lack of anticipated interactions drive it to move in a random fashion, resulting in the increase of its distance to the target over time. While doing so, however, it progressively refines its representations as it observes the effects of its random actions. As a result, approximately at $t_1 = 56$, the representations are accurate enough to enable the agent to actually plan its motion towards the target. Around $t_2 = 112$, the agent has reached very close to the target, and continues to refine its representations by moving around it, being able, however, to restrain its motion in a small region. At $t_3 = 155$ the agent is assigned a new target, hence the steep increase in the distance readings. In the samples to follow, however, it is clear that its motion towards the target is now more controlled and directed than before, as the distance constantly drops, even though not at a constant rate. At approximately $t_4 = 259$ the agent has reached the area of the target and behaves in a similar fashion as in (t_2, t_3), maintaining a reasonably small distance to the target. At $t_5 = 381$, the agent is assigned yet another target; this time, its motion until it reaches it at $t_6 = 453$ is almost linear, and the distance drops at a constant rate. In conjunction with the agent restricting its motion to an even smaller region of the target after t_6, the above results indicate the agent's capability to progressively refine its beliefs about the effects of its own actions and, eventually, put them to actual, deliberate and effective use as a means to achieve its goal.

5 Conclusions

We have presented an example as an application of the proposed framework, where agents create grounded representational structures based on their interaction with the environment. In the proposed model, autonomy and the resulting intelligence is not considered as an extra module, but as an asset emerging from the agent's functionality for interaction. It has to be noted that relevant attempts have been made also in [19], [20] and [21] with a much more specific world for the agent to interact with and with many more hand-coded aspects of its functionality. Specifically, what has been achieved in this paper is that the logic of semiotic sign processes provides a tool for the integration of perception and action under functional representations, which refer to certain environmental states of affairs as subjective anticipations of possible outcomes of the interaction of the agent with the respective instance of the environment. While the way differentiations of the environment and the respective indications for possible outcome of an interaction (i.e. anticipations/beliefs) are formed in a way compatible with the one suggested in the proposed theoretical framework – all under the guidance of a certain goal which is implicitly defined through a graded satisfaction signal, as in any biological agent – the number of introduced abstraction levels (sign levels – more developed signs and anticipations which result in more complex generalisations) and the criterion for their introduction are inevitably hard-coded. Nevertheless, keeping in mind that the aim of this paper was to set the basis for the designing of an IVA which will support, up to a certain satisfying level, the fundamental properties of autonomy, in different types of dynamically changing environments, we have demonstrated that the suggested architecture is on the right track, while the abovementioned difficulties remain as a significant challenge in any relevant attempt.

References

1. Aylett, R., Luck, M.: Applying Artificial Intelligence to Virtual Reality: Intelligent Virtual Environments. Applied Artificial Intelligence 14(1), 3–32 (1999)
2. Thalmann, D.: Control and Autonomy for Intelligent Virtual Agent Behaviour. In: Vouros, G.A., Panayiotopoulos, T. (eds.) SETN 2004. LNCS (LNAI), vol. 3025, pp. 515–524. Springer, Heidelberg (2004)
3. Gillies, M., Ballin, D.: Integrating Autonomous Behavior and User Control for Believable Agents. In: 3rd International Joint Conference on Autonomous Agents and Multi Agent Systems (AAMAS 2004), pp. 336–343 (2005)
4. Dinerstein, J., Egbert, P.K.: Fast multi-level adaptation for interactive autonomous characters. ACM Transactions on Graphics 24(2), 262–288 (2005)
5. Gratch, J., Rickel, J., Andre, E., Badler, N., Cassell, J., Petajan, E.: Creating interactive virtual humans: Some assembly required. IEEE Intelligent Systems 17(4), 54–63 (2002)
6. Kasap, Z., Magnenat-Thalmann, N.: Intelligent Virtual Humans with Autonomy and Personality: State-of-the-Art. Intelligent Decision Technologies 1(1-2), 3–15 (2007)
7. Harnad, S.: The Symbol Grounding Problem. Physica D 42, 335–346 (1990)
8. Janlert, L.E.: Modeling change: The frame problem. In: Pylyshynn, Z.W. (ed.) The robots dilemma: The frame problem in artificial intelligence. Ablex, Norwood (1987)

9. Rickel, J., Johnson, W.L.: Animated Agents for Procedural Training in Virtual Reality: Perception, Cognition, and Motor Control. Applied Artificial Intelligence 13, 343–382 (1999)
10. Ziemke, T.: Rethinking Grounding. In: Riegler, P., von Stein (eds.) Understanding Representation in the Cognitive Sciences. Plenum Press, New York (1999)
11. Arnellos, A., Spyrou, T., Darzentas, J.: Towards the Naturalization of Agency based on an Interactivist Account of Autonomy. New Ideas in Psychology (Forthcoming, 2008)
12. Collier, J.: Autonomy in Anticipatory Systems: Significance for Functionality, Intentionality and Meaning. In: Dubois, D.M. (ed.) The 2nd Int. Conf. on Computing Anticipatory Systems. Springer, New York (1999)
13. Arnellos, A. Spyrou, T., Darzentas, J.: Emergence and Downward Causation in Contemporary Artificial Agents: Implications for their Autonomy and Some Design Guidelines. Cybernetics and Human Knowing (Forthcoming, 2008)
14. Ruiz-Mirazo, K., Moreno, A.: Basic Autonomy as a Fundamental Step in the Synthesis of Life. Artificial Life 10, 235–259 (2004)
15. Bickhard, M.H.: Representational Content in Humans and Machines. Journal of Experimental and Theoretical Artificial Intelligence 5, 285–333 (1993)
16. Bickhard, M.H.: Autonomy, Function, and Representation. Communication and Cognition — Artificial Intelligence 17(3-4), 111–131 (2000)
17. Peirce, C.S.: The Essential Peirce. Selected Philosophical Writings, vol. 1(1867–1893). Indiana University Press, Bloomington, Indianapolis (1998)
18. Arnellos, A., Spyrou, T., Darzentas, J.: Dynamic Interactions in Artificial Environments: Causal and Non-Causal Aspects for the Emergence of Meaning. Systemics, Cybernetics and Informatics 3, 82–89 (2006)
19. Tani, J., Nolfi, S.: Learning to perceive the world as articulated: An approach for hierarchical learning in sensory-motor systems. In: Proceedings of the fifth international conference on simulation of adaptive behavior. MIT Press, Cambridge (1998)
20. Vogt, P.: The emergence of compositional structures in perceptually grounded language games. Artificial Intelligence 167(1-2), 206–242 (2005)
21. Roy, D.: Semiotic Schemas: A Framework for Grounding Language in the Action and Perception. Artificial Intelligence 167(1-2), 170–205 (2005)

A Sparse Regression Mixture Model for Clustering Time-Series

K. Blekas, N. Galatsanos, and A. Likas

Department of Computer Science, University of Ioannina, 45110 Ioannina, Greece
{kblekas,galatsanos,arly}@cs.uoi.gr

Abstract. In this study we present a new sparse polynomial regression mixture model for fitting time series. The contribution of this work is the introduction of a smoothing prior over component regression coefficients through a Bayesian framework. This is done by using an appropriate Student-t distribution. The advantages of the sparsity-favouring prior is to make model more robust, less independent on order p of polynomials and improve the clustering procedure. The whole framework is converted into a maximum a posteriori (MAP) approach, where the known EM algorithm can be applied offering update equations for the model parameters in closed forms. The efficiency of the proposed sparse mixture model is experimentally shown by applying it on various real benchmarks and by comparing it with the typical regression mixture and the K-means algorithm. The results are very promising.

Keywords: Clustering time-series, Regression mixture model, sparse prior, Expectation-Maximization (EM) algorithm.

1 Introduction

Clustering is a very interesting and challenging research problem and a wide spectrum of methodologies has been used to address it. Probabilistic mixture modeling is a well established model-based approach for clustering that offers many advantages. One such advantage is that it provides a natural platform to evaluate the quality of the clustering solution [1], [2]. Clustering time-series is a special case of clustering in which the available data have one or both of the following two features: first they are of very large dimension and thus conventional clustering methods are computationally prohibitive, and second they are not of equal length and thus conventional clustering methods cannot straightforwardly be applied. In such cases it is natural initially to fit the available data with a parametric model and then to cluster based on that model. Different types of functional models have been used to for such data. Among them polynomial and spline regression are the most commonly used models [3] and have been successfully applied to a number of diverse applications, ranging from gene clustering in bioinformatics to clustering of cyclone trajectories, see for example [4] [5], [6] and [7].

Sparse Bayesian regression is methodology that has received a lot of attention lately, see for example [8], [9], [10] and [11]. Enforcing sparsity is a fundamental

J. Darzentas et al. (Eds.): SETN 2008, LNAI 5138, pp. 64–72, 2008.

machine learning regularization principle and lies behind some well known subjects such as *feature selection*. The key idea behind sparse priors is that we can obtain more flexible inference methods by employing models with many more degrees of freedom than could uniquely be adapted given data. In particular, the target of sparse Bayesian regression is to impose a heavy tail priors to the coefficients of the regressor. Such prior will zero out the coefficients that are not significant and maintain only a few large coefficients that are considered significant based on the model. The main advantage of such models is that they address the problem of model order selection which is a very important problem in many model based applications including regression. If the order of the regressor model is too large it overfits the observations and does not generalize well. On the other hand if it is too small it might miss trends in the data.

In this paper we present a sparse regression mixture model for clustering time-series data. It is based on treating the regression coefficients of each component as Gaussian random variables, and sequentially the inverse of their variance as Gamma hyperpriors. These two hierarchical priors constitute the Student-t distribution which has been proved to be very efficient [8]. Then, a maximum a posteriori expectation maximization algorithm (MAP-EM) [12], [2] is applied to learn this model and cluster the data. This is very efficient since it leads to update rules of model parameters in closed form during the M-step and improves data fitting.

The performance of the proposed methodology is evaluated using a variety of real datasets. Comparative results are also obtained using the classical K-means algorithm and also the typical regression mixture model without the sparse prior. Since the ground truth is already known, we have used the percentage of correct classification for evaluating each method. As experimentally have shown, the main advantage of our method is through sparsity property to achieve more flexibility and robustness with better solutions.

In section 2 we present the simple polynomial regression mixture model and how the EM algorithm can be used for estimating its parameters. The proposed sparse regression method is then given in section 3 describing the sparse Student-t prior over the component regression coefficients. To assess the performance of the proposed methodology we present in section 4 numerical experiments with known benchmarks. Finally, in section 5 we give our conclusions and suggestions for future research.

2   Regression Mixture Models

Suppose the set of N time-series data sequences $Y = \{y_{il}\}_{i=1,\ldots,N}^{l=1,\ldots,T}$, where l denotes the temporal index that corresponds to time locations t_l. It must be noted that although during the present description of the regression model it is assumed that all y_i sequences are of equal length, this can be easily changed. In such case, each y_i for $i = 1, \ldots, N$ is of variable length T_i. This corresponds to the general case of the model.

To model time-series y_i we use p-order polynomial regression on the time range $t = (t_1, \ldots, t_T)$ with an additive noise term given by

$$y_i = X\beta + e_i \,, \tag{1}$$

where X is the Vandermonde matrix, i.e.

$$X = \begin{pmatrix} 1 & t_1 & \ldots & t_1^p \\ \vdots & \vdots & \ldots & \vdots \\ 1 & t_T & \ldots & t_T^p \end{pmatrix}$$

and β is the $p+1$-vector of regression coefficients. Finally, the error term e_i is a T-dimensional vector that is assumed to be Gaussian and independent over time, i.e. $e_i \sim \mathcal{N}(0, \Sigma)$ with a diagonal covariance matrix $\Sigma = diag(\sigma_1^2, \ldots, \sigma_T^2)$. Thus, by assuming $X\beta$ deterministic, we can model the joint probability density of the sequence y with the normal distribution $\mathcal{N}(X\beta, \Sigma)$.

In this study we consider the problem of clustering time-series, i.e. the division of the set of sequences y_i with $i = 1, \ldots N$ into K clusters, where each cluster will contain sequences of the same generation mechanism (polynomial regression model). To this direction, the regression mixture model is a useful generative model that can be used to capture heterogeneous sources of time-series. This can be described by the following probability density function:

$$f(y_i|\Theta) = \sum_{j=1}^{K} \pi_j p(y_i|\theta_j) \,, \tag{2}$$

which has a generic and powerful meaning in model-based clustering. Following this scheme, each sequence is generated by first selecting a source j (cluster) according to probabilities π_j and then by performing sampling based on the corresponding regression relationship with parameters $\theta_j = \{\beta_j, \Sigma_j\}$ as described by the normal density function $p(y_i|\theta_j) = \mathcal{N}(y_i|X\beta_j, \Sigma_j)$. Moreover, the unknown mixture probabilities satisfy the constraints: $\pi_j \geq 0$ and $\sum_{j=1}^{K} \pi_j = 1$.

Based on the above formulation, the clustering problem becomes a maximum likelihood (ML) estimation problem for the mixture parameters $\Theta = \{\pi_j, \theta_j\}_{j=1}^{K}$, where the log-likelihood function is given by

$$L(Y|\Theta) = \sum_{i=1}^{N} \log\{\sum_{j=1}^{K} \pi_j \mathcal{N}(y_i|X\beta_j, \Sigma_j)\} \,. \tag{3}$$

The Expectation-Maximization (EM) algorithm [12] is an efficient framework for solving likelihood estimation problems for mixture models. It performs iteratively two steps: The E-step, where the current posterior probabilities of samples to belong to each cluster are calculated:

$$z_{ij}^{(t)} = P(j|y_i, \Theta^{(t)}) = \frac{\pi_j^{(t)} \mathcal{N}(y_i|X\beta_j^{(t)}, \Sigma_j^{(t)})}{f(y_i|\Theta^{(t)})} \,, \tag{4}$$

and the M-step, where the maximization of the expected value of the complete log-likelihood is performed. This leads to the following updated rules for the mixture parameters [4], [3]:

$$\pi_j^{(t+1)} = \frac{\sum_{i=1}^{N} z_{ij}^{(t)}}{N} , \tag{5}$$

$$\beta_j^{(t+1)} = \left[\sum_{i=1}^{N} z_{ij}^{(t)} X^T \Sigma_j^{-1\,(t)} X \right]^{-1} X^T \Sigma_j^{-1\,(t)} \sum_{i=1}^{N} z_{ij}^{(t)} y_i , \tag{6}$$

$$\sigma_{jl}^{2\,(t+1)} = \frac{\sum_{i=1}^{N} z_{ij}^{(t)} (y_{il} - [X\beta_j^{(t+1)}]_l)^2}{\sum_{i=1}^{N} z_{ij}^{(t)}} , \tag{7}$$

where $[.]_l$ indicates the l-th component of the T-dimensional vector that corresponds to location t_l. After convergence of the EM, the association of the N observable sequences y_i with the K clusters is based on the maximum value of the posterior probabilities. The generative polynomial regression function is also obtained per each cluster, as expressed by the $(p+1)$-dimensional vectors of the regression coefficients $\beta_j = (\beta_{j0}, \beta_{j1}, \ldots, \beta_{jp})^T$.

3   Sparse Regression Mixture Models

An important issue, when using the regression mixture model is to define the order p of the polynomials. The appropriate value of p depends on the shape of the curve to be fitted. Polynomials of smaller order lead to underfitting, while large values of p may lead to curve overfitting. Both cases may lead to serious deterioration of the clustering performance as also verified by experimental results.

The problem can be tackled using some regularization method that penalizes large order polynomials. An elegant statistical method for regularization is the Bayesian approach. This technique assumes a large value of the order p and imposes a prior distribution $p(\beta_j)$ on the parameter vectors $(\beta_{j0}, \beta_{j1}, \ldots, \beta_{jp})^T$ of each polynomial.

More specifically, the prior is defined in a hierarchical way as follows:

$$p(\beta_j|\alpha_j) = \mathcal{N}(\beta_j|0, A_j^{-1}) = \prod_{k=0}^{p} \mathcal{N}(\beta_k|0, \alpha_{jk}^{-1}) \tag{8}$$

where $\mathcal{N}(\mu|0, \Sigma)$ is the normal distribution and A_j is a diagonal matrix containing the $p+1$ elements of the hyperparameter vector $\alpha_j = [\alpha_{j0} \ldots \alpha_{jp}]$.

In addition a Γ prior is also imposed on the hyperparameters α_{jk}:

$$p(\alpha_j) = \prod_{k=0}^{p} Gamma(\alpha_{jk}|a,b) \propto \prod_{k=0}^{p} \alpha_{jk}^{a-1} e^{-b\alpha_{jk}}, \tag{9}$$

where where a and b denote parameters that are a priori set to near zero values.

The above two-stage hierarchical prior on α_j is actually a Student-t distribution and is called *sparse* ([8]), since it enforces most of the values α_{jk} to be large, thus the corresponding β_{jk} are set zero and eliminated from the model. In this way the complexity of the regression polynomials is controlled in an automatic and elegant way and overfitting is avoided. This prior has been successfully employed in the Relevance Vector Machine (RVM) model [8].

In order to exploit this sparse prior we resort to the MAP approach where the log-likelihood of the model (Eq. 3) is augmented with a penalty term that corresponds to the logarithm of the prior $p(\beta_j)$.

$$L(Y|\Theta) = \sum_{i=1}^{N} \log\{\sum_{j=1}^{K} \pi_j \mathcal{N}(y_i|X\beta_j, \Sigma_j)\} + \sum_{j=1}^{K} \log p(\beta_j|\alpha_j) + \sum_{j=1}^{K} \log p(\alpha_j) \tag{10}$$

where the parameter vector Θ is augmented to include the parameter vectors a_j: $\Theta = \{\pi_j, \beta_j, \Sigma_j, \alpha_j\}_{j=1}^{K}$.

Maximization of the MAP log-likelihood with respect to the parameters Θ is again achieved using the EM algorithm. At each EM iteration t, the computation of the posteriors z_{ij} in the E-step is again performed using Eq. (4). The same happens in the M-step for the update of the parameters π_j and Σ_j which is performed using Eq. (4) and Eq. (6). The introduction of the sparse prior affects the update of the parameter vectors β_j which is now written as:

$$\beta_j^{(t+1)} = \left[\sum_{i=1}^{N} z_{ij}^{(t)} X^T \Sigma_j^{-1(t)} X + A_j^{(t)}\right]^{-1} X^T \Sigma_j^{-1(t)} \sum_{i=1}^{N} z_{ij}^{(t)} y_i, \tag{11}$$

while the update of the hyperparameters α_{jk} is given by:

$$\alpha_{jk}^{(t+1)} = \frac{2a-1}{\beta_{jk}^{(t+1)} + 2b} \tag{12}$$

In the last equation the values of a and b have been set to 10^{-4}.

It must be noted that at each M-step, in order to accelerate convergence, it possible to iteratively apply the update equations for β_j and α_{jk} more than once. In our experiments two update iterations were carried out.

4   Experimental Results

We have made experiments on a variety of known benchmarks in order to study the performance of the proposed sparse regression mixture model, referred as

Sparse RM. Comparative results were obtained with the typical regression mixture model, that will be referred next as *Simple RM.* Both methods were initialized with the same strategy. In particular, at first K time-series are randomly selected form the dataset for initializing the polynomial coefficients β_j of the K components of the mixture model, following the simple least-square fit solution. Then, the log-likelihood function value is calculated after performing one step of the EM algorithm. One hundred (100) such different one-EM-step executions are made and the parameters of the model that capture the maximum likelihood value are finally used for initializing the EM algorithm.

Table 1. The description of the five datasets used in our experimental study

Dataset	Number of classes (K)	Size of dataset (N)	Time series length (T)
CBF	3	930	128
ECG	2	200	96
Gun problem	2	200	150
Synthetic control	6	600	60
Trace	4	200	275

In Table 1 we present some characteristics (the size and the number of classes) of the five (5) real datasets we have used in our experimental study. In particular, we have selected five (5) datasets for evaluating our method [13]:

- The *Cylinder-Bell-Funnel (CBF)* dataset contains time series from three different classes generated by three particular equations, see [14].
- The *ECG* dataset characterized by underlying patterns of periodicity.
- The *Gun problem* comes from a video surveillance domain that gives the motion streams from the right hand of two actors.
- The *Synthetic control* dataset which comes from monitoring and control of process environments.
- The *Trace dataset* which is a synthetic dataset designed to simulate instrumentation failures in nuclear plant.

More details on these benchmarks can be found at [13].

The obtained results from the comparative study on these benchmarks are summarized in Figure 1. For each one of the five problems we present a diagram with the accuracy of both methods for various values of polynomial order p. Note that we present here the mean values of the correct classification percentage as obtained from twenty (20) runs per order value. Furthermore, these diagrams illustrate (grey straight lines) the performance of the $K - means$ algorithm, where the time series are treated as feature vectors. It must be noted that these results are published in [13] and correspond to the best solution found after 10 different runs of the K-means.

As it is obvious from the diagrams of Figure 1 the typical regression model (Simple RM) deteriorate the clustering performance, especially in cases of large

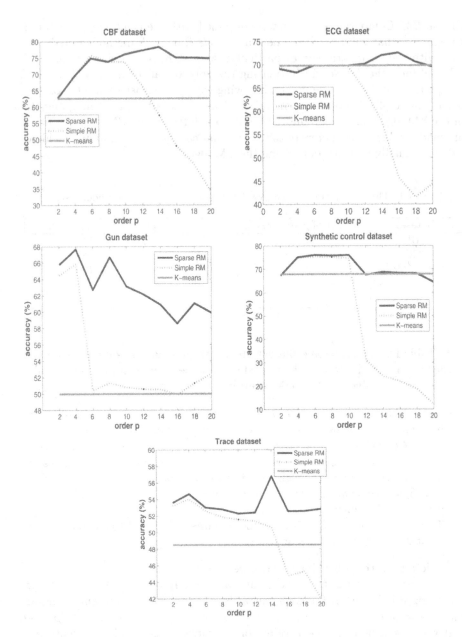

Fig. 1. Comparative results from the experiments on five benchmarks. The mean value of the correct classification for each one of the three methods is illustrated for different values of the polynomial order p.

polynomial order p values, leading in strong overfitting in all experimental datasets. On the other hand, the proposed sparse regression mixture model has the ability to overcome this disadvantage with the use of the sparsity potential, and to maintain better performance throughout the range of the polynomial order p. Therefore, it seems that the availability of sparsity makes the clustering approach independent on the polynomial order. Furthermore, that is more interesting is that in most cases, such as in the CBF, the ECG and the Trace datasets (Figure 1), the best clustering solution was obtained for large values of order p, where the typical RM completely fails. This is of great advantage, since it may lead to improve the model generalization ability. Thus, we recommend the use of sparse regression mixture model with a large polynomial order value (e.g. $p = 15$) and allow the model to select the most useful among the regression coefficients and sets to zero the rest of them.

5   Conclusions and Future Work

In this paper we presented an efficient methodology for clustering time-series, where the key aspect of the proposed technique is the sparsity of the regression polynomials. The main scheme applied here is the polynomial regression mixture model. Adding sparsity to the polynomial coefficients introduces a regularization principle that allows to start from an overparameterized model and force regression coefficients close to zero if they are not useful. Learning in the proposed sparse regression mixture model is treated in a maximum a posteriori (MAP) framework that allows the EM algorithm to be effectively used for estimating the model parameters. This has the advantage of establishing update rules in closed form during the M-step and thus data fitting can be done very efficiently. Experiments on difficult datasets have demonstrated the ability of the proposed sparse models to improve the performance and the robustness of the typical regression mixture model.

Our further research on this subject is mainly focused on three directions. In particular, we can alternatively use another design matrix X for regression (Eq. 1), apart from the simplest Vandermonde matrix in the case of polynomial. Following the Relevance Vector Machine (RVM) approach [8], different types of Kernel matrices can be examined, such as Gaussian Kernel. On the other hand, we are planning to examine also the possibility of using another type of more advantageous sparse priors, such as those presented at [10], [11] that have recently applied to general linear sparse models. The third target of our future work is to eliminate the dependence of the proposed regression mixture model on the initialization. Experiments have shown that in some cases there is a significant dependence on initializing model parameters especially on the regression parameters β_{jk}. A possible solution is to design an incremental procedure for learning a regression mixture model by adopting successful schemes that have already been presented in the case of classical mixture models [15]. Finally, we are planning to study the performance of the proposed methodology and its extensions in computer vision applications, such as visual tracking problems and object detection in a video surveillance domain [16], [17].

References

1. Bishop, C.M.: Pattern Recognition and Machine Learning. Springer, Heidelberg (2006)
2. McLachlan, G.M., Peel, D.: Finite Mixture Models. John Wiley & Sons, Inc., New York (2001)
3. Gaffney, S.J., Smyth, P.: Curve clustering with random effects regression mixtures. In: Bishop, C.M., Frey, B.J. (eds.) Proc. of the Ninth Intern. Workshop on Artificial Intelligence and Statistics (2003)
4. DeSarbo, W.S., Cron, W.L.: A maximum likelihood methodology for clusterwise linear regression. Journal of Classification 5(1), 249–282 (1988)
5. Chudova, D., Gaffney, S., Mjolsness, E., Smyth, P.: Mixture models for translation-invariant clustering of sets of multi-dimensional curves. In: Proc. of the Ninth ACM SIGKDD Intern. Conf. on Knowledge Discovery and Data Mining, Washington, DC, pp. 79–88 (2003)
6. Gaffney, S.J.: Probabilistic curve-aligned clustering and prediction with regression mixture models. Ph.D thesis, Department of Computer Science, University of California, Irvine (2004)
7. Blekas, K., Nikou, C., Galatsanos, N., Tsekos, N.V.: A regression mixture model with spatial constraints for clustering spatiotemporal data. Intern. Journal on Artificial Intelligence Tools (to appear)
8. Tipping, M.E.: Sparse Bayesian Learning and the Relevance Vector Machine. Journal of Machine Learning Research 1, 211–244 (2001)
9. Zhong, M.: A Variational method for learning Sparse Bayesian Regression. Neurocomputing 69, 2351–2355 (2006)
10. Schmolck, A., Everson, R.: Smooth Relevance Vector Machine: A smoothness prior extension of the RVM. Machine Learning 68(2), 107–135 (2007)
11. Seeger, M.: Bayesian Inference and Optimal Design for the Sparse Linear Model. Journal of Machine Learning Research 9, 759–813 (2008)
12. Dempster, A.P., Laird, N.M., Rubin, D.B.: Maximum Likelihood from incomplete data via the EM algorithm. J. Roy. Statist. Soc. B 39, 1–38 (1977)
13. Keogh, E., Xi, X., Wei, L., Ratanamahatana, C.A.: The ucr time series classification/clustering homepage (2006), www.cs.ucr.edu/~eamonn/timeseriesdata/
14. Keogh, E.J., Pazzani, M.J.: Scaling up Dynamic Time Warping for Datamining Applications. In: 6th ACM SIGKDD International Conference on Knowledge Discovery & Data Mining, pp. 285–289 (2000)
15. Vlassis, N., Likas, A.: A greedy EM algorithm for Gaussian mixture learning. Neural Processing Letters 15, 77–87 (2001)
16. Williams, O., Blake, A., Cipolla, R.: Sparse Bayesian Learning for Efficient Visual Tracking. IEEE Trans. on Pattern Analysis and Machine Intelligence 27(8), 1292–1304 (2005)
17. Antonini, G., Thiran, J.: Counting pedestrians in video sequences using trajectory clustering. IEEE Trans. on Circuits and Systems for Video Technology 16(8), 1008–1020 (2006)

Human Distress Sound Analysis and Characterization Using Advanced Classification Techniques

Charalampos Doukas[1] and Ilias Maglogiannis[1,2]

[1] Dep. of Information & Communication Systems Engineering, University of the Aegean, Samos, Greece
{doukas,imaglo}@aegean.gr
[2] Dep. of Biomedical Informatics, University of Central Greece, Lamia, Greece
imaglo@ucg.gr

Abstract. The analysis of sounds generated in close proximity of a subject can often indicate emergency events like falls, pain and other distress situations. This paper presents a system for collecting and analyzing sounds and speech expressions utilizing on-body sensors and advanced classification techniques for emergency events detection. A variety of popular classification and meta-classification algorithms have been evaluated and the corresponding results are presented.

Keywords: Human sound analysis, sound classification, meta-classification, Pervasive Application.

1 Introduction

Medical telemonitoring at home, a pervasive telemedicine application intended for intelligent and assistive environments, is an interesting solution compared to health facility institutions for the patients since it offers a medical surveillance in a familiar atmosphere for the patient and can reduce the costs of medical treatment. Within this context, the telemonitoring of both human physiological data and status comprehension is interesting for the purpose of emergency event detection. In the case of elderly people living on their own, there is a particular need for monitoring their behavior. The first goal of surveillance is the detection of major incidents such as falls, or long periods of inactivity or any other kind of distress situations that would require aid. The early detection of such events is an important step to alert and protect the subject, so that serious injury can be avoided. Motion data acquired from accelerometers and audiovisual content from the patient's environment have been proposed as practical, inexpensive and reliable methods for monitoring ambulatory motion in elderly subjects and for the detection or the prediction of falls. Additional sound analysis from the subjects' surrounding environment has been used in order to detect emergency incidents or call for help.

Advanced processing (e.g., filtering and feature selection) and classification of the aforementioned data are required for the proper diagnosis of related emergency incidents.

This paper presents the initial implementation of a human status awareness system based on sound data analysis. On-body sensors capture environment sounds and

J. Darzentas et al. (Eds.): SETN 2008, LNAI 5138, pp. 73–84, 2008.

transmit them wirelessly to a monitoring unit that undertakes the sound analysis task and results in a decision for a patient status. The rest of the paper is organized as follows; Section 2 discusses related work in the context of patient activity interpretation. Section 3 describes the proposed architecture and Section 4 presents details regarding the acquisition of the patient movement and sound data using sensors. Section 5 discusses the data processing and classification methods. Section 6 presents the initial evaluation results and Section 7 concludes the paper.

2 Related Work

Although the concept of patient activity and status recognition with focus on distress events detection is relatively new, there exists significant related research work, which may be retrieved from the literature ([1]-[29]). Information regarding the patient movement and activity is frequently acquired through visual tracking of the patient's position. In [6] and [15] overhead tracking through cameras provides the movement trajectory of the patient and gives information about user activity on predetermined monitored areas. Unusual inactivity (e.g., continuous tracking of the patient on the floor) is interpreted as a fall. Similarly, in [10] omni-camera images are used in order to determine the horizontal placement of the patient's silhouettes on the floor (case of fall). Success rate for fall detection is declared at 81% for the latter work. Head tracking is used in [13] in order to follow patient's movement trajectory with a success rate of fall detection at 66.67%. Environmental sound processing is used in [24]-[29] for monitoring the overall behavior of patients and detection of alarm incidents.

The aforementioned activities detection methods based on audio and visual information of the user require capturing equipment and thus are limited to indoor environment usage. In addition, some of the methods require also the a-priori knowledge of the area structure (e.g., obstacles, definition of floor, etc.), or user information (e.g., height in [10]). A different approach for collecting patient activity information is the use of sensors that integrate devices like accelerometers, gyroscopes, contact sensors and microphones. The decrease of sensors size and weight, in conjunction with the introduction of embedded wireless transceivers allows their pervasive placement on patients and the transmission of the collected movement and audio information to monitoring units wirelessly. The latter approach is less depended on the patient and environmental information and can be used for a variety of applications for user activity recognition ([1], [3], [9]).

Regarding fall detection using sound processing, most of the related work focuses on collecting and analyzing sound data captured from the patient's environment. In [25] - [27] authors present a sound analysis system for special sounds detection and association with events related to specific patient activities or events that might indicate need for first aid (e.g., falls, glass breaking, call for help, etc.). Sound event detection and feature extraction is performed through Discrete Wavelet Transformation (DWT) whereas classification to predefined events or vocal expressions is performed through a Gaussian Mixture Model (GMM) technique. In [28] Mel Frequency Cepstral Coefficients (MFCC) are used in order to detect a variety of sound signatures of both distressful and normal events. The examined sounds are categorized into classes according to their corresponding average magnitude levels that emerge from the application of Fourier Transform on the sound signal. Ceptral coefficients are used as features fed into a GMM model for proper classification. Accuracy of proper classification achieves 91.58% according to the authors. The latter methods needs special equipment to be installed on site and can be used exclusively in indoor environments. Additionally it requires much effort for isolating sounds generated within patient's proximity from other sounds.

3 System Architecture

The proposed system aims at collecting sound events of the subject's surrounding environment and analyzing the latter for generating status awareness of the patient. The system thus follows the architecture illustrated in Fig. 1. Special sensor nodes with networking capabilities are required for collecting and transmitting sound data. These sensors can be attached on several locations on the subject's body. A monitoring node is required for collecting the aforementioned data and performing required processing in order to enable an estimation of the human status. Two major modules are utilized for the latter estimation; a data processing module and a classification module. The first one consists of a filtering sub - module, responsible for filtering the incoming data (e.g. removing background noise, etc.) and a feature extraction sub - module for selecting and extracting appsropriates features from the filtered data (see Section 5 for details). These features are the used by the second module which utilizes a number of features for building a classification train model and then for providing estimations regarding the patient's status based on the incoming sounds. When a distress event is detected medical or treatment personnel can be alerted and the corresponding event data are provided.

The monitoring node can be a personal computer or a PDA device equipped with networking capabilities for alerting medical personnel regarding the patient's status and sending the related data.

4 Sound Acquisition Details

Sensor data acquisition may be accomplished through on-body (wearable) network. On body networks or Wireless Personal Area Networks (WPANs) are defined within the IEEE 802.15 standard. The most prominent protocols for pervasive telemonitoring systems such as the proposed system are Bluetooth and ZigBee (IEEE 802.15.4 standard). In our study we selected the ZigBee protocol for a number of reasons:

- Low cost and very low power consumption:
- Low complexity that makes the protocol ideal for integration on sensor nodes.
- Higher range compared to Bluetooth (up to 100 meter).
- Can be used for automatic creation of mesh networks and
- Contains built-in security measures.

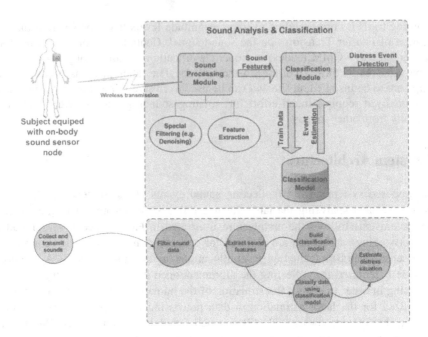

Fig. 1. Architecture of the presented system for distress sound analysis and classification. The lower part of the figure contains a data flow diagram illustrating the main procedures and information exchange at the different modules.

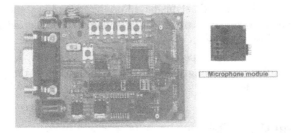

Fig. 2. The SARD ZigBee node. The node acts as both receiver and transmitter. The RS232 interface provides connectivity with the monitoring device (e.g., a PC or a PDA with appropriate interface adapter) when the node is used as receiver. One or more microphone modules can be attached on the node.

The sensor node used in the initial implementation of the proposed system consists of a SARD ZigBee node [2] (see Fig. 2). The latter contains a 2.4 GHz wireless data transceiver RF reference design with printed circuit antenna, which provides all hardware required for a complete wireless node using IEEE 802.15.4 (ZigBee) [17] packet structure. It includes an RS232 port for interface with a personal computer or a PDA, background debug module for in-circuit hardware debug, four switches and LEDs for control and monitoring, a low-power, low-voltage MCU (MicroController Unit) with 60KB of on-chip Flash which allows the user flexibility in establishing wireless data networks. A separate SensiNode [31] board has been also attached containing a microphone and additional sensors like illumination and temperature sensors.

For the acquisition of sound data one node is preferably placed on subject's chest. The latter position has been proven more convenient for the collection of both body fall sounds and distress speech expressions.

5 Sound Processing and Feature Extraction

The presented work uses spectrogram analysis that is based on short-time-Fourier Transform (STFT) on the signal for the detection of sounds generated by the fall of the human body and for the detection of vocal stress in speech expressions indicating distress events. Given a signal x(t) and its Fourier Transform $X(\tau, \omega)$ the STFT is:

$$STFT\{x()\} \equiv X(\tau,\omega) = \int_{-\infty}^{\infty} x(t)\omega(t-\tau)e^{-j\omega}dt \qquad (1)$$

The spectrogram is respectively given by the magnitude of the STFT of one function:

$$spectrogram(t,\omega) = |STFT(t,\omega)|^2 \qquad (2)$$

In the most usual format of a spectrogram, the horizontal axis represents time, the vertical axis is frequency, and the intensity of each point in the image represents amplitude of a particular frequency at a particular time (see Fig. 3). Based on conducted experiments the relative amplitude of a signal and the peak frequency at a given time can give a successful indication of a patient fall sound as captured by the on-body sensor microphone; Body falls generate low frequency sounds with high amplitude. Using a threshold of >90% for relative signal amplitude and <200 Hz for peak frequency, the differentiation of a fall sound against other sounds is possible. More precisely, over a series of 20 sound samples containing both body fall sounds and

Fig. 3. Spectrogram analysis illustration of a sound sample containing human distress sounds and background noise. Strongly dashed lines indicate time segments where distress sounds have been recorded whereas softly dashed lines indicate background noise respectively.

background noise (e.g., radio, object falls, etc) the detection of the fall was possible for the 80% of them. Different types of floor (i.e. wood, cement and flooring tile) were also used. The presented method has low computational complexity and can be easily integrated on sensor devices for real time sound processing. The unsuccessful detection of fall sounds was observed in cases where high amplitude sounds were generated in close proximity to the microphone. The same analysis has been applied on vocal sounds in an effort to detect distress expressions.

6   Utilizing Advanced Classification Techniques

Several advanced classification techniques have been utilized in order to build proper models for proper characterization of the body fall sound and the distress speech expressions. The algorithms used were: Bayes Networks, Naïve Bayes, Naïve Bayes Multunomial, Support Vector Machines, Logistic regression, Multilayer perceptron, Nearest Neighbor and K-Nearest Neighbor, Neural Networks, PART, NBTree, and SimpleCart. In addition, the following meta-classifiers have also been used:

- AdaBoost [33] : Class for boosting a nominal class classifier using the Adaboost M1 method. Only. Often dramatically improves performance, but sometimes overfits.
- Classification via regression [34] : Class for doing classification using regression methods. Class is binarized and one regression model is built for each class value.
- CVparameterSelection [35]: Class for performing parameter selection by cross-validation for any classifier.
- RandomSubSpace [36]: This method constructs a decision tree based classifier that maintains highest accuracy on training data and improves on generalization accuracy as it grows in complexity. The classifier consists of multiple trees constructed systematically by pseudorandomly selecting subsets of components of the feature vector, that is, trees constructed in randomly chosen subspaces.
- NestedDichotomies [37]: A meta-classifier for handling multi-class datasets with 2-class classifiers by building a random class-balanced tree structure.
- Dagging [38]: This meta-classifier creates a number of disjoint, stratified folds out of the data and feeds each chunk of data to a copy of the supplied base classifier. Predictions are made via majority vote, since all the generated base classifiers are put into the Vote meta-classifier. Useful for base classifiers that are quadratic or worse in time behavior, regarding number of instances in the training data. Usually Support Vector Machines are used as base classifiers.
- ThresholdSelector [40]: A meta-classifier that selecting a mid-point threshold on the probability output by a Classifier. The midpoint threshold is set so that a given performance measure is optimized. Currently this is the F-measure. Performance is measured either on the training data, a hold-out set or using cross-validation. In addition, the probabilities returned by the base learner can have their range expanded so that the output probabilities will reside between 0 and 1 (this is useful if the scheme normally produces probabilities in a very narrow range).

For the performance assessment of the aforementioned techniques the weka [40] tool has been used. The following section contains details regarding the creation of classification models and the evaluation procedure.

7 Evaluation Results

This section describes the experimental setup of the classification evaluation and presents the corresponding results. Two separate experiments have been conducted involving the acquisition and analysis of sounds generated by human body falls, and distress sounds respectively. Two individuals of average size and height equipped with the sensor nodes described in section 4 have been used for the collection of sounds. During the first experiment a total number of 245 sounds have been collected; half of them considered actual body fall on a number of different floor types (wooden, cement, etc.) and the rest of them were sounds utilized as noise (e.g., background noise, object fall, glass break, etc.). The second experiment involved the acquisition of 280 speech expressions; 180 containing distress speech expressions that could be associated with emergency cases and 100 containing irrelevant sounds (e.g., background noise, normal speech, etc.)

The sounds features extracted and utilized in the classification phase were the maximum frequency, the peak frequency, the amplitude and 6 randomly selected frequencies around the area of the signal where the relative amplitude was at 100%. The latter features were derived from the analyzed sound signal over a certain time window of 5 seconds for each sound sample. A total number of 19 classification models have been created based on the classification algorithms discussed in section 6 and the weka tool. The datasets described previously were used for the creation of the corresponding classification models and a 20-fold cross validation over the latter was used for the performance evaluation of each model. A number of statistical analysis performance metrics have been used: the accuracy, the kappa statistic [40] (which measures the agreement between two raters who each classify N items into C mutually exclusive categories), the mean absolute error of the false classified instances versus the correctly ones, the root mean square error of the latter, the relative absolute error and the root relative squared error. Tables 1 and 2 present the evaluation results of the discussed classification algorithms for the cases of distress speech and fall sounds classification respectively.

As indicated by the evaluation results, the majority of the algorithms achieve high accuracy results. Performance differences between some of the evaluated algorithms with high accuracy are illustrated through the ROC analysis curves. In the case of meta-classifiers and distress speech classification it is interesting to notice that Classification via Regression has higher accuracy than the RandomSubSpace (98.5765 % against 97.8648) but the latter one seems to have a better True False (TF) rate.

From the specific results it is hard to distinguish a better performance of the meta-classifiers against the simple ones. AdBoost meta-classifier and NBTree simple classifier managed both to identify correctly all sound speech expressions. It is also interesting to see that popural classification techniques like Support Vector Machines

Table 1. Evaluation results for the different classification algorithms used on distress speech expressions classification. Accuracy, Kappa statistic, absolute mean error, Root Mean Square Error (RMSE), relative absolute error and Root relative squared error are utilized as performance metrics.

Classification Algorithm	Correctly Classified Instances	Kappa statistic	Mean absolute error	Root mean squared error	Relative absolute error	Root relative squared error
BayesNet	98.58%	0.9392	0.0176	0.1169	7.81%	34.97%
NaiveBayes	97.51%	0.8923	0.0246	0.1337	10.92%	40.00%
Logistic	94.31%	0.7389	0.0957	0.2251	42.44%	67.32%
MultiLayerPerceptron	98.58%	0.9378	0.0212	0.0984	9.39%	29.43%
SVM	90.75%	0.4736	0.0925	0.3042	41.02%	90.98%
IB1 (Nearest Neighbor)	92.53%	0.5547	0.0747	0.2734	33.14%	81.76%
IBK (K Nearest)	92.53%	0.5547	0.0779	0.2724	34.53%	81.46%
NNge	94.66%	0.7094	0.0534	0.231	23.67%	69.10%
PART	99.29%	0.9674	0.0112	0.0851	4.95%	25.44%
NBTree	100%	1	0.0072	0.0219	3.20%	6.55%
SimpleCart	98.58%	0.9331	0.0242	0.1193	10.73%	35.67%
AdaBoost	100	1	0.0043	0.0211	1.9087	6.3142
ClassificationViaRegression	98.5765	0.9347	0.0728	0.1562	32.2821	46.7101
CVParameterSelection	87.1886	0	0.2256	0.3343	100	100
RandomSubSpace	97.8648	0.9044	0.0449	0.1238	19.8779	37.0152
NestedDichotomies	98.5765	0.9331	0.0239	0.1222	10.6124	36.549
Dagging	87.9004	0.1895	0.127	0.3054	56.3066	91.349
ThresholdSelector	92.8826	0.6958	0.1016	0.228	45.0493	68.195

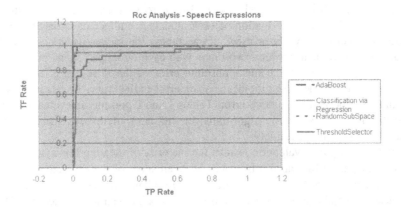

Fig. 4. Roc Curve for illustrating the performance of an indicative sample of the assessed meta-classifiers over the distress speech sound classification

(SVM) managed to achieve less accuracy (90.75% in speech classification and 81.22% in fall sounds respectively) than techniques like Bayes Networks (98.58% in speech classification and 97.55% in fall sounds respectively).

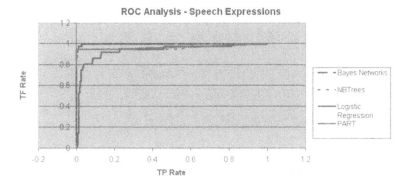

Fig. 5. Roc Curve for illustrating the performance of an indicative sample of the assessed classifiers over the distress speech sound classification

Table 2. Evaluation results for the different classification algorithms used on fall sounds classification. Accuracy, Kappa statistic, absolute mean error, Root Mean Square Error (RMSE), relative absolute error and Root relative squared error are utilized as performance metrics.

Classification Algorithm	Correctly Classified Instances	Kappa statistic	Mean absolute error	Root mean squared error	Relative absolute error	Root relative squared error
BayesNet	97.55%	0.9484	0.0417	0.1461	8.77%	29.98%
NaiveBayes	75.92%	0.4627	0.2273	0.4477	47.83%	91.87%
NaiveBayesMultiNomial	68.57%	0.329	0.3152	0.5585	66.33%	114.59%
Logistic	83.67%	0.6438	0.245	0.3573	51.57%	73.32%
MultiLayerPerceptron	95.51%	0.906	0.0789	0.194	16.60%	39.81%
SVM	81.22%	0.587	0.1878	0.4333	39.52%	88.91%
IB1 (Nearest Neighbor)	97.14%	0.9395	0.0286	0.169	6.01%	34.68%
IBK (K Nearest)	97.14%	0.9395	0.0326	0.1684	6.86%	34.55%
NNge	93.88%	0.8728	0.0612	0.2474	12.89%	50.77%
PART	99.59%	0.9914	0.0043	0.0627	0.90%	12.87%
NBTree	100%	1	0.0069	0.0241	1.45%	4.95%
SimpleCart	97.96%	0.9568	0.0244	0.1392	5.13%	28.56%
AdaBoost	93.0612	0.8501	0.1388	0.2394	29.2101	49.1244
ClassificationViaRegression	95.5102	0.9034	0.1497	0.2182	31.5116	44.7648
CVParamterSelection	61.2245	0	0.4751	0.4874	100	100
RandomSubSpace	99.5918	0.9914	0.085	0.1426	17.8912	29.2673
NestedDichotomies	99.5918	0.9914	0.0056	0.0642	1.1682	13.1699
Dagging	77.9592	0.4862	0.2551	0.4467	53.6906	91.6592
ThresholdSelector	79.1837	0.5641	0.2625	0.3713	55.239	76.184

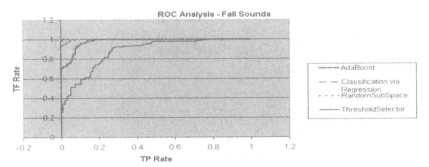

Fig. 6. Roc Curve for illustrating the performance of an indicative sample of the assessed meta-classifiers over the fall sounds classification

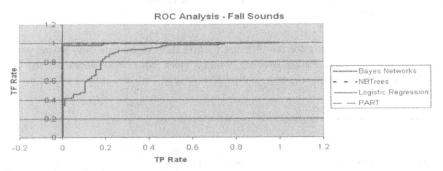

Fig. 7. Roc Curve for illustrating the performance of an indicative sample of the assessed classifiers over the fall sounds classification

8 Conclusions

In this paper we propose and evaluate a methodology for the analysis and classification of sounds and speech expressions for emergency events detection. A novel method for sound feature extraction, capable of implementation on mobile devices, has been presented. A number of state of the art classification methods has been evaluated against the successful detection of distress speech expressions and fall sounds. The initial results are very promising and indicate a significant proof for the concept of the proposed approach. It is in our future plans to combine the sound samples with information from the visual channel and with data collected by on-body wearable sensors (biosignals, accelerometers etc.) and additionally use more sensors for better redundancy.

References

[1] Noury, N., Herve, T., Rialle, V., Virone, G., Mercier, E., Morey, G., Moro, A., Porcheron, T.: Monitoring behavior in home using a smart fall sensor and position sensors. In: Proc. of 1st Annual International Conference on Microtechnologies in Medicine and Biology, October 2000, pp. 607–610 (2000)

[2] Noury, N.: A smart sensor for the remote follow up of activity and fall detection of the elderly. In: Proc. of 2nd Annual International Conference on Microtechnologies in Medicine and Biology, May 2002, pp. 314–317 (2002)

[3] Prado, M., Reina-Tosina, J., Roa, L.: Distributed intelligent architecture for falling detection and physical activity analysis in the elderly. In: Proc. of the 24th Annual IEEE EMBS Conference, October 2002, pp. 1910–1911 (2002)

[4] Fukaya, K.: Fall detection sensor for fall protection airbag. In: Proc. of the 41st SICE Annual Conference, August 2002, pp. 419–420 (2002)

[5] Sixsmith, A., Johnson, N.: A smart sensor to detect the falls of the elderly. IEEE Pervasing Computing 3(2), 42–47 (2004)

[6] Nait-Charif, H., McKenna, S.J.: Activity summarisation and fall detection in a supportive home environment. In: Proc. of the 17th International Conference on Pattern Recognition ICPR 2004, August 2004, pp. 323–236 (2004)

[7] Hwang, J.Y., Kang, J.M., Jang, Y.W., Kim, H.C.: Development of novel algorithm and real-time monitoring ambulatory system using Bluetooth module for fall detection in the elderly. In: Proc. of the 26th Annual International Conference of the IEEE Engineering in Medicine and Biology Society, pp. 2204–2207 (2004)

[8] Luo, S., Hu, Q.: A dynamic motion pattern analysis approach to fall detection. In: Proc. of the 2004 IEEE International Workshop on Biomedical Circuits and Systems, December 2004, pp. 1–8 (2004)

[9] Wang, S., Yang, J., Chen, N., Chen, X., Zhang, Q.: Human activity recognition with user-free accelerometers in the sensor networks. In: Proc. of International Conference on Neural Networks and Brain, October 2005, pp. 1212–1217 (2005)

[10] Miaou, S.-G., Sung, P.-H., Huang, C.-Y.: A Customized Human Fall Detection System Using Omni-Camera Images and Personal Information. In: Proc. of 1st Transdisciplinary conference on Distributed Diagnosis and Home Healthcare, pp. 39–42 (2006)

[11] Alwan, M., Rajendran, P.J., Kell, S., Mack, D., Dalal, S., Wolfe, M., Felder, R.: A Smart and Passive Floor-Vibration Based Fall Detector for Elderly. In: Proc. of 2nd Information and Communication Technologies Conference, ICTTA 2006, April 2006, pp. 1003–1007 (2006)

[12] Allen, F.R., Ambikairajah, E., Lovell, N.H., Celler, B.G.: An Adapted Gaussian Mixture Model Approach to Accelerometry-Based Movement Classification Using Time-Domain Features. In: Proc. of 28th Annual International Conference of the IEEE Engineering in Medicine and Biology Society, August 2006, pp. 3600–3603 (2006)

[13] Rougier, C., Meunier, J., St-Arnaud, A., Rousseau, J.: Monocular 3D Head Tracking to Detect Falls of Elderly People. In: Proc. of 28th Annual International Conference of the IEEE Engineering in Medicine and Biology Society, August 2006, pp. 6384–6387 (2006)

[14] Cao, X.B., Chen, D., Qiao, H., Xu, Y.W.: An Evolutionary Support Vector Machines Classifier for Pedestrian Detection. In: Proc. of 2006 IEEE International Conference on Intelligent Robots and Systems, October 2006, pp. 4223–4227 (2006)

[15] Bart, J., Rudi, D.: Context aware inactivity recognition for visual fall detection. In: Proc. of 2006 Pervasive Health Conference and Workshops, December 2006, pp. 1–4 (2006)

[16] Gaura, E.I., Rider, R.J., Steele, N., Naguib, R.N.G.: Neural-network compensation methods for capacitive micromachined accelerometers for use in telecare medicine. IEEE Transactions on Information Technology in Biomedicine 5(3), 248–252

[17] http://www.zigbee.org

[18] http://www.freescale.com

[19] Christianini, N., Shawe-Taylor, J.: An introduction to support vector machines. Cambridge University Press, Cambridge (2000)

[20] Schölkopf, B.: Statistical learning and kernel methods, http://research.Microsoft.com/~bsc

[21] Chang, C.-C., Lin, C.-J.: LIBSVM: a library for support vector machines (2001), http://www.csie.ntu.edu.tw/~cjlin/libsvm

[22] Sasiadek, J.Z., Khe, J.: Sensor fusion based on fuzzy Kalman filter. In: Proc. of 2nd International IEEE Workshop on Robot Motion and Control, October 2001, pp. 275–283 (2001)

[23] Grewal, M.S., Andrews, A.P.: Kalman Filterinf., 2nd edn. John Wiley & Sons, Chichester (2001)

[24] Virone, G., Istrate, D., Vacher, M., Noury, N., Serignat, J.F., Demongeot, J.: First Steps in Data Fusion between a Multichannel Audio Acquisition and an Information System for Home Healthcare. In: Proc. of the 25th Annual International Conference of the IEEE in Engineering in Medicine and Biology Society, September 2003, vol. 2, pp. 1364–1367 (2003)

[25] Istrate, D., Castelli, E., Vacher, M., Besacier, L., Serignat, J.-F.: Information extraction from sound for medical telemonitoring. IEEE Transaction on Information Theory in Biomedicine 10(2), 264–274 (2006)

[26] Vacher, M., Istrate, D., Serignat, J.F.: Sound Detection and Classification through transient models using Wavelet Coefficient Trees. In: 12th Eurasip European Signal Processing Conference (September 2004)

[27] Istrate, D., Vacher, M., Serignat, J.F.: Generic Implementation of a Distress Sound Extraction System for Elder Care. In: Proceedings of the 28th IEEE EMBS Annual International Conference, New York City, USA, August 30-September 3, pp. 3309–3312 (2006)

[28] Laydrus, N.C., Ambikairajah, E., Celler, B.: Automated Sound Analysis System for Home Telemonitoring using Shifted Delta Cepstral Features. In: 15th International Conference on Digital Signal Processing, July 2007, pp. 135–138 (2007)

[29] Nguyen, C.P., Pham, T.N.Y., Eric, C.: Toward a Sound Analysis System for Telemedicine. In: Wang, L., Jin, Y. (eds.) FSKD 2005. LNCS (LNAI), vol. 3614, pp. 352–361. Springer, Heidelberg (2005)

[30] Sonogram Speech and Sound processing tool, http://www.christoph-lauer.de/

[31] SensiNode, http://www.sensinode.com

[32] Alwan, M., Dalal, S., Mack, D., Kell, S.W., Turner, B., Leachtenauer, J., Felder, R.: Impact of Monitoring Technology in Assisted Living: Outcome Pilot. IEEE Trans. on IT in Biomedicine 10(1), 192–198

[33] Freund, Y., Schapire, R.E.: Experiments with a new boosting algorithm. In: Thirteenth International Conference on Machine Learning, San Francisco, pp. 148–156 (1996)

[34] Frank, E., Wang, Y., Inglis, S., Holmes, G., Witten, I.H.: Using model trees for classification. Machine Learning 32(1), 63–76 (1998)

[35] Kohavi, R.: Wrappers for Performance Enhancement and Oblivious Decision Graphs. Department of Computer Science, Stanford University (1995)

[36] Ho, T.K.: The Random Subspace Method for Constructing Decision Forests. IEEE Transactions on Pattern Analysis and Machine Intelligence 20(8), 832–844 (1998)

[37] Dong, L., Frank, E., Kramer, S.: Ensembles of Balanced Nested Dichotomies for Multiclass Problems. In: Jorge, A.M., Torgo, L., Brazdil, P.B., Camacho, R., Gama, J. (eds.) PKDD 2005. LNCS (LNAI), vol. 3721, pp. 84–95. Springer, Heidelberg (2005)

[38] Ting, K.M., Witten, I.H.: Stacking Bagged and Dagged Models. In: Fourteenth international Conference on Machine Learning, San Francisco, CA, pp. 367–375 (1997)

[39] Witten, I.H., Frank, E.: Data Mining: Practical machine learning tools and techniques, 2nd edn. Morgan Kaufmann, San Francisco (2005)

[40] Carletta, J.: Assessing Agreement on Classification Tasks: The Kappa Statistic. Computational Linguistics 22(2), 249–254 (1996)

A Genetic Programming Environment for System Modeling

Efstratios F. Georgopoulos[1,3], George P. Zarogiannis, Adam V. Adamopoulos[1,2],
Anastasios P. Vassilopoulos[4], and Spiridon D. Likothanassis[1]

[1] Pattern Recognition Laboratory, Dept. of Computer Engineering & Informatics,
University of Patras, 26500, Patras, Greece
[2] Medical Physics Laboratory, Department of Medicine, Democritus University of Thrace,
68100, Alexandroupolis, Greece
[3] Technological Educational Institute of Kalamata, 24100, Kalamata, Greece
[4] Composite Construction Laboratory (CCLab), Ecole Polytechnique Fédérale de Lausanne
(EPFL), Station 16, CH-1015 Lausanne, Switzerland
sfg@teikal.gr, adam@med.duth.gr,
anastasios.vasilopoulos@epfl.ch, likothan@cti.gr

Abstract. In the current paper we present an integrated genetic programming
environment with a graphical user interface (GUI), called jGPModeling. The
jGPModeling environment was developed using the JAVA programming language, and is an implementation of the steady-state genetic programming algorithm. That algorithm evolves tree based structures that represent models of
input – output relation of a system. During the design and implementation of
the application, we focused on the execution time optimization and tried to limit
the bloat effect. In order to evaluate the performance of the jGPModeling environment, two different real world system modeling tasks were used.

Keywords: Genetic Programming, Evolutionary Algorithms, System Modeling,
MEG modeling, fatigue modeling.

1 Introduction

The problem of discovering a mathematical expression that describes the operation of
a physical or artificial system using empirically observed variables or measurements
is a very common and important problem in many scientific areas. Usually, the observed data are noisy and sometimes missing. Also, it is very common, that there is no
known mathematical way to express the relation using a formal mathematical way.
These kinds of problems are called modeling problems, symbolic system identification problems, black box problems, or data mining problems [14].

Most data-driven system modeling or system identification techniques assume an
a-priori known model structure and focus mainly to the calculation of the model parameters' values. But what can be done when there is no *a-priori* knowledge about the
model structure?

Genetic programming (GP) is a domain-independent problem-solving technique in
which computer programs are evolved to solve, or approximately solve, problems. Genetic programming is a member of a broad family of techniques called Evolutionary

J. Darzentas et al. (Eds.): SETN 2008, LNAI 5138, pp. 85–96, 2008.

Algorithms. All these techniques are based on the Darwinian principle of reproduction and *survival of the fittest* and are similar to the biological genetic operations such as crossover and mutation. Genetic programming addresses one of the central goals of computer science, namely automatic programming; which is to create, in an automated way, a computer program that enables a computer to solve a problem [1], [2].

In GP the evolution operates on a population of computer programs of varying sizes and shapes. GP starts with an initial population of thousands or millions of randomly generated computer programs, composed of the available programmatic ingredients and then applies the principles of biological evolution to create a new (and often improved) population of programs. The generation of this new population is done in a domain-independent way using the Darwinian principle of *survival of the fittest,* an analogue of the naturally-occurring genetic operation of sexual recombination (crossover), and occasional mutation [3]. The crossover operation is designed to create syntactically valid offspring programs (given closure amongst the set of programmatic ingredients). GP combines the expressive high-level symbolic representations of computer programs with the near-optimal efficiency of learning of Holland's genetic algorithm. A computer program that solves (or approximately solves) a given problem often emerges from this process [3].

GP for the problem of system modeling doesn't need an a priori knowledge of the model structure like other techniques do. GP evolves a system model from scratch. It doesn't use a predefined model structure or a bank of model structures and search just for the parameter values that fit better the data, like other data driven methods. Instead, GP creates an initial population of models and evolves those using genetic operators in order to find the mathematical expression that best fit the data of the given system. That is, GP searches simultaneously for the model structure and parameters. Genetic Programming has been applied successfully in a number of system modeling problems like the ones in [12], [13], [14], [15], [16], [17], [18], [19], [20].

In the current paper we present an integrated GP environment with a graphical user interface (GUI), called jGPModeling. The jGPModeling environment was developed using the JAVA programming language, and is an implementation of the steady-state genetic programming algorithm. That algorithm evolves tree based structures that represent models of input – output relation of a system. During the design and implementation of the application, we focused on the execution time optimization and tried to limit the bloat effect, which refers to the continuously increasing and expanding tree size. In order to evaluate the performance of the jGPModeling environment, two different real world system modeling tasks were used. First, we tried to find a model of the MagnetoEncephaloGram (MEG) recordings of epileptic patients. MEG recordings where obtained using a Superconductive QUantum Interference Device (SQUID) which is installed in Medical Physics Laboratory, in the General Hospital of Alexandroupolis, Greece. Second, we tried to model the behavior of a composite material subject to stress tests. This material is a fiberglass laminate, which is a typical material for wind turbine blade construction.

The rest of the paper is organized as follows. Section 2 is an introduction to Genetic Programming, while section 3 presents the jGPModeling environment. The modeling experiments are presented in section 4. Finally, section 5 discusses the concluding remarks.

2 The GP Algorithm

In the jGPModeling environment it is implemented a steady state GP, which is a variation of the classic GP algorithm and in contrast with that, it doesn't make use of the generational evolutionary scheme. The Steady state GP was selected because the great multiprocessing capabilities that exhibits. The steady state GP algorithm generates models by executing the following steps, [3]:

(1) Generates an initial population of randomly constructed models (these are random compositions of the programmatic ingredients of the problem). Ever model is represented as a tree. The size of each model is limited to a maximum number of points (i.e. total number of functions and terminals) or a maximum depth of the model tree. The models in this initial population will almost always have very poor performance. Nonetheless, some individuals will turn out to be somewhat more fit than others. These differences in performance are then exploited by genetic programming.

(2) Selects, in random, a subset of the population for a tournament (tournament selection).

(3) Evaluates the members of this subset and assign it a fitness value. Typically, each model runs over a number of different fitness cases so that its fitness is measured as a sum or an average over a variety of representative different situations. For example, the fitness may be measured in terms of the sum of the absolute value of the differences between the output produced by the model and the desired output (i.e., the Minkowski distance) or the square root of the sum of the squares (i.e., Euclidean distance). These sums are taken over a sampling of different fitness cases. The fitness cases may be chosen at random or may be chosen in some structured way (e.g., at regular intervals) [3].

(4) Selects the tournament winners according to their fitness values.

(5) Applies genetic operators of mutation and crossover on the winners.

    a. *Mutation*: Create a new model from an existing one by mutating a randomly chosen part of it. One mutation point is randomly and independently chosen and the sub-tree occurring at that point is deleted. Then, a new sub-tree is grown at that point using the same growth procedure as was originally used to create the initial random population - this is only one of the many different ways that mutation operation can be implemented. For more details about mutation implementation see [2].

    b. *Crossover*: Create two new models from two existing ones by genetically recombining randomly chosen parts of them using the crossover operation applied at a randomly chosen crossover point within each model. Because entire sub-trees are swapped, the crossover operation always produces syntactically and semantically valid models, as offspring, regardless of the choice of the two crossover points. Because models are selected to participate in the crossover operation based on their fitness, crossover allocates future trials to regions of the search space whose models contain parts from promising models [3]. (For details about crossover implementation see [2]).

(6) Replaces in the population the tournament losers by the winner offspring.

(7) If the termination criterion is not fulfilled returns to step 2.

(8) Returns the best ever found individual (model) of the population as the result.

3 The jGPModeling Environment

The jGPModeling environment was implemented in the Java programming language because of the many advantages that this language offers, like: object orientation,

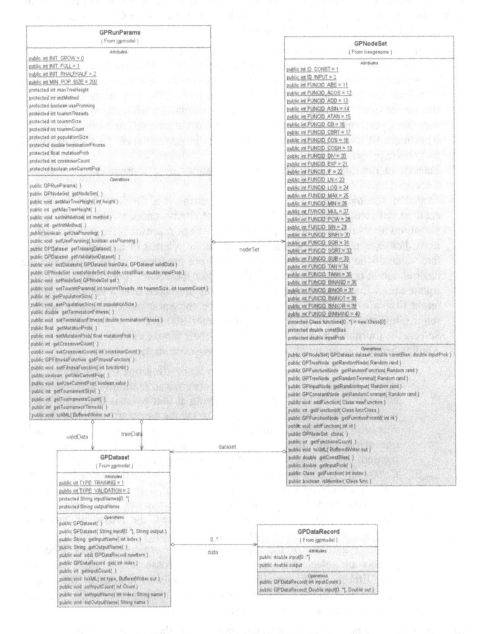

Fig. 1. Basic classes of the jGPModeling Environment

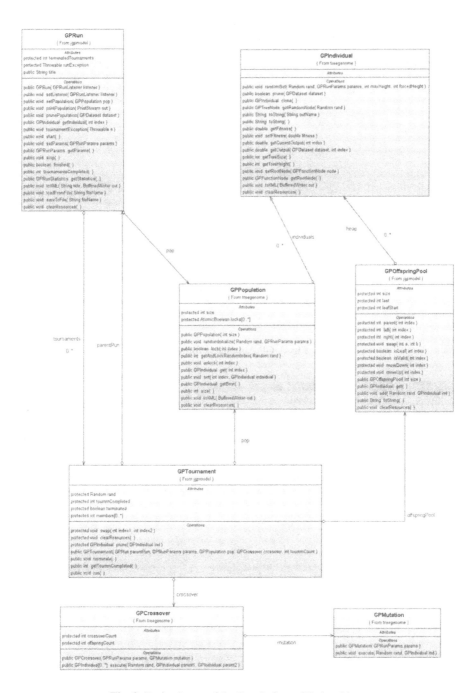

Fig. 2. Basic classes of the Steady State GP algorithm

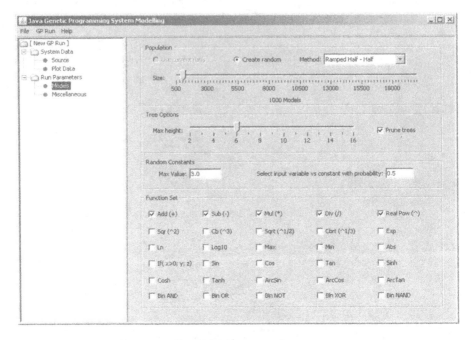

Fig. 3. GP Parameters form

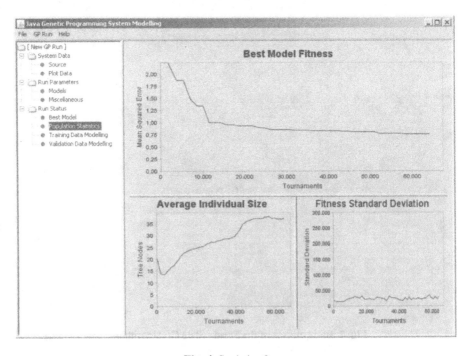

Fig. 4. Statistics form

freeware compilers and Integrated Development Environments, portability, multi-threading, multiprocessing capabilities, a huge number of internet communities working on Java, freeware and open source libraries for graphics, network programming, multithreading, security etc. As IDE it was used the SUN NetBeans environment (www.netbeans.org). In order to speed up the GP execution the implemented algorithm uses multithreading, so that it can be benefited from the multiprocessing capabilities of the modern multi-core processors. Moreover, there are implemented caching and loop unroll techniques, as while as techniques to control the bloat phenomenon that is very common in GP implementations.

Figure 1 depicts the basic classes of the jGPModeling application, which are the *GPDataRecord:* holds the input output data of one training/testing example, the *GPDataSet:* holds the whole training and testing set, the *GPNode:* is the class of the tree representation of the models and the *GPRunParams*: holds the algorithms' parameter values. Figure 2 depicts the basic classes of the GP algorithm, which are the *GPIndividual*: the class of the population individuals (models), the *GPMutation*: the mutation class, the *GPCrossover*: the crossover class, the *GPPopulation*: the population class, the *GPOffspring*: the class of the offspring stack and is used in the tournament operation, the *GPTournament*: the class of the tournament operation and the *GPRun*: a "master" class responsible for the operation of the whole steady state GP algorithm. Figures 3 and 4 depict some (from the many) forms of the jGPModeling environment's GUI. The GUI is a user friendly and fully operational user interface where the user can insert all of the GP algorithm's parameters, the data files (training and test) from the problem and observe the evolution process through a number of different online graphs. At the end of the evolutionary process the tool prints the best produced model in a mathematical expression form.

4 Modeling Experiments

The jGPModeling environment was tested on two system modeling problems. The first one is the modeling of the MagnetoEncephaloGram (MEG) of epileptic patients and the second one is the modeling of fatigue behavior of composite laminates.

4.1 MEG

Brain dynamics can be evaluated by recording the changes of the neuronal electric voltage, either by the electroencephalogram (EEG), or by the magnetoencephalogram (MEG). The EEG recordings represent the time series that match up to neurological activity as a function of time. On the other hand the MEG is generated due to the time varying nature of the neuronal electric activity, since time-varying electric currents generate magnetic fields. EEG and MEG are considered to be complementary, each one carrying a part but not the whole of the information related to the underlying neural activity [4]. Thus, it has been suggested that the EEG is mostly related to the inter-neural electric activity, whereas the MEG is mostly related to the intra-neural activity. MEG recordings are obtained using Superconductive QUantum Interference Devices (SQUIDs) that are very sensitive magnetometers, capable to detect and record biomagnetic fields of the order of fT ($=10^{-15}$ T).

Table 1. The Genetic Programming parameters for MEG modeling

Population Size:	3.000
Max tree depth:	5
Constants' range:	[-3, 3]
Function Set:	+, -, *, /, ^, ^2, ^3, ^1/2, ^1/3, Exp
Fitness evaluation function:	Mean Squared Error
Tournament Size;	4
Crossover trials:	2
Mutation Probability:	0.75

Table 2. Model found by GP for MEG modeling

x(t) = (((((1.01041 * x(t-5)) + (x(t-1) – x(t-5))) + ((x(t-3) – x(t-4)) / -1.10131)) + (((x(t-4) – x(t-7)) + (x(t-2) – x(t-1))) / -1.75476)) - ((((x(t-2) * 1.01041) * 1.00921) + (x(t-6) – x(t-5))) + ((x(t-6) – x(t-5)) – x(t-1))))

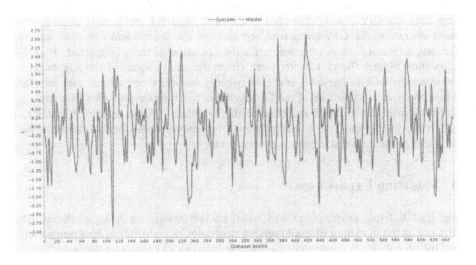

Fig. 5. Performance of the model found for MEG modeling problem on the test set

To investigate, model and even more, to predict neural and brain dynamics in epilepsy many efforts have been made in the past. Neural network models have been developed to elucidate the underlying neural and brain modeling of epileptogenesis [5] and propagation of epileptic seizures [6]; dimensionality calculations of the EEG [7] and the MEG [8] of epileptic patients have been performed, considering the emerging brain dynamics with respect to the Theory of Nonlinear Systems and Chaos for the analysis of the *strange attractors* that are observed; even more, intelligent signal processing methods have been applied, including the use artificial neural networks (ANN) for EEG and MEG signal prediction, as well as, the development of Nonlinear Autoregressive Moving Average with external inputs (NARMAX) to model of these signals [9]. All these methods conclude to a common point, that is, in epileptic brain dynamics, a phase transition does occur, leading form higher to lower dimensional dynamics, emerging self-organization and synchronicity at the neural, as well as, at the systemic level. To our knowledge, none of the previous studies considered the task of modeling epileptic brain dynamics with the use of GP. In this case the task is to find a mathematical expression that connects the current MEG value (x(t)) with some former ones (x(t-1), x(t-2), etc). Table 2 depicts a model evolved by the GP algorithm for the parameters of table 1, while figure 5 the performance of the best produced model.

4.2 Fatigue Life of Composite Laminates

Many experimental data sets can be found in the literature on the fatigue behavior of composite laminates (see [10] and [11] for details). The database used here, contains fatigue data of composite materials tested under several fatigue loading conditions of constant amplitude (see [10] for details). All of the examined materials are fiberglass-polyester and fiberglass-epoxy laminates, typical materials used in wind turbine rotor blade construction. Specifically, we consider the fatigue data from coupons cut at several on- and off-axis angles from a multidirectional laminate of the stacking sequence $[0/(\pm45)_2/0]_T$. Two unidirectional laminates of 100% aligned fibers, with a weight of 700 g/m^2 and two stitched ±45, in the middle, of 450 g/m^2, 225 g/m^2 in each off-axis direction. The material is glass polyester and the laminates were fabricated by hand lay-up. Coupons were cut, by a diamond saw wheel, at several on- and off-axis directions, 0°, on-axis, and 15°, 30°, 45°, 60°, 75° and 90° off-axis directions. In total, 257 valid test results were obtained and comprise the constant amplitude fatigue data base. Fatigue data were used for the determination of 17 S-N curves at various on- and off-axis loading directions, under four different stress ratios ($R=\sigma_{min}/\sigma_{max}$), namely, R=10 (C-C), R=-1 (T-C), R=0.1 and R=0.5 (T-T). At least three coupons were tested at each one of the four or five stress levels that were pre-assigned for the determination of each S-N curve, according to the material under examination. Thus, 12-18 coupons were tested for the determination of each one of the 17 S-N curves. Details about the specified material, preparation and testing procedures can be found in [10]. Here, the task is to calculate a mathematical expression of the number of cycles to failure N as a function of R and σ_{max}. To our knowledge, this is the first time that Genetic Programming is used for modeling the fatigue behavior of composite laminates. Table 4 depicts a model evolved by the GP algorithm for the parameters of table 3, while figure 6 depicts the performance of this model.

Table 3. The Genetic Programming parameters for Material #2 modeling

Population Size:	1.500
Max tree depth:	7
Constants' range:	[-3, 3]
Function Set:	+, -, *, /, ^, ^2, ^3, ^1/2, ^1/3, Exp, If
Fitness evaluation function:	Mean Squared Error
Tournament Size:	4
Crossover trials:	3
Mutation Probability:	0,9

Table 4. Model found by GP for Material #2 modeling

logN = (((((normsmax - ((normsmax ^ 0.75353) ^ (normsmax / 0.05539))) ^ IF(((normR ^ 0.58267) + normsmax) > 0 ; (normsmax ^ (normsmax ^ 0.01139)) ; 0.34968)) + 0.50471) ^ exp(((((normR * 1.63523) ^ (normR - 1.58772)) ^ IF((0.55180 - normR) > 0 ; (normR ^ 0.90228) ; (-3.00786 / normsmax))) ^ (IF((normsmax + -0.68915) > 0 ; 5.15377 ; (1.33572 - normR)) ^ 2)))) / IF(cbrt(((((normR ^ normsmax) + 0.50471) / cbrt(normR)) + IF(normR > 0 ; (normsmax * -3.00786) ; (0.50471 * normsmax)))) > 0 ; (((((normsmax - normR) ^ 3) ^ 3) + ((normsmax / (normsmax ^ 0.01055)) / (0.98240 ^ (normR ^ normsmax)))) * exp((((normsmax - -0.04625) ^ (0.55180 ^ normR)) ^ 3))) ; ((IF(IF(normR > 0 ; 0.75353 ; normsmax) > 0 ; ((1.65546 ^ normR) * (0.55180 ^ normR)) ; ((normsmax /0.05539) / 1.33182)) ^ IF((sqrt(normR) + normsmax) > 0 ; normsmax ; normR)) + 4639731933.59830)))

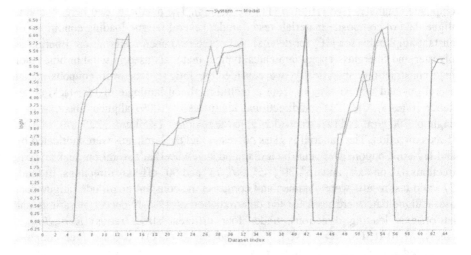

Fig. 6. Performance of the model found for fatigue modeling problem on the test set

5   Conclusions

Observing the models of the real systems obtained by jGPModeling, it is confirmed what is already known in the community of artificial intelligence and machine learning. GP is an algorithm capable to solve difficult problems evolving complicated structures of programs. From the two very different types of systems that have been modeled with great success it is getting obvious that the jGPModeling application can be used in a broad spectrum of problems in different scientific areas. Also, it is important to note that the rapid development of very fast computer hardware encourages the use of techniques like GP, and is expected that these kinds of techniques will be used more and more for solving difficult problems in the future.

Now, concerning some future directions of the presented jGPModeling environment these could be:

- The application of jGPModeling tool to other system modeling task in various scientific areas, with the help of experts in sectors that exist such type problems.
- The further improvement of the tool in terms of speed and memory management.

- The implementation of more genetic operators and enhancements in the basic GP algorithm, like the Automatically Defined Functions [2], the use of some grammar for the syntax of models and etc.
- Using as a base the multiprocessing abilities of jGPModeling environment we shall work on the direction of adding new features in order jGPModeling environment to be able to run in a distributed computer network. Thus, the tool taking advantage of the large computational power of these distributed systems it will be able to tackle difficult modeling tasks.

Thus, the jGPModeling environment proved to be a user friendly and powerful GP tool with great expansion capabilities that we intend to improve in the near future.

References

1. Koza, J.R.: Genetic Programming: A Paradigm for Genetically Breeding Populations of Computer Programs to Solve Problems Technical Report STAN-TR-CS 1314, Stanford University (1990)
2. Koza, J.R.: Genetic programming: on the programming of computers by means of natural selection. MIT Press, Cambridge (1992)
3. Koza, J.R.: Genetic Programming. In: Williams, J.G., Kent, A. (eds.) Encyclopedia of Computer Science and Technology, vol. 39 (suppl. 24), pp. 29–43. Marcel Dekker, New York (1998)
4. Dassios, G., Fokas, A.S., Hadjiloizi, D.: On the complementarity of EEG and MEG. Inverse Problems 23, 2541–2549 (2007)
5. Anninos, P.A., Tsagas, N., Adamopoulos, A.: A brain model theory for epilepsy and its treatment: experimental verification using SQUID measurements. In: Cotterill, R.M.J. (ed.) Models of brain function, pp. 405–422. Cambridge University Press, Cambridge (1989)
6. Jouny, C.C., Frabaszczuk, P.L., Bergey, G.K.: Complexity-based analysis of the dynamic of propagation of epileptic seizures. In: Proceedings of the 2nd International IEEE EMBS Conference on Neural Engineering, pp. 155–157 (2005)
7. Tsoutsouras, V.G., Iliopoulos, A.C., Pavlos, G.P.: Non-linear analysis and modelling of EEG in patients with epilepsy. In: 19th Panhellenic Conference on Nonlinear Science and Complexity, Thessaloniki, Greece, July 10-22 (2006)
8. Adamopoulos, A., Pavlos, G., Anninos, P., Tsitsis, D., Rigas, A.: Chaotic analysis of biomagnetic measurements of human brain, NATO ASI on Chaotic Synamics – Theory and Practice, Patras, Greece, July, 11-20 (1991)
9. Adamopoulos, A.V.: Intelligent adaptive modelling and identification of MEG of epileptic patients. WSEAS Transactions on Biology and Biomedicine 3(2), 69–76 (2006)
10. Philippidis, T.P., Vassilopoulos, A.P.: Complex stress state effect on fatigue life of GRP laminates. Part I, experimental. Int. J. Fatigue 24, 813–823 (2002)
11. Nijssen, R.P.L.: OptiDAT – fatigue of wind turbine materials database, http://www.kc-wmc.nl/optimat_blades/index.htm
12. Winkler, S.: Identifying Nonlinear Model Structures Using Genetic Programming. Diploma Thesis, Institute of Systems Theory and Simulation, Johannes Kepler University Linz, Austria (2004)
13. Winkler, S., Affenzeller, M., Wagner, S.: New Methods for the Identification of Nonlinear Model Structures Based Upon Genetic Programming Techniques. In: Proceedings of the 15th International Conference on Systems Science, vol. 1, pp. 386–393. Oficyna Wydawnicza Politechniki Wroclawskiej (2004)

14. Winkler, S., Affenzeller, M., Wagner, S.: Identifying Nonlinear Model Structures Using Genetic Programming Techniques. In: Cybernetics and Systems 2004, pp. 689–694. Austrian Society for Cybernetic Studies (2004)
15. Madár, J., Abonyi, F.: Szeifert: Genetic Programming for the Identification of Nonlinear Input-Output Models (accepted, 2005)
16. Madár, J., Abonyi, J., Szeifert, F.: Genetic Programming for System Identification. In: Intelligent Systems Design and Applications (ISDA) (2004)
17. Willis, M.J., Hiden, H.G., Marenbach, P., McKay, B., Montague, G.A.: Genetic programming: An introduction and survey of applications. In: Second International Conference on Genetic Algorithms in Engineering Systems, pp. 314–319 (1997)
18. Tsang, E.P.K., Butler, J.M., Li, J.: EDDIE Beats the Bookies. Journal of SOFTWARE – Practice and Experience 28(10) (1998)
19. Alex, F., Stechert, A.: Evolving Nonlinear Predictive Models for Lossless Image Compression with Genetic Programming. In: Genetic Programming 1998: Proceedings of the Third Annual Conference, pp. 95–102. Morgan Kaufmann, Wisconsin (1998)
20. Werner, J.C., Fogarty, T.C.: Genetic programming applied to Collagen disease & thrombosis. South Bank University, London (2001)

Mining Gene Expression Profiles and Gene Regulatory Networks: Identification of Phenotype-Specific Molecular Mechanisms

Alexandros Kanterakis[1], Dimitris Kafetzopoulos[2], Vassilis Moustakis[1,3], and George Potamias[1,*]

[1] Foundation for Research & Technology – Hellas (FORTH), Institute of Computer Science, Vassilika Vouton, P.O. Box 1385, 71110 Heraklion, Greece
[2] Foundation for Research & Technology – Hellas (FORTH), Institute of Molecular Biology & Biotechnology, Vassilika Vouton, P.O. Box 1385, 71110 Heraklion, Greece
[3] Technical University of Crete, Department of Production Engineering and Management, Chania, Greece
{kantale,moustaki,potamias}@ics.forth.gr, kafetzo@imbb.forth.gr

Abstract. The complex regulatory mechanisms of genes and their transcription are the major gene regulatory steps in the cell. Gene Regulatory Networks (GRNs) and DNA Microarrays (MAs) present two of the most prominent and heavily researched concepts in contemporary molecular biology and bioinformatics. The challenge in contemporary biomedical informatics research lies in systems biology – the linking of various pillars of heterogeneous data so they can be used in synergy for life science research. Faced with this challenge we devised and present an integrated methodology that 'amalgamates' knowledge and data from both GRNs and MA gene expression sources. The methodology, is able to identify phenotype-specific GRN functional paths, and aims to uncover potential gene-regulatory 'fingerprints' and molecular mechanisms that govern the genomic profiles of diseases. Initial implementation and experimental results on a real-world breast-cancer study demonstrate the suitability and reliability of the approach.

Keywords: Data mining, Bioinformatics, Biomedical informatics, Microarrays, gene expression profiling, Gene regulatory networks. Bresat cancer.

1 Introduction

With the completion of the sequencing of the human genome we entered the *post-genomic* age. The main focus in genomic research is switching from sequencing to the understanding of how genomes are functioning. Now we need high-quality annotation of all the *functionally* important genes and the variations within them that contribute to health and disease.

Contemporary post-genomics bioinformatics research seeks for methods that not only combine the information from dispersed and heterogeneous data sources but distil the knowledge and provide a systematic, genome-scale view of biology [1]. The

---

* Corresponding author.

J. Darzentas et al. (Eds.): SETN 2008, LNAI 5138, pp. 97–109, 2008.
© Springer-Verlag Berlin Heidelberg 2008

advantage of this approach is that it can identify emergent properties of the underlying molecular system as a 'whole' – an endeavor of limited success if targeted genes, reactions or even molecular pathways are studied in *isolation* [2]. Individuals show different phenotypes for the same disease as they respond differently to drugs, and sometimes the effects are unpredictable. Many of the genes examined in early clinico-genomic studies were linked to single-gene traits, but future advances engage the elucidation of multi-gene determinants of drug response. Differences in the individuals' background DNA code but mainly, differences in the underlying gene regulation mechanisms alter the expression or function of proteins being targeted by drugs, and contribute significantly to variation in the responses of individuals.

Gene Regulatory Networks (GRNs) and DNA Microarrays (MAs) present two of the most prominent and heavily researched concepts in contemporary molecular biology and bioinformatics. GRNs model the interfering relations among gene products during the regulation of the cell function. The real future of bioinformatics and bio-medical informatics lies in systems biology – the linking of various pillars of hetero-geneous data so they can be used in synergy for life science research. So, the biggest challenge right now is the integration and analysis of disparate data. The aim is to accelerate our understanding of the *molecular mechanisms* underlying genetic variations and to produce targeted individualized therapies.

Faced with the aforementioned challenges we devised and present an integrated methodology that 'amalgamates' knowledge and data from both GRNs and MA gene expression sources. The methodology aims to uncover potential gene-regulatory 'fin-gerprints' and molecular mechanisms that govern the genomic profiles of diseases.

In the next section we present the background of MA technology and gene expres-sion data mining, as well as the background to GRNs and their analysis, accompanied with the intrinsic problems related to MAs and GRNs when studied in isolation. In the third section we present our combined GRNs/MA mining methodology in detail. In the next section we present experimental results and discuss on the application of the meth-odology on a real-world breast cancer microarray gene expression study. In the last section we summarize our contribution and point to on-going and future research work.

2   Background: Microarrays and Gene Regulatory Networks

Recent post-genomic studies aim towards the integration of complementary method-ologies into a holistic picture of the molecular mechanisms underlying cell function. In this context *system biology* approaches take the lead [1]. The rationale behind this integrative approach is simple: while no dataset can comprehensively define a cellular pathway or response, several complementary data sets may be integrated in order to reveal such pathways or, provide insight into a molecular process that remains cov-ered when a single data source is utilized. For example, MA gene expression profiling and analysis may be used to identify genes differentially regulated at the level of transcription. However, it cannot be used to identify the underlying regulation mecha-nism. But if they considered in union with GRNs they may provide an indication of both transcriptional and regulatory events. So, a more comprehensive and holistic view of the genes and the regulatory mechanisms driving a given cell response may be revealed [3].

Microarrays. With the recent advances in MA technology [4], [5], the potential for molecular diagnostic and prognostic tools seem to come in reality. A number of pioneering studies have been conducted that profile the expression-level of genes for various types of cancers [6], [7]. The aim is to add molecular characteristics to the classification of diseases so that diagnostic procedures are enhanced and prognostic predictions are improved. By measuring transcription levels of genes in an organism under various conditions we can build up *gene expression profiles* that characterize the dynamic functioning of each gene in the genome. Gene expression data analysis depends on gene-expression data mining aiming towards: (a) the investigation and modeling of a biological process that leads to predicted results, and (b) the detection of underlying *hidden-regularities* in biological data [8]. Possible prognostic genes for disease outcome, including response to treatment and disease recurrence are then selected to compose the targeted disease *molecular signature* of (or, *gene-markers* for) .

Gene Regulatory Networks. GRNs are the "on"–"off" switches (acting as rheostats) of a cell operating at the gene level. GRNs are network structures that depict the interaction of DNA segments during the transcription of the genes into mRNA. They dynamically orchestrate the level of expression for each gene by controlling whether that gene will be transcribed into RNA. A simple GRN may consist of one or more input *signalling pathways* and it can be viewed as a cellular input-output device (see Figure 1).

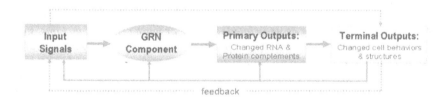

Fig. 1. A general layout of signalling molecular pathways: their components and functions

The prominent and vital role of GRNs in the study of various biology processes is a major sector in contemporary biology research, with a big number of thorough studies being published [9]. From a computational point of view GRNs can be conceived as analogue biochemical computers that regulate the level of expression of target genes [10]. The network by itself acts as a mechanism that determines cellular behavior where the nodes are genes and edges are functions that represent the molecular reactions between the nodes. These functions can be perceived as *Boolean* functions, where nodes have only two possible states - "on" and "off", and the whole network represented as a simple *directed graph* [11]. Most of the relations in known and established GRNs have been derived from laborious and extensive laboratory experiments.

Most of the current efforts in GRN research focus on the *reconstruction* of GRNs from MA gene expression data. The various approaches include: Boolean Networks [12], Bayesian Networks [13], differential equations and steady-state models [14], statistical and probabilistic [15], and data mining methods [16], [17]. Finally the integration of these methods with visualization techniques and software tools plays an important role in the field [18], [19].

2.1 The Limitations of Microarrays and Gene Regulatory Networks

MA experiments involve more variables (genes) than samples (patients). This fact, leads to results with poor biological significance. Simon et al. published a very strict criticism on common pitfalls on microarray data mining [20]. It is an established fact that different methods produce gene-marker lists that are strikingly different [21]. On the other hand, very few methods of gene regulatory inference are considered superior to the others mainly because of the intrinsically noisy property of the data, 'the curse of dimensionality', and the unknown 'true' underlying networks. Cell, tissues, organs, organisms or any other biological systems defined by evolution are essentially complex physicochemical systems. They consist of numerous complex and dynamic networks of biochemical reactions and signalling interactions between active cellular components. This cellular complexity has made it difficult to build a complete understanding of cellular machinery to achieve a specific purpose [22].

To circumvent this complexity, microarrays, biology knowledge and biology networks can be combined in order to document and support the detected and predicted interactions. A very nice review on the intersections between high-throughput technology and molecular networks could be found in [23] where, the validation of our approach, with respect to amalgamating GRNs and MAs, could be traced: *"when analyzed separately, datasets obtained from these (our note: GRNs and MAs) experiments cannot explain the whole picture. Intersection of the experimental data with the interaction content of the networks has to be used in order to provide the closest possible view of the activated molecular machinery in a cell. As all objects on the networks are annotated, they can be associated with one or more cellular functions, such as apoptosis, DNA repair, cell cycle check points and fatty acid metabolism"* (p.1747 in [23]). Recently, specific GRNs were utilized in order to document and validate the findings of MA experiments. For example Anisimov in [24], studied the neural pathway of stem cells as they differentiate to neural cells, in order to understand what pathway the neural cells follow during their development and degeneration. Special tools like GenMapp [25] and Pathway Explorer [26] have been used to identify the subparts of targeted GRNs that have been up- or down-regulated in the gene expression profiles of specific samples. In particular, Kulterer et al., following a somewhat similar to our approach, visualise on specific molecular pathways the status of differentiating genes during developmental (time-series) events (during osteogenesis) [27]. But a clear differentiation of the underlying molecular mechanisms for the different developmental phases could not be easily traced. In our approach we identify, visualise and trace-down the status of genes that clearly differentiate between different disease phenotypes. In any case, the advances and tools that each discipline carries can be integrated in a holistic and generic perspective so that the chaotic complexity of biology networks can be traced down.

Our methodology utilises known and well-established GRNs and 'amalgamate' them with the gene expression profiles assigned to specific categories or, *phenotypes*. The power of known and well established known GRNs rests on their underlying biological validity. This is in contact to the artificially reconstructed GRNs the molecular validity of which is to be explored. In other words, the artificial reconstruction of GRNs aims towards exploratory endeavours - to uncover new regulations, e.g., "which genes regulate other genes?" In contrast, our approach tries to take advantage and link established biological and molecular knowledge (as presented by established GRNs) with clinical knowledge, as presented by real phenotypic characterisation of diseases (e.g., histo-pathological measurements). The novelty of the approach relies on the *identification of characteristic GRN functional subparts that match, differentiate and putatively characterise different phenotypes.* The approach could be considered as orthogonal to approaches that map and visualize the gene expression profile of single samples on molecular pathways [28] or, highlight metabolic pathways of differentially expressed genes [29]. Our aim is to identify phenotype-related molecular mechanisms that uncover and putatively govern the genomic status of disease classes (status, responses etc.) The target is an integrated methodology that ease the identification of *individualised* molecular profiles and supports diagnostic, prognostic and therapeutic decision making needs in the context of the rising evidence-based individualised (or personalised) genomic medicine environment [30].

3 Combing MAs and GRNs: The Methodology

The genetic regulatory relationship we model assumes that a change in the expression of a regulator gene modulates the expression of a target gene. A change in the expression of these genes might change dramatically the behaviour of the whole network. The identification and prediction of such changes is a challenging task in bioinformatics.

3.1 Decomposition of GRNs into Functional Paths

Known and established GRNs, despite of their possible errors and misfits, they employ sufficient biological evidence. Existing GRNs databases provide us with widely utilized networks of proved molecular validity. Online public GRN repositories contain a variety of information: the GRN graphs (with their XMLs); links to online bio-databanks via the respective GRN's nodes (genes) and edge (regulation), meta-data and various bio-related

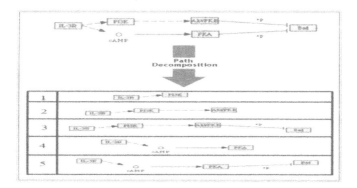

Fig. 2. Path decomposition. Top: A target part of the KEGG cell-cycle GRN; Bottom: The five decomposed paths for the targated GRN part - all possible routes (paths) taking place during network regulaion machinery

annotations. In the current implementation we utilize the KEGG repository of pathways (http://www.genome.jp/ kegg/). KEGG provides a standardized format representation (KGML, http://www.genome.jp/kegg/xml/). It also encompasses and models a variety of regulation links (please refer to http://www.genome.jp/kegg/document/help_ pathway.html).

Our methodology relies on a novel approach for GRN utilization and processing that takes into account all possible *interpretations* of the network. The different GRN interpretations correspond to the different *functional paths* that can be followed during the regulation of a target gene. It unfolds into the following steps: (i) Different GRNs are downloaded from the KEGG repository and cashed locally; (ii) With an XML parser we obtain all the internal network semantics, i.e., genes (nodes) and relations (regulation types, as shown in Figure 2); and (iii) With a specially devised path finding process, all possible and *functionally-valid* network paths are identified and extracted (see Figure 2). Each functional path is annotated with the possible valid gene values according to Kauffman's principles that follow a binary setting, i.e., each gene in a functional path can be either 'ON' or 'OFF'. The extracted and annotated paths are stored in a database that acts as a repository for future reference.

Semantics and Interpretation of GRN Functional Paths. According to Kauffman [13], the following functional gene-regulatory semantics apply: (a) the network is a directed graph with genes (inputs and outputs) being the graph nodes and the edges between them representing the casual (regulatory) links between them; (b) each node can be in one of the two states: 'on' or 'off'; (c) for a gene, 'on' corresponds to the gene being expressed (i.e., the respective substance being present); and (d) time is viewed as proceeding in discrete steps - at each step, the new state of a node is a Boolean function of the prior states of the nodes with arrows pointing towards it.

Since the regulation edge connecting two genes defines explicitly the possible values of each gene, we can set all possible state-values that a gene may take in a path (see Figure 3). Thus, each extracted path contains not only the relevant sub-graph but the state-values ('ON' or 'OFF') of each gene as well. The only requirement concerns the assumption that for a path being functional, the path should be '*active*' during the GRN regulation process. In other words we assume that all genes in a path are

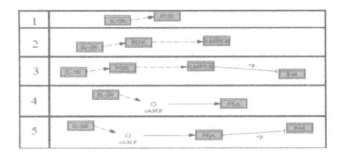

Fig. 3. Identification of functional paths: 'ON' (red) and 'OFF' (blue) values are assigned to genes according to their regulation to form the corresponding (the decomposed paths of figure 3 are shown). For exmple, th eregulatory relation IL-3R ---> PI3K is active only if: (i) IL-3K is 'ON', otherwise the regulatory relation could not be activated; and (ii) PI3k is also 'ON' – the indirect activation relation '--->' from IL-3R forces it to become active.

functionally active. For example assume the functional path A → B ('→' is an activation/expression regulatory relation). If gene A is on an 'OFF' state then, gene B is not allowed to be on an 'ON' state - B could become 'ON' only and only if it is activated/expressed by another gene in a different functional path, e.g., C → B). If we had allowed non-functional genes to have arbitrary values then the significant paths would be more likely to be 'noisy' rather than of biological importance.

3.2 Matching GRN Functional-Paths with MA Gene expression Profiles

The next step is to locate microarray experiments and respective gene expression data where we expect the targeted GRN to play an important role. For example the cell-cycle and apoptosis GRNs play an important role in cancer studies and in the respective experiments dealing with tumor progression. With a gene expression/functional-path *matching* operation, the valid and most prominent GRN functional paths are identified.

In order to combine and match GRN paths with microarray data the respective gene expression values should be transformed into two (binary) states - "on" and "off". Microarray gene expression discretization is a popular method to indicate vigorously the expressed (up-regulated/"on") and not-expressed (down-regulated/"off") genes. We employ an information-theoretic process for the *binary transformation* of gene expression values. The process pre-supposes the categorization of the input samples into two categories (or, phenotypes), and is based. A detailed description of the followed discretization process may be found in [8].

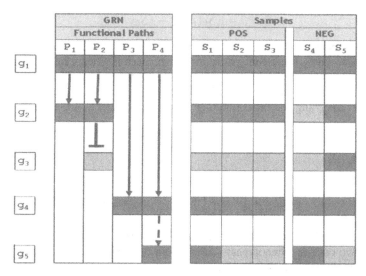

Fig. 4. Matching GRN functional paths with samples' gene expression profiles; 'red' and 'light-green' coloring represents 'on' and 'off' states of genes, respectively, '━▶' and '━◀' represent activation/expression and inhibition relations between genes, respectively)

GRN and MA gene expression data matching aims to *differentiate* GRN functional paths and identify the most prominent of them with respect to the gene expression profiles of the input samples. In other words, the quest is for those functional paths that exhibit *high matching scores* for one of the samples' class/phenotype and low matching scores for the other. This is a paradigm shift from mining for genes with differential expression, to mining for subparts of GRN with differential function.

The procedure for differential path identification is inherently simple. Assume five input samples $\{S_1, S_2, S_3\}$ and $\{S_4, S_5\}$ assigned to (phenotypic) class 'POS' and 'NEG', respectively, and engages five genes $\{g_1, g_2, g_3, g_4, g_5\}$. Furthermore assume that we are about to explore (target) an artificial GRN being decomposed into four functional-paths $\{P_1, P_2, P_3, P_4\}$ (see Figure 4).

In order to match the gene expression profile of a sample with a specific path we consider just the genes engaged in the path. Our notion of matching is realized as a *consistency* relation between functional GRN paths and gene expression profiles. In other words, we care just for those GRN 'parts' which represent 'causal' *putative molecular mechanisms* that govern the expression status of genes for specific samples and phenotypes. For example, sample S_1 matches perfectly path P_1 – even if this path engages just two of the total five genes. In contrast, S_5 does not match perfectly path P_4 – even if both share the same ('on') state for genes g_1 and g_4, they differ in the state of gene g_5 ('on' in path P_4 and 'off' in sample S_5). In general, for each path we compute the number of samples being consistent with each class. Suppose that there are S_1 and S_2 samples belonging to the first and second class, respectively. Assume that path P_i is consistent with $S_{i;1}$ and $S_{i;2}$ samples form the first and second class, respectively. The formula below, computes the *differential power* of the specific path with respect to the two classes.

$$dp(P_i) = \left| (S_{i;1} / S_1) - (S_{i;2} / S_2) \right|$$

Ranking of paths according to the above formula provides the most differentiating and prominent GRN functional paths for the respective phenotypic classes. These paths *uncover* putative molecular mechanisms that govern specific phenotypes (disease, disease states, drug responses etc). For the example of figure 3 the following path/samples matching and differentiating power matrix may be computed:

	S_1	S_2	S_3	S_4	S_5	POS	NEG	*dp*
P_1	✓	✓	✓	✗	✓	1.0	0.5	0.5
P_2	✓	✓	✓	✗	✗	1.0	0.0	1.0
P_3	✓	✓	✓	✓	✓	1.0	1.0	0.0
P_4	✓	✗	✗	✓	✗	0.3	0.5	0.2

Examining the matrix we note that path P_2 (shaded) differentiates perfectly between the two classes (phenotypes) – it is a path that matches just class 'POS' samples and none from class 'NEG'. So, a direct conclusion is that a putative molecular mechanism governing class 'POS' is represented by functional path P_2. Of course, in the case of not-perfect matches there will be exceptions, i.e., functional paths shared by both phenotypes. In this case the ranking of paths (by their *dp* power) could help to identify the most prominent, with respect to the targeted phenotype, paths.

The introduced *dp* formula can be enriched so that longer consistent paths acquire stronger power. It can also be relaxed so that 'consistent' presents a continuous, rather than a Boolean indicator. Finally we may introduce 'unknown' values for missing and erroneous gene expression values.

4   Experimental Results and Discussion

We applied the presented methodology on a real-world cancer related microarray study performed in the context of the Prognochip project [31], [32]. The study concerns the gene expression profiling of breast-cancer (BRCA) patients. In this paper we target the *Estrogen Receptor* (ER) phenotypic categories of the engaged patient samples namely, 'ER+' and 'ER-' for patients that exhibit positive and negative ER statuses, respectively. ER status is a very-well known and studied histopathology factor with big prognostic value, and constitutes the target of many microarray studies. The study concerns the gene expression profiles of 17 and 9 'ER+' and 'ER-' patients, respectively, and targets the expression profile of 34772 genes.

Targeting ER-Phenotype Discriminant Genes. In an effort to identify such mechanisms we applied the presented methodology on the aforementioned BRCA microarray dataset targeting various GRNs. We first identified the most discriminant genes that differentiate between the ER+ and ER- patients. For that, we utilised a gene-selection process implemented in the MineGene system [8], [33]. We were able to identify four genes that perfectly discriminate between the two phenotypes. Based on

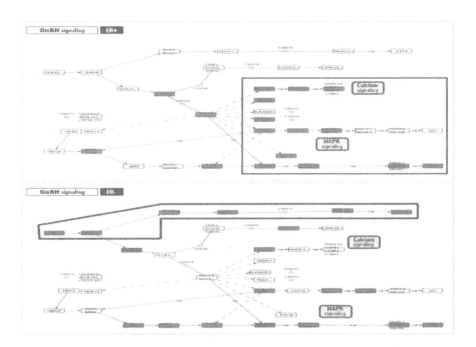

Fig. 5. The highly differentiating ER+ (upper-part) and ER- (bottom-part) GnRH-signaling paths

a literature search we identified two regulatory molecular pathways that engage these genes namely: the *GnRH* (Gonadotropin Releasing Hormone) and the *Apoptosis* signalling regulatory pathways.

We then applied the presented methodology on these GRNs with all available patients' gene expression profiles and targeting the four discriminant genes.

(a) *GnRH-signalling*: A total of 392 functional paths were identified after parsing. Each of these paths matches with a different degree the expression profiles of the input samples. Moreover, each of these paths exhibit different *dp* figures with respect to the two phenotypic categories. We focused on the paths with the highest differential power for one category and zero power for the other, i.e., paths matching just and only one of the phenotypic categories. We were able to identify 46 and 34 such paths for the ER+ and ER- phenotypic categories, respectively. Then, we isolated from these paths all the genes with their corresponding functional status (i.e., "on", "off"). Utilising the KGML-ED pathway visualization tool (http://kgml-ed.ipk-gatersleben.de/Introduction.

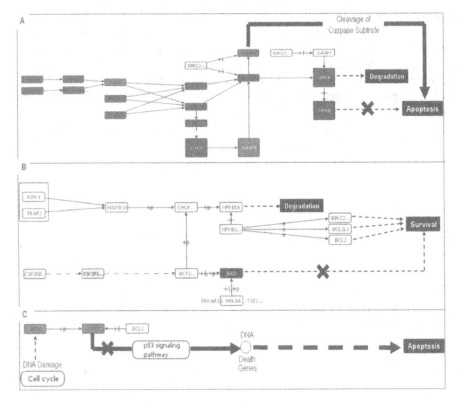

Fig. 6. The highly differentiating ER+/ER- Apoptosis paths – Notation: 'red rectangle with black text: ER+ "on" gene; 'rectangle with red text: ER- "on" genes; 'blue rectangle with white text': ER- "off" gene; 'red square with white text': ER+ and ER- "on" gene; and 'blue square with white text': ER+ and ER- "off" gene

html), we mapped and visualised these genes on the targeted GnRH GRN. Results are shown in Figure 5 where, the corresponding differentiating paths are noted with bold surrounding for each of the phenotypic categories (upper-part for ER+ and bottom-part for ER-). These paths match and cover most of the sample cases and present putative regulatory cell-signalling mechanisms that govern the targeted phenotypes. It should be evident that there are two clearly separate molecular mechanisms (governed by potential different genotypes) underlying the ER+ vs. ER- phenotypes (surroundings in figure 5).

(b) *Apoptosis*: Following the same line of analysis, a total of 407 functional paths were identified, with 89 matching just the ER- phenotype and 45 matching just the ER+ one (see Figure 6 and the respective legend for an explanation of the graph notations). From the upper part (A) of the figure one can easily trace the ER+ functional routes guiding to apoptotic events ('➜' signs); in contrast to the ER- case where these routes are blocked ('✗' sign). The finding is consistent with the fact that ER+ BRCA patient cases show good prognostic behaviour, i.e., apoptotic or, programmed cell-death is taking place. In the middle-part (B) of the figure it can be seen that for the ER- case the 'survival' of genes route is blocked, guiding to potential anti-apoptotic and uncontrolled cell-proliferation events – a finding consistent with the bad prognosis of ER- BRCA cases. In the bottom-part (C) of the figure the two different TP53-related mechanisms that differ between the two phenotypes could be identified and traced; the ER+ routes guide to apoptosis; in contrast to the ER- case where the route to apoptosis is blocked by the up-regulation of the BCL2 gene and subsequent down-regulation of TP53. The TP53 gene is a well-known and crucial factor for cancer progression with its blocking guiding to tumourgenesis and metastatic behaviours. So, we may state that the findings uncover and are consistent with established clinical and biological knowledge.

5 Conclusions

We have presented an integrated methodology for the combined mining of both GRN and MA gene-expression data sources. In the heart of the methodology rests the decomposition of GRNs into all possible functional paths, and the matching of these paths with samples' gene-expression profiles. Initial implementation and experimental results on a real-world breast-cancer study demonstrate the suitability and reliability of the approach.

Among other, our on-going and immediate research on the field include: (a) further experimentation with various real-world microarray studies and different GRN targets (accompanied with the evaluation of results form molecular biology experts); (b) extension of path-decomposition to multiple GRNs; (c) elaboration on more sophisticated path/gene-expression sample matching formulas and operations; (d) incorporation of different gene nomenclatures in order to cope with microarray experiments from different platforms and gene encodings; and (e) porting of the whole methodology in a Web-Services oriented workflow environment.

References

1. Ideker, T., Galitski, T., Hood, L.: A new approach to decoding life: systems biology. Annu. Rev. Genomics Hum. Genet. 2, 343–372 (2001)
2. Collins, F.S., Green, E.D., Guttmacher, A.E., Guyer, M.S.: 'A Vision for the Future of Genomics Research. Nature 422(6934), 835–847 (2003)
3. Kumar, A.: Teaching Systems Biology: An Active-learning Approach. Cell Biol. Educ. 4(4), 323–329 (2005)
4. Friend, H.F.: How DNA microarrays and expression profiling will affect clinical practice. Br. Med. J. 319, 1–2 (1999)
5. Bassett, D.E., Eisen, M.B., Boguski, M.S.: Gene expression informatics: it's all in your mine. Nature Genetics 21(suppl. 1), 51–55 (1999)
6. Golub, T.R., et al.: Molecular classification of cancer: class discovery and class prediction by gene expression monitoring. Science 286, 531–537 (1999)
7. van 't Veer, L.J., et al.: Gene Expression Profiling Predicts Clinical Outcome of Breast Cancer. Nature 415, 530–536 (2002)
8. Potamias, G., Koumakis, L., Moustakis, V.: Gene Selection via Discretized Gene-Expression Profiles and Greedy Feature-Elimination. In: Vouros, G.A., Panayiotopoulos, T. (eds.) SETN 2004. LNCS (LNAI), vol. 3025, pp. 256–266. Springer, Heidelberg (2004)
9. Bower, J.M., Bolouri, H.: Computational Modeling of Genetic and Biochemical Networks. Computational Molecular Biology Series. MIT Press, Cambridge (2001)
10. Arkin, Ross, J.: Computational functions in biochemical reaction networks. Biophys J. 67(2), 560–578 (1994)
11. Kauffman, S.A.: The Origins of Order: Self-Organization and Selection in Evolution. Oxford Univ. Press, New York (1993)
12. Akutsu, T., Miyano, S., Kuhara, S.: Identification of genetic networks from a small number of gene expression patterns under the Boolean network model. In: Pac. Symp. Biocomput., pp. 17–28 (1999)
13. Imoto, S., Goto, T., Miyano, S.: Estimation of genetic networks and functional structures between genes by using Bayesian networks and nonparametric regression. In: Pac. Symp. Biocomput., pp. 175–186 (2002)
14. Kimura, S., et al.: Inference of S-system models of genetic networks using a cooperative co-evolutionary algorithm. Bioinformatics 21(7), 1154–1163 (2005)
15. Segal, E., et al.: Module networks: identifying regulatory modules and their condition-specific regulators from gene expression data. Nat. Genet. 34(2), 166 (2003)
16. Guanrao, P.L., Almasri, E., Dai, Y.: Rank-based edge reconstruction for scale-free genetic regulatory networks. BMC Bioinformatics 9, 75 (2008)
17. Daisuke, T., Horton, P.: Inference of scale-free networks from gene expression time series. Journal of Bioinformatics and Computational Biology 4(2), 503–514 (2006)
18. Milenkovic, T., Lai, J., Przulj, N.: GraphCrunch: a tool for large network analyses. BMC Bioinformatics 9, 70 (2008)
19. Wu, C.C., Huang, H.C., Juan, H.F., Chen, S.T.: GeneNetwork: an interactive tool for reconstruction of genetic networks using microarray data. Bioinformatics 20(18), 3691–3693 (2004)
20. Simon, R., Radmacher, M.D., Dobbin, K., McShane, L.M.: Pitfalls in the Use of DNA Microarray Data for Diagnostic Classification. Journal of the National Cancer Institute 95(1), 14–18 (2003)
21. Pan, W.: A comparative review of statistical methods for discovering differentially expressed genes in replicated microarray experiments. Bioinformatics 18(4), 546–554 (2002)

22. Kitano, H.: Systems biology: a brief overview. Science 295(5560), 1662–1664 (2002)
23. Kwoh, K., Ng, P.Y.: 'Network analysis approach for biology. Cell. Mol. Life Sci. 64, 1739–1751 (2007)
24. Anisimov, S.V., Christophersen, N.S., Correia, A.S., Li, J.-Y., Brundin, P.: NeuroStem Chip: a novel highly specialized tool to study neural differentiation pathways in human stem cells. BMC Genomics 8, 46 (2007)
25. Doniger, S.W., Salomonis, N., Dahlquist, K.D., Vranizan, K., Lawlor, S.C., Conklin, B.R.: MAPPFinder: using Gene Ontology and GenMAPP to create a global gene-expression profile from microarray data. Genome Biology 4(1), R7 (2003)
26. Mlecnik, B., Scheideler, M., Hackl, H., Hartler, J., Sanchez-Cabo, F., Trajanoski, Z.: PathwayExplorer: web service for visualizing high-throughput expression data on biological pathways. Nucleic Acids Res. 33, W633-W637 (2005)
27. Kulterer, B., et al.: Gene expression profiling of human mesenchymal stem cells derived from bone marrow during expansion and osteoblast differentiation. BMC Genomics 8, 70 (2007)
28. Weniger, M., Engelmann, J.C., Schultz, J.: Genome Expression Pathway Analysis Tool – Analysis and visualization of microarray gene expression data under genomic, proteomic and metabolic context. BMC Bioinformatics 8, 179 (2007)
29. Grosu, P., Townsend, J.P., Hartl, D.L., Cavalieri, D.: Pathway Processor: A Tool for Integrating Whole-Genome Expression Results into Metabolic Networks. Genome Research 12(7), 1121–1126 (2002)
30. Evans, W.E., Relling, M.V.: Moving towards individualized medicine with pharmacogenomics. Nature 429, 464–468 (2004)
31. Kafetzopoulos, D.: The Prognochip Project: Transcripromics and Biomedical Informatics for the Classification and Prognosis of Breast Cancer. ERCIM News 60 (2005)
32. Analyti, A., Kondylakis, H., Manakanatas, D., Kalaitzakis, M., Plexousakis, D., Potamias, G.: Integrating Clinical and Genomic Information through the PrognoChip Mediator. In: Maglaveras, N., Chouvarda, I., Koutkias, V., Brause, R. (eds.) ISBMDA 2006. LNCS (LNBI), vol. 4345, pp. 250–261. Springer, Heidelberg (2006)
33. Potamias, G., Kanterakis, A.: Supporting Clinico-Genomic Knowledge Discovery: A Multi-strategy Data Mining Process. In: Antoniou, G., Potamias, G., Spyropoulos, C., Plexousakis, D. (eds.) SETN 2006. LNCS (LNAI), vol. 3955, pp. 520–524. Springer, Heidelberg (2006)

MOpiS: A Multiple Opinion Summarizer

Fotis Kokkoras[1], Efstratia Lampridou[1], Konstantinos Ntonas[2],
and Ioannis Vlahavas[1]

[1] Department of Informatics, Aristotle University of Thessaloniki,
54 124 Thessaloniki, Greece
{kokkoras,elamprid,vlahavas}@csd.auth.gr
[2] University of Macedonia Library & Information Center,
Egnatias 156, 54 006 Thessaloniki, Greece
kntonas@gmail.com

Abstract. Product reviews written by on-line shoppers is a valuable source of information for potential new customers who desire to make an informed purchase decision. Manually processing quite a few dozens, or even hundreds, of reviews for a single product is tedious and time consuming. Although there exist mature and generic text summarization techniques, they are focused primarily on article type content and do not perform well on short and usually repetitive snippets of text found at on-line shops. In this paper, we propose MOpiS, a multiple opinion summarization algorithm that generates improved summaries of product reviews by taking into consideration metadata information that usually accompanies the on-line review text. We demonstrate the effectiveness of our approach with experimental results.

Keywords: text summarization, opinion mining, product reviews.

1 Introduction

The Web has changed the way people express their opinion. They can now easily discuss and express their views about everything, generating in this way huge amounts of on-line data. One particular on-line activity that generates such data is on-line shopping. Modern successful on-line shops and product comparison sites allow consumers to express their opinion on products and services they purchased. Although such information can be useful to other potential customers, reading and mentally processing quite a few dozens or hundreds of reviews for a single product are tedious and time consuming.

Mature text summarization systems can provide a shortened version of a (rather long) text, which contains the most important points of the original text [1]. These summaries are produced based on attributes (or features) that are usually derived empirically, by using statistical and/or computational linguistics methods. The values of these attributes are derived from the original text and the summaries typically have 10%-30% of the size of the original text [2].

Although there are text summarizers available that perform well on article type content, the discreteness of the on-line reviews suggests that alternative techniques

J. Darzentas et al. (Eds.): SETN 2008, LNAI 5138, pp. 110–122, 2008.

are required. On-line reviews are usually short and express only the subjective opinion of each reviewer. The power of these reviews lies behind their large number. As more and more reviews for a specific product or service are becoming available, possible real issues or weaknesses of it are revealed as they are reported by more users. The same holds for the strong features of it.

The problem with all these written opinions is that it takes much time for someone to consult them. Sometimes it is even impossible to read them all due to their large number. Mining and summarizing customer reviews is the recent trend in the, so called, research field of *opinion mining*. Unlike traditional summarization, opinion (or review) summarization mines the features of the product on which the customers have expressed their opinions and tries to identify whether the opinions are positive or negative [3, 4, 5]. It does not summarize the reviews by selecting a subset or rewriting some of the original sentences.

Popular on-line shops such as newegg.com or product comparison portals such as pricegrabber.com, contain already categorized reviews (with pros and cons) that contain additional metadata such as the familiarity of the user with the domain of the product, the duration of ownership at the time of the review, the usefulness of the review to other users, etc. Such augmented reviews can help us decide what is better to include in a summary. For example, they might provide hints for the reliability of the reviewer.

In this paper, we identify cases of such valuable metadata and propose MOpiS, a novel summarization approach for multiple, metadata augmented, product reviews. We work with the reviews at the sentence level. We first create a dictionary of the domain and then score the available sentences using a simple statistical method. We then utilise the available metadata of each review to increase or decrease this score in a weighted fashion. At the end we provide a redundancy elimination step to improve the quality of the summary produced.

We also present experimental results, which provide strong evidence for the validity of our claims. The summarization algorithm we propose outperforms two commercial, general purpose summarizers and a naive version of our approach that all ignore such metadata.

The rest of the paper is organized as follows: Section 2 presents related word, while Section 3 identifies potential metadata that can serve our approach and describes the way we collect all these data (review text and metadata). Our summarization algorithm is described in detail in Section 4, while Section 5 includes our experimental results and discussion about them. Finally, Section 6 concludes the paper and gives insight for future work.

2   Related Work

Most of the related research work in review summarization focuses on the problem of identifying important product features and classifying a review as positive or negative for the product or service under consideration.

Hu and Liou in [4] mine the features of the product on which the customers have expressed their opinions and decide whether the opinions are positive or negative. They do not summarize the reviews by selecting a subset neither rewrite some of the original sentences from the reviews to capture the main points, as in the classic text summarization.

Morinaga et al. [5, 6] collect statements regarding target products using a general search engine and then extract opinions from them, using syntactic and linguistic rules derived by human experts from text test samples. They then mine these opinions to find statistically meaningful information.

In [7], Dave et al. use information retrieval techniques for feature extraction and scoring. They use their system to identify and classify (as positive or negative) review sentences in web content.

Nguyen et al. in [8] classify the sentences of restaurant reviews into negative and positive and then categorize each sentence into predefined types, such as food and service. From each type, both a negative and a positive review are selected for the summary.

OPINE is an unsupervised information extraction system presented in [9], which extracts fine-grained features and opinions from on-line reviews. It uses a relaxation-labelling technique to determine the semantic orientation of potential opinion words, in the context of the extracted product features and specific review sentences.

None of the approaches mentioned above take advantage of the additional available metadata of each review to improve the efficiency of the task, whether they perform summarization or opinion categorization. To the best of our knowledge, our approach is novel and outlined as follows:

- We weigh the importance of the available metadata by using a multicriteria approach and use web content extraction and a simple statistical approach to build (once) a dictionary of the domain.
- We rank the sentences of multiple reviews on the basis of the frequency of their words and the dictionary, and then adjust their importance to some degree (weighted adjustment) by considering the available metadata.
- We select the sentences for the final summary eliminating redundancy at the same time.

3 Metadata Identification and Extraction

3.1 Metadata Identification

We have examined the review facilities provided by many popular on-line shops, for additional information that accompanies the review text and that can potentially contribute to a summarization task. The features we located are presented in Table 1.

Since the features in the list we built do not all exist in every e-shop, we focused on providing a usage methodology that can be followed even in the absence of any of these metadata. It is obvious though that, in such a case, some performance degradation is expected. Our main concern was to keep it graceful.

Note also that our approach is based on reviews that are already categorised by the reviewers, by providing separate positive and negative comments. As a result we discriminate between positive (*pros*) and negative (*cons*) reviews. If the reviews are not categorised, then an opinion categorization step is required. This is future work for our case. Finally, we consider parameters with calculated values (*Respectability* in

Table 1). Such metadata, if required, can be calculated by using web content extraction techniques.

Table 1. Useful and common metadata, accompanying product reviews in e-shops

Field	Possible Values
Tech Level (of the reviewer)	*average, somewhat high, high*
Ownership Duration (of the product under review)	*a few days, about week, a few weeks, a few months, a year, more that a year*
Usefulness (of the review)	*"n out of m people found this review helpful"* number of people (n) who vote this review useful out of the total number of people (m) who voted either for or against the review
Respectability (of the reviewer)	this is a calculated metadata: percentage value, equal to the average usefulness of all the reviews this user has made

3.2 Metadata Extraction

Unfortunately, the data required for the summarization task usually resides in proprietary databases and is considered inaccessible for automated processing. The reviews are only available in HTML pages generated automatically from page templates and database content. The only way to gather arbitrary such semi-structured data is to use web content extraction techniques.

For the web data extraction task we developed ΔEiXTo [11], a general purpose, web content extraction tool which consists of two separate components:

- GUI ΔEiXTo a graphical application that is used to visually build, test, fine-tune, execute and maintain extraction rules (wrappers), and
- ΔEiXTo executor, an open source Perl application that screen scrapes desired web content based on extraction rules created with GUI ΔEiXTo.

Data extracted with ΔEiXTo can be saved in various formats, suitable for further processing, including XML and RSS. Additionally, both components can be easily scheduled to run automatically and extract desired content. Some kind of cooperative extraction (between two or more wrappers) is also possible with ΔEiXTo. The detailed presentation of ΔEiXTo is beyond the scope of this paper and will be done in the near future.

4 The MOpiS Summarization Algorithm

In this section we present MOpiS, the proposed Multiple Opinion Summarization algorithm. MOpiS works at the sentence level. The available positive and negative comments from the reviews of a product are aggregated to form the positive and the negative sum, respectively. Then, each sum is partitioned into individual sentences from which we remove the stop words, the punctuation, the numbers and the symbols. As a result, a *Pros* and a *Cons* sentences set is produced.

Besides the review text, our approach uses an automatically generated dictionary, containing certain keywords related to the domain of the product in question. The dictionaries (one for each product category) are produced once by a Perl script that processes a large amount of reviews on products of the domain in question. The exact dictionary generation procedure is the following:

- Extract review data for 50 products using ΔEiXTo (we collected a few thousands reviews for each domain).
- For each domain, create a single text file containing the pros and cons part of the review data.
- Remove the stop words (articles, prepositions, pronouns), as well as 500 quite common English words.
- For each word calculate the frequency of occurrence and keep the 150 most frequent words.

Finally, we identify which of the metadata of Table 1 are present in our reviews and use ΔEiXTo to extract them. The extraction takes place in the same task that collects the review text (ΔEiXTo is capable of extracting many fields at the same time).

Thus, for each product p for which we want to summarize the reviews, our algorithm takes as input a set with the review data of p (either *Pros* or *Cons*), the dictionary D of the domain and $k \leq 4$ sets of metadata. The summarization algorithm is described next.

4.1 The Scoring Procedure

4.1.1 Text Contribution

The main concept of the scoring procedure is that each sentence should be given a score depending on the importance of the words that it contains, but also on the additional metadata of the review that it belongs to. For each sentence i, we calculate an initial score R_i based on the text and then adjust this score according to the metadata presented. This is expressed with equation (1) in which w_j is a factor which defines the importance we give to this metadata category.

$$S_i = R_i + R_i \cdot \sum_{j=1}^{k} w_j \qquad (1)$$

For each sentence, the R_i parameter in equation (1) is calculated on the basis of the importance of the words the sentence contains. Each word v_l of the sentence contributes to the score its frequency of occurrence f_{v_l}, unless this word belongs to the dictionary D, in which case its contribution is doubled. This is depicted in equation (2).

By doubling the contribution of dictionary words to the initial score of a sentence, we increase the probability to have this sentence in the final summary, as the more dictionary words it contains the more important it is considered.

$$R_i = \sum_{v_l \notin D} f_{v_l} + 2 \cdot \sum_{v_l \in D} f_{v_l} \qquad (2)$$

4.2 Metadata Contribution

Let us now define the way the metadata of each review contribute to the total score S_i (equation 1) of each sentence. Since the various metadata fields are of different nature, those considered more important should contribute more to S_i, that is, their w value should be greater. For the task of assigning proper values to the factors w_j, we used the Analytic Hierarchy Process (AHP [10]), which provides a methodology to estimate consistent weight values for criteria, according to the subjective importance we give to these criteria. This importance values are selected from a predefined (by AHP) scale between 1 and 9.

Particularly and according to [10], we considered:

- the ownership duration to be "very little more important" than the technology level of the user (importance 2 in AHP),
- the usefulness of the review to be "a little more important" than the ownership duration (importance 3 in AHP), "more important" than the technology level of the reviewer (importance 4 in AHP) and "very little more important" than the respectability of the reviewer (importance 2 in AHP),
- the respectability of the reviewer to be "very little more important" than the duration of ownership (importance 2 in AHP) and "a little more important" than the technology level of the reviewer (importance 3 in AHP).

With these considerations we were able to define the pairwise comparison matrix required by the AHP for the calculation of initial weight values w'_j. This is a 4x4 matrix if all four possible metadata categories of **Table 1** are used, 3x3 if one is omitted, etc. We also calculated the consistency criterion, as described in AHP. This metric provides evidence that we made no contradicting assumptions on the importance we assigned to the metadata categories.

In each case, the calculated w'_j provides good initial values for the w_j of equation (1). To provide further flexibility based on the values of the metadata category under consideration, we allow the replacement of w'_j with a function of $g(d, w'_j)$, where d is some function of the metadata value in hand.

For example, say that w_1 corresponds to tech level and we wish to give more credit to reviews from users of high tech level (a rational decision). We can define function $g(high, w'_1)$ as in equation 3.

$$w_1 = \begin{cases} w'_1 & TechLevel = high \\ 0 & otherwise \end{cases} \tag{3}$$

4.3 Redundancy Elimination

When the sentence scoring is over, *MOpiS* enters into its final step which is the elimination of redundant sentences. This step tries to prevent the inclusion of many sentences that have the same meaning with sentences that are already into the final summary.

First, the sentence S_i with the highest rank is chosen. However, if the sentence is quite long (we used a threshold of 30 words) it is rejected and the next sentence is chosen. This rejection arises due to our observation that very long sentences were somehow artificially lengthy, because the reviewer did not obeyed common syntactic rules.

When the sentence with the highest score S_u is selected, it is removed from the ranked list and is added to the final summary. At the same time the score S_i of all of the rest sentences ($i \neq u$) in the ranked list is readjusted according to equation (4):

$$S_i' = S_i - \sum_{\forall v_l \notin D} f_{v_l} - 2 \cdot \sum_{\forall v_l \in D} f_{v_l} \quad i \neq u \tag{4}$$

In equation (4), v_l is a word of sentence S_u which is already selected for the summary, and f_{v_l} is its initially calculated frequency of occurrence. The rest of the symbols are as defined in equation (2).

Actually, the score of each of the rest sentences is decreased for every word that has been given bonus before, but now already appears in the summarization text. Thus, the recurrence of concepts in the summarization text is reduced.

This selection-readjustment procedure is repeated for the next sentence in the top of the ranked list until the desired number of sentences is added into the summary.

The whole task described in Section 4 is performed once for the *pros* summary and a second time for the *cons* summary. The only difference in these two "runs" is the initial set of sentences.

5   Experimental Results

5.1   Case Study A

We extracted 1587 review records for 9 different products belonging to 3 different product categories (3 randomly selected products from each category) from newegg.com, one of the most successful on-line stores, where each review is organized in the way presented in Fig. 1.

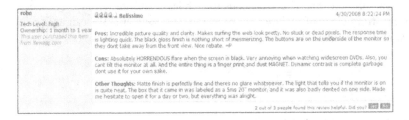

Fig. 1. A typical review record at newegg.com

We used a single extraction rule capable of performing a sequence of page fetches (by following "Next Page" links) and capturing all reviews and data fields under interest. A total of 160 web pages were processed. The amount of the extracted data is summarized in Table 2.

Table 2. The dataset used

Domain:	Monitors			Printers			CPU Coolers		
Models:	A	B	C	A	B	C	A	B	C
#Reviews:	218	130	358	124	86	86	293	126	166

In particular, each review contains positive comments (*pros*), negative comments (*cons*), how familiar is the user with the related technology (*tech level*), the duration of ownership of the product (*ownership*) and the *usefulness* of the review to other users.

Using AHP and the importance values we discussed in Section 4.2, we calculated the following initial values: $w'_1=0.14$, $w'_2=0.24$ and $w'_3=0.62$ (w'_1 for *Tech Level*, w'_2 for *Ownership Duration* and w'_3 the *Usefulness* of the review).

We further adjusted w_j using equation (3) for w_1, equation (5) for w_2 and equation (6) for w_3.

$$w_2 = \begin{cases} w'_2 & Ownership = more\ than\ a\ year \\ 0 & otherwise \end{cases} \qquad (5)$$

$$w_3 = g(\delta_v, w'_3) = \left(\frac{1}{1+e^{-0.2 \cdot \delta_v}} - \frac{w'_3}{1.24} \right) \cdot 1.24 \qquad (6)$$

In regard to the factor w_3 of the usefulness of each review, a sigmoid function was used (equation (6)) to adjust w'_3 according to the difference δ_v between the positive and negative votes of a review. This favors reviews that were found useful by most users and penalizes reviews that were not considered useful by the majority of users.

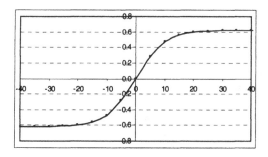

Fig. 2. The sigmoid function $g(\delta_v, w'_3)$ that modulates the factor w_3 according to δ_v

The rest parameters of equation (6) were decided on the need to vary w_3 between w'_3 and $-w'_3$ (the value calculated with AHP) and move the plateau of g away from values of $|\delta_v|<20$, because we observed that the majority of δ_v values lies in this range. Fig. 2 displays the way w_3 depends on δ_v, through $g(\delta_v, w'_3)$.

As mentioned in Section 4.2, we used the AHP because it provides a methodology to check the consistency of the subjective importance values we assigned to each of the criteria. We applied this methodology to the importance values we assigned and found them to be consistent.

Besides *MOpiS*, we also used two well known, commercial summarizers, the *Copernic* [12] and the *TextAnalyst* [13]. Both are general purpose summarizers. This means that they work better with lengthy, article style texts. Reviews on the other hand are usually not so lengthy, they are many and some of them have almost the same meaning. Additionally we calculated how the *MOpiS* algorithm performed when we ignore the second addendum of equation (1), that is, ignore the metadata contribution – we call this version *naive MOpiS*.

Copernic produces document summaries by detecting the concepts of the text and then extracting sentences that reflect these concepts. It mostly uses statistical methods to identify the concepts. Additional important words cannot be inserted by the user, as the concepts extracted are considered to be the keywords required.

TextAnalyst can analyze unstructured text and create a semantic network from it. The semantic network is utilized to score the individual sentences. The system collects those sentences that have a semantic weight greater than a certain adjustable threshold value. It is possible to define an external dictionary of concepts but early tests with the dictionaries we had created led to reduced performance. Therefore no dictionary was set for *TextAnalyst*.

The results of our experiments are summarized in Table 3. We adjusted all systems so as to create a summary of 10 sentences for *pros* and 10 sentences for *cons*. The numbers in parenthesis are the performance of *naive MOpiS*. It is obvious that inclusion of metadata information in the way we suggested, improves the summary (to a degree of about 16% in our experiments), confirming our initial hypothesis. Precision and recall measures are average values that were calculated on the basis of three human-generated summaries. These individuals were provided only with the text of the reviews (without the additional metadata) and the variation in their judgment was les than 3.1%.

It is also obvious that the other two summarizers, although quite sophisticated without any doubt, do not perform very well with this kind of data (many short reviews with overlapped information).

Regarding *TextAnalyst*, the blank cells at recall and precision in Table 3 are due to our inability to adjust the system so as to produce summary of the desired length. In those cases, the summary contained either too many or too few sentences. Consequently, it was not comparable with the summary of *MOpiS* and *Copernic*.

Further investigation of the resulted summaries revealed some interesting facts. The most recent reviews for monitor B were from customers that owned the product more than a year. All of them complained about severe malfunctions after one year of possession (this was also the warranty period). Moreover, it was said that when warranty was over, service was no longer provided by the company. Although such facts were not reported by the majority of the reviewers, these two aspects were depicted in our summary, as they came from reviews with long duration of ownership that were subsidized by our algorithm. They were not mentioned though by neither *Copernic* and *TextAnalyst* nor the *naive MOpiS*.

The contribution of the usefulness of a review is also distinct. By increasing the score of a sentence belonging to a useful review and decreasing it in the opposite case

Table 3. Experimental Results for newegg.com

			MOpiS (naive MOpiS)		Copernic		TextAnalyst	
			Recall	**Precision**	**Rec**	**Prec**	**Rec**	**Prec**
Monitors	A	Pros	90.9 (90.9)	70 (70)	60	60	45.3	30
		Cons	75 (62.5)	70 (50)	25	30	62.5	60
	B	Pros	100 (77.8)	90 (80)	100	60	66.7	70
		Cons	88.8 (66.7)	70 (60)	75	60	33.3	70
	C	Pros	100 (100)	90 (80)	72.7	60	-	-
		Cons	88.9 (66.7)	80 (60)	60	40	-	-
Printers	A	Pros	85.7 (85.7)	70 (70)	62.5	40	-	-
		Cons	87.5 (62.5)	60 (40)	50	40	50	40
	B	Pros	100 (100)	60 (40)	83.3	60	66.7	40
		Cons	87.5 (37.5)	70 (40)	75	60	50	70
	C	Pros	87.5 (75)	80 (70)	87.5	60	62.5	50
		Cons	100 (100)	70 (70)	71.4	70	50	60
CPU Coolers	A	Pros	100 (100)	70 (60)	66.7	70	-	-
		Cons	100 (80)	80 (70)	60	60	60	62.5
	B	Pros	83.3 (83.3)	100 (100)	66.7	80	50	90
		Cons	100 (75)	70 (60)	60	60	25	10
	C	Pros	75 (75)	70 (50)	75	70	-	-
		Cons	100 (80)	50 (60)	100	70	80	40
	Average:		**91.7 (78.8)**	**73.3 (62.8)**	**69.5**	**58.3**	**54**	**53.3**

(equation (6)), significant sentences were kept in the summary while those with no importance were excluded. Because of that, the occurrence of false information in the summary due to malicious reviews is highly unlikely, as those reviews get negative votes of usefulness by the other users. For instance, the following review from monitor A gathered 21 negative votes and 0 positive for being useful:

Monitor had a sticker on it "Certified for Windows Premium", but when I tried to install the software it said "This software does not work with Vista". I phoned <company> - they refused to send me replacement software that will work with Vista!

This sentence was selected by *TextAnalyst* as, despite its meaning, it contains important words. *MOpiS* decreased its score by setting w_3=-0.60 in equation (1). Similarly, *naive MOpiS* selected a sentence from an abusive review that was voted down by the users. None such sentence was selected by *MOpiS*, resulting in summary of better quality.

On the contrary, reviews that received many positive votes are considered more useful and likely to hold important information, so their sentences are given precedence. This is also a way of not depending exclusively on statistical methods, because important statements may not have a high word frequency.

For example, in printer A, there were reviews complaining about the printer being reset in Japanese. Human summarization can easily identify this as a negative aspect in spite of the low frequency (it was not mentioned by many reviews). *MOpiS'* summary reported it though, because of the high usefulness of the reviews. None of the other systems tracked it down.

Moreover, the redundancy elimination aspect of *MOpiS* performed well. Repetition on concepts was minimal or absent since it does not select the highly rated sentences but readjusts the score of all the rest sentences according to the one that was selected for the summary. In *TextAnalyst* however, the repetition of the concepts presented was evident. Actually, in one case, its summary included two identical sentences, coming from a review that was submitted twice! Redundancy elimination also helped *naive MOpiS* to outperform the two commercial summarizers.

It was also observed that, the special nature of the review data affects the performance of plain text summarizers like *Copernic*. Although its function is based on statistical methods, its results were affected to a great degree by the structure of the text. When the same data (aggregated reviews) had different order, different summary was generated.

Finally, we used *MOpiS* in another summarization task worth mentioned. In one case, we were asked to verify if there were problems reported regarding the operation of a RAID controller in a certain computer motherboard under a certain operating system. We summarized 142 negative comments (cons) and this "rumor" was reflected in the summary. Neither *Copernic* nor *TextAnalyst* verified it though.

5.1 Case Study B

We conducted another experiment, this time on a different web site, the pricegrabber.com. This site does not contain the amount of reviews of newegg.com, since it directs the buyer to retailer e-shops for the final transaction – it seems the buyers prefer to review the product at the retailer's site. We extracted data for two printers A and B (27 and 33 records respectively).

Reviews in pricegrabber.com are quite short and contain *strengths* and *weaknesses* instead of *pros* and *cons*. They do not include the *tech level* of the reviewer but provide access to other reviews of the same person. As a result, we decided to create a calculated *Respectability* value by averaging the usefulness values of his reviews. The initial w_j factors were calculated as: $w'_1 = 0.16$ (for ownership), $w'_2 = 0.3$ (for respectability) and $w'_3 = 0.54$ (for usefulness). We further adjusted those values like we did in case study A. We used equation (5) for *ownership*, a sigmoid function like equation (6) for *usefulness* and $w_3 = 0.006 * respectability - w'_3$ for *respectability*.

Due to the small length of the reviews, some sights of saturation were observed. *Copernic* and *TextAnalyst* combine sentences to create new. As a result, they packed

Table 4. Experimental results for pricegrabber.com

			MOpiS (naive MOpiS)		Copernic		TextAnalyst	
			Recall	**Precision**	**Rec**	**Prec**	**Rec**	**Prec**
Printers	A	Pros	100 (100)	90 (60)	100	80	83.3	90
		Cons	100 (100)	80 (70)	100	60	-	-
	B	Pros	100 (100)	90 (60)	100	70	-	-
		Cons	83.3 (83.3)	70 (70)	100	70	100	80
	Average:		**95.8 (95.8)**	**82.5 (65)**	**100**	**58.3**	**91.65**	**85**

many small reviews into long sentences, including in this way almost all the initial reviews. They couldn't prevent though the repetition of the same fact many times in their reviews. Repetition was minimal in *MOpiS* which also adapted well to the different kind of metadata of pricegrabber.com.

6 Conclusions

In this paper, we proposed a novel, multi review/opinion summarization algorithm that is based not only on the text of the review but on additional review metadata. We detected four such frequently found metadata and based on AHP, we consistently defined how important we consider them in a useful review, in relation to each other. We used these importance values to define weights that control the way these metadata contribute to our review scoring procedure.

The additive nature of our algorithm allows it to adapt to review sites with any subset of the set of metadata we detected. Moreover, we allow custom modulation of the initial calculated weight to give bonus or penalize certain values for the metadata field of the review. Finally, the redundancy elimination step reduces concept repetition in the final summary.

Our experimental results demonstrated the usefulness of this metadata inclusion by means of improved precision and recall metrics. This is clearly demonstrated by the improved performance of *MOpiS* compared to *naive MOpiS*.

Consequently, our next step is to remove the requirement for categorized reviews (pros/strengths and cons/weaknesses) since there exist many sites which do not discriminate between pros and cons, but rather have a single, mixed review.

A limited version of *MOpiS* (3 product categories from newegg.com), is available online at http://deixto.csd.auth.gr/newegg/newegg.html, for real-time demonstration.

References

1. Mani, I.: Automatic Summarization. John Benjamins Publishing Company, Amsterdam (2001)
2. Mani, I., Maybury, M.T.: Advances in Automatic Text Summarization. MIT Press, Cambridge (1999)
3. Liu, B.: Web Data Mining. Springer, Heidelberg (2007)
4. Hu, M., Liu, B.: Mining and summarizing customer reviews. In: Proceedings of the 10th ACM SIGKDD International Conference on Knowledge Discovery and Data Mining, SIGKDD 2004, pp. 168–177 (2004)
5. Hu, M., Liu, B.: Mining Opinion Features in Customer Reviews. In: Proceedings of the 19th National Conference on Artificial Intelligence (AAAI 2004), San Jose, USA (2004)
6. Morinaga, S., Yamanishi, K., Tateishi, K., Fukushima, T.: Mining Product Reputations on the Web. In: Proceedings of the 8th ACM SIGKDD International Conference on Knowledge Discover and Data Mining, KDD 2002, pp. 341–349 (2002)
7. Dave, K., Lawrence, S., Pennock, D.N.: Mining the Peanut Gallery: Opinion Extraction and Semantic Classification of Product Reviews. In: Proceedings of the 12th International World Wide Web Conference, WWW 2003, pp. 451–460 (2003)

 8. Nguyen, P., Mahajan, M., Zweig, G.: Summarization of Multiple User Reviews in the Restaurant Domain. Technical Report, Microsoft Research, MSR-TR-2007-126 (2007)
 9. Popescu, A.M., Etzioni, O.: Extracting product features and opinions from reviews. In: Proceedings of the Conference on Human Language Technology and Empirical Methods in Natural Language Processing, Vancouver, Canada, pp. 339–346 (2005)
10. Saaty, T.L.: Decision Making for Leaders: The Analytic Hierarchy Process for Decisions in a Complex World. RWS Publications, Pittsburgh (1999)
11. ΔEiXTo web data extraction tool, http://deixto.csd.auth.gr
12. Copernic Summarizer, http://www.copernic.com
13. TextAnalyst, http://www.megaputer.com/textanalyst.php

Human Behavior Classification Using Multiple Views

Dimitrios I. Kosmopoulos, Panagiota Antonakaki, Konstandinos Valasoulis,
Anastasios Kesidis, and Stavros Perantonis

NCSR Demokritos, Institute of Informatics and Telecommunications
{dkosmo,ganton,kvalas,akesidis,sper}@iit.demokritos.gr

Abstract. We present our current results in understanding the behavior of humans moving an a plane using multiple cameras. We exploit the merits of camera registration based on homography estimation to extract position on a 2D map. We use the output of HMM and SVM classifiers to model and extract human behaviour from both the target trajectory and the target short term activity (i.e., walking, running, abrupt motion etc). The proposed approach is verified experimentally in an indoor environment. The current results with a single moving agent are presented.

1 Introduction

Motion analysis in video, and particularly human behaviour understanding, has attracted many researchers [1], mainly because of its fundamental applications in video surveillance, video indexing, virtual reality and computer-human interfaces. One of the most challenging problems in computer vision is to automatically model and recognize human behaviour, for example in a visual surveillance context, thus reducing human intervention. Such a system will monitor a location and will automatically detect, categorize and recognize human behaviours, calling for human attention only when necessary. With the growing number of cameras set in many public areas, this kind of approach appears more appealing. The research in the area of behaviour understanding concentrates mainly on the development of methods for analysis of visual data in order to extract and process information about the behavior of physical objects in a scene. Many methods have been proposed, as we present in the next section, the common problems are occlusions in crowded scenes and lack of well defined spatial information in case of 2D tracking. In case of 3D tracking through multiple cameras the problem is the high system complexity. Based on the remark that abnormal events are rare, unpredictable and can not be described by a single model, the goal of our method is to approach the problem by modelling the normal behaviours. Since multiple criteria can be taken into consideration when an observer is trying to decide whether an observed behaviour is normal or not, such as rare activities, or abnormal sequence of usual activities, or unusual trajectories, we implemented a system that applies these criteria. To overcome problems like occlusion, view variance and high complexity we have developed a system with

J. Darzentas et al. (Eds.): SETN 2008, LNAI 5138, pp. 123–134, 2008.
© Springer-Verlag Berlin Heidelberg 2008

multiple cameras, which capitalises on homography estimation (thus avoiding the laborious camera calibration and 3D reconstruction procedures) and views from different cameras. In our method, we assume that all targets move on the same plane, the ground plane and that the ground plane can be viewed from all cameras. The information that we use is that of the object's blob, which lead us to determine the "ground points", i.e., the projections of target blobs on the ground plane. Having extracted the sequence of "ground points" we define the trajectory in 2D, while most of the previous works use object trajectory as a sequence of consecutive locations of the object's centroid on a coordinate system or in the image. Additionally we separate the problem of trajectory classification from the classification of the "short term actions" that can be localised at a certain position, e.g., walking, running, abrupt motion, inactive (standing still), active (body movement in the same position). For this purpose we use features extracted from the blob (optical flow and speed) and the relative pose to the camera. The features that we extract are relatively simple and thus computationally efficient.

The rest of the paper is organized as follows: in section 2 we present the previous work done, which will help to evaluate the innovations of the proposed approach; section 3 provides an overview of the proposed architecture; section 4 explains briefly how we compute the principal axis and how this is projected on the ground plane; section 5 describes the short term behaviour representation and classification, while section 6 describes the classification for trajectories; in section 7 we provide the experimental results and section 8 concludes this paper.

2   Related Work

The framework for behaviour understanding systems in the related literature is divided in layers as displayed in Fig. 1. The first low level tasks such as motion

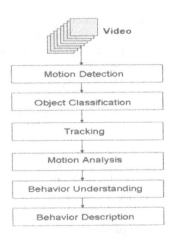

Fig. 1. The main framework for video surveillance systems

segmentation, object classification and tracking are not going to be developed in this work. The tracker provides the system with verified data referring human motion, and we present methods that take a step further and analyze that motion providing behaviour description. In our work, we not only focus on the higher level of behaviour description, but also on fusion of data from many cameras, trying to overcome some of the difficulties discussed before, such as occlusion and view dependence. As mentioned earlier, we examine both the type of motion (or short term activity) and the target trajectory to model normal behavior and detect abnormal events. In the related literature these two are mostly treated separately.

Motion based techniques are normally used for short term action classification, without considering the target trajectories. These calculate features of the motion itself and perform recognition of behaviors based on them. There are many methods for this kind of recognition, for example Bobick et al. in [2] use Motion Energy Images (MEIs) and Motion History Images (MHIs) to classify "aerobic" - type exercises. Taking this work another step further Weinland et al. in [3], focus on the extraction of motion descriptors analogous to MHIs, called motion history volumes, from multiple cameras and their classification into primitive actions. [4], compute optical flow of the object of interest to recognize short term behaviours (e.g., walking, running, fighting) in a nearest neighbour framework.

There are several other methods that use the target trajectory for behaviour classification using the centroid of the target blob. These methods, however, ignore the importance of what the person is doing (short term activity). They also extract trajectories in 2D images thus having problems with view dependency, occlusions etc.

Hidden Markov Models are highly applicable to behavior recognition using trajectories, e.g., [5], [6], [7], [8], due to their transition and emission probabilities, to their automatically training, their simplicity, their computational efficiency and mainly because motion can be viewed as piece-wise stationary signal or a short-term stationary signal. There exist several image-based techniques, which model the motion in the pixel level no matter if it results from local or global motion. In [9] the foreground pixels are clustered using fuzzy k-means clustering to model normal patterns. The trajectories are clustered hierarchically using spatial and temporal information and then each motion pattern is represented with a chain of Gaussian distributions. Coupled hidden Markov models were used for modeling interactions between actors, [10]. Xiang and Gong [11], use Dynamically Multi-Linked Hidden Markov Models to model actions and interactions between persons. Abstract Hidden Markov Models are used by Nguyen et al. in [12] to deal with noise and duration independence, while Wang et al. in [13] use Conditional Random Fields for behaviour recognition in order to be able to model context dependence in behaviours. In [14] a feature vector composed of features giving position and target state is used and the normal behaviours' representations are extracted through clustering.

The aforementioned methods seem attractive in cases of crowd, however, the correspondence to real world activities is not intuitive. Our approach differs in the sense that we decouple the position (trajectory) from the target state claiming that these can be, in many cases, separate problems and addressing them separately may help to reduce the problem dimensionality. Moreover, in contrast to those methods we are able to say if the abnormality is due to abnormal trajectory or abnormal short term activity.

In behaviour understanding, few works depend on homography. Park et al. in [15] use homography to extract object features, and with spatio-temporal relationships between people and vehicle tracks to extract semantic information from interactions calculated from relative positions. Ribeiro et al. in [16], estimate homography that allows the system to have an orthographic view of the ground plane to eliminate perspective distortion for a single camera. Then they calculate features in order to classify the data in four activities (Active, Inactive, Walking, Running), however no trajectory information is employed.

In the literature referenced above in order to extract features that can be used for classification it has been assumed that the targets move almost vertically to the camera z-axis or within a range that is small compared to the distance from the camera, so their size variation is small. Furthermore, the assumption that humans are planar objects, so that homography-based image rectification can be possible, may be true when the cameras are close to being vertical to the common plane, e.g., cameras viewing from high ceilings, but is definitely wrong in the general case.

In our work in contrast to the aforementioned ones we compute simple features that represent short term activity without making assumptions about camera position or relative pose of target to the camera. We also provide a novel representation of an agent's behavior by modeling both short term activity and trajectory on a 2D projection map. This representation bypasses the computationally intensive 3D world representation. It is closer to the human perception compared to pure image-based techniques due to separate handling of the two information sources.

Furthermore, in this work we present a novel classification scheme, which combines the intuition of a human expert with automated learning, by using a binary tree with SVM classifiers.

3 System Overview

The proposed system processes video streams from several cameras with overlapping fields of view (ground plain) as displayed in Fig.2. From each camera we take an image sequence (video) and then we extract some low level features from the foreground objects resulting from optical flow and relative pose to each camera. The foreground objects result from a background subtraction process using mixtures of Gaussian distributions [17]. In parallel to this process we project on the ground plane the points of each camera-specific blob using the homography matrix, which has been calculated offline. The maxima that result from that projection give the target position on the ground plane, from which the trajectory and the speed can be easily calculated. The target speed along with

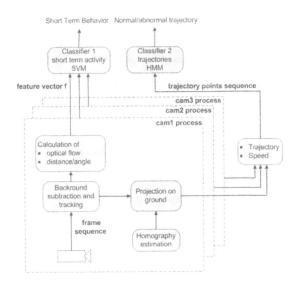

Fig. 2. The system architecture with 3 cameras

the view-specific features are input to a classifier to extract short term action. The features from all cameras are used to create a feature vector to decide the short term action as perceived by all cameras. In parallel to the aforementioned classification we also classify the target trajectory using the currently calculated target position. It is assumed that the classifiers for short term action and trajectories have been trained in a supervised learning fashion. If either the short term action or the trajectory are found not to be consistent with the "normal" patterns this is highlighted to the user. More details on the individual subsystems are given in the following sections.

4   Target Localization Via Homogrpahies

The (planar) homographies are geometric entities whose role is to provide associations between points on different planes. Assume that the scene viewed by a camera comprises a predominant plane, the ground. Assume as well that a coordinate system is attached, so that a point on the plane is expressed as: $\mathbf{P}_\pi = (X, Y, 1)^T$. Finally, assume that the coordinates of this point on the camera plane is $\mathbf{P}_c = (x, y, 1)^T$. The homography $\mathbf{H}$ is a 3x3 matrix which, relates $\mathbf{P}_\pi$ and $\mathbf{P}_c$ as follows:

$$\mathbf{P}_\pi = \mathbf{H} \cdot \mathbf{P}_c \qquad (1)$$

We assume that the target moves on a predominant ground plane. Multiple cameras view the target. The homographies between the cameras and the views project subject information to the ground plane. Synergy of information from all views into the ground plane allows for increased accuracy in the localization

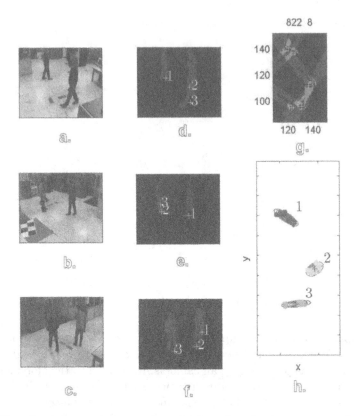

Fig. 3. View from three cameras and extraction of projections on ground plane. In a,b,c the views, in d, e, f the points extracted from background subtraction (several occlusions are present), in g the maxima of the accumulator corresponding to feet positions (represented as crosses) and in h the positions in ground plane.

of the subject. Our approach is related to [18]. In that work, the homographies between each view and the ground plane are calculated. Subsequently, foreground likelihoods in all views are computed. The pixels belonging to the foreground blobs are projected to the ground plane using (1). The projected likelihood maps are multiplied, and a synergy map is applied. The maxima of the map correspond to ground point position of the viewed target on the ground plane, that is, the position where the feet touch the ground. The projection from each camera casts a "shadow" on the ground plane as depicted in Figure 3. The maximum of the accumulator corresponds to the position of the subjects on the ground plane.

5   Short Term Activity Classification

The goal of this work is to separate the classification of short term activity from trajectory abnormalities. To this end we define and extract features separately as described in the following. The short term activity is defined as the activity

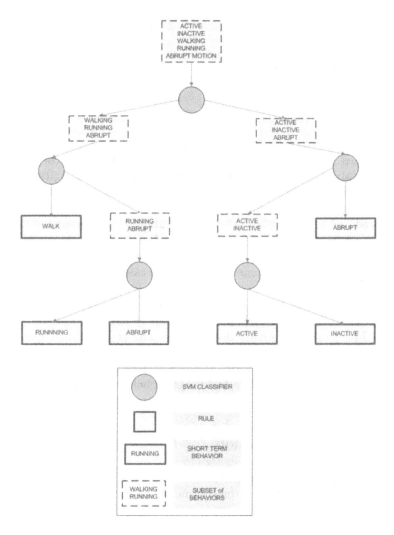

Fig. 4. The classification scheme used for short term behaviors

that takes place locally in the global coordinate system and within a short time period, e.g., a few frames. It is associated with types of motion like walking, running, standing still, body motion while no translation takes place on the global coordinate system (active), abrupt motion. In the present work, we use features coming from trajectory points on the common coordinate system (on the ground plane), the optical flow of the target (i.e. the pattern of apparent motion of targets in the scene caused by the relative motion between an observer and the target) and the relative pose to the camera. From the trajectory points we can extract the target speed (e.g., walking vs running), while from the optical flow are able to differentiate motion types when the target does not move significantly in the global coordinate system (staying still vs moving abruptly). The relative pose to the camera is an important feature since the optical flow may give low values for targets that are far away or move in parallel to the camera axis as opposed to targets that move vertically to the camera axis and are close. The short term activity is thus represented by the following feature vector (for three cameras):

$$\mathbf{f} = (a_1, s_1, d_1, c_1, a_2, s_2, d_2, c_2, a_3, s_3, d_3, c_3, v) \tag{2}$$

where a_i is the mean optical flow in the target bounding rectangle for two consecutive frames (calculated according to the method [19]), s_i the corresponding standard deviation, d_i the distance of the target from the camera, c_i the cosine of the target motion related to the camera and v the velocity norm on the ground plane respectively.

The speed and pose calculation are computationally inexpensive. The latter is given by considering the (x,y) position of each camera and the recent target positions. The mean optical flow requires more computational time but it can be calculated accurately in real time by limiting its calculation in the foreground regions.

For the classification of each short term behavior type we could have used a multiclass classifier, however we had better results by incorporating the "expert" knowledge in the form of a binary decision tree. This is depicted in figure 4. We have set a simple threshold for the speed (speed is calculated quite accurately) to separate active, inactive from running and walking (by definition). Then we have used binary SVM classifiers to find the exact behaviors as will be mentioned in section 7.

6   Trajectory Classification

In order to classify trajectory classifiers able to handle time series may be used. One of the most popular ones is the continuous Hidden Markov Model. In our case each (x,y) ground position is supposed to be the observation vector. Given a continuous Hidden Markov Model, which models each observation as a mixture of distributions, e.g., Gaussians it is possible to classify in real time the trajectory up to the current moment as normal or abnormal, based on a set of training parameters which have been learnt offline by using an EM algorithm see, e.g.,[20]. The problem is stated as follows: if the forward variable

$$a_t(i) = P(O_1...O_t, q_t = S_i/\lambda) \tag{3}$$

is the probability of the partial observation sequence $O_1 ...O_t$ and state S_i at time t given the model λ, then the $a_t(i)$ is calculated inductively by the following:

$$\alpha_1(i) = \pi_i b_i(O_1), 1 \le i \le N \tag{4}$$

where N is the number of states and π_i the state priors, $b_i(O_1)$ the probability of observation O_1 at $t=1$, given that we are in state i

$$\alpha_{t+1}(j) = [\sum_{i=1}^{N} \alpha_t(i)a_{ij}]b_j(O_{t+1}), 1 \le t \le T - 1, 1 \le j \le N \qquad (5)$$

where a_{ij} the transition probability from state i to state j and $b_j(O_{t+1})$ the probability of observation O_{t+1} at $t+1$ given that we are in state j. Then the desired probability is the sum of terminal probabilities:

$$P(O_1, ...O_T|\lambda) = \sum_{i=1}^{N} \alpha_T(i) \qquad (6)$$

The observation probability is given by:

$$b_j(\mathbf{O}) = \sum_{m=1}^{M} c_{jm} N(\mathbf{O}, \mu_{jm}, \Sigma_{jm}) \qquad (7)$$

where c_{jm} the probability that the sample is drawn from the m-component and mu_{jm} and Σ_{jm} the mean vector and the covariance matrix of the m-component in the j-state.

7   Experiments

For our experiments in our lab, we have installed three cameras as displayed in figure 5. We have defined a scenario of moving in a room following a certain path for entering and exiting. A barrier does not allow entering from a certain side, so when someone visits areas which are not neighboring due to the barrier (non consecutive states according to the HMM) there is abnormal activity. Similarly when areas are visited in the wrong order (e.g., entering from the exit or exiting from the entrance) according to the modeled state order this also abnormal.

To evaluate our system we have used 64 single-person sequences with several behaviors, which produced approximately 3600 vectors. We applied a eleven-fold validation method, i.e., eleven sequences have been evaluated while we have trained with the rest ones by applying all possible combinations. We trained a HMM with 10 states. The precision/recall rates are given in Table 1. The log likelihood has been normalised according to trajectory duration and a threshold was experimentally set in order to find outliers.

Table 1. Evaluation of trajectory classification. The number of sequences, number of vectors, the precision and recall rates are given.

Label	Seq Nr	Vect Nr	Precision	Recall
normal	42	3029	97.36	88.09
abnormal	22	677	80.76	95.45

Fig. 5. The test configuration. Example of a normal trajectory and of an abnormal one. In the latter the target goes over the barrier.

Table 2. Confusion matrix for Short Term Behaviors

STB	active	inactive	walk	run	abrupt
active	765	22	0	0	0
inactive	5	532	0	0	0
walk	25	0	985	20	75
run	8	0	25	532	33
abrupt	5	0	23	45	425

For the short term behaviors we have used the classification scheme described in section 5. Additional sequences have been used. The "normal" short term activity is defined to be "walking", "active" (moving but not 2D changing coordinates), "inactive", (staying still) and the abnormal activity is defined as "running" or "abrupt motion". The characterization of each vector was done by human operators. The binary SVM classifier with radial basis function kernel has been used (c=0.1, gamma=2). The results are given in Table 2. In most cases the short term behaviors are correctly classified and the results seem promising enough to pursue the multi-agent scenarios.

8   Conclusions

In our work we have provided a richer representation of an agent's behavior compared to most current methods by modeling both short term activity and trajectory on a 2D projection map. Through the proposed representation scheme we have bypassed the computationally intensive 3D world representation. Furthermore, our model is closer to the human perception compared to pure image-based techniques due to separate handling of the two information sources.

The experiments were performed with a single moving target, though the method can be generalised to multiple moving targets, which may occlude each other, due to the use of multiple cameras. Our next steps include the extension of the proposed work for behavior modeling of multiple persons under occlusions and for the modeling cooperation between agents.

References

1. Moeslund, T.B., Hilton, A., Krüger, V.: A survey of advances in vision-based human motion capture and analysis. Comput. Vis. Image Underst. 104(2), 90–126 (2006)
2. Bobick, A.F., Davis, J.W.: The recognition of human movement using temporal templates. IEEE Trans. Pattern Anal. Mach. Intell. 23(3), 257–267 (2001)
3. Weinland, D., Ronfard, R., Boyer, E.: Free viewpoint action recognition using motion history volumes. Comput. Vis. Image Underst. 104(2), 249–257 (2006)
4. Efros, A.A., Berg, A.C., Mori, G., Malik, J.: Recognizing action at a distance. In: IEEE International Conference on Computer Vision, Nice, France, pp. 726–733 (2003)
5. Bregler, C., Malik, J.: Learning appearance based models: Mixtures of second moment experts. In: Mozer, M.C., Jordan, M.I., Petsche, T. (eds.) Advances in Neural Information Processing Systems, vol. 9, p. 845. The MIT Press, Cambridge (1997)
6. Ivanov, Y.A., Bobick, A.F.: Recognition of visual activities and interactions by stochastic parsing. IEEE Trans. Pattern Anal. Mach. Intell. 22(8), 852–872 (2000)
7. Bashir, F.I., Qu, W., Khokhar, A.A., Schonfeld, D.: Hmm-based motion recognition system using segmented pca. In: ICIP, vol. (3), pp. 1288–1291 (2005)
8. Sukthankar, G., Sycara, K.: Robust recognition of physical team behaviors using spatio-temporal models. In: Proceedings of Workshop on Modeling Others from Observations (MOO 2005) (2006)
9. Hu, W., Xiao, X., Fu, Z., Xie, D., Tan, T., Maybank, S.: A system for learning statistical motion patterns. IEEE Transactions on Pattern Analysis and Machine Intelligence 28(9), 1450–1464 (2006)
10. Brand, M., Oliver, N., Pentland, A.: Coupled hidden markov models for complex action recognition. In: CVPR 1997: Proceedings of the 1997 Conference on Computer Vision and Pattern Recognition (CVPR 1997), Washington, DC, USA, p. 994. IEEE Computer Society, Los Alamitos (1997)
11. Xiang, T., Gong, S.: Beyond tracking: Modelling activity and understanding behaviour. Int. J. Comput. Vision 67(1), 21–51 (2006)
12. Nguyen, N., Bui, H., Venkatesh, S., West, G.: Recognising and monitoring high-level behaviours in complex spatial environments. In: Proceedings of the IEEE International Conference on Computer Vision and Pattern Recognition (CVPR 2003) (2003)
13. Wang, T., Li, J., Diao, Q., Hu, W., Zhang, Y., Dulong, C.: Semantic event detection using conditional random fields. In: CVPRW 2006: Proceedings of the 2006 Conference on Computer Vision and Pattern Recognition Workshop, Washington, DC, USA, p. 109. IEEE Computer Society, Los Alamitos (2006)

14. Xiang, T., Gong, S.: Unsupervised video behaviour profiling for on-the-fly anomaly detection. IEEE Transactions on Pattern Analysis and Machine Intelligence (2008)
15. Park, S., Trivedi, M.M.: Analysis and query of person-vehicle interactions in homography domain. In: VSSN 2006: Proceedings of the 4th ACM international workshop on Video surveillance and sensor networks, pp. 101–110. ACM, New York (2006)
16. Ribeiro, P., Santos-Victor, J.: Human activities recognition from video: modelling, feature selection and classification architecture. In: Proc. Workshop on Human Activity Recognition and Modelling (HAREM 2005), pp. 61–70 (2005)
17. Zivkovic, Z., van der Heijden, F.: Efficient adaptive density estimation per image pixel for the task of background subtraction. Pattern Recogn. Lett. 27(7), 773–780 (2006)
18. Khan, S.M., Shah, M.: A multiview approach to tracking people in crowded scenes using a planar homography constraint. In: ECCV, vol. (4), pp. 133–146 (2006)
19. Lucas, B.D., Kanade, T.: An iterative image registration technique with an application to stereo vision (ijcai). In: Proceedings of the 7th International Joint Conference on Artificial Intelligence (IJCAI 1981), April 1981, pp. 674–679 (1981)
20. Rabiner, L.R.: A tutorial on hidden markov models and selected applications in speech recognition. Proceedings of the IEEE 77(2), 257–286 (1989)

A Hamming Maxnet That Determines all the Maxima

Konstantinos Koutroumbas

Institute for Space Applications and Remote Sensing,
National Observatory of Athens,
Metaxa and V. Paulou, Palaia Penteli,
152 36 Athens, Greece
koutroum@space.noa.gr

Abstract. In this paper the problem of the determination of the maximum among the M members of a set of positive real numbers S is considered. More specifically, a version of the Hamming Maxnet is proposed that is able to determine all maxima of S, in contrast to the original Hamming Maxnet and most of its variants, which can not deal with multiple maxima in S. A detailed convergence analysis of the proposed network is provided. Also, the proposed version is compared with other variants of the Hamming Maxnet via simulations.

Keywords: Neural Networks, Hamming Maxnet, Cognitive modelling, Perception.

1 Introduction

A common problem for several applications (e.g. data mining) is the selection of the maximum among M (usually positive) numbers x_i, $i = 0, \ldots, M - 1$. Several recurrent and non-recurrent techniques have been proposed for facing this problem. Among the most well known recurrent techniques is the so called Hamming MaxNet (HMN). This has originally been proposed in [6] and has been studied in [2], [3], [5], [8].

HMN mimics the heavy use of lateral inhibition evident in biological neural networks, such as the Limulus eye [12]. It consists of two layers of nodes, each one consisting of M nodes (see fig. 1(a)). Each node in the first layer takes as input one of the x_i's and is connected with all other nodes in this layer including itself. The dynamics of the first layer of HMN are described by the following equation

$$x_i(t + 1) = f(x_i(t) - \varepsilon \sum_{j=0,\ j\neq i}^{M-1} x_j(t)), \quad i = 0, \ldots, M - 1, \tag{1}$$

where $x_i(t)$ denotes the state of the i node at the t iteration. $f(x)$ is the ramp function which equals x, if $x \geq 0$ and 0 otherwise, and ε is chosen such that $0 < \varepsilon < 1/(M-1)$ [3]. Note that it is assumed that $x_i(0) \geq 0$ and $\max x_i(0) > 0$,

J. Darzentas et al. (Eds.): SETN 2008, LNAI 5138, pp. 135–147, 2008.

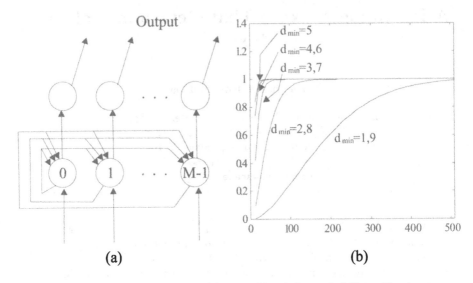

Fig. 1. (a) The Hamming Maxnet. (b) The plot of the probability of having two or more stored vectors having minimum Hamming distance from the input vector, with respect to M, for various values of $d_{\min}$.

$i = 0, \ldots, M - 1$. Intuitively speaking, each node inhibits all the rest and, at the end all nodes converge to 0 except the one with the maximum initial value, which remains positive. Each one of the nodes in the second layer of HMN takes input from the corresponding node of the first layer and outputs 1 if its input is positive and 0 otherwise. Clearly the HMN can be used in any framework where the determination of the maximum among M numbers need to be determined. In addition, it can be used either if its input is discrete-valued (as is the associative memory case) or continuous-valued.

Two problems of HMN (see eg. [3]) are its slow convergence and its inability to identify the maximum if two or more nodes have maximum initial value. In this case all nodes with maximum initial value tend asymptotically to 0. Approaches to cope with this problem are discussed in [3], [5] at the cost of very slow convergence in the former approach and at the cost of increased hardware complexity in the latter approach.

Several variations of the original HMN have been proposed eg. [12], [9] (Improved HMN), [10] (GEMNET), [11] (HITNET), [4] (HMN1, HMN2), [1], [7], [13]. Although most of them achieve significant convergence acceleration, they cannot deal with the case where two or more nodes have maximum initial value.

In this paper a variation of HMN is given that deals with the problem of multiple initial maxima. Before we proceed any further, let us see how often this problem arises. Let us consider the associative memory application. In this case M n-dimensional binary vectors e_i, $i = 0, \ldots, M - 1$ are stored and for a given input vector y, the similarity between y and each one of the e_i's is computed as $x_i = n - d(\boldsymbol{y}, \boldsymbol{e}_i)$, where $d(\boldsymbol{y}, \boldsymbol{e}_i)$ is the Hamming distance between $\boldsymbol{y}$ and $\boldsymbol{e}_i$, that is the number of places where $\boldsymbol{y}$ and $\boldsymbol{e}_i$ differ. Clearly $x_i \in \{0, 1, \ldots, n\}$, $i = 0, 1, \ldots, M - 1$. Then x_i's are passed to the HMN in order to identify the maximum.

The appearance of two or more nodes of HMN with maximum initial value implies that there are two or more stored vectors $\boldsymbol{e}_i$'s that lie at (Hamming) distance $d_{\min}$ from the input vector $\boldsymbol{y}$, where $d_{\min} = \min_{i=0,\ldots,M-1} d(\boldsymbol{y}, \boldsymbol{e}_i)$. Choosing independently M $\boldsymbol{e}_i$'s out of the set of the 2^n n-dimensional binary vectors, it is easy to see that, among the 2^n possible n-dimensional vectors, the number of vectors that lie at distance $d_{\min}$ from $\boldsymbol{y}$ is $C(n, d_{\min})$, where $C(n, m) = n!/(n - m)!m!$. Thus, the problem may be stated as:

"Among the 2^n n-dimensional binary vectors, there are $p = C(n, d_{\min})$ vectors that lie at distance $d_{\min}$ from $\boldsymbol{y}$ [1]. Selecting M out of the 2^n vectors, what is the probability that at least two vectors lie at distance $d_{\min}$ from $\boldsymbol{y}$?"

Let X be the random variable that corresponds to the selected vectors that lie at distance $d_{\min}$ from $\boldsymbol{y}$. Using the hypergeometric distribution we have

$$P(X = i) = \frac{C(p, i)C(2^n - p, M - i)}{C(2^n, M)}. \tag{2}$$

Then, it is

$$P(X > 1) = \sum_{i=2}^{\min\{p,M\}} P(X = i) = \sum_{i=2}^{\min\{p,M\}} \frac{C(p, i)C(2^n - p, M - i)}{C(2^n, M)}. \tag{3}$$

In fig. 1(b) a plot of $P(X > 1)$ with respect to M is given, with $d_{\min}$ varying from 1 to 9. In this case $n = 10$, $2^n = 1024$ and M increases up to 512. It is not difficult to see that as M increases, the probability of having two or more vectors $\boldsymbol{e}_i$ at distance $d_{\min}$ from $\boldsymbol{y}$ increases and tends to 1. Note that the sharpest increase occurs at $d_{\min} = n/2$, the value of $d_{\min}$ for which $C(n, d_{\min})$ becomes maximum. More specifically, for $M \geq 100$ and $2 \leq d_{\min} \leq 8$, $P(X > 1)$ is very close to 1.

The above analysis indicate that the HMN and most of its variants fail often in the framework of the associative memory application, for which it has originally been proposed (unless $M \ll 2^n$). Similar situations arise in other areas where x_is are discrete-valued.

Thus, handling the case of multiple nodes with maximum initial value in the HMN is of paramount importance especially in cases where the competing numbers $x_i(0)$ are discrete-valued.

In the next section a novel version of the HMN that resolves the above problem, called *HMN-Multiple-Maxima (HMN-MM)* is described and its convergence analysis is carried out. In section 3, the proposed method is compared via simulations with other known variants of the HMN, in terms of convergence speed. Finally, section 4 contains the concluding remarks.

---

[1] For the sake of simplicity of the notation, the dependence of p from $d_{\min}$ is not shown.

2 The HMN-Multiple-Maxima

2.1 Description

Let
$$S = \{x_0, x_1, \ldots, x_{M-1}\},$$

with $x_i \geq 0$, $i = 0, \ldots, M-1$ and $\max_{j=0,\ldots,M-1} x_j > 0$. The problem considered here is the determination of the one or more maxima of S. In the sequel, without loss of generality, we assume that

$$x_0 = x_1 = \ldots = x_{q-1} > x_q \geq x_{q+1} \geq \ldots \geq x_{M-1}. \tag{4}$$

In the proposed scheme the parameter ε of the original HMN[2] is replaced by $\varepsilon(t) - 1/M(t)$, where $M(t)$ is the number of the nonzero nodes at the t iteration. The term $\varepsilon(t)$ is computed as

$$\varepsilon(t) = \left(\frac{1}{M(t)} \sum_{i=0}^{M(t)-1} \varepsilon_i^{2^k}(t)\right)^{1/2^k}, \tag{5}$$

where k is a positive integer [3] and $\varepsilon_i(t)$ is determined as

$$\varepsilon_i(t) = \frac{x_i(t)}{\sum_{m=0}^{M(t)-1} x_m(t)}, \quad i = 0, 1, \ldots, M-1. \tag{6}$$

It is easy to verify that $\varepsilon_i(t) \in [0, 1]$. In addition, $\varepsilon(t)$ is an increasing sequence with respect to k and tends to $\max_{i=0,\ldots,M-1}\{\varepsilon_i(t)\}$, as $k \to +\infty$ (see appendix).

The dynamics of the proposed version of HMN are described by the equation

$$x_i(t+1) = f(x_i(t) - (\varepsilon(t) - \frac{1}{M(t)}) \sum_{j=0,\ j\neq i}^{M(t)-1} x_j(t)), \quad i = 0, 1, \ldots, M-1, \tag{7}$$

or

$$x_i(t+1) = f((1 + (\varepsilon(t) - \frac{1}{M(t)}))x_i(t) - (\varepsilon(t) - \frac{1}{M(t)}) \sum_{j=0}^{M(t)-1} x_j(t)), \quad i = 0, \ldots, M-1,$$
$$\tag{8}$$

It is worth noting that, unlike most of the variants of the HMN, in the present case $\varepsilon(t)$ is determined taking into account the values of the nodes at time t.

2.2 Convergence Analysis

The HMN-MM converges in a finite number of iterations to a state where all the nodes are zero except the nodes with maximum initial value, which are stabilize

---

[2] From now on we focus only the 1st layer of HMN.

[3] For the sake of clarity of notation, the dependence of $\varepsilon(t)$ from k is not shown.

to a positive number. The above statement is proved via the series of lemmas and theorems given below.

First, we prove that the factor $\varepsilon(t) - 1/M(t)$ is positive as long as there are positive nodes with less than maximum initial value at the t iteration. This is vital, since otherwise no lateral inhibition takes place.

Lemma 1: It is

$$\varepsilon(t) - \frac{1}{M(t)} > 0, \tag{9}$$

for $k \geq 1$, as long as at the t iteration there are positive nodes with less than maximum initial value.

Proof: Let $k = 1$. Then eq. (5) becomes $\varepsilon(t) = (\frac{1}{M(t)} \sum_{i=0}^{M(t)-1} \varepsilon_i^2(t))^{1/2}$. Substituting to eq. (9) and after some elementary manipulations, we have

$$\sum_{i=0}^{M(t)-1} \varepsilon_i^2(t) > \frac{1}{M(t)}$$

Utilizing the Cauchy-Schwartz inequality, we have

$$\left(\sum_{i=0}^{M(t)-1} \varepsilon_i(t) \right)^2 < \left(\sum_{i=0}^{M(t)-1} \varepsilon_i^2(t) \right)\left(\sum_{i=0}^{M(t)-1} 1 \right) = M(t) \sum_{i=0}^{M(t)-1} \varepsilon_i^2(t). \tag{10}$$

The inequality is strict since, taking into account (6) and hypothesis, not all ε_i's are equal to each other. The last inequality gives

$$\sum_{i=0}^{M(t)-1} \varepsilon_i^2(t) > \frac{1}{M(t)}\left(\sum_{i=0}^{M(t)-1} \varepsilon_i(t) \right)^2. \tag{11}$$

Thus, it suffices to prove that

$$\frac{1}{M(t)}\left(\sum_{i=0}^{M(t)-1} \varepsilon_i(t) \right)^2 \geq \frac{1}{M(t)}.$$

Taking into account the definition of $\varepsilon_i(t)$'s and after some algebra we obtain $1 \geq 1$ which obviously holds.

Thus, we have proved the claim for $k = 1$. Taking into account that the sequence defined by (5) is increasing with respect to k (see proposition 1 in the appendix), the claim holds also for $k > 1$. $\qquad \qquad QED$

The next lemma states that if all nodes with less than maximum initial value are zero, the factor $\varepsilon(t) - 1/M(t)$ becomes zero. This is important for the stabilization of the nodes with maximum initial value.

Lemma 2: If at a specific iteration t each node is either equal to 0 or equal to a positive number R, then $\varepsilon(t) - 1/M(t) = 0$, where $M(t)$ is the number of nodes that are equal to R.

Proof: For a nonzero node, say i, at the t iteration, eq. (6) gives $\varepsilon_i(t) = 1/M(t)$. Substituting to eq. (5), the claim follows. QED

Lemma 3: If $\exists t_0$ such that $x_i(t_0) = 0$ for some i, then $x_i(t) = 0$, $\forall t \geq t_0$.

Proof: Applying eq. (7), for $t = t_0$ we have that the argument of f is negative. Thus $x_i(t_0 + 1) = 0$. By induction the claim follows. QED

Lemma 4: $x_i(t)$'s are non-increasing sequences.

Proof: If $x_i(t + 1) = 0$, the claim holds since $x_i(t + 1) = 0 \leq x_i(t)$.

Assume now that $x_i(t + 1) > 0$. Taking into account (a) the definition of f, (b) the fact that $x_m(t) \geq 0$, for $m = 0, \ldots, M - 1$ and (c) the fact that $\varepsilon(t) - 1/M(t) \geq 0$ (lemmas 1 and 2), it is

$$x_i(t + 1) = x_i(t) - (\varepsilon(t) - \frac{1}{M(t)}) \sum_{m=0, m \neq i}^{M(t)-1} x_m(t) \leq x_i(t).$$

QED

The following lemma states that the initial ordering of the nodes is not violated in the subsequent iterations.

Lemma 5: If $x_l(0) > x_j(0)$, then

$$(a) \; \Delta x_{lj}(t) \equiv x_l(t) - x_j(t) = \Pi_{r=0}^{t-1}(1 + (\varepsilon(r) - \frac{1}{M(r)}))(x_l(0) - x_j(0)), \quad (12)$$

as long as $x_l(t)$ and $x_j(t)$ are positive.
  (b) $\Delta x_{lj}(t)$ is increasing as long as $x_l(t)$ and $x_j(t)$ are positive
  (c) $x_l(t) > x_j(t)$ as long as $x_l(t) > 0$.

Proof: (a) Taking into account eq. (7) the claim follows by induction.
  (b) This is a direct consequence of part (a) of the lemma and lemma 1, which implies that $(1 + (\varepsilon(r) - 1/M(r))) > 1$, for $r = 0, \ldots, t - 1$.
  (c) If $x_j(t) = 0$, $x_l(t) > 0 = x_j(t)$.

Suppose now that $x_j(t) > 0$. Then from part (a) of the lemma and hypothesis, the claim follows. QED

The next lemma states that two nodes with the same initial value will have the same value for all the subsequent iterations.

Lemma 6: If $x_i(0) = x_j(0) = a$, then
  (a) $x_i(t) = x_j(t)$, $\forall t$ and
  (b) $\varepsilon_i(t) = \varepsilon_j(t)$, $\forall t$.

Proof: (a) Utilizing eq. (7) the claim follows by induction.
  (b) The claim follows from the definition of ε_i's and (a). QED

Let $\Lambda_m = \lim_{t \to \infty} x_m(t)$, $m = 0, \ldots, M - 1$. The next three theorems guarantee that the network converges in a finite number of iterations to a state where all nodes with less than maximum initial value are zero, while all nodes with maximum initial value stabilize to the same positive number.

Theorem 1: $\Lambda_j = 0$, for all nodes with less than maximum initial value, provided that the variable k involved in the definition of $\varepsilon(t)$ (see eq. (5)) is chosen such that

$$\lambda_k(t) \equiv \varepsilon_l(t) - \varepsilon(t) < \frac{\sum_{m=0}^{M-1}(x_l(0) - x_m(0))}{M \sum_{m=0}^{M-1} x_m(0)} - c, \tag{13}$$

for $t \geq 1$, where c is a small positive constant chosen such that the right hand side of the previous inequality is positive and l is a node with maximum initial value.

Proof: Suppose that $\Lambda_j > 0$ for $j = 0, \ldots, M - 1$. Since the sequences $x_j(t)$'s are decreasing, we have $x_j(t) \geq \Lambda_j > 0$, $\forall t \geq 1$, and, thus, $M(t) = M$, $\forall t$.

From lemma 5(b), it is

$$\Lambda_l \geq \Lambda_j > 0. \tag{14}$$

In the sequel, based on the above assumptions we prove two results that will be utilized in the sequel. First we show that the inequality

$$\varepsilon_l(t) - \frac{1}{M} > \frac{\sum_{m=0}^{M-1}(x_l(0) - x_m(0))}{M \sum_{m=0}^{M-1} x_m(0)}. \tag{15}$$

is valid. Indeed, taking into account the definition of $\varepsilon_l(t)$, we have:

$$\varepsilon_l(t) - \frac{1}{M} = \frac{x_l(t)}{\sum_{m=0}^{M(t)-1} x_m(t)} - \frac{1}{M} = \frac{\sum_{m=0}^{M-1}(x_l(t) - x_m(t))}{M \sum_{m=0}^{M-1} x_m(t)}.$$

Utilizing the arguments given in the proof of lemma 4 under the assumption that all $x_j(t)$'s are positive, we conclude that the sequences $x_j(t)$'s are strictly decreasing. This fact, combined with lemma 5(a) proves eq. (15).

Next we prove that $\varepsilon(t) - \frac{1}{M}$ is bounded below by a positive constant. Proposition 2 in the appendix guarantees that $\lim_{k\to\infty} \lambda_k(t) = 0$. Choosing a value k_1 for k such that eq. (13) is satisfied and taking into account eq. (15) and the definition of $\lambda_k(t)$ we have

$$\varepsilon(t) - \frac{1}{M} = \varepsilon_l(t) - \frac{1}{M} - \lambda_k(t) > c > 0. \tag{16}$$

After the proof of (15) and (16), we proceed with the main part of the proof of the theorem. By the definition of f it follows that

$$x_j(t+1) = x_j(t) - (\varepsilon(t) - \frac{1}{M}) \sum_{m=0, m\neq j}^{M-1} x_m(t). \tag{17}$$

Let $x_j(t) = \Lambda_j + \delta_j(t)$, with $\delta_j(t) > 0$, for $t \geq 1$. Substituting to eq. (17) it is

$$x_j(t+1) = \Lambda_j + (\delta_j(t) - (\varepsilon(t) - \frac{1}{M}) \sum_{m=0, m\neq j}^{M-1} x_m(t)). \tag{18}$$

Note that $\sum_{m=0,m\neq j}^{M-1} x_m(t) > \sum_{m=0,m\neq j}^{M-1} \Lambda_m(t) > 0$, since $x_m(t)$'s are strictly decreasing. In order the term in the parenthesis at the right hand side in the above equation to be negative, it suffices

$$\delta_j(t) < \frac{\sum_{m=0}^{M-1}(x_l(0) - x_m(0))}{M \sum_{m=0}^{M-1} x_m(0)} \sum_{m=0,m\neq j}^{M-1} \Lambda_m \equiv Q. \tag{19}$$

Clearly Q is positive. Let $0 < \delta < Q$. Then, taking into account the fact that $x_j(t)$ is strictly decreasing, there exists t_0 such that $\Lambda_j < x_j(t_0) < \Lambda_j + \delta$ implies that $x_j(t_0 + 1) < \Lambda_j$ (as eqs. (18) and (19) guarantee), which is a contradiction.

Assume now that for $r(< M-q)$ nodes with less than maximum initial value it is $\Lambda_j = 0$. Taking into account the ordering hypothesis (see eq. (4)) and lemma 4, we conclude that these nodes will necessarily be the last r nodes, i.e. the nodes $M - r, \ldots, M - 1$. Applying the above arguments to the network taking into account only the first $M - r$ nodes, we reach again a contradiction. Thus, we have prove that $\Lambda_j = 0$, for all nodes with less than maximum initial value, provided that the variable k is chosen such that eq. (13) is satisfied. QED

Theorem 2: All nodes with less than maximum initial value attain their zero limit after a finite number of iterations, provided that k satisfies eq. (13).

Proof: Suppose that all nodes with less than maximum initial value remain positive forever, i.e. $x_j(t) > 0$, $\forall t$. Once more, $M(t) = M$, $\forall t$. Also, note that (16) is also valid in this context.

Let l be a node with maximum initial value. Taking into account lemma 5 and the fact that $x_j(t)$'s are strictly decreasing (as is discussed in the proof of theorem 1), we have that $x_l(t) - x_j(t) > x_l(0) - x_j(0) \equiv N > 0$. Therefore $\Lambda_l - \Lambda_j \geq N$. But from theorem 1 it is $\Lambda_j = 0$. Thus $\Lambda_l \geq N > 0$.

In order to be

$$x_j(t) < (\varepsilon(t) - \frac{1}{M}) \sum_{m=0,m\neq j}^{M-1} x_m(t) \tag{20}$$

it suffices $x_j(t) < cN$. Taking into account the fact that $x_j(t)$ is strictly decreasing, there exists t_0 such that $0 < x_j(t_0) < cN$, and $x_j(t_0 + 1) = 0$ (as a consequence of the definition of f), which is a contradiction.

Assume now that $r(< M - q)$ nodes with less than maximum initial value attain the zero value after a finite number of iterations. Taking into account the ordering hypothesis (see eq. (4)) and lemma 4, we conclude that these nodes will necessarily be the last r nodes, i.e. the nodes $M - r, \ldots, M - 1$. Applying the above arguments to the network taking into account only the first $M - r$ nodes, we reach again a contradiction. Thus, we have prove that all nodes with less than maximum initial value attain the zero value after a finite number of iterations, provided that the variable k is chosen such that eq. (13) is satisfied.
QED

Theorem 3: All nodes with maximum initial value stabilize to a positive number after a finite number of iterations.

Proof: Let t_0 be the first iteration where all nodes with non maximum initial value become zero. We prove first that $x_i(t_0) > 0$, $i = 0, \ldots, q-1$. From proposition 3 (see appendix), we have that $\varepsilon(t_0 - 1) \leq x_i(t_0 - 1)/\sum_{j=0}^{M-1} x_j(t_0 - 1)$. Utilizing this fact we have

$$A = (1 + (\varepsilon(t_0 - 1) - \frac{1}{M(t_0 - 1)}))x_i(t_0 - 1) -$$

$$(\varepsilon(t_0 - 1) - \frac{1}{M(t_0 - 1)}) \sum_{j=0}^{M(t_0-1)-1} x_j(t_0 - 1) \geq$$

$$(1 + (\varepsilon(t_0-1) - \frac{1}{M(t_0 - 1)}))x_i(t_0-1) - x_i(t_0-1) + \frac{1}{M(t_0 - 1)} \sum_{j=0}^{M(t_0-1)-1} x_j(t_0-1) =$$

$$(\varepsilon(t_0 - 1) - \frac{1}{M(t_0 - 1)})x_i(t_0 - 1) + \frac{1}{M(t_0 - 1)} \sum_{j=0}^{M(t_0-1)-1} x_j(t_0 - 1).$$

As a consequence of lemmas 1 and 5, the first term in the last expression is positive, while the second is non-negative. Thus $A > 0$ and, taking into account eq. (8), we have that $x_i(t_0) > 0$.

Lemma 6 implies that $x_0(t_0) = \ldots = x_{q-1}(t_0) = R > 0$. Also, lemma 2 implies that $\varepsilon(t_0) - 1/M(t_0) = 0$. Thus, from eq. (7) we have $x_i(t_0 + 1) = x_i(t_0) = R$, $i = 0, \ldots, q - 1$. By induction, it can be easily shown that $x_i(t) = x_i(t_0) = R$, $i = 0, \ldots, q - 1$, $\forall t \geq t_0$. Thus, the nodes with maximum initial value stabilize to a positive value after a finite number of iterations. QED

3   Simulation Results

In this section, a comparison of the HMN-MM, with the original HMN, the improved HMN, the GEMNET, the HITNET, the HMN1 and HMN2 is carried out, in terms of convergence speed. In each of the data sets that follow, there are at least two maxima. Clearly, the last six methods fail in this case to identify any of them. Specifically, all the nodes of these networks with maximum initial value converge asymptotically to zero. Thus, we externally test at each iteration whether or not all the nodes with non-maximum initial value are zero.

The above methods are compared for various values of M, that is $M = 10, 50, 100, 500, 1000, 5000, 10000$ and for the cases where the members of S stem from (a) the uniform $U(0,1)$ distribution, (b) the peak distribution, where all numbers stem from the uniform distribution $U(0,1)$ except the maxima which are set equal to 2 and (c) the gaussian distribution $N(0,1)$. In each of the resulting data sets we determine the maximum of S and we set x_0 and x_1 equal to this value [4]. For each value of M and for each distribution, 100 different samples for

---

[4] This is an artificial way of producing data sets containing at least two maxima.

Table 1. Number iterations required for the determination of the maxima of S, when x_i's stem from the uniform distribution $U(0,1)$. For GEMNET and HITNET the γ parameter is set to 2. Note that in some cases, HITNET failed to determine the maximum due to over-inhibition. Finally, note that for the first four networks an additional test has been embedded at each iteration in order to check whether all nodes with non-maximum initial value are zero.

	Original HMN	Modified HMN	GEMNET $\gamma = 2$	HITNET $\gamma = 2$	HMN1 $k = 6$	HMN2 $k = 6$	HMN-MM $k = 6$
$M = 10$	6.1	3.14	2.46	2.46	1.22	2.63	30.84
$M = 50$	37.5	5.43	4.70	3.91	1.99	2.36	24.65
$M = 100$	75.4	6.44	5.66	4.31	2.22	3.56	22.52
$M = 500$	419.1	8.77	8.00	4.94	3.15	4.52	34.57
$M = 1000$	799.4	9.71	9.01	5.09	3.33	4.74	35.39
$M = 5000$	4004.0	12.00	11.29	5.43	4.20	5.63	35.53
$M = 10000$	8324.2	13.10	12.29	5.61	4.71	6.16	38.29

Table 2. Number iterations required for the determination of the maxima of S, when x_i's stem from the peak distribution in (0,1), with the peak value equal to 2. See also caption of table 1. γ and k are parameters for the corresponding methods.

	Original HMN	Modified HMN	GEMNET $\gamma = 2$	HITNET $\gamma = 2$	HMN1 $k = 6$	HMN2 $k = 6$	HMN-MM $k = 6$
$M = 10$	1.98	1.92	1.83	1.82	1	1	2.15
$M = 50$	7.03	3.00	3.00	2.03	1	1	2.00
$M = 100$	11.02	3.29	3.10	2.03	1	1	2.00
$M = 500$	29.58	4.91	4.79	2.00	1	1	2.00
$M = 1000$	43.99	5.00	5.00	2.00	1	1	2.00
$M = 5000$	104.60	6.00	6.00	2.00	1	1	2.00
$M = 10000$	149.54	7.00	7.00	2.00	1	1	2.00

Table 3. Number iterations required for the determination of the maxima of S, when x_i's stem from the gaussian distribution $N(0,1)$. See also caption of table 1.

	Original HMN	Modified HMN	GEMNET $\gamma = 2$	HITNET $\gamma = 2$	HMN1 $k = 6$	HMN2 (k=6)	HMN-MM $k = 6$
$M = 10$	3.7	2.24	1.39	2.46	1.72	2.68	12.61
$M = 50$	23.7	3.96	3.05	3.91	1.95	2.69	10.57
$M = 100$	43.7	4.55	3.58	4.31	2.00	2.56	7.89
$M = 500$	248.0	6.34	5.41	4.94	2.14	2.95	11.54
$M = 1000$	490.1	7.07	6.17	5.09	2.31	3.05	9.99
$M = 5000$	2545.6	8.80	7.82	5.43	2.59	3.45	13.14
$M = 10000$	4659.7	9.33	8.42	5.61	2.58	3.30	11.27

S are drawn and all the methods are performed for each one of them. Then, the number of iterations for the 100 different samples for each value of M and for each distribution are averaged. From these simulations the following conclusions can be drawn

(i) The HMN-MM, requires more iterations for convergence than the modified HMN the GEMNET, the HITNET, the HMN1 and the HMN2 in the case where the data stem from the uniform or the normal distribution.

(ii) The HMN-MM, the HITNET, the HMN1 and HMN2 require less iterations for convergence than all the rest methods considered above, in the case where the data stem from the peak distribution (however, HMN1 and HMN2 requires a single iteration for the determination of the maximum).

(iii) In all cases, the HMN-MM requires significantly less iterations than the original HMN.

(iv) The number of required iterations in the HMN-MM does not necessarily increase with M.

In addition, the HMN-MM does not exhibit the danger of over-inhibition [5] that is present in the GEMNET and the HITNET.

4   Concluding Remarks

In this paper a variant of the Hamming Maxnet, called HMN-Multiple-Maxima (HMN-MM), that is able to determine all the maxima of a set of M positive numbers S, is proposed. Specifically, the parameter ε in the original HMN is replaced by the difference $\varepsilon(t) - 1/M(t)$. In addition, the value of $\varepsilon(t)$ is not determined in terms of the number of nodes or in terms of the number of nonzero nodes at each iteration. Instead, the values themselves of the nodes at each iteration are used in order to determine $\varepsilon(t)$. Specifically, for each node the value $\varepsilon_i(t)$ is determined as the state of the i-th node at iteration t divided by the summation of the states of all the nodes of the network at the same iteration. Then, $\varepsilon(t)$ is defined as a number very close to the maximum of $\varepsilon_i(t)$'s (see eq. (5)).

A detailed convergence analysis of the proposed scheme has been carried out, showing its ability to determine all the maxima that occur in S, in contrast to the original HMN and most of its variants which fail in the case where more than one maxima occur in S. In terms of the number of iterations, the above method is significantly faster than the original HMN. Also, it is slower than the modified HMN, the GEMNET, the HITNET, the HMN1 and the HMN2 in the case where the members of S stem from the uniform or the normal distribution and it is slightly faster when the members of S stem from the peak distribution.

In addition, the proposed scheme does not include any parameters that depend on the distribution of the numbers of S, which in practice is unknown, as is the case with GEMNET and HITNET. Also, no over-inhibition can occur in

---

[5] That is the case where the nodes with maximum initial value are forced to 0.

the HMN-MM. On the other hand, each iteration in the proposed version is computationally more demanding than the other methods except HMN1 and HMN2.

Acknowledgement. The author would like to thank Dr A. Rondogiannis for his support in the subject of the hypergeometric distribution.

References

1. Chen, C.-M., Hsu, M.-H., Wang, T.-Y.: A fast winner-take-all neural network with the dynamic ratio. Journal of Information Science and Engineering 18, 187–210 (2002)
2. Floreen, P.: The Convergence of Hamming Memory Networks. IEEE Trans. on Neural Networks 2(4) (1991)
3. Koutroumbas, K., Kalouptsidis, N.: Qualitative Analysis of the Parallel and asynchronous modes of the Hamming Network. IEEE Trans. on Neural Networks 5(3), 380–391 (1994)
4. Koutroumbas, K.: Accelerating the Hamming Maxnet. In: Vouros, G.A., Panayiotopoulos, T. (eds.) SETN 2004. LNCS (LNAI), vol. 3025, pp. 338–347. Springer, Heidelberg (2004)
5. Koutroumbas, K., Kalouptsidis, N.: Generalized Hamming Networks and applications. Neural Networks 18(7), 896–913 (2005)
6. Lippmann, R.P.: An Introduction to Computing with Neural Nets. IEEE ASSP Magazine 4(2) (April 1987)
7. Meilijson, I., Ruppin, E., Sipper, M.: A single-iteration threshold Hamming Network. IEEE Transactions on Neural Networks 6, 261–266 (1995)
8. Sum, J., Tam, P.K.S.: Note on the Maxnet dynamics. Neural Computation 8(3), 491–499 (1996)
9. Yadid-Pecht, O., Gur, M.: A biologically-inspired improved Maxnet. IEEE Transactions on Neural Networks 6(3), 757–759 (1995)
10. Yang, J.F., Chen, C.M., Wang, W.C., Lee, J.Y.: A general mean based iteration winner-take-all neural network. IEEE Transactions on Neural Networks 6(1), 14–24 (1995)
11. Yang, J.F., Chen, C.M.: Winner-take-all neural networks using the highest threshold. IEEE Transactions on Neural Networks 11(1), 194–199 (2000)
12. Yen, J.C., Chang, S.: Improved winner-take-all neural network. Electronics Letters, 662–664 (1992)
13. Yen, J.C., Chang, F.J., Chang, S.: A new winners-take-all architecture in artificial neural networks. IEEE Transactions on Neural Networks 5(5), 838–843 (1994)

Appendix

Proposition 1: The following sequence is increasing

$$A_k = \left(\frac{1}{M} \sum_{i=0}^{M-1} a_i^{2^k}\right)^{\frac{1}{2^k}}, \quad k = 0, 1, 2, \ldots.$$

Proof: It is

$$A_{k-1} = (\frac{1}{M} \sum_{i=0}^{M-1} a_i^{2^{k-1}})^{\frac{1}{2^{k-1}}}.$$

Raising both sides of the above equality to the 2^k-th power and using the Cauchy-Schwartz inequality, we obtain

$$A_{k-1}^{2^k} = (\frac{1}{M} \sum_{i=0}^{M-1} a_i^{2^{k-1}})^2 \le \sum_{i=0}^{M-1} (a_i^{2^{k-1}})^2 \sum_{i=0}^{M-1} \frac{1}{M^2} = \frac{1}{M} \sum_{i=0}^{M-1} a_i^{2^k}, \qquad (21)$$

with the equality being hold only when all a_i's are equal to each other. Raising now to the $1/2^k$-th power, we obtain

$$A_{k-1} \le (\frac{1}{M} \sum_{i=0}^{M-1} a_i^{2^k})^{\frac{1}{2^k}} \equiv A_k. \qquad (22)$$

Thus, the claim is proved.   *QED*

Proposition 2: A_k converges to the maximum of a_i's.

Proof: A_k can be written as

$$A_k = (\frac{1}{M})^{1/2^k} (\sum_{i=0}^{M-1} a_i^{2^k})^{1/2^k}.$$

Since $(1/M)^{1/2^k}$ is a subsequence of $(1/M)^{1/n}$ and $(\sum_{i=0}^{M-1} a_i^{2^k})^{1/2^k}$ is a subsequence of $(\sum_{i=0}^{M-1} a_i^n)^{1/n}$, we have that

- $\lim_{k \to +\infty} (\frac{1}{M})^{1/2^k} = 1$
- $\lim_{k \to +\infty} (\sum_{i=0}^{M-1} a_i^{2^k})^{1/2^k} = \max_{i=0,\dots,M-1} a_i.$

Combining these facts the claim follows.   *QED*

Proposition 3: A_k is less than the maximum of a_i's, for all k, provided that not all a_i's are equal to each other.

Proof: Defining $\lambda = \max_{i=0,\dots,M-1} a_i$, it is

$$A_k = (\frac{1}{M} \sum_{i=0}^{M-1} a_i^{2^k})^{\frac{1}{2^k}} < (\frac{1}{M} \sum_{i=0}^{M-1} \lambda^{2^k})^{\frac{1}{2^k}} = (\lambda^{2^k})^{\frac{1}{2^k}} = \lambda. \qquad (23)$$

                                                                QED

Item-Based Filtering and Semantic Networks for Personalized Web Content Adaptation in E-Commerce

Panayiotis Koutsabasis and John Darzentas

University of the Aegean,
Department of Product and Systems Design Engineering,
Hermoupolis, Syros, Greece, GR-84100
{kgp,idarz}@aegean.gr

Abstract. Personalised web content adaptation systems are critical constituents of successful e-commerce applications. These systems aim at the automatic identification, composition and presentation of content to users based on a model about their preferences and the context of interaction. The paper critically reviews related work in the field and presents an integrated approach for the design of personalization and web content adaptation in e-commerce that places emphasis on item-based collaborative filtering and on short-term, dynamic user models represented as semantic networks. The proposed approach for personalised web content adaptation can provide different types of interesting recommendations taking into account the current user interaction context in a computationally inexpensive way. It is also respectful of user personal information and unobtrusive with respect to user feedback.

Keywords: personalisation, web content adaptation, item-based collaborative filtering, user model, semantic networks, e-commerce.

1 Introduction

Over the last decade it has been realised that personalised web content adaptation can significantly enhance user experience in e-commerce. According to Bunt et al. [4] personalised content adaptation "*involves identifying the content most relevant to a given user and context (jointly referred to as the interaction context), as well as how this content should be organized*". The large number of approaches, techniques and systems available makes it difficult for designers to identify and develop personalization and web content adaptation components that address specific e-commerce application requirements.

The recent book entitled The Adaptive Web [3] reviews several approaches for modelling technologies (including user profiles, data mining and document modelling), adaptation technologies (including content-based, collaborative and hybrid information filtering), applications (including e-commerce, e-learning, etc.) and challenges (including group collaboration, privacy and usability engineering). Furthermore, Adomavicius and Tuzhilin [1] present a thorough review of algorithms and techniques for content-based, collaborative and hybrid filtering approaches and identify extensions in a number of directions including improvement of understanding of

J. Darzentas et al. (Eds.): SETN 2008, LNAI 5138, pp. 148–159, 2008.

users and items, incorporation of contextual information, support for multi-criteria ratings, and provision of more flexible and less intrusive types of recommendations. In addition, Montaner et al [19] analyse 37 different systems from the perspective of user profile generation, maintenance and exploitation approaches and techniques and conclude that *"the combination of modelling particular user preferences, building content models and modelling social patterns in intelligent agents seems to be an ideal solution"*.

Personalised web content adaptation has grown into a large research field that attracts scientists from different communities such as: hypertext, user modelling, machine learning, natural language generation, information retrieval, intelligent tutoring systems and cognitive science [3]. The aims of the paper are to (a) review related work in the field and (b) to present the architectural design of personalised web content adaptation system for an e-commerce application that hosts virtual product exhibitions in the area of traditional folk art. The architecture is generic and may be applied to any e-commerce system. In addition, the proposed approach is novel regarding the combinatory employment of item-based filtering and semantic networks for representing short-term, dynamic user models that are generated without user intervention.

2 Related Work

Related work in the field is reviewed according to the following important design considerations that are relevant to any type of personalised web content adaptation system: types of data sources, goals for data mining, user profile generation and representation and recommendation approaches. The review concludes with outlining important research challenges.

2.1 Types of Data Sources

Data sources for personalised web content adaptation can be divided into two main groupings: structured, such as databases or metadata-encoded information and unstructured, such as plain text. The principle data source of an e-commerce application concerns items (or products), which are usually stored in a structured database that can provide automatic responses to user queries. Other types of data sources include: retailer information, user demographic information (which may be part of the user profile), usage statistics (which is typically constructed off-line via data mining), and so on. All these types of data sources are typically structured databases.

In the case of unstructured data sources, there is a need to provide structure to enable search; this usually happens in a vector representation [19] (or "bag of words"), with each word being assigned a weight using methods such as tf*idf (term frequency times inverse document frequency)[1] [20]. To improve recall, techniques like removing stop words and stemming (keeping only the root of the word) are used. The similarity with other documents or profiles is usually measured with the cosine metric (of the

---

[1] Terms frequently appearing in a document (tf), but rarely in other documents (idf), are more likely to be relevant to the topic of the document.

angle between the two vectors). There are many other approaches to represent documents and calculate similarity; a useful review is given in [15].

2.2 Goals for Data Mining

Data mining, and particularly web usage mining has emerged as an important component of contemporary personalisation and content adaptation systems. Web usage mining can be defined as *"the automatic discovery and analysis of patterns in clickstream and associated data collected or generated as a result of user interactions with web resources on one or more web sites"* [18]. According to Mobasher [17] the data mining cycle includes the three phases of: data collection and pre-processing, pattern discovery and evaluation, and applying the discovered knowledge in real-time.

The goals and implementation of data mining differ depending on the specific application. The primary data sources used in web usage mining are the server log files, which typically include web server access logs and application server logs. Of particular interest to web personalisation and adaptation is the data collection and pre-processing phase of data mining which includes data fusion (from multiple logs), data cleaning, page view identification, sessionization (distinguishing user actions or episodes of a single interaction session), episode identification and finally data transformation that includes the aggregation and integration to e-commerce application databases [17].

2.3 User Profile Generation and Representation

Most personalisation and content adaptation systems make use of user profiles. User profiles may include demographic information and may also represent the interests or preferences of either a group of users or a single person [10]. User profiles may be created explicitly when users directly specify their preferences or implicitly by user agents that monitor user activity. User profiles can be dynamic or static, depending on whether their content changes over time. Furthermore, user profiles may be long-term or short-term depending on whether they are maintained for a long period of time in the system and they reflect preferences that do not change frequently over time or vise-versa.

According to Gauch et al. [10] user profiles are generally represented as sets of weighted keywords, semantic networks and concept-based profiles. User profiles represented as weighted keywords are updated by web documents that are interesting to users. When user profiles are represented as semantic networks, the nodes correspond to keywords or bags of words and the arcs to weights: the bigger the weight, the more relevant the connected nodes. In concept-based profiles, the nodes represent abstract topics considered interesting to the user, rather than specific words or sets of related words as in semantic networks. In general there are many variations in the implementations of user profiles based on the above approaches. Recently, ontologies have been employed to represent user profiles on the basis of domain classifications, however with interest in search engines in general [9] or in particular engines for research papers [16].

2.4 Recommendation Aproaches

Recommendation approaches (also referred as information filtering and recommender systems) are usually classified in the following categories [2]:

- Content-based: recommendations are made on the basis of estimating the similarity of new items with those that the user has preferred (purchased, positively reviewed, etc.) in the past. Pure content-based approaches suffer from that they cannot produce novel recommendations to users because the user profile is the only source of information about user preferences and does not change frequently. Another drawback is that in many cases users are required to explicitly provide their preferences.
- Collaborative: recommendations are made on the basis of estimating the similarity of the current user's profile with other users; items that other users prefer will be recommended to the current user. Pure collaborative filtering algorithms are computationally expensive and will not work until other users have provided items with ratings.
- Hybrid: recommendations are made combining both aforementioned approaches in various ways [5].

Other approaches, that are usually combined with content-based and collaborative recommenders include:

- Cluster models [24] that maintain profiles for clusters of users that have similar preferences. Recommended items are pulled from other users belonging in the same cluster.
- Demographic filtering approaches [20] that suggest items taking into account demographic information like age, occupation, education, nationality, place of living, etc.
- Knowledge-based: that recommend items based on domain-dependent inferences about user needs and preferences [5].
- Item-based filtering [13][14], which matches each of the user's preferred items to other similar items and then combines those similar items into a recommendation list. According to [21], item-based techniques first analyze the user-item matrix to identify relationships between different items, and then use these relationships to indirectly compute recommendations for users.

2.5 Challenges for Personalised Web Content Adaptation

Offline vs. Real-Time Computation. Contemporary personalisation and web content adaptation systems typically use a two-stage process for generating recommendations. The first stage is carried out offline, where web usage data from previous interactions is collected and mined and a model for making recommendations is created [17]. The second stage is carried out in real-time as users start interacting with the web site. During interaction, the model for making recommendations for the particular user is retrieved and data from the current user session is embedded into the model. Web content adaptation systems must perform most time-consuming calculations offline in order to perform fast and scale up when large amounts of data are made available.

Explicit vs. Implicit User Information Collection. Many recommender systems are intrusive in the sense that they require explicit feedback from the user and often at a significant level of user involvement [1]. Explicit user information collection typically happens through HTML forms (e.g. to directly specify preferences, to provide demographic information, etc.), as well as when users are asked to provide ratings, reviews and opinions for items. Obviously, these methods assume that users will be willing to provide this information in advance. However, related studies have shown that this assumption is unrealistic: users are reluctant to provide personal information to computer applications in general [12] mainly due to privacy concerns; they are even laconic when they provide queries [8]! On the other hand, Gauch et al. [10] identify a number of implicit user feedback techniques: browser cache, proxy servers, browser agents (e.g. the Google toolbar), desktop agents (e.g. Google desktop), web logs and search logs (based on cookies). Each technique has advantages and disadvantages with the most prominent that most still require some user intervention.

The 'Cold-Start Problem' in Generating Recommendations. Nearly all approaches for item recommendation suffer from the 'cold-start' problem, which refers to the fact that there may not be (interesting) recommendations to make to users, when there are few user profiles or ratings available. For example, in a content-based recommendation approach there may be no (interesting) recommendations for new users, unless they explicitly provide feedback about their preferences. There are many workarounds to this problem, like introducing a knowledge-based component that does not rely on historical information, to boost the process [7].

Contextual Web Content Adaptation. The large majority of recommender systems make their recommendations based only on the user and item information and do not take into consideration additional contextual information that may be crucial in some applications [1]. For example, a user that likes alternative rock music may log in to their favourite music recommendation system in order to buy a gift to a friend that likes classical music. The system should take into account this particular context of use and recommend accordingly, however this requires that it can scale up well during a particular user session.

Multi-Criteria Ratings. Most of the current recommender systems deal with single criterion ratings. However in some applications it seems quite useful to incorporate multi-criteria ratings into recommendation methods. For example, a scientific book may be reviewed on novelty, soundness, use of language, etc.; a movie on plot, direction, photography, and so on. According to Adomavicius and Tuzhilin [1] multi-criteria ratings have not yet been examined in the recommender systems literature.

Evaluation. There are not established approaches for evaluation of personalisation and web content adaptation [11]. The goals and measures of evaluations differ significantly across studies ranging from accuracy and relevance to item coverage and usability; thus the selection (or rejection) of some evaluation measures is an issue. In addition, some recommendation techniques and algorithms have been designed with specific computational requirements in mind (e.g. many collaborative filtering algorithms have been designed specifically for data sets with many more users than items); which results to satisfactory results in only some cases of data sets [14].

3   The Design of Personalised Web Content Adaptation for Virtual Product Exhibitions in E-Commerce

This section presents the architectural design of personalised web content adaptation system for an e-commerce application that hosts virtual product exhibitions in the area of traditional folk art.

Traditional folk art includes a huge variety of products that ranges from traditional clothing and decoration to artefacts of everyday life such as kitchen appliances and jewellery. These products often have a high commercial value and address a large market comprising clients who want to decorate public and private spaces (such as hotels, offices, restaurants, etc.), tourists, collectors, and interested industries such as the packaging and film industry. Of particular interest to this project are ceramics that have a multitude of uses ranging in packaging, decoration and household artefacts.

The traditional folk art sector is fragmented consisting of many SMEs that design and manufacture artefacts on the basis of self-learning and family tradition. The virtual exhibition of traditional folk art products will provide new opportunities for retailers in the sector to promote their products through the web. The traditional folk art sector typically addresses 'foreigners', i.e. customers that are not accustomed to the tradition of others, and in this respect the global market.

The aims of the personalised web content adaptation system are to complement the dynamic page segments of the e-commerce application[2] in order to:

- Provide interesting recommendations about items that are relevant to the current user interaction context, thus enabling contextual content adaptation;
- Provide a number of different types of recommendations based on different information filters;
- Integrate complementary approaches for recommendations thus constituting a hybrid recommendation system, in an attempt to address the "cold-start" problem in recommender systems;
- Ensure that the recommendation algorithms are computationally inexpensive;
- Ensure that customers interact with the e-commerce system unobtrusively of requests related to providing their preferences or other personal information;

The design of the proposed web content adaptation system is presented in the next section in terms of the conceptual architecture and its basic components.

3.1   Conceptual Architecture of Personalised Web Content Adaptation

The conceptual architecture of the proposed personalised web content adaptation system is depicted in Fig. 1. The main ideas of the proposed system are to combine item-based recommendations which are computationally inexpensive and sensitive to user interaction context with user profiling/modelling represented as a short-term, semantic network to enable fast memory-based operations. In addition, the user is not burdened by requests to provide explicit personal information, which is collected and

---

[2] Except for the item recommendations function, the e-commerce application is deterministic [22], i.e. it returns the same set of items to all customers with the same query at a certain time.

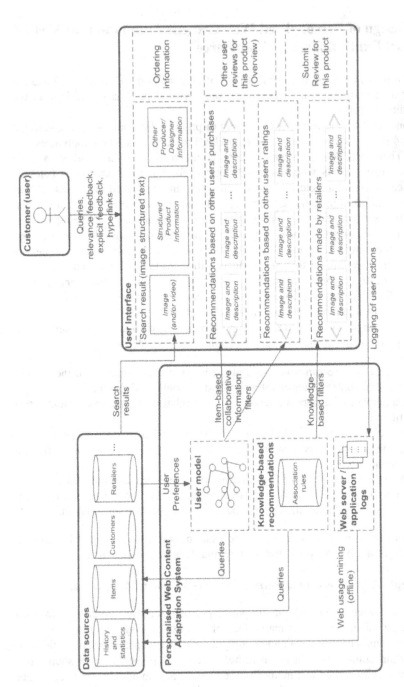

Fig. 1. Conceptual architecture of the personalised web content adaptation system for virtual product exhibitions in e-commerce

analysed by a data mining component offline. Finally, a knowledge-based component is proposed that uses association rules related to item properties to aid the item-based recommendation in order to get around the cold-start problem.

User Interface. The web user interface is dynamically generated after a user has selected an item from the pool of search results (a draft wireframe of the user interface is depicted as part of Fig. 1). The user interface is composed of a set of fragments. The 'search result' fragment presents detailed information about the item selected, which can consist of related image(s), video(s), and structured text description. There are fragments from which users can provide ordering information (e.g. a shopping basket, a wish-list etc.) and see ratings and reviews for this item provided by other users. Furthermore, users can submit their own ratings and reviews for the item available – to enable relevance feedback for producing recommendations. All the aforementioned fragments are retrieved directly from the e-commerce application databases.

In addition, there are recommendation fragments, which provide certain types of recommendations to customers based on different information filters. These fragments are dynamically generated from by the user model in the case of item-based collaborative information filters and from the knowledge-based component in the case of knowledge-based recommendations. These fragments present recommendations of other items to the user that can be selected to continue the navigation of the e-commerce application.

Item-Based Collaborative Filtering. Item-based collaborative filters recommend items that have been purchased or rated by other users in corresponding user interface fragments. If there is not (reliable) knowledge about recommendations, these page fragments are simply not added to the user interface (e.g. for a new user).

Item-based collaborative filtering happens in two-stages. The first stage calculates offline the similarity of each item in the e-commerce database with all other items on the basis of users' preferences (e.g. purchases and ratings, obviously with a different level of importance). More specifically, in our design the algorithm first identifies all items that have been purchased or rated by a particular customer. Then, the algorithm records these preferences for all customers. After recording all relationships between items (jointly purchased and/or rated with other items by customers), the similarity of items can be computed in various ways: for example, Sarwar et al. [21] propose cosine-based similarity, correlation-based similarity and adjusted cosine similarity for item-based collaborative filtering in particular. We are to initially implement weight and similarity metrics according the vector space model (tf*idf weighting scheme and cosine metric similarity), which is the most widely tested and used. The algorithm is similar to [14], with the difference that we also keep track of rated items for user recommendations. The complexity of the algorithm is $O(I^{2*}U)$ in the worst case (i.e. if an item has been purchased or rated by nearly all customers!), where I is the number of items and U the number of users, however in practice it is much closer to $O(I*U)$ [14].

User Model as a Semantic Network. The second stage of item-based collaborative filtering happens in real-time by creating the user model and employing this during interaction. In our design of item-based filtering, the user model is a semantic network that is retrieved in memory each time a user selects an item from the pool of search results. According to [15] *"a semantic network is formed by a directed graph, whose nodes are organized in hierarchic structures, while the arcs connecting them represent the binary relations between them, such as relations 'is-a' and 'part-of'"*.

In our design, the semantic network consists of a central node which is the selected item (the 'search result' in Fig. 1.) and it is connected with nodes that are related items to this item (and have been identified by the item-based collaborative filtering algorithm offline, as described above). We define two types of relations: "purchased together" and "rated" (by other users). Each relation also holds a degree of similarity that can be used to prioritise the presentation to the user. An example of a simple semantic network of this type that shows related nodes at one level of depth is depicted in Fig. 2. Obviously the semantic network can be retrieved and built in memory in further depth, if required.

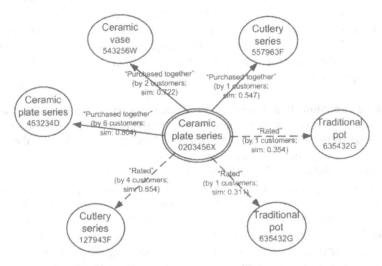

Fig. 2. An example of a simple semantic network whose nodes are items related with "purchased together" and "rated" relationships

Whenever the user selects an item from the pool of search results, the semantic network for this item is retrieved from the data sources and constructed in memory. Further to this, summary information about related items is retrieved and displayed to the user according to the type of relationship of the selected item. Each type of relationship corresponds to an information filter. The degree of similarity of each neighbouring node-item to the selected item is taken into account to prioritise the order of presentation of the recommended items. The complexity of the algorithm is then $O(I*\log I)$, i.e. the computational cost to retrieve an item I from the sorted index.

Knowledge-based Recommendations. Knowledge-based recommendations are employed to address to some extent the cold start problem in recommender systems. This type of recommendation is designed so that it relies on association rules that are inserted by knowledge engineers at the retailer site, when adding new items (products) to the data sources. For the purposes of this project, creation of digital content will take place and this content will then be inserted to the e-commerce system. During this phase of the project, the e-commerce system backend will require from retailers to associate related products.

Web Usage Mining. The proposed approach also incorporates web usage mining, mainly with regard to data collection and pre-processing that result to updating structured databases with user behavioural data such as user queries made and items browsed. Although this data is not particularly related to the current design of the personalised web content adaptation system, it will be required for further development, for example to let the user see previously browsed items or create related information filters.

3.2 Other Related Systems

The proposed personalisation and web content adaptation system is currently under development, and thus it cannot be evaluated by a user-centred approach. Therefore, a number of related systems are presented that have similar design ideas.

The most widely known system that uses item-based collaborative filtering is Amazon.com to generate several types of recommendations. In [14] it is reported that *"our algorithm's online computation scales independently of the number of customers and number of items in the product catalog, produces recommendations in realtime, scales to massive data sets, and generates high quality recommendations"*. It needs to be noted that several aspects of the Amazon.com approach to recommendation are not known, such as the user profile–item matching technique and techniques used to find similar users [19].

Another system that has implemented item-based algorithms for evaluation purposes is MovieLens, which is a movie recommender. The experimental study of item-based algorithms in MovieLens suggests that *"item-based algorithms provide dramatically better performance than user-based algorithms, while at the same time providing better quality than the best available user-based algorithms"* [14].

With respect to the employment of semantic networks to the representation of user profiles, Gauch et al. [10] review a number of systems that build semantic networks either with keywords as nodes and similarity weights as arcs, or bags of words that correspond to concept-nodes that are related to other concept-nodes again via weighted arcs. In our approach we extend those ideas by incorporating items in the nodes (that can be seen at the level of abstraction of either a bag of words or structured information) and by connecting those items with both similarity weights and types of relations that correspond to information filters. We also anticipate that our approach for representing the semantic network can be further extended to include more types of relations, e.g. such as "items browsed" by other users.

4 Summary and Future Work

The paper first presented a critical review of related work in the area of personalised web content adaptation. Important research challenges in the field include those of: offline vs. real-time computation, explicit vs. implicit user information collection, the cold-start problem in producing recommendations, contextual web content adaptation, multi-criteria ratings and evaluation of the personalisation and web content adaptation process.

The main result of the paper is however the architectural design of an integrated approach to personalised web content adaptation that addresses the first four of the aforementioned challenges. The conceptual architecture of the proposed system was discussed mainly in terms its two distinguishing features of item-based collaborative filtering and the dynamic creation of short-term user models that are represented as semantic networks.

The proposed system is under development and will be integrated to an open source e-commerce platform for hosting virtual product exhibitions of traditional folk art products.

References

1. Adomavicius, G., Tuzhilin, A.: Toward the Next Generation of Recommender Systems: A Survey of the State-of-the-Art and Possible Extensions. IEEE Transactions on Knowledge and Data Engineering 17(6) (2005)
2. Balabanovic, M., Shoham, Y.: Fab: Content-Based, Collaborative Recommendation. Communications of the ACM 40(3), 66–72 (1997)
3. Brusilovsky, P., Kobsa, A., Nejdl, W. (eds.): Adaptive Web 2007. LNCS, vol. 4321, pp. 409–432. Springer, Heidelberg (2007)
4. Bunt, A., Carenini, G., Conati, C.: Adaptive Content Presentation for the Web. In: Brusilovsky, P., Kobsa, A., Nejdl, W. (eds.) Adaptive Web 2007. LNCS, vol. 4321, pp. 409–432. Springer, Heidelberg (2007)
5. Burke, R.: Hybrid Recommender Systems: Survey and Experiments. UMUAI 12(4), 331–370 (2002)
6. Burke, R.: Knowledge-based Recommender Systems. In: Kent, A. (ed.) Encyclopedia of Library and Information Systems, vol. 69 (sup. 32) (2000)
7. Burke, R.: Hybrid Web Recommender Systems. In: Brusilovsky, P., Kobsa, A., Nejdl, W. (eds.) Adaptive Web 2007. LNCS, vol. 4321, pp. 377–408. Springer, Heidelberg (2007)
8. Carroll, J.M., Rosson, M.-B.: The paradox of the active user. In: Carroll, J.M. (ed.) Interfacing thought, pp. 80–111. MIT Press, Cambridge (1987)
9. Gauch, S., Speretta, M., Pretschner, A.: Ontology-based User Profiles for Personalised Search. In: Sharman, R., Kishore, R., Ramesh, R. (eds.) Ontologies: A Handbook of Principles, Concepts and Applications in Information Systems. Springer, Heidelberg (2007)
10. Gauch, S., Speretta, M., Chandramouli, A., Micarelli, A.: User Profiles for Personalized Information Access. In: Brusilovsky, P., Kobsa, A., Nejdl, W. (eds.) Adaptive Web 2007. LNCS, vol. 4321, pp. 54–89. Springer, Heidelberg (2007)
11. Herlocker, J.L., Konstan, J.A., Terveen, L.G., Riedl, J.T.: Evaluating Collaborative Filtering Recommender Systems. ACM Transactions on Information Systems (TOIS) 22(1), 5–53 (2004)

12. Jansen, B.J., Spink, A., Saracevic, T.: Real life, real users, and real needs: A study and analysis of user queries on the Web. Information Processing and Management 36(2), 207–227 (2000)
13. Karypis, G.: Evaluation of Item-Based Top-N Recommendation Algorithms. Technical Report, CS-TR-00-46, Computer Science Dept., University of Minnesota (2000)
14. Linden, G., Smith, B., York, J.: Amazon.com Recommendations: Item-to-Item Collaborative Filtering. IEEE Internet Computing, 76–80 (January-February 2003)
15. Micarelli, A., Sciarrone, F., Marinilli, M.: Web Document Modelling. In: Brusilovsky, P., Kobsa, A., Nejdl, W. (eds.) Adaptive Web 2007. LNCS, vol. 4321, pp. 155–192. Springer, Heidelberg (2007)
16. Middleton, S.E., Shadbolt, N.R., De Roure, D.C.: Ontological User Profiling in Recommender Systems. ACM Transactions on Information Systems 22(1), 54–88 (2004)
17. Mobasher, B.: Data Mining forWeb Personalization. In: Brusilovsky, P., Kobsa, A., Nejdl, W. (eds.) Adaptive Web 2007. LNCS, vol. 4321, pp. 90–135. Springer, Heidelberg (2007)
18. Mobasher, B.: Web usage mining and personalization. In: Singh, M.P. (ed.) Practical Handbook of Internet Computing. CRC Press, Boca Raton (2005)
19. Montaner, M., Lopez, B., De La Rosa, J.: A Taxonomy of Recommender Agents on the Internet. Artificial Intelligence Review 19, 285–330 (2003)
20. Pazzani, M.: A Framework for Collaborative, Content-Based, and Demographic Filtering. Artificial Intelligence Rev., 393–408 (December 1999)
21. Salton, G., McGill, M.: Introduction to Modern Information Retrieval. McGraw-Hill, New York (1983)
22. Sarwar, B., Karypis, G., Konstan, J., Riedl, J.: Item-Based Collaborative Filtering Recommendation Algorithms, Item-based collaborative filtering recommendation algorithms. In: Proceedings of the 10th International World Wide Web Conference (WWW10), Hong Kong, May 1-5 (2001)
23. Tanudjaja, F., Mui, L.: Persona: A Contextualized and Personalized Web Search. In: Proceedings of the 35th Hawaii International Conference on System Sciences – 2002. IEEE, Los Alamitos (2002)
24. Ungar, L., Foster, D.: Clustering Methods for Collaborative Filtering. In: Proc. Workshop on Recommendation Systems. AAAI Press, Menlo Park (1998)

Fuzzy Representation and Synthesis of Concepts in Engineering Design

Vassilis C. Moulianitis[1,2], Nikos A. Aspragathos[2], and Argiris J. Dentsoras[2]

[1] Department of Product and System Design Engineering,
University of the Aegean, 84100, Ermoupolis, Syros, Greece
[2] University of Patras, Mechanical Engineering and Aeronautics Department,
26500, Patras, Greece
{moulian,asprag,dentsora}@mech.upatras.gr

Abstract. In the present paper a new mathematical fuzzy-logic-based formulation of the design objects and the rules that govern a design problem during the conceptual design phase is presented.. A procedure for the automatic generation of degrees of satisfaction of the design specifications for each feasible solution - subjected to design constraints - is introduced. A table containing the satisfaction degrees is used for the derivation of the set of all possible synthesized solutions. The determination of this set, which is a subset of the set of the synthesised solutions, is based on a suitable partition of the Euclidean space. An illustrative example of a knowledge based system for the conceptual design of grippers for handling fabrics is presented. The advantages of this model are revealed via a comparison with previous implementations of the conceptual design phase based on crisp production rules or certainty factors.

Keywords: Conceptual Design, Fuzzy Logic.

1 Introduction

It was only by the middle of the 20th century when design became the focus point of many researchers aiming to the formulation of a rigorous and scientific basis for what had been an intense human intellectual activity throughout the centuries. Since then, design philosophies, models, methods and techniques have been developed [1, 2, 3] and different approaches have been worked out concerning the nature of the design. The design models have tended to fall within two main classes: prescriptive and descriptive. Prescriptive models have been used to look at the design process from a global perspective covering the procedural steps, and have been based on either the design process itself or on the product attributes. The descriptive models are related to the designer' s activities during the design process. Recent advances in the field of artificial intelligence (AI) offer the scientific basis to clarify and articulate concepts of design so as to establish an advanced framework for the design research, towards the so called computational models. A survey on the application of soft computing (SC) techniques in engineering design is presented in [4]. Design activities encompass a spectrum from routine design, through variant design to creative design of new artifacts. While the routine design is possible to be computable to a great extent, it is very difficult to model adequately the creative design.

J. Darzentas et al. (Eds.): SETN 2008, LNAI 5138, pp. 160–172, 2008.

The main activities taking place in the conceptual design phase are the concept generation and evaluation. In the conceptual design phase, human's description of the design objects as well as their relation are qualitative, linguistic and subjective, so it is natural to use fuzzy logic approach to model them. Furthermore, if the design problem is ill-defined, crisp production rules may be inadequate to handle the fuzziness of the design problem in the conceptual design phase. The fuzzy set theory provides a viable tool for solving such problems [5]. According to Zimmermann [6], one of the main reasons for the use of the fuzzy set theory is the natural means provided for the communication between the designer and the computer. In addition, the storage of the vague and uncertain portion of the knowledge expressed by rules is much more realistic when fuzzy sets are used instead of crisp concepts and symbolism. Finally, the uncertainty in the conclusions that arises as a natural consequence of the uncertainty in the hypothesis can be easily overcome by applying aggregation procedures. Recently, efforts have been made to model the conceptual design phase by applying fuzzy logic. The formulation of a design process model is very important step towards the development of knowledge-based systems for the conceptual design. A number of design models of the Quality Function Deployment (QFD) technique and of the evaluation process based on fuzzy logic appeared in the relevant literature [7, 8, 9, 10, 11].

In [7], the problem of representation and assessment of conflicting requirements related to the use of House of Quality (HOQ) is solved using fuzzy logic. An application of the fuzzy sets theory with QFD technique in the field of construction is presented in [11]. The system supports the evaluation of buildable designs through adapting matrices of conventional HOQ. Fuzzy numbers are used to represent the linguistic and imprecise nature of decisions and judgments of buildable designs, and a fuzzy inference mechanisms is established to process design-relevant information. Fuzzy logic is used to prioritize, categorize or translate customer requirements and design requirements in QFD providing more reliable results than the conventional approach [8, 10]. Fuzzy logic allows a design team to reconcile tradeoffs among the various performance characteristics representing the customer satisfaction as well as the inherent fuzziness [9]. However, QFD is not based on a solid mathematical foundation and the results of this technique are very difficult to be proven [7]. In addition, fuzzy logic in the above papers solves problems that arise in the execution of QFD and do not support the mathematical foundation of this technique.

Fuzzy logic calculus is used to represent and handle imprecise design information [12, 13, 14]. The method of imprecision (MoI) consists of seven axioms and handles design problems as multicriteria evaluation problems modeled with fuzzy numbers. In this method, design specifications are represented with numbers and not with linguistic variables. In [15] an outranking preference model based on the possibility theory is developed to select the "best" design concepts from a number of alternatives. The alternatives are compared in pairs using a fuzzy outranking relation.

The present paper introduces a mathematical model of the conceptual design phase based on fuzzy logic. It contributes towards a better understanding of the concept selection during the generation process in the conceptual design phase where the most critical decisions are made with the design objects still kept to an abstract level and it provides a mathematical base for the development of knowledge based systems. The design objects are represented by fuzzy variables and the relationships among them as fuzzy rules. A table containing the degree of satisfaction among the design specifications and feasible solutions is used to formulate the set of synthesized potential final solutions from a knowledge base of simple solutions. The final solutions are found with a suitable resolve of a Euclidean space where the alternative solutions lie.

2 Fuzziness in the Design Process

According to Ullman [16], conceptual design is the second phase in a six-phase product design process. In the first phase, two main tasks must be accomplished; (a) to understand the design problem and, (b) to plan the design process. In the conceptual design phase, mainly concepts are generated and evaluated. In the third phase, detailed design is performed followed by the product evaluation, while the production planning is concurrently integrated.

Usually, the solutions are generated by applying forward, data-driven strategies, which presupposes the satisfaction of the posed design specifications.

In the early design phases, the design objects are usually represented by linguistic variables, sketches, draft drawings, etc, while in the following phases the objects are quantified and the subsequent decision making is based on quantitative processes. In the conceptual design phase the design objects should be kept as abstract as possible and at the same level of abstraction. Consequently, these objects cannot be represented by quantitative variables and analytical formulas or arithmetic procedures. In order to develop a knowledge-based system to be used in the conceptual design process, linguistic variables and production rules must be defined. Since human's description of the design objects and of the relationships among them are qualitative, fuzzy, subjective and depending, to a great extent, on expertise, it seems convenient to use a fuzzy logic approach to model them. For example, imprecise and vague terms such as "Poor", "Tolerable" that represents linguistic concepts for design solutions used in the conceptual design phase can be represented in a mathematically well-defined way which simulates the human's information processing. Furthermore, if the entities in the design domain are not deterministically defined, simple production rules may be inadequate to handle the fuzziness of the domain in the conceptual design phase. The fuzzy set theory provides a viable tool for solving problems in ill-defined domains [5].

The present work contributes towards the development of new systems where the designer has the main role in the determination and verification of the solutions using the following capabilities:

- The development of the knowledge base that describes attributes of design objects and contains rules expressed in a well-defined formulated way. The knowledge about the domain is obtained by using the well established knowledge acquisition techniques.
- The final solution space is determined considering the limits defined by the designer.
- The designer can verify the feasibility of the alternative solutions.

In the present paper the design objects are formulated using fuzzy set theory. The knowledge acquisition and the development of the system' s knowledge base is not discussed since it has been already analyzed in previous papers [17, 18, 19]. The design objects are introduced in a generalized way together with the determination mechanisms of alternative final solutions to the problem.

3 Generation of Solutions

In this section, the design objects and the procedures taking place for the generation of solutions in the conceptual design phase are presented. The design objects are represented using fuzzy set theory and their relationships by fuzzy production rules. In addition, the procedure for the automatic determination of the degrees of satisfaction of the design specifications for each feasible solution subjected to the design constraints is presented. This procedure is compared to the heuristic methods that can be alternatively used. In addition, the designer's contribution throughout this process is discussed.

The block diagram shown in fig. 1 illustrates the design objects and the procedures taken place for the generation of solutions in the conceptual design phase. In every design problem, a set of design specifications have to be defined by the designer that must be fulfilled simultaneously by one or more feasible solutions constrained by a set of design constraints. Let the design specifications, design constraints and feasible solutions be members of the sets R_d, C_d and S respectively. The designer defines sets R_d and C_d, which are then used by the fuzzy inference mechanism to produce the table that contains the degrees of satisfaction (DoS Table) of the design specifications by the feasible solutions. The set **CFS** (Combined or synthesized Feasible Solutions) contains all the synthesized members of the S. Every member of **CFS** is a synthesis of one or more members of S that are also present in the DoS Table. The degrees of satisfaction of the synthesized objects are determined using a maximum operator. The set **FS** (Final Solutions) is a subset of **CFS**, which under the conditions specified by the designer, represents the set of the final design solutions. The generation of **FS** is achieved through suitable partitioning of hypercube, formed by the members of **CFS**. In the following paragraphs the structure and organization of the domain knowledge is presented.

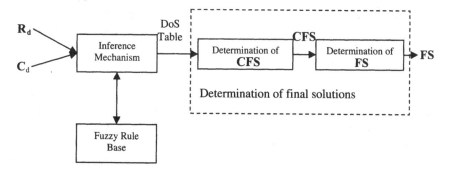

Fig. 1. Block diagram of the fuzzy generation in the conceptual design

3.1 Formulation of the Design Objects

The design problem requires the satisfaction of a subset R_d of the specification defined by the designer. This set is a subset of set $T = \{Spec\_1, Spec\_2, ..., Spec\_a_1\}$, where a_1 is the number of specifications, that represents all possible design specifications in a particular domain. In this case, design specifications manifest themselves as a set of needs that must be fulfilled by one or more operating principles. The members of **T** are inserted in the system in order to define its operational domain. For example, the set **T** can be the set of the tasks that a class of grippers can accomplish when handling non-rigid materials or the functions that a product must accomplish, e.g. the deflection of water and mud by a mudguard of a bicycle in the Splashgard design example [9].

The set of design specifications $\mathbf{R_d}$ is defined as:

$$\mathbf{R_d} = \{R_{d,i} \mid i = 1,2,...,n_1\} \subseteq \mathbf{T} \tag{1}$$

The cardinality n_1, $(n_1 \le a_1)$, and the members of $\mathbf{R_d}$ are selected by the designer.

Let $\mathbf{DC} = \{\mathbf{Constraint_1}, \mathbf{Constraint_2},..., \mathbf{Constraint_{a_2}}\}$ be the set of the universes of the design constraints for the specific problem domain. Every design constraint included in **DC** is represented by a collection of fuzzy sets A_i^k, for $k = 1,2,3,...,k_A$ where k_A is the number of the fuzzy sets which are partitions of the universe $\mathbf{Constraint_i}$ for $i = 1,2,...,a_2$, expressed by means of membership functions [20]:

$$A_i^k \in F(\mathbf{Constraint_i}), \mu_{A_i^k}(x): \mathbf{Constraint_i} \to [0,1],$$
$$\text{for } k = 1,2,...,k_A \text{ and } i = 1,2,...,a_2 \tag{2}$$

Where $F(\cdot)$ denotes a family of fuzzy sets, and x the members of each set $\mathbf{Constraint_i}$. The number of fuzzy sets for each constraint may be different. In the present paper, without loss of generality and for reasons of simplicity, it is assumed that they are the same for every fuzzy variable.

The designer specifies the constraints that form a set:

$$\mathbf{C_d} = \{C_{d,i} \mid C_{d,i} \in \mathbf{DC}, i = 1,2,...,n_2\} \tag{3}$$

Where n_2, $(n_2 \le a_2)$, is the number of constraints specified by the designer for the specific design problem. For example, in the case of a gripper design for handling non-rigid materials, both material's and environment's characteristics are design constraints that are included in **DC**. In the case of the Splashgard example [9], the quantities of the water and the mud that the mudguard must handle are constraints to this design problem.

Since the designer specifies the constraints, a vector $I = [R_{d,i} \quad \underline{x}] = [R_{d,i} \quad x_1 \quad x_2 \quad ... \quad x_{n_2}] \in \mathbf{R_d} \times \mathbf{C_d}^{n_2}$ is the input vector to the design system shown in fig. 1. The elements number of I is equal to $n_2 + 1$ and every member of $\underline{x} = [x_1 \quad x_2 \quad ... \quad x_{i_1} \quad ... \quad x_{n_2}]$ is defined for each constraint in its universe of discourse $\mathbf{Constraint_{i_1}} = [b_1, b_2]$ where $b_1, b_2 \in \mathfrak{R}$, $b_1 < b_2$ or it can be defined by one linguistic variables from those used for the fuzzy sets A_i^k for $k = 1,2,3,...,k_A$, and $i_1 = 1,2,...,n_2$. For example, the thickness of a non-rigid material can be defined by a numerical value, e.g. 0.1 mm, or, more naturally, by a linguistic variable such as "low thickness".

The set of feasible solutions manifest themselves as a set of operating principles:

$$S = \{S_1, S_2,...S_j,...,S_m\}, j = 1,2,...,m \tag{4}$$

that must satisfy the set of design specifications R_d. Every operating principle satisfies every member of R_d to a different degree. So, every member of S can be defined as a set of ordered pairs as follows:

$$S_j = \left\{ \left\langle \bar{c}_j, i \right\rangle \mid \bar{c}_j \in [0,1], i = 1,2,...,n_1 \right\} \tag{5}$$

Where, i denotes the design specification $R_{d,i}$, and $\bar{c}_j$ the crisp degree of satisfaction with $j = 1,2,...,m$. For reasons of simplicity the ordered pair $\left\langle \bar{c}_j, i \right\rangle$ is written as c_{ij}.

Let **DoC** be the universe of discourse of the degrees of satisfaction c_{ij} that is represented by a collection of fuzzy sets B_i^k, for $k = 1,2,3,...,k_B$ which are partitions of the space **DoC**, expressed by the following membership functions [20]:

$$B_i^k \in F(\textbf{DoC}); \mu_{B_i^k}(c_{ij}): \textbf{DoC} \rightarrow [0,1], \text{ for } k = 1,2,...k_B,$$
$$i = 1,2,...,n_1 \text{ and } j = 1,2,...,m \tag{6}$$

In the present paper and for reasons of simplicity, the space **DoC** is considered to be the closed interval $[0,1]$. The upper bound of **DoC** means that the specific operating principle satisfies fully the design specifications while the lower bound does not satisfies it at all. The vector $O = \begin{bmatrix} c_{i1} & c_{i2} & ... & c_{im} \end{bmatrix} \in S_1 \times S_2 \times ... \times S_m$ has m elements and represents the output of the inference mechanism in the form of satisfaction degrees. For every $R_{d,i}$ a degree of satisfaction is determined for every S_j, showing the degree of satisfaction of every operating principle S_j to the specification $R_{d,i}$.

3.2 Formulation of the DoS Table and K_s

In this section, the formulation of the degrees of satisfaction table and the mapping among the design objects are presented. The vector I is the input to the inference mechanism that derives the vector O, which is the defuzzified aggregated output of a set of fuzzy rules, and derives the rows of the DoS Table. This procedure is repeated n_1 times, where n_1 is the cardinality of the R_d.

As it is referred in section 3, the DoS Table maps the degree of satisfaction of every design specification by a feasible solution. The design specifications and the design constraints contribute to the determination of the satisfaction degrees. These degrees represent the confidence of the designer concerning the ability of a specific solution to satisfy the specification and the constraints. The mapping among those design objects is represented formally by the following relation:

$$K_s : R_d \times C_d^{n_2} \rightarrow S_1 \times S_2 \times ... \times S_m \tag{7}$$

The application of K_s derives the rows of the DoS Table where every column represents the set S_j for $j = 1,2,...,m$. The K_s relation expresses the knowledge about the problem in the earlier design phases and it can be suitably represented by production rules, since they can process linguistic variables. Production rules capture the design knowledge in a domain and give a novice access to expertness. The formulation of K_s using simple production rules and rules with degrees of satisfaction was presented in [17, 18, 19]. In the present paper, the $\mathbf{K_s}$ relation is formulated using fuzzy logic and is expressed as a set of fuzzy conjunctive rules of the form:

$$\text{i,j-Rule: If } R_{d,i} \text{ is } Spec\_i \text{ and } x_1 \text{ is } A_1^k \text{ and } x_2 \text{ is } A_2^k \text{ and } \dots \text{ and } x_{n_2} \text{ is }$$
$$A_{n_2}^{\ k} \text{ Then } c_{ij} \text{ is } B_{ij}^{\ k} \text{ (for } j = 1,2,...,m \text{ and } k = 1,2,3,...)$$
(8)

where $x_i \in \mathbf{Constraint_i}$ for $i = 1,2,...,n_2$. For reasons of simplicity, k indices are the same for the elements of the relation (8). In the developed system, k index represents different fuzzy sets and follows the formulation of the relations (2) and (6). If p_1 rules of the form shown in (8) are fired, then the membership function $\mu_{c_{ij}}(c_{ij})$ describing the overall output is [20]:

$$\mu_{c_{ij}}(c_{ij}) = \underset{p}{aggreg}\left(compos\left(1, \mu_{A_1^k}(x_1), ..., \mu_{A_{n_2}^k}(x_{n_2}), \mu_{p,j-Rule}(1, x_1, ..., x_{n_2}, c_{ij})\right)\right),$$
$$k = 1,2,3,...$$
(9)

Where 1 denotes the truth of "$R_{d,i}$ is $Spec\_i$" and $\underline{x} = \begin{bmatrix} x_1 & x_2 & ... & x_{n_2} \end{bmatrix}$ is the vector of fuzzy inputs for the fuzzy constraints, $\mu_{p,j-Rule}(1, x_1, x_2, ..., x_{n_2}, c_{ij})$ is the fuzzy implication of the p,j-Rule for $p = 1,2,...,p_1$ and $j = 1,2,...,m$. Composition extracts the output of the rules using the input vector and aggregation combines all the results of the rules. Implication, composition and aggregation can be defined in various ways [20]. The final result for $\overline{c}_{ij}$ appears in the DoS matrix is determined by a defuzzification process [20]:

$$\overline{c}_{ij} = defuzz\left(\mu_{c_{ij}}(c_{ij}), c_{ij}\right)$$
(10)

Where *defuzz* represents a specific defuzzification method. In the system developed for the conceptual design of grippers for handling fabrics, the Mamdani's implication, the max-min composition, the maximum aggregation and the centroid defuzzification method have been used [20]. The main reason of using these methods is the simplicity of construction of these connectives. The minimum operator determines the maximum value for the membership functions of all T-norms, while the maximum operator the minimum value of all S-norms providing the most tolerant of the intolerant strategies for aggregation. The centroid defuzzification method determines the representative of the set that presents the least error among the members of the set.

3.3 Definition of the Final Solutions Set CFS

Let **CFS** be the set of the synthesized feasible solutions, where $\mathbf{CFS} \subset \mathbf{S}^w$, where is $w = \min(n_1, m)$. The **CFS** is constructed by the composition of the elements of the **S**. The number of the elements that are synthesized depends on the cardinality of the $\mathbf{R_d}$ or the number of operating principles shown in DoS Table. If the number of specifications are less than the number of operating principles, and taking into account that every operating principle can satisfy at least one specification, then $\mathbf{CFS} \subset \mathbf{S}^{n_1}$. In the case where the number of operating principles is less than the cardinality of the $\mathbf{R_d}$, then every operating principle must satisfy more than one design specifications, so $\mathbf{CFS} \subset \mathbf{S}^m$. Every member of $\mathbf{CFS}$ is defined as a set of degrees of satisfaction such as:

$$FS_{j_1} = \left\{ \overline{fc}_{ij_1} \mid \overline{fc}_{ij_1} = \max_i \left(c_{ij} \right) \in [0,1], i, j = 1,2,..,n_1, \overline{c}_{ij} \in S_j, S_j \in \mathbf{S} \right\}$$

$$j_1 = 1,2,...,n_c, n_c = \sum_{a=1}^{\min(n_1,m)} \binom{m}{a}$$

(11)

In (11), the upper bound of the variable a depends on the number of design specifications and feasible solutions. If the number of specifications is smaller than the number of operating principles, then the upper bound of a is the number of design specifications. In the case where the number of operating principles is lower than the cardinality of the $\mathbf{R_d}$, the upper bound of a is equal to the number of operating principles. If $n_1 > m$, then there are members where the $n_1 - m$ members of FS_{j_1} are the identity element of $\mathbf{S}$ defined as:

$$S_0 = \left\{ \overline{c}_{i0} \mid \overline{c}_{i0} = 0, i = 1,2,...,n_1 \right\}$$

(12)

In addition, one member of $\mathbf{S}$ can be a final solution if it satisfies all the elements of $\mathbf{R_d}$ synthesized with $n_1 - 1$ identity elements. The S_0 operating principle is a mathematical invention that supports the synthesis. The members of S_0 does not affect the degrees of satisfaction of the synthesized operating principles since 0 is the identity element of the maximum operation used in (11). The set of final solutions $\mathbf{FS}$ is a subset of $\mathbf{CFS}$.

3.4 Determination of the Final Solutions Set FS

Every member of $\mathbf{CFS}$ is considered as a vector in a n_1-th dimensional Euclidean space, which is a hypercube $[0,1]^{n_1}$. In order to obtain the final solutions to the problem, belonging in $\mathbf{FS}$, a measure is defined which is the Euclidean norm $\|\cdot\|$, of a vector $FS_{j_1} = \left(\overline{fc}_{1j_1}, \overline{fc}_{2j_1}, ..., \overline{fc}_{n_1 j_1} \right)$, [21]:

$$\left\| FS_{j_1} \right\| = \sqrt{\left(\sum_{i=1}^{n_1} \left| \overline{fc}_{ij_1} \right|^2 \right)}$$

(13)

In this case the norm is considered as the index, used for the solution evaluation. In this space the maximum norm corresponds to the vector E_{n_1}, which has all its elements equal to 1:

$$\left\| E_{n_1} \right\| = \left\| (1,1,...,1) \right\| = \sqrt{n_1}$$

(14)

The norm $\|\cdot\|$ defines a relation that maps every FS_{j_1} to a subset of $\mathfrak{R}$:

$$\|\cdot\| : FS_{j_1} \rightarrow \left[0, \left\| E_{n_1} \right\| \right], j_1 = 1,2,...,n_c$$

(15)

The members of $\mathbf{FS}$ are determined by the following simple rule:

$$\text{if } \left\| FS_{j_1} \right\| \geq \ell_1 \text{ then the } FS_{j_1} \in \mathbf{FS}$$

(16)

where ℓ_1 is a limit defined as:

$$\ell_1 = f_1 \|E_{n-1}\|$$

$$f_1 \in \left(1, \frac{\|E_n\|}{\|E_{n-1}\|}\right] \tag{17}$$

The heuristic search methods presented in [18, 19] can be used to determine the **FS** in an easier way. These methods do not search the entire **CFS** and should be used when the cardinal number of **CFS** is too high. For example, if the cardinality of R_d and **S** are 5 and 20 respectively then the cardinality of **CFS** is 21699 as it is determined by (11). The heuristic search methods checks at least 20 members of **CFS** and in the worst case 15504 members (71,45% of **CFS**).

Additionally, the synthesized operating principles are new operating principles and their feasibility must be checked by the designer. The feasibility of the proposed solution cannot be check automatically by the proposed method, since the available knowledge is synthesized and there are no previous tests or simulations. The synthesized solutions of the **CFS** represent the capability of the system to provide new alternative solutions.

4 An illustrative Example

A knowledge-based system (KBS) for the design of grippers for handling non-rigid materials is developed as an application. The knowledge used for it can be found in [17] and has been reformulated to adapt to the introduced fuzzy logic representation. All the sets and the rules have been defined using fuzzy sets and the KBS was run in MATLAB with the Fuzzy Logic Toolbox [22].

A lot of operational concepts for grippers in apparel industry have been elaborated so far. In the past, each gripper was dedicated to one task. The current trend is to manufacture grippers which, under the same concept, may accomplish more than one task. The existing grippers do not present the same efficiency for different tasks.

Consider the case where the gripper has to accomplish two tasks contained in **T** which have five members. The designer specifies the following two tasks as design specifications $R_d = \{Apply\_tension, Separation\}$. The design constraints to this domain are $DC = \{Thickness, Porosity, Roughness, Density, Molecular\_weight\}$ For reasons of simplicity, every design constraint presents a normalized universe of discourse $Constraint_i = [0,1]$ with three triangular membership functions; $A_i^1 = Low$, $A_i^2 = Medium$ and $A_i^3 = Medium$. In this case $C_d \equiv DC$ and the inputs for the constraints specified by the designer are $\underline{x} = \{0.5 \quad 0.1 \quad High \quad Medium \quad High\}$ Material's thickness and porosity are defined with crisp values where the first denotes a medium value and the second a low one respectively. If the designer is not able to determine the crisp values then the linguistic variables that corresponds to the fuzzy sets can be used.

$\mathbf{S} = \{Clamp, Vacuum, Elec\,/\,sive, Adhesive, Velcro, Pinch, Freezing, Pin, Airjet\}$ is the set of the gripper's operating principles, where the order of appearance in the $\mathbf{S}$ denotes the pointer j of the S_j member for $j = 1,2,...,9$ (e.g. $S_4 = Adhesive$).

The universe of discourse of the degrees of satisfaction c_{ij} of every operating principle is defined to be the set $\mathbf{DoC} = [0,1]$ with three triangular membership functions: $B^1 = Poor$, $B^2 = Tolerable$, $B^3 = Good$.

In the following, three rules for the case of $R_{d,2} = Separation$ and $S_2 = Vacuum$ that are stored in the knowledge base are shown:

If $R_{d,2}$ is *Separation* and x_2 is Low Then c_{22} is Good.

If $R_{d,2}$ is *Separation* and x_2 is Medium Then c_{22} is Tolerable.

If $R_{d,2}$ is *Separation* and x_2 is High Then c_{22} is Poor.

Since x_2 has a low value the first two rules will fire. The other constraints do not affect the S_2 operating principle and are omitted. The porosity of the material affects strongly the vacuum operating principles. If the material is very porous then the vacuum gripper cannot separate a single ply form a stack and, consequently, this task cannot be satisfied by this operating principle. Using max-min composition, maximum aggregation and centroid defuzzification the DoS Table is filled and the results are shown in table 1. The **CFS** has $n_c = 45$ elements. The 9 elements shown in table 1 and the 6 elements shown in table 1 are 15 elements of this set. In order to avoid further elaborating and evaluating bad solutions of the **CFS**, the search algorithm presented in [18] with $l_1 = 1.01$ and $l_2 = 0.1$ has been used to obtain the **FS**. The elements of this set are shown in table 2. None of the operating principles presented in table 1 can accomplish successfully both tasks. Clamp operating principle is the only operating principle that accomplishes to a successful degree the "apply tension" task, while the other six operating principles presented in table 2 can accomplish successfully the separation task. The synthesis of the clamp operating principle with each one of the six operating principles presented in table 2 can accomplish both tasks. In previous implementations of the KBS there was only one solution, the "Clamp" gripper, [18]. As it is shown in table 1, "Clamp" gripper presents tolerable values for each specification, a fact that is moderately acceptable. The fuzzy implementation provides six alternative solutions that are better than the unique solution, since the designer can choose the best by applying a proper evaluation process.

Table 1. DoS Matrix

Design Specifications	Feasible Solutions								
	S_1	S_2	S_3	S_4	S_5	S_6	S_7	S_8	S_9
$R_{d,1}$	0.58	0.16	0.17	0.16	0.15	0.16	0.16	0.21	0.16
$R_{d,2}$	0.58	0.84	0.58	0.82	0.16	0.83	0.82	0.82	0.83

The fuzzy implementation of the KBS compared to the previous one [18] seems to be more complex; a problem that is relieved by the mathematical formulation presented in this work. The intrinsic fuzziness is handled efficiently and the results obtained are better since an aggregated procedure is used. As it is shown in the present example, the designer can handle the knowledge in a natural way. The design objects in the concept generation and evaluation are represented with linguistic variables while, in previous implementations, the designer had to assign arithmetic variables to the concepts and criteria, a task considered as difficult at least for novice designers. Since the results are numeric the designer is able to determine easier the best solution. The arithmetic variables that the designer must determine are the limits of the Euclidean space that affect the quantity and quality of solutions.

Table 2. Final solutions to the problem

	FS_1	FS_2	FS_3	FS_4	FS_5	FS_6
Synthesized Operating principles	Clamp Vacuum	Clamp Adhesive	Clamp Pinch	Clamp Freezing	Clamp Pin	Clamp Airjet
$\overline{fc}_{1,j_1}$	0.58	0.58	0.58	0.58	0.58	0.58
$\overline{fc}_{2,j_1}$	0.84	0.82	0.83	0.82	0.82	0.83
Norm	1.03	1.01	1.02	1.01	1.01	1.02

5 Conclusions

In the present paper, a formal mathematical base for qualitative reasoning based on fuzzy logic for the conceptual design phase is presented. The design objects are represented via fuzzy variables and their relationships as fuzzy rules. Using fuzzy logic the intrinsic fuzziness, inherent in the conceptual design phase, is efficiently handled. The determination of final solutions to the design problem is based on a suitable resolving of a Euclidean space where the alternative feasible solutions are represented as vectors. The benefits of this model are summarized in the following:

- A formal mathematical base for qualitative reasoning in the conceptual design phase is introduced.
- The intrinsic fuzziness appearing during the design process is handled using fuzzy logic.

These benefits are shown in the example of the conceptual design of grippers for handling fabrics, where comparisons with previous implementations of the KBS are also presented. The advantages of the introduced prescriptive design model may contribute towards a more systematic and automated design process as well as towards the development of more effective and intelligent design systems that could provide substantial help to the designer throughout the design process.

Acknowledgments. University of Patras is partner and University of Aegean is associate partner of the EU-funded FP6 Innovative Production Machines and Systems (I*PROMS) Network of Excellence.

References

1. Evduomwan, N.F.O., Sivaloganathan, S., Jebb, A.: A survey of design philosophies, models, methods and systems. Proceedings of the Institution of Mechanical Engineers, Part B: Journal of Engineering Manufacture, B4 210, 301–320 (1996)
2. Finger, S., Dixon, J.R.: A Review of Research in Mechanical Engineering Design. Part II: Representations, Analysis, and Design for the Life Cycle. Research in Engineering Design 1, 121–137 (1989)
3. Finger, S., Dixon, J.R.: A Review of Research in Mechanical Engineering Design. Part I: Descriptive, Prescriptive, and Computer-Based Models of Design Processes. Research in Engineering Design 1, 51–67 (1989)
4. Saridakis, K.M., Dentsoras, A.J.: Soft computing in engineering design – A review Advanced Engineering Informatics, vol. 22, pp. 202–221 (2008)
5. Feng, C.X., Li, P.-G., Liang, M.: Fuzzy mapping of requirements onto functions in detail design. Computer-Aided Design 33, 425–437 (2001)
6. Zimmermann, H.-J.: Fuzzy Set Theory-And Its Applications. Kluwer Academic Publishers, USA (1996)
7. Temponi, C., Yen, J., Amos Tiao, W.: House of quality: A fuzzy logic-based requirements analysis. European Journal of Operational Research 117, 340–354 (1999)
8. Lin, C.-T.: A fuzzy logic-based approach for implementing quality function deployment. Smart Engineering System Design 5, 55–65 (2003)
9. Kim, K.-J., Moskowotz, H., Dhingra, A., Evans, G.: Fuzzy multicriteria models for quality function deployment. European Journal of Operational Research 121, 504–518 (2000)
10. Yan, W., Chen, C.-H., Khoo, L.-P.: An integrated approach to the elicitation of customer requirements for engineering design using picture sorts and fuzzy evaluation. AIEDAM 16, 59–71 (2002)
11. Yang, Y.Q., Wang, S.Q., Dulaimi, M., Low, S.P.: A fuzzy quality function deployment system for buildable design decision-making. Automation in Construction 12, 381–383 (2003)
12. Wood, K.L., Antonsson, E.K.: Computations with Imprecise Parameters in Engineering Design:Background and Theory. Journal of Mechanisms, Transmission and Automation in Design 111, 616–625 (1989)
13. Vanegas, L.V., Labib, A.W.: Application of new fuzzy-weighted average (NFWA) method to engineering design evaluation. Int. J. Prod. Res. 39, 1147–1162 (2001)
14. Scott, M.J., Antonsson, E.K.: Aggregation functions for engineering design trade-offs. Fuzzy Sets and Systems 99, 253–264 (1998)
15. Wang, J.: Ranking engineering design concepts using a fuzzy outranking preference model. Fuzzy sets and systems 119, 161–170 (2001)
16. Ullman, D.G.: The mechanical design process. McGraw-Hill Inc., New York (1992)
17. Moulianitis, V.C., Dentsoras, A.J., Aspragathos, N.A.: A knowledge-based system for the conceptual design of grippers for handling fabrics. AIEDAM 13, 13–25 (1999)
18. Moulianitis, V.C., Dentsoras, A.J., Aspragathos, N.A.: A Search Method in Knowledge-Based Systems using Euclidean Space Norm - An Application to Design of Robot Grippers. In: AIENG 1998, Galway, Ireland, pp. 247–260 (1998)

19. Moulianitis, V.C., Dentsoras, A.J., Aspragathos, N.A.: A Heuristic Method for Knowledge-Based Conceptual Design of Robot Grippers based on the Euclidean Space Inner Product. In: CACD 1999, Lancaster, UK, pp. 37–48 (1999)
20. Jamshidi, M., Vadiee, N., Ross, T.J.: Fuzzy Logic and Control (Software and hardware applications). PTR Prentice Hall, USA (1993)
21. Kreyszig, E.: Introductory functional analysis with applications. John Wiley & Sons, Canada (1989)
22. Fuzzy Logic Toolbox™2 User's Guide. The MathWorks, Inc. (2008)

Phonotactic Recognition of Greek and Cypriot Dialects from Telephone Speech

Iosif Mporas[1], Todor Ganchev[2], and Nikos Fakotakis[2]

[1,2] Artificial Intelligence Group, Wire Communications Laboratory,
Dept. of Electrical and Computer Engineering, University of Patras,
26500 Rion, Greece
imporas@upatras.gr, {tganchev,fakotaki}@wcl.ee.upatras.gr

Abstract. In the present work we report recent progress in development of dialect recognition system for the Standard Modern Greek and Cypriot dialect of Greek language. Specifically, we rely on a compound recognition scheme, where the outputs of multiple phone recognizers, trained on different European languages are combined. This allows achieving higher recognition accuracy, when compared to the one of the mainstream phone recognizer. The evaluation results reported here indicate high recognition accuracy - up to 95%, which makes the proposed solution a feasible addition to existing spoken dialogue systems, such as voice banking applications, call routers, voice portals, smart-home environments, e-Government speech oriented services, etc.

Keywords: Dialect recognition, phone recognition.

1 Introduction

The globalization tendency in the last two decades has forced the speech technology community to turn to the development of multilingual functional systems. Specifically, multilingual speech recognition and synthesis, which will enable automatic speech translation, will become increasingly important [1]. The corner-stone of multilingual speech applications is Language Identification (LID) that is the task of automatic recognition of the language of a speech signal. The role of LID is essential for multilingual functionalities, such as spoken dialog systems (e.g. info-kiosks, voice banking, e-Government, voice portals, etc) that support a group of languages, spoken document retrieval and human-to-human communication systems (e.g. call routers, speech-to-speech translation) [1]. Due to the high importance of LID, intensive efforts have been devoted to the development of this technology. This has led to significant progress, which has been made in the last few years [2].

One of the challenging research tasks related to LID is Dialect Identification (DID). Similarly to LID, in DID task, a system is supposed to identify correctly one among different dialects of a given language from a spoken utterance. The DID functionality is crucial for spoken dialogue systems, when there is need the speech recognition engine to be adapted to the speaking style and manner of articulation of users originating from areas that speak different dialects. Dialect adaptation facilitates higher speech recognition performance, when compared to the baseline performance

J. Darzentas et al. (Eds.): SETN 2008, LNAI 5138, pp. 173–181, 2008.

without adaptation. Various techniques have been proposed for addressing the challenges of the DID task. Most of them bear similarity to corresponding techniques used in LID tasks, from which they were originally inspired. However, generally speaking, DID is considered as a more difficult task [3], compared to LID, due to the intrinsic similarities among the dialects within a specific language.

Specifically, in the DID task various sources of information, which is encoded in the different levels of spoken language can be utilized for successful discrimination among dialects. This discrimination can be performed on various levels, such as the acoustic level (e.g. spectral information), the prosodic level (e.g. prosody), the phonotactic level (e.g. language models) and the lexical information [4]. Regarding the acoustic level, spectral information of the speech signal is extracted through speech parameterization techniques and further fed to powerful classification algorithms such as Gaussian mixture models [3], support vector machines [5] and neural networks [6]. At the prosodic level, duration of phonetic units [4, 7, 8, 9] and rhythm [10] have been exploited. Lexical information also has been proved as a useful source when recognizing dialects or languages [11, 12].

To this end, the most successfully applied approach for both LID and DID is the phonotactic approach [13, 14]. In the phonotactic approach, the speech signal is decomposed to its corresponding phone sequence, and further fed to dialect-specific language models. The language model with the maximum probability score indicates the recognized dialect. The decomposition of the speech waveform to phonetic sequence can be performed using a single phone recognizer followed by the target language models (PRLM), or using parallel phone recognizers (PPRLM).

In the present work, we investigate the task of automatic recognition between the two major dialectal categories of Greek language, namely the Standard Modern Greek and the Cypriot. We develop and evaluate various configurations of a DID system, which follow the phonotactic approach. Specifically, we utilize six different language-dependent phone recognizers and investigate several configurations for fusion of their outputs.

The remainder of this paper is organized as follows: In Section 2, we describe the main differences between the Standard Modern Greek and Cypriot dialect and briefly outline the characteristics of the speech corpora used for the purpose of performance evaluation. In Section 3, we offer a detailed description of the architecture of the DID system that has been developed. The experimental setup is explained in Section 4. Section 5 is devoted to analysis of the experimental results. Finally, Section 6 offers summary and conclusion.

2 Greek and Cypriot Speech Corpora

The technical characteristics of the Cypriot dialect vary from the Standard Modern Greek in the phonology, morphology, vocabulary and syntax. In the phonological level, where the approach employed here falls in, the basic difference is that in the Standard Modern Greek there is no variation in vowel lengths. This means that there is no phonemic distinction between long and short vowels. However, in the Cypriot dialect there are expanded phonemic consonant length distinctions. Detailed description of the variations between Greek and Cypriot can be found in [15].

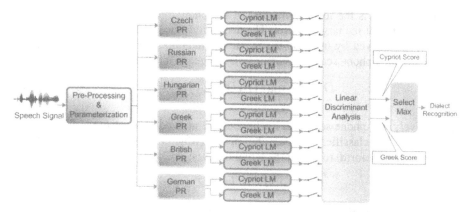

Fig. 1. DID system for Greek – Cypriot dialects. Phone recognizers and language models are denoted as PR and LM respectively.

There exist two speech corpora that capture the Standard Modern Greek and Cypriot dialect: SpeechDat(II) FDB5000 Greek database [16], and the Orientel Cypriot Greek database [17]. Both databases include spontaneous answers to prompted questions. The recorded utterances consist of isolated and connected digits, natural numbers, money amounts, yes/no answers, dates, application words and phrases, phonetically rich words and phonetically rich sentences.

The SpeechDat(II) FDB5000 Greek database consists of prompted speech recordings from 5000 native Greek speakers (both males and females), recorded over the fixed telephone network of Greece. The speech waveforms were sampled at 8 KHz and stored in 8-bit A-law format.

The Orientel Cypriot Greek database consists of recordings of 1000 Cypriot native speakers. Following the conventions of SpeechDat project, the recordings were collected over the fixed telephone network of Cyprus. The speech waveforms were sampled at 8 KHz and stored in 8-bit A-law format.

3 System Description

The DID system presented in this section follows the PPRLM approach. In Fig. 1, we outline the general architecture of the system. As the figure shows, the unlabelled speech signal (i.e. of unknown dialect) is initially pre-processed and parameterized. Afterwards, the feature vector sequence, F, is forwarded in parallel to $N=6$ phone recognizers. Each phone recognizer decomposes the feature vector sequence to a phone sequence, O_n, $n=1,..., 6$. Each one of the phone sequences consists of the corresponding phone labels that constitute the nth phone recognizer. For every one of the computed phone sequences the likelihood against the corresponding dialect-dependent n-gram language model L_{nl} is computed:

$$P_{nl} = P(O_n \mid L_{nl}) \ . \tag{1}$$

where $n=1,..., N$ is the identity number of the phone recognizer, and $l=1, 2$, is the identity number of the target dialect. The decision D about the unknown input speech waveform is derived by utilizing the maximum-likelihood criterion on the scores of all the computed phone sequences against the language models:

$$D = \arg \max_{n,l} \{P(O_n \mid L_{nl})\}.$$ (2)

Alternatively, the phone sequence likelihoods are utilized as input I to a discriminant classifier C. The classifier will map the input I (which corresponds to the unknown input speech waveform) to one of the target dialects:

$$C : I \rightarrow \{D_l\}, \ l = 1, 2 \ .$$ (3)

where D_l is the decision for the unknown speech waveform either to belong the Standard Modern Greek or to the Cypriot dialect.

3.1 Phone Recognizers

The system studied in the present work utilizes six parallel phone recognizers. Each recognizer was trained on the phone set of one language, namely on Czech, Russian, Hungarian, Greek, British English and German.

Specifically, the Czech, Russian and Hungarian phone recognizers were trained on the SpeechDat-E database [18]. These recognizers utilize Mel-scale filter bank energies and temporal patterns of critical-band spectral densities. Phone posterior probabilities are further computed using neural networks. Further details concerning these phone recognizers are available in [19].

The remaining three phone recognizers, namely the Greek, British English and German, were trained on the SpeechDat(II)-FDB databases [20], utilizing the HTK toolkit [21]. Each phone is modeled by a 3-state left-to-right hidden Markov model (HMM). They utilize the 12 first Mel frequency cepstral coefficients together with the 0th coefficient. Phone models are trained using the base feature vector as well as the first and second derivative coefficients. Each state of the HMMs is modeled by a mixture of eight continuous Gaussian distributions. The size of the training data for the Greek phone recognizer was almost 6 hours of speech recordings. The British English phone recognizer was trained by employing approximately 21 hours of speech. Finally, the German phone recognizer was trained utilizing nearly 14 hours of speech. All phone models in these three recognizers are context-independent.

3.2 Language Models

For the construction of the language models we exploited the CMU Cambridge Statistical Language Modeling (SLM) Toolkit [22]. Every phonetic sequence was modeled by an m-gram language model. For each dialect and for each output of a phone recognizer, one 3-gram ($m=3$) language model was trained, resulted to a total of $6x2=12$ language models.

3.3 Fusion of PRLMs

The output scores of each PRLM were forwarded as input to a linear discriminant classifier. The classifier makes the final decision about the dialect that the unknown speech waveform belongs to, by fusing the outputs of the parallel tokenizers (e.g. phone recognizers). Besides the case where we fuse the outputs of all six phone recognizers, in Section 5 we also investigate fusion of other subsets of PRLM scores.

4 Experimental Setup

For the present evaluation of dialect recognition performance, we utilized recordings from the Greek and Cypriot speech corpora, described in Section 2. Specifically, we used 20,000 speech files (10,000 from each dialect). The experiments were carried out by dividing the speech data to 80% (8,000 files per dialect) for training the language models, 10% (1,000 files per dialect) for training the linear discriminant classifier and 10% (1,000 files per dialect) for evaluating the identification accuracy. All types of utterances mentioned in Section 2 have been utilized in the present evaluation. The experimental procedure was implemented by performing 5-fold cross validation. This experimental setup takes advantage of the available data while keeping non-overlapping training and testing datasets.

Two types of errors can occur during the dialect recognition process: The first one called a false negative error (i.e. *miss* to recognize the correct target) occurs when the true target is falsely rejected as being a non-target. The second type called a false positive error occurs when a tryout from a non-target is accepted as if it comes from the target dialect. The latter error is also known as a *false alarm*, because a non-target trial is accepted as a target one. We use the miss probability and false alarm probability as indicators of the DID performance. These two error types are utilized in the Detection Error Trade-off (DET) [23] plots shown in the experimental results section. The DET plots utilize the scores computed for the Cypriot and Greek dialect models, and therefore show the system performance in multiple operation points of the DID system.

5 Experimental Results

Following the experimental protocol outlined in Section 4, we evaluated the performance of the individual language-specific phone recognizers as well as several score fusion schemes. Specifically, in Fig. 2 we present the DET curves for the Russian (RU), Hungarian (HU), Czech (CZ), Greek (GR), British English (BR), German (GE) phone recognizers and the score-level fusion (FU) of all the six PRLM outputs. As the figure presents, the British English and German phone recognizers exhibited superior performance when compared to the remaining phone recognizers. We can explain this observation with the significantly larger amounts of training data used for building these models. On the other hand, the performance of the Czech, Russian and Hungarian phone recognizers is significantly inferior, when compared to the Greek one, mainly because of the differences in the phone sets among these languages and the Greek dialects.

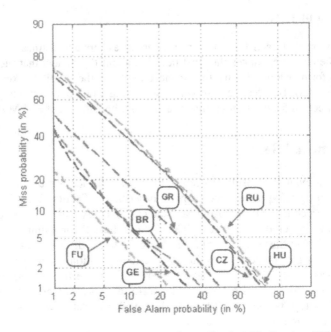

Fig. 2. DET plots of the individual phone recognizers: Czech (CZ), Russian (RU), Hungarian (HU), Greek (GR), British English (BR), German (GE) and the fusion of all of them (FU)

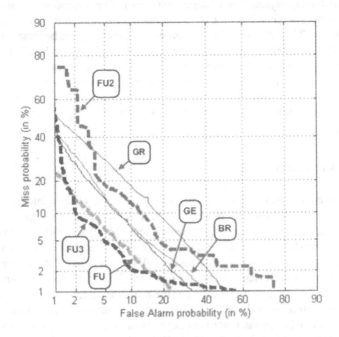

Fig. 3. DET plots for different fusion schemes: FU – all phone recognizers; FU3 – fusion of the top-3 (GE, BR, GR) phone recognizers; and as FU2 – of the top-2 (GE, BR) scorers. Plots of the individual phone recognizers: Greek (GR), British English (BR), German (GE)

Furthermore, the fusion results for different sets of phone recognizers are presented in Fig. 3. Specifically, the fusion result for all phone recognizers (FU) is contrasted to these obtained for the fusion of the top-3 (FU3) scorers (GE, BR, GR), and for the top-2 (FU2) scorers (GE, BR). As the DET plots present, the FU2 doesn't offer any advantage, when compared to the individual Greek and British English phone recognizers. Furthermore, adding the Greek phone recognizer to the top-2 set, significantly improves the recognition accuracy. As Fig. 3 presents, the top-3 fusion scheme, FU3, exhibits the highest recognition accuracy at the area of Equal Error Rate (EER), i.e. where the probability of missing a target trial is equal to the probability of accepting a non-target trial as originating from the target. Finally, the fusion of all six phone recognizers FU offers more balanced performance in all areas of the DET plot.

In particular, as presented in the corresponding DET-plot, the FU fusion scheme guarantees satisfactory recognition accuracy for the low-miss probability, low-false alarm and the EER operating points of the Greek-Cypriot dialect identification system. Such balanced performance makes the FU fusion scheme more predictable, and thus, a more attractive choice for a wide range of applications. However, for applications where the preferred operating point is near the EER the FU3 fusion scheme maximizes the dialect recognition performance. Finally, the FU3 fusion scheme could be the option of interest when the computational demands and memory requirements are important factors and is desirable to be bounded.

6 Conclusion

The present-day speech recognition technology relies on large speech corpora, which serve in the creation of statistical models of speech. However, for many European languages and dialects such extensive resources are not available or the existing data are of very limited size, mainly due to the high expenses for collecting spoken corpora. As a consequence, the citizens speaking these languages or dialects are not in position that would allow them to comfortably take advantage of the modern technology, since they often experience difficulties when assessing public domain spoken dialogue systems.

In the present contribution, we presume that a dialect recognition component will make the spoken interaction application aware to the dialect spoken by the user. This information would enable for selection of dialect-specific speech recognition settings, which would improve the overall task completion rate in spoken interaction, and as a result will contribute for improved quality of service.

Specifically, evaluating different configurations of our dialect identification component, we study a fusion scheme that exploits a number of phone recognizers, trained from significant amounts of speech corpora.

The use of a compound phone recognition scheme improves dialect identification rates, when compared to the use of the mainstream Greek phone recognizer alone.

In conclusion, we would like to summarize that the reported effort led to successful development of compound dialect recognition component, which is capable to achieve recognition accuracy above 95%, and which has the potential to facilitate for improved speech recognition of Cypriot speakers.

Acknowledgments. The authors would like to thank the anonymous reviewers for their useful comments and corrections.

References

1. Schultz, T., Kirchhoff, K.: Multilingual Speech Processing. Academic Press, Elsevier (2006)
2. Martin, A.F., Le, A.N.: NIST 2007 Language Recognition Evaluation. In: Odyssey 2008 - The Speaker and Language Recognition Workshop ISCA Tutorial and Research Workshop, (2008)
3. Torres-Carrasquillo, P.A., Gleason, T.P., Reynolds, D.A.: Dialect Identification using Gaussian Mixture Models. In: Odyssey 2004 - The Speaker and Language Recognition Workshop, ISCA Tutorial and Research Workshop, pp. 297--300 (2004)
4. Tong, R., Ma, B., Li, H., Chng, E.S.: Integrating Acoustic, Prosodic and Phonotactic Features for Spoken Language Identification. In: 2006 IEEE International Conference on Acoustics, Speech and Signal Processing (ICASSP), pp. 205--208 (2006)
5. Campbell, W. M., Singer, E., Torres-Carrasquillo, P.A., Reynolds, D.A.: Language Recognition with Support Vector Machines. In: Odyssey 2004 - The Speaker and Language Recognition Workshop, ISCA Tutorial and Research Workshop, pp. 285--288 (2004)
6. Braun, J., Levkowitz, H.: Automatic Language Identification with Perceptually Guided Training and Recurrent Neural Networks. In: 5th International Conference on Spoken Language Processing (ICSLP), pp. 3201--3205 (1998)
7. Ghesquiere, P.J., Compernolle, D.V.: Flemish Accent Identification based on Formant and Duration Features. In: 2002 IEEE International Conference on Acoustics, Speech and Signal Processing (ICASSP), pp. 749--752 (2002)
8. Lin, C.Y., Wang, H.C.: Fusion of Phonotactic and Prosodic Knowledge for Language Identification. In: 9th International Conference on Spoken Language Processing (ICSLP), pp. 425--428 (2006)
9. Hazen, T., Zue, V.: Segment-based Automatic Language Identification. J. of the Acoustic Society of America, (101) 4, pp. 2323--2331 (1997)
10. Farinas, J., Pellegrino, F., Rouas, J.L., Andre-Obrecht, R.: Merging Segmental and Rhythm Features for Automatic Language Identification. In: 2002 IEEE International Conference on Acoustics, Speech and Signal Processing (ICASSP), pp. 753--756 (2002)
11. Huang, R., Hansen, J.: Dialect/Accent Classification via Boosted Word Modeling. In: 2005 IEEE International Conference on Acoustics, Speech and Signal Processing (ICASSP), pp. 585--588 (2005)
12. Campbell, W.M., Richardson, F., Reynolds, D.A.: Language Recognition with Word Lattices and Support Vector Machines. In: 2007 IEEE International Conference on Acoustics, Speech and Signal Processing (ICASSP), pp. 425--428 (2007)
13. Zissman, M.: Comparison of Four Approaches to Automatic Language Identification. J. IEEE Transactions on Speech and Audio Processing, 4 (1), pp. 31--44 (1996)
14. Tsai, W.H., Chang, W.W.: Chinese Dialect Identification using an Acoustic-Phonotactic Model. In: 6th European Conference on Speech Communication and Technology (EUROSPEECH), pp. 367--370 (1999)
15. Κοντοσόπουλος, Ν.Γ.: Διάλεκτοι και ιδιώματα της Νέας Ελληνικής. Εκδόσεις Γρηγόρη (1994)

16. Chatzi, I., Fakotakis, N., Kokkinakis, G.: Greek speech database for creation of voice driven teleservices. In: 5th European Conference on Speech Communication and Technology (EUROSPEECH), pp.1755--1758 (1997)

17. Kostoulas, T., Georgila, K.: Orientel Cypriot Greek Database. V.2.0, (2007)

18. Pollak, P., Cernocky, J., Boudy, J., Choukri, K., Van den Heuvel, H., Vicsi, K., Virag, A., Siemund, R., Majewski, W., Staroniewicz, P., Tropf, H.: SpeechDat(E) - Eastern European Telephone Speech Databases. In: XLDB Workshop and Satellite Event to LREC Conference on Very Large Telephone Speech Databases (2000)

19. Schwarz, P., Matejka, P., Cernocky, J.: Towards Lower Error Rates in Phoneme Recognition. In: Sojka, P., Kopecek, I. Pala K. (eds.): TSD 2004, LNAI 3206, pp. 465–472, Springer, Heidelberg (2004)

20. Hoge, H., Draxler, C., Van den Heuvel, H., Johansen, F.T., Sanders, E., Tropf, H.S.: SpeechDat Multilingual Speech Databases for Teleservices: Across the Finish Line. In: 6th European Conference on Speech Communication and Technology (EUROSPEECH), pp. 2699--2702 (1999)

21. Young, S., Evermann, G., Gales, M., Hain, T., Kershaw, D., Moore, G., Odell, J., Ollason, D., Povey, D., Valtchev, V., Woodland, P.: The HTK Book (for HTK Version 3.3), Cambridge University (2005)

22. Clarkson, P.R., Rosenfeld, R.: Statistical Language Modeling Using the CMU-Cambridge Toolkit. In: 5th European Conference on Speech Communication and Technology (EUROSPEECH), pp. 2707--2710 (1997)

23. Martin, A., Doddington, G., Kamm, T., Ordowski, M., Przybocki, M.: The DET curve in assessment of detection task performance. In: 5th European Conference on Speech Communication and Technology (EUROSPEECH), vol.4. pp.1895--1898 (1997)

A "Bag" or a "Window" of Words for Information Filtering?

Nikolaos Nanas and Manolis Vavalis

Centre for Research and Technology - Thessaly (CE.RE.TE.TH), Greece
{n.nanas,mav}@cereteth.gr

Abstract. Treating documents as bag of words is the norm in Information Filtering. Syntactic and semantic correlations between terms are ignored, or in other words, term independence is assumed. In this paper we challenge this common assumption. We use Nootropia, a user profiling model that uses a sliding window approach to capture term dependencies in a network and a spreading activation process to take them into account for document evaluation. Experiments performed based on TREC's routing guidelines demonstrate that given an adequate window size the additional information that term dependencies encode, results in improved filtering performance over a traditional bag of words approach.

1 Introduction

Information Filtering (IF) is concerned with the problem of providing a user with relevant information, based on a tailored representation of the user's interests, called "profile". Content-based document filtering, in particular, deals with text documents and in this case the user profile is an abstraction directly derived from their content. IF is traditionally approached with user profiling models that treat documents as "bag of words" and ignore any syntactic or semantic correlations between terms. In other words, term dependencies are ignored. Although a lot of effort has been put in capturing and exploiting term dependencies in Information Retrieval (IR) and other domains, this is not true for research in IF.

In this paper we challenge the common term independence assumption in IF research and show that the additional information that term dependencies provide can be successfully incorporated in a user profile for effective IF. This is done with Nootropia[1], a user profiling model that uses a network of terms, incorporating term dependencies, to represent a user's interests and a spreading activation process for document evaluation. To identify term dependencies, we take the context of terms into account using a sliding window approach. A window is a span of contiguous words in a document's text and its size is an important parameter that defines the kind of term correlations that one can identify. A small window of a few words, typically no more than three, is called "local context" and is appropriate for identifying adjacent, syntactic correlations

---

[1] Greek word for: "an individual's or a group's particular way of thinking, someone's characteristics of intellect and perception'.'

J. Darzentas et al. (Eds.): SETN 2008, LNAI 5138, pp. 182–193, 2008.

between terms, such as compounds. "Global context" on the other hand, is defined by a larger window (more than three terms) that may incorporate several sentences, or even the complete document [1]. We will refer to the latter, special case, as "document context". The problem with document context is that it ignores differences in the semantic content of different document sections. Two terms within the same document are associated even if they appear in separate sections about possibly unrelated topics. More generally, when global context is used the captured dependencies are caused by both syntactic and semantic correlations between terms. We argue and support experimentally that in terms of filtering performance, using global context is superior to both local context and the traditional bag of words approach. It is therefore shown that, in contrast to the current practices, term dependencies can be effectively incorporated in IF.

The rest of the paper is organised as follows. We start with a review of existing approaches to term dependence representation, mainly within IR research. It is made clear that a lot of effort has been put in tackling the issue, but the proposed solutions are not directly applicable to IF. Nootropia, which is described in section 3, is one of the few user profiling models that incorporates term dependencies and the first serious attempt to highlight their importance for IF. We concentrate on Nootropia's comparative evaluation using TREC 2001 routing guidelines (section 4). Nootropia is compared to a baseline profile, which ignores term dependencies, with positive results (section 5). Experiments have been performed for various window sizes and the results indicate that the best filtering performance is achieved when term dependencies are captured using global context. We conclude with a summary and pointers for future work (section 6).

2   Related Work

It is generally acknowledged that the term independence assumption is false and a lot of research work has focused on capturing and exploiting term dependencies, especially within IR. One way to take into account term dependencies is by extending the Vector Space Model, which inherently ignores them. Various methods for transforming the vector space so that term dependencies are captured have been proposed. These include the Generalised Vector Space Model [2], Latent Semantic Indexing (LSI) [3] and the use of term context vectors [4].

Probabilistic models, which are also popular in IF, can in principle incorporate term dependencies, but these are usually ignored because they significantly increase the amount of parameters that have to be estimated. For n terms there are 2^n term pairs and so calculating all possible dependencies becomes impractical. Various remedies to this problem exist. van Rijsbergen proposes the use of an appropriately constructed "dependence tree" [5], derived using the Maximum Spanning Tree (MST) technique, that allows the computation to focus on the most significant of dependencies between terms in the same document. Similarly, but for dependencies between terms in the same sentence, the MST has also been used in [6]. Other proposed solutions to the above problem include the Chow expansion [7] and the Bahadur-Lazarsfeld expansion [8].

Inference networks are also probabilistic model that have been successfully applied to IR. They are used to calculate the probability that a document satisfies a user's query. They comprise three layers of nodes: document nodes, representational nodes and query nodes. Typically, inner-layer links are excluded and hence term dependencies are ignored. To implicitly capture term dependencies, [9] have generalised the basic inference network using an additional layer of nodes that capture text representations which can include phrases, or compound terms. [10] propose the automatic construction of a thesaurus using a term similarity measure on a "collocation map". A collocation map is a type of inference network that encodes the statistical associations between terms in a given document collection.

Like inference networks, spreading activation models and neural networks (NNs) are based on a network structure where documents, terms and queries are represented as nodes [11]. They differ in that links model statistical, rather than strictly probabilistic relations between terms. Typically, links between terms and between documents are not used and therefore term dependencies are again ignored. In spreading activation models for IR, document evaluation takes place by assigning an initial energy to the user's query node, which subsequently flows through the network towards the document nodes. A document's relevance is calculated as the final amount of energy it receives. Only in cases where the activation of documents is allowed to spread back to the term nodes, activating terms which are not included in the initial query, does the associative nature of the spreading activation process allow dependencies between terms in the same document to be taken into account [12]. This basic architecture has also been adopted by NN approaches [13]. Nevertheless, NNs have additional means for capturing term dependencies. For example, [14] adds links between terms based on thesaurus relations. Furthermore, in non-linear NNs, hidden layers capture implicitly the dependencies between terms in the same documents. This kind of NNs have already been applied to IR [15].

There are also methods for the automatic construction of hierarchical networks that explicitly capture topic-subtopic relations between terms. These networks are usually referred to as "Concept", "Topic" or "Subject" hierarchies, and can be applied for the organisation, summarisation and interactive access to information. One method for the automatic construction of a concept (subsumption) hierarchy is through the use of subsumption associations between terms [16]. Subsumption hierarchies exploit term cooccurences within the document context and combine them with evidence from the complete document set, in order to identify topic-subtopic relations between terms. Lexical correlations are not explicitly taken into account. In contrast, lexical correlations support the construction of so called "Lexical Hierarchies" [17]. Local context is used to identify lexical compounds which are then combined with statistics provided by the complete set of extracted compounds, in order to structure the terms into a concept hierarchy. Lexical hierarchies exclude correlations between terms that do not appear close enough in the text.

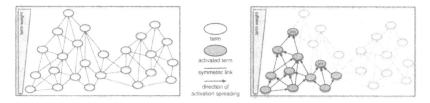

Fig. 1. Hierarchical Term Network: (left) deactivated, (right) activated

Clearly a lot of work has already been done in taking into account term dependencies when retrieving, classifying, or organising information. However, the above models are not directly applicable to IF. IR deals with relatively static document collections and only temporary queries. Consequently, the relevant models concentrate on capturing term dependencies for document, rather than query representation. In other words, term dependence representation is integrated in the document indexing process. This is true for all approaches that are based on the vector space model and all network approaches that include a layer of document nodes. However, IF is concerned with a stream of documents, or more generally with dynamic document collections, where indexing is not always possible.

This may explain why, in most cases, IF systems are based on user profiling models that ignore term dependencies, like the common vector space model [18,19], probabilistic classifiers [20] and linear NNs with no hidden units [21]. Exceptions are the systems described in [22] and [23], where the user profile is an associative graph that captures syntactic correlations between terms in the same sentence. These two systems demonstrate that capturing term dependencies in IF is feasible, but they have failed to attract the attention of the IF community, possibly due to the lack of experimental evidence in support of this design choice.

3   Nootropia

Nootropia is a user profiling model that represents a new stance towards IF, challenging the common term independence assumption. IF is not treated as a specialisation of IR, but as an independent research domain with its own specifications. The user profile is no longer just a persistent query, but a much richer information structure. Documents are not just bag of words, but word sequences, where the context of words is a rich source of additional information. The model is inspired from the work described in the previous section, but modifies it appropriately to account for the particularities of IF.

More specifically, in Nootropia the user profile is a term network that represents terms as nodes and their correlations as links (fig. 1 (left)). Both nodes and links are weighted. A term's weight measures statistically the term's importance in respect to the user's interests. Various methods can be used to calculate this weight. Here we adopt Information Gain (IG) that has exhibited the best performance in comparative experiments reported in [24]. The weight of the link between two terms measures their statistical dependence.

The profile is built based on a set of documents that the user has specified as interesting. Initially, after stop word removal and stemming, terms in the documents are weighted and those with weight over a certain threshold are extracted to become profile terms. To identify correlations between the extracted terms a sliding window approach is adopted. As already discussed the size of the window is a significant parameter, which defines a term's context and thus the kind of correlations being recognised. We experimented both with local and global contexts, but we avoided document context for reasons already explained in the introduction. Two extracted terms are linked if they appear together within the sliding window. The weight w_{ij} $(0 < w_{ij} \leq 1)$ of the link between two extracted terms t_i and t_j $(t_i \neq t_j)$ is calculated using formula 1, which extends the one adopted by [10] with the additional distance parameter $1/d_{ij}$. In this way, a link's weight is a combined measure of the statistical dependence caused by both semantic and syntactic correlations between the linked terms. The weight of the link between two terms is large whenever the two terms appear appear frequently together within the window and/or close to each other. The link identification and weighting process generates out of the extracted terms an associative graph similar to those in [10,22]. Finally the graph is transformed by ordering terms according to decreasing weight. The resulting ordered network (fig. 1) complies , according to [25], with many of the design principles set by [16] for the generation of a concept hierarchy using subsumption and reflects, according to [26], the general topics of interest to the user.

$$w_{ij} = \frac{fr_{ij}^2}{fr_i \cdot fr_j} \cdot \frac{1}{d_{ij}} \tag{1}$$

where:
   fr_{ij} is the number of times t_i and t_j cooccur
      within the sliding window
   fr_i and fr_j are respectively the number of occurrences
      of t_i and t_j in the user specified documents
   d_{ij} is the average distance between t_i and t_j,
      within the sliding window

Document evaluation is based on a spreading activation process. When confronted with a new document, a window of the same size used to construct the profile, slides through the document's text. For each position of the window, profile terms that appear within it are activated (in fig. 1 (right) shadowed nodes depict activated terms). The initial activation of a profile term equals its weight in the document. Here we consider the simplest case of binary document indexing and the initial activation is equal to one. Subsequently, each activated term disseminates part of its activation to activated terms higher in the hierarchy that it is linked to. This process follows the sequence that the hierarchy imposes. It starts from the activated term lower in the hierarchy and moves up the hierarchy order until the highest activated term is reached and no further energy is disseminated. The amount of activation that is disseminated between two activated terms is proportional to the weight of the corresponding link. The current window's relevance score $S(w)$ is then calculated as the weighted sum of the final activation of activated profile terms (equation 2). The process is repeated for each new position of the sliding window. The document's overall score is calculated as the sum of the scores of individual window positions, normalised to the logarithm of the number of terms in the document.

$$S(w) = \sum_t w_t \cdot A_t^f \tag{2}$$

where:

    t is an activated term
    w_t the term's weight
    A_t^f the term's final activation

The use of a sliding window during document evaluation is a significant improvement over previous implementations of the model, which treated documents being evaluated as bag of words, for reasons of computational efficiency [26]. Here we sacrifice some efficiency over representational accuracy. When a document being evaluated is treated as a bag of words, two correlated profile terms can be mistakenly activated even if they appear far apart in the document's text. On the contrary, with the current implementation two terms that have been linked because they appear together within the same context (window) are only jointly activated when they reappear within the same sort of context (local or global) in the evaluated document. For example, if a window comprising three contiguous terms (local context) is used when the profile is built and the lexical compound "knowledge management" is identified, then using the same window during document evaluation ensures that the profile terms "knowledge" and "management" are only going to be activated when they reappear within a window of size 3 and not in separate and possibly unrelated sections of the document's text.

With the above spreading activation process, document evaluation takes into account the term dependencies that the profile represents. How much a term contributes to the document's relevance does not depend only on its weight, but also on the term's links to other terms and their weights. A document's relevance increases when it activates connected and not just isolated terms. Document evaluation can also incorporate additional evidence derived from the pattern of network activation [26]. Spreading activation is performed efficiently because it exploits the direction imposed by the hierarchy and avoids the costly iterations of existing spreading activation approaches [22].

Although not within the scope of the current work, it can be formally proved that the profile's network stores more information than a simple vector comprising the same weighted terms. In summary, it can be shown that the document evaluation process can realise a deterministic mapping from a profile network to a weighted vector consisting of the network's terms with weights equal to their final activation. In fact, more than one network configurations can be mapped deterministically to the same weighted vector. It follows, that the infinity of weighted networks is larger than the infinity of weighted vectors and hence our profile's network structure can store more information than a vector based profile, comprising the same terms. In other words, it has the potential to more accurately represent the user's interests.

4   Experimental Methodology

We adopt for our experiments the routing guidelines of the 10th Text Retrieval Conference (TREC 2001). TREC's filtering track has been the most well established standard for the evaluation of IF systems. It adopts the Reuters Corpus Volume 1 (RCV1), an archive of 806,791 English language news stories that have been manually categorised according to topic, region, and industry sector[2]. The TREC-2001 filtering track is based on 84 out of the 103 RCV1 topic categories. Furthermore, it divides RCV1 into 23,864 training stories and a test set comprising the rest of the stories. TREC's routing sub-task concentrates on a static filtering problem[3]. For each topic a profile is built based on the complete relevance information and any non-relevance related information provided by the training set. The profile is then tested against the complete test set and its output is the top 1000 ranked documents. It is then evaluated by calculating the Average Uninterpolated Precision (AUP) of this list. AUP combines precision and recall in a single evaluation measure. It is defined as the sum of the precision value at each point in the list where a relevant document appears, divided by the total number of relevant documents. We have only deviated from the routing guidelines by experimenting with the first 42 out of the complete set of 84 topic categories. This was done to allow for a larger number of experiments to be performed. Furthermore, we did not exploit the topic statement or statistics derived from corpora other than RCV1.

To test the effect of term dependencies on filtering performance, we experimented with two types of profiles. The first is our baseline profile, which ignores term dependencies. It is a weighted keyword vector and performs document evaluation using the inner product. The second type is based on the proposed Nootropia model. For a specific topic, both types comprise the same set of weighted terms, which are extracted from the training documents, after stop word removal and stemming, using IG and a threshold equal to 0.001[4]. Therefore any difference in the performance of the two profile types is due to the additional information that links encode in the case of Nootropia. We also compare Nootropia to the average TREC participant to provide an indication of its level of performance.

In addition, we wish to test the effect of context in capturing term dependencies, and so in the case of Nootropia, links are generated with a variety of window sizes, which vary from local to global context. In particular, we experimented with windows of size 3, 5, 10 and 20. A profile built with a specific window size uses the same window during document evaluation. Furthermore, to investigate

---

[2] http://about.reuters.com/researchandstandards/corpus/index.asp

[3] For more details see: http://trec.nist.gov/data/filtering/T10filter_guide.htm

[4] This value was chosen based on preliminary experiments which are not reported here.

Table 1. Experimental Results

Topic	baseline	Nootropia (window size) 3	5	10	20	TREC average
1	0.00411	0.00832	0.01297	0.01799	0.01859	0.014278
2	0.06523	0.03996	0.04226	0.04910	0.05066	0.048389
3	0.00199	0.00345	0.00490	0.01040	0.01189	0.0105
4	0.04526	0.00856	0.01094	0.01895	0.02130	0.051556
5	0.01630	0.00582	0.00645	0.00745	0.00737	0.023722
6	0.25454	0.23252	0.24959	0.25964	0.23729	0.171944
7	0.05148	0.02507	0.04185	0.04609	0.04441	0.033833
8	0.07111	0.06984	0.06939	0.06746	0.06719	0.063278
9	0.22180	0.05655	0.06294	0.09045	0.11752	0.166667
10	0.17447	0.17535	0.17577	0.17465	0.15157	0.151944
11	0.01566	0.00634	0.00696	0.01464	0.01560	0.032944
12	0.09203	0.07586	0.08712	0.09939	0.10674	0.067111
13	0.02327	0.03055	0.03241	0.03288	0.02918	0.024
14	0.03933	0.04004	0.04035	0.04104	0.03735	0.037667
15	0.14136	0.13757	0.14143	0.15978	0.14860	0.094333
16	0.01739	0.01913	0.02027	0.02124	0.02118	0.017389
17	0.00767	0.00972	0.01003	0.01067	0.00973	0.013056
18	0.10542	0.14591	0.14695	0.13925	0.10619	0.090333
19	0.01015	0.04370	0.04348	0.03771	0.03626	0.037389
20	0.00164	0.00183	0.00470	0.01042	0.01093	0.030056
21	0.08214	0.17947	0.11543	0.17491	0.17569	0.173222
22	0.03741	0.05233	0.05382	0.05289	0.04205	0.031611
23	0.20605	0.10822	0.11794	0.13429	0.12725	0.130333
24	0.12723	0.05306	0.08323	0.12163	0.13633	0.084167
25	0.09008	0.08993	0.08946	0.09011	0.08890	0.054222
26	0.09913	0.09897	0.09869	0.09923	0.09802	0.065611
27	0.07562	0.07771	0.07785	0.07936	0.07965	0.054389
28	0.06643	0.09118	0.09423	0.09589	0.09668	0.0665
29	0.01810	0.00210	0.00299	0.00891	0.01944	0.0225
30	0.21695	0.11652	0.17083	0.24095	0.22404	0.237556
31	0.13870	0.15352	0.15262	0.14660	0.13611	0.101944
32	0.14435	0.16887	0.16711	0.16300	0.14930	0.110722
33	0.14725	0.02991	0.04103	0.07714	0.03668	0.197667
34	0.12055	0.10776	0.11631	0.14653	0.16226	0.137556
35	0.03620	0.00229	0.00361	0.00645	0.01187	0.121556
36	0.00004	0.01715	0.01939	0.01988	0.02868	0.123944
37	0.27648	0.22105	0.21450	0.25002	0.28970	0.260056
38	0.04782	0.06022	0.06057	0.06053	0.05943	0.038056
39	0.03696	0.02876	0.03383	0.03738	0.03720	0.031056
40	0.20316	0.28971	0.28930	0.28785	0.28503	0.213722
41	0.29913	0.33744	0.34223	0.33377	0.33140	0.254
42	0.00000	0.00000	0.00000	0.00000	0.00000	0.136611
average	0.09119	0.08148	0.08466	0.09373	0.09203	0.09113
> baseline		19	21	28	25	
> TREC		21	23	30	26	

if the strength of links affects filtering performance, experiments were conducted with profiles that maintain a percentage of the strongest generated links. These percentages are 10%, 20%, 40% and 100% for the complete set of links.

5 Experimental Results

Table 1 summarises the results for the various window sizes. The table has 7 columns. The first is the topic numbers and the second the AUP score that the baseline, vector-based profile achieves for each topic. Columns three to six show Nootropia's AUP scores for the four window sizes and the 100% of generated links. Finally, column seven concentrates the average AUP score over all TREC 2001 participants[5]. The last two rows count the number of topics for which Nootropia outperforms the baseline profile and the TREC average. Note that the analytical results for the various percentages of maintained links are not presented here due to space limitation. They are summarised instead in table 2 which presents the macro averages for Nootropia, the baseline profile and the average TREC participant.

---

[5] Many thanks to Ellen Voorhees for providing the anonymised TREC results.

Table 2. Summary

Window Size	Link Percentage					TREC average
	baseline	10	20	40	100	
3	0.091191	0.081787	0.08177	0.081802	0.081482	0.091128
5	0.091191	0.084585	0.084561	0.084613	0.08466	0.091128
10	0.091191	0.093614	0.093581	0.093644	0.093726	0.091128
20	0.091191	0.091363	0.091527	0.092	0.09203	0.091128

Nootropia's comparison to the baseline reveals that there are window sizes for which taking into account term dependencies improves filtering performance. For sizes 3 and 5, Nootropia performs worse than the baseline in terms of macro average (table 2). In the first case, it performs better than the baseline for 19 out of 42 topics and for window size 5, it is better for half of the topics (table 1). However, as the window size increases the situation reverses. For sizes 10 and 20 Nootropia performs clearly better than the baseline in terms of macro average, irrespective of percentage of maintained links (table 2). For window size 10, it is better than the baseline for 28 out of 42 topics and for window size 20 for 25 topics (table 1).

Clearly global context produces better filtering results than local context. This can be due to a number of reasons. One possibility is that the syntactic correlations that one can identify using local context are not enough by themselves. Semantic correlations between terms that appear within the same semantic unit, but not necessarily close to each other in text are also important. Another possibility is that the drawback of local over global context arises during document evaluation. A small window size constraints the number of profile terms that can be jointly activated. On the other hand, it should be noted that Nootropia exhibits the best performance for window size 10 and not for 20 (table 2). This is an indication that while local context should be avoided, very large window sizes are not necessarily preferable either. We expect that as the window size increases towards document context, filtering performance will drop for reasons already explained. Of course more experiments with larger window sizes are necessary to support this hypothesis.

A sliding window of size 8 has also been applied successfully for concept-based query modelling in information retrieval, where the sliding window is used to define the context of concepts [27]. There is also independent evidence from Statistical Natural Language Processing that a larger window size is needed to allow term dependencies to have an impact. [28] report experiments which aimed to establish how much evidence, in terms of running text, is required to show that terms do not distribute homogeneously. Their findings demonstrate that a window size of 5 is not sufficient for statistically significant evidence to occur, to defeat the null hypothesis of homogeneous term distribution, except for very rare words. A window size of 100, on the other hand, is sufficient even for function words to show heterogeneous distributions in some collections. This supports our claim that the effects of term dependency will only surface in larger window sizes. Their experiments also suggest that ideal window size may be language and collection dependent.

Another interesting finding is that for window sizes 5, 10 and 20 the best performance is achieved when all links are maintained (table 2). Despite the large number of possible links, using a sliding window for document evaluation prevents the saturation of the activated profile network and all links can contribute to increased performance. In addition, the results indicate that irrespective of link percentage, there is consistency in the topics for which Nootropia outperforms the baseline profile. This leads to the conclusion that the beneficiary effect of links for these topics is not accidental. It derives either from statistical characteristics of these topics, or more likely their semantic properties. Although further analysis is required, this finding makes us believe that there are domains for which taking into account term dependencies is particularly important.

Finally, Nootropia's performance compares favourably with the average TREC participant. For window sizes 10 and 20, Nootropia performs better than the average TREC participant for 30 and 26 topics, respectively. Given that various improvements to the model are possible, including calibration of link weights and incorporation in the document evaluation process of additional evidence derived from the pattern of network activation, this level of performance is satisfactory and a good starting point for further work.

6 Summary and Future Work

It is generally acknowledged that the term independence assumption is false and a lot of effort has been put in capturing and exploiting term dependencies when retrieving or organising information. In IF however, documents are typically treated as bag of words and any syntactic or semantic correlations between terms are ignored. In this paper we challenged this common practice with Nootropia, a user profiling model that uses a sliding window approach to identify and measure term dependencies, represents them in a network of terms and takes them into account for document evaluation with a spreading activation process. Experiments performed based on TREC 2001 routing guidelines show that a large enough window size (global context), which allows both syntactic and semantic term correlations to be incorporated, results in improved filtering performance over both small window sizes (local context) and a bag of words approach that ignores term dependencies all together. We have stressed that, since we compared profiles that comprised the same set of terms, these differences in performance are only due to the additional information that links encode.

The promising findings complement previous experimental results highlighting the importance of representing term dependencies for multi-topic IF with a single user profile [26]. Term dependence representation allows the profile to concentrate on the most relevant out of the complete set of possible term combinations. It increases the profile's information storage capacity and renders it resistant to dimensionality problems. So in principle, Nootropia can incorporate a large number of different types of features that describe the content or usage of information items. Nodes in the network can be features that are automatically extracted from a variety of media. They can correspond to the users that have viewed or purchased an information item, or to the tags that have been assigned to it. Complemented with a biologically inspired process for profile adaptation [29], the above property allows for an integrated solution to adaptive content-based and collaborative filtering with a single profile structure. This is a fascinating interdisciplinary research direction that we are currently pursuing.

References

1. Ide, N., Veronis, J.: Word sense disambiguation: The state of the art. Computational Linguistics 24, 1–40 (1998)
2. Wong, S.K.M., Ziarko, W., Wong, P.C.N.: Generalized vector space model in information retrieval. In: Annual International ACM SIGIR Conference on Research and Development in Information Retrieval, pp. 18–25. ACM Press, New York (1985)
3. Deerwester, S., Dumais, S.T., Landauer, G.W., Hashman, R.: Indexing by latent semantic analysis. Journal of the American Society for Information Science 41, 391–407 (1990)
4. Billhardt, H., Borrajo, D., Maojo, V.: A context vector model for information retrieval. Journal of the American Society for Information Science and Technology 53, 236–249 (2002)
5. van Rijsbergen, C.J.: A theoretical basis for the use of co-occurrence data in information retrieval. Journal of Documentation 33, 106–199 (1977)
6. Nallapati, R., Allan, J.: Capturing term dependencies using a language model based on sentence trees. In: 11th International Conference on Information and Knowledge Management (CIKM 2002), pp. 383–390. ACM Press, New York (2002)
7. Lee, C., Lee, G.G.: Probabilistic information retrieval model for a dependency structured indexing system. Information Processing and Management 45, 161–175 (2005)
8. Losee, R.M.: Term dependence: Truncating the bahadur-lazarsfeld expansion. Information Processing and Management 30, 293–303 (1994)
9. Turtle, H., Croft, W.B.: Inference networks for document retrieval. In: 13th Annual International ACM SIGIR Conference on Research and Development in Information Retrieval, pp. 1–24 (1990)
10. Park, Y.C., Choi, K.S.: Automatic thesaurus construction using bayesian networks. Information Processing and Management 32, 543–553 (1996)
11. Cunningham, S., Holmes, G., Littin, J., Beale, R., Witten, I.: Applying connectionist models to information retrieval. In: Amari, S., Kasobov, N. (eds.) Brain-Like Computing and Intelligent Information Systems, pp. 435–457. Springer, Heidelberg (1997)
12. Belew, R.K.: Adaptive information retrieval: using a connectionist representation to retrieve and learn about documents. In: Belkin, N., Rijsbergen, C. (eds.) 12th Annual International ACM SIGIR Conference on Research and Development in Information Retrieval, pp. 11–20. ACM Press, New York (1989)
13. Wilkinson, R., Hingston, P.: Using the cosine measure in a neural network for document retrieval. In: 14th Annual Internation ACM SIGIR conference on Research and Development in Information Retrieval, pp. 202–210. ACM Press, New York (1991)
14. Mothe, J.: Search mechanisms using a new neural network model comparison with the vector space model. In: Intelligent Multimedia Information Retrieval Systems and Management (RIAO 1994), pp. 275–294 (1994)

15. Wong, S.K.M., Cai, Y.J., Yao, Y.Y.: Computation of term associations by a neural network. In: 16th Annual International ACM SIGIR Conference on Research and Development in Information Retrieval, pp. 107–115. ACM Press, New York (1993)
16. Sanderson, M., Croft, B.W.: Deriving concept hierarchies from text. In: 22nd Annual Internation ACM SIGIR Conference on Research and Development in Information Retrieval, Berkeley, California, United States, pp. 206–213. ACM Press, New York (1999)
17. Anick, P., Tipirneri, S.: The paraphrase search assistant: Terminological feedback for iterative information seeking. In: Hearst, M., Gey, F., Tong, R. (eds.) 22nd Annual International ACM SIGIR Conference on Research and Development in Information Retrieval, pp. 153–159 (1999)
18. Widyantoro, D.H., Ioerger, T.R., Yen, J.: An adaptive algorithm for learning changes in user interests. In: ACM/CIKM 1999 Conference on Information and Knowledge Management, Kansas City, MO, pp. 405–412 (1999)
19. Mostafa, J., Mukhopadhyay, S., Palakal, M., Lam, W.: A multilevel approach to intelligent information filtering: model, system, and evaluation. ACM Transactions on Information Systems (TOIS) 15, 368–399 (1997)
20. Mladeni'c, D.: Using text learning to help web browsing. In: 9th International Conference on Human-Computer Interaction (HCI International 2001), New Orleans, LA, pp. 893–897 (2001)
21. Menczer, F., Belew, R.: Adaptive information agents in distributed textual environments. In: 2nd International Conference on Autonomous Agents, Minneapolis, MN, pp. 157–164 (1998)
22. Sorensen, H., O' Riordan, A., O' Riordan, C.: Profiling with the informer text filtering agent. Journal of Universal Computer Science 3, 988–1006 (1997)
23. McElligott, M., Sorensen, H.: An evolutionary connectionist approach to personal information filtering. In: 4th Irish Neural Networks Conference 1994, University College Dublin, Ireland, pp. 141–146 (1994)
24. Nanas, N., Uren, V., Roeck, A.D.: A comparative evaluation of term weighting methods in information filtering. In: 4th International Workshop on Natural Language and Information Systems (NLIS 2004), pp. 13–17 (2004)
25. Nanas, N., Uren, V., De Roeck, A., Domingue, J.: Building and applying a concept hierarchy representation of a user profile. In: 26th Annual International ACM SIGIR Conference on Research and Development in Information Retrieval, pp. 198–204. ACM press, New York (2003)
26. Nanas, N., Uren, V., De Roeck, A., Domingue, J.: Multi-topic information filtering with a single user profile. In: 3rd Hellenic Conference on Artificial Intelligence, pp. 400–409 (2004)
27. Bruza, P.D., Song, D.: Inferring query models by information flow analysis. In: Proceedings of the 11th International ACM Conference on Information and Knowledge Management (CIKM 2002), pp. 260–269 (2002)
28. Roeck, A.D., Sarkar, A., Garthwaite, P.H.: Defeating the homogeneity assumption. In: 7th International Conference on the Statistical Analysis of Textual Data (JADT), pp. 282–294 (2004)
29. Nanas, N., De Roeck, A.: Autopoiesis, the immune system and adaptive information filtering. Natural Computing (2008)

A Fuzzy Clustering Algorithm to Estimate the Parameters of Radial Basis Functions Neural Networks and Its Application to System Modeling

Antonios D. Niros and George E. Tsekouras[*]

University of the Aegean, Department of Cultural Technology and Communication,
Sapfous 5, 81100 Mytilini, Greece
Tel.:+30 2251036631; fax:+30 2251036609
gtsek@ct.aegean.gr

Abstract. We propose a two-staged fuzzy clustering algorithm to train radial basis function neural networks. The novelty of the contribution lies in the way we handle the input training data information between the two stages of the algorithm. The back-propagation method is employed to optimize the network parameters. The number of hidden nodes is determined by the iterative implementation of the fuzzy clustering and the back-propagation. Simulation results show that the methodology produces accurate models compared to the standard and more sophisticated techniques reported in the literature.

Keywords: Radial Basis Function Neural Networks; Fuzzy clustering; Back-propagation.

1 Introduction

Radial basis function neural networks (RBFNN) form a class of artificial neural networks, which has certain advantages over other types of neural networks including better approximation capabilities, simple network structure, and faster learning [1-4]. RBFNN training algorithms are split into two basic categories: (i) algorithms where the number of hidden nodes is predetermined and (ii) algorithms that involve structure selection mechanisms [5].

The training of Gaussian type RBFNNs is based on the estimation of three kinds of parameters: centers, widths and connecting weights between neurons. Commonly, the estimation of the above parameters is carried out using two-staged learning strategies. In the first stage, cluster analysis is implemented to calculate the appropriate values of the centers and the widths, while in the second stage supervised optimization procedures, such as gradient descent, are involved in the optimal estimation of the connecting weights [6-10].

So far, fuzzy clustering appears to be a very efficient tool to estimate the parameters of a Gaussian type RBFNN, since it is able to detect meaningful structures in the available data set, adopting Gaussian membership functions in the first place [11, 12]. This

---

[*] Corresponding author.

J. Darzentas et al. (Eds.): SETN 2008, LNAI 5138, pp. 194–204, 2008.

paper proposes a novel method to train Gaussian type RBFNNs by using a special designed fuzzy clustering algorithm. More specifically, the proposed algorithm uses the fuzzy c-means to group all the training data into a number of clusters. Then, each cluster center is assigned a weight value, which is determined by the fuzzy cardinality of the respective cluster. The set of these cluster centers along with their weights is further clustered using the weighted fuzzy c-means [13]. Thus, unlike the approach developed in [14], in the proposed algorithm the contribution of each cluster center to the final fuzzy partition is decided by the respective weight since two different cluster centers of different sizes should not weight equally with respect to the final partition. By elaborating the fuzzy clusters, produced by the weighted fuzzy c-means, initial estimations of the centers and the widths are obtained. These estimations are used for the calculation of the connecting weights between the hidden nodes. Finally, the network parameters are further optimized by using the back-propagation algorithm.

The paper is organized as follows: the proposed algorithm is presented in Section 2. The numerical examples can be found in Section 3. Finally, the paper concludes in Section 4.

2 The Proposed Algorithm

The proposed algorithm utilizes three learning phases and its flow sheet is depicted in Fig. 1. Before proceeding with the theoretical description of each step, we briefly overview the whole algorithmic scheme and discuss certain design issues.

Phase 1 utilizes three steps to extract initial conditions for the centers and the widths parameters of the network nodes. In the first step, the available data set is preprocessed using the well-known fuzzy c-means. In the second step, the clusters centers obtained by the fuzzy c-means are considered to be the new data set, which is further clustered by using the weighted fuzzy c-means algorithm. The justification of the above choices is given later on in this section. In the third step, based on the resulting fuzzy partition, we perform an initial estimation of the centers and the widths. Phase 2 is concerned with the initial parameter estimation of the neuron connecting weights, which is accomplished by the least squares method. Finally, Phase 3 performs a back-propagation optimization procedure to fine tune all the RBFNN parameters. The above phases are iteratively applied, where in each iteration the number of nodes increases by one until the approximation accuracy lies within acceptable levels. At this point, we have to stress on that the number of RBFNN nodes is equal to the number of clusters produced by the weighted fuzzy c-means. Thus, starting with two clusters (nodes) we finally obtain the number of nodes for which the network's performance is accurate. This remark directly implies that the final number of nodes will be small but yet reliable enough to produce an efficient RBFNN.

To better understand the relation between the design steps, we analyze in more details the main criteria used here. The first criterion is referred to the choice of using the fuzzy c-means. The fuzzy c-means has been recognized as a very efficient tool to discover structures hidden in the data. Since the RBFNN nodes' parameters are strongly depend on these structures, its use in the beginning of the algorithm will provide certain benefits related to the selection of a "good" starting point for the training process. The fuzzy c-means is used to preprocess the training data and therefore, this step is not concerned with designing a rigorous fuzzy c-means scheme. The only

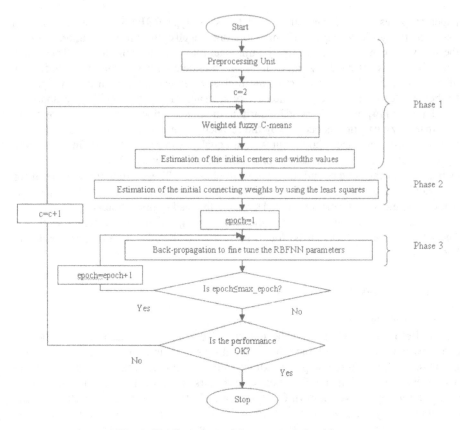

Fig. 1. The flow sheet of the proposed algorithm

restriction here is that the number of clusters is large enough, so that the training instances in the weighted fuzzy c-means, that follows, will be adequate to obtain a credible clustering result.

The second criterion refers to the weighted fuzzy c-means. Similar approaches [14] utilized the classical fuzzy c-means to elaborate the cluster centers produced by the preprocessing unit. Apparently, the implementation of the fuzzy c-means assumes that all these centers are of equal significance. However, this may be not true, because two different initial clusters may contain different number of training instances. Thus, by applying the fuzzy c-means we delete all the information, which relates the original data set to the above centers. Therefore, the real data structure may not be discovered. To resolve this problem we assign weights to the centers, and we elaborate them using the weighted version of the fuzzy c-means. Each of the above weights is defined as the normalized cardinality of the respective initial cluster. In this way, all the necessary information related to the original data set is taken into account during the whole model's design process.

The third criterion concerns the use of the back-propagation algorithm. The basic back-propagation algorithm often finds local minimum. It is also sensitive to its design parameters. However, many researchers [8, 13, 14] use back-propagation because of two main reasons: (a) back-propagation exhibits a fast convergence, and (b) providing good initial conditions to the back-propagation algorithm it is able to obtain very good results. Hence, the key point of all these approaches, including our approach here, is to provide the back-propagation algorithm with good initial conditions.

2.1 Phase 1

Let $X = \{x_1, x_2, ..., x_N\}$ be a set of N unlabeled feature data vectors with $x_k \in \Re^p$ $(1 \le k \le N)$, and n $(2 \le n \le N)$ be a number of fuzzy clusters. Given that the membership function of the k-th vector to the i-th cluster is denoted as $u_{ik} = \{u_i(x_k), 1 \le i \le c, 1 \le k \le n\}$ and the fuzziness parameter been denoted as m, the fuzzy c-means minimizes the following objective function,

$$J_m = \sum_{k=1}^{N} \sum_{i=1}^{n} (u_{ik})^m \| x_k - v_i \|^2 \tag{1}$$

under the next constraint,

$$\sum_{i=1}^{n} u_{ik} = 1 \tag{2}$$

The cluster centers and the membership degrees that solve the above optimization problem are given as,

$$v_i = \frac{\sum_{k=1}^{N} u_{ik}^m x_k}{\sum_{k=1}^{N} u_{ik}^m} \quad , \quad 1 \le i \le n \tag{3}$$

$$u_{ik} = \frac{1}{\sum_{j=1}^{c} \left(\frac{\| x_k - v_i \|_A}{\| x_k - v_j \|_A} \right)^{\frac{2}{m-1}}} \tag{4}$$

Then, the fuzzy c-means consists of iteratively applying the above two equations. In the next step, the cluster centers v_i $(1 \le i \le n)$ are considered the new data set. Because the vector v_i is referred to a cluster of the original data set, its weight of significance is defined as follows,

$$\rho_i = \frac{\aleph(U^i)}{\sum_l^n \aleph(U^l)} \tag{5}$$

where $\aleph(U^i)$ is the fuzzy cardinality of the i-th cluster,

$$\aleph(U^i) = \sum_{k=1}^{N} u_{ik} \quad (1 \le i \le n) \tag{6}$$

Then, the objective function of the weighted fuzzy c-means is,

$$J_\rho = \sum_{i=1}^{n} \sum_{j=1}^{c} \rho_i \, (\mu_{ji})^m \, \| v_i - v_j \|^2 \tag{7}$$

where c is the number of the final clusters, which coincides with the number of RBFNN hidden nodes, v_j $(1 \le j \le c)$ are the centers of the final clusters, $m \in (1, \infty)$ is a factor to adjust the membership degree weighting effect, and ρ_i is the weight of significance that is assigned to v_i.

The problem is to minimize J_ρ under the following constraint,

$$\sum_{j=1}^{c} \mu_{ji} = 1 \qquad \forall i \tag{8}$$

The final prototypes and the respective membership functions that solve this constrained optimization problem are given by the next equations,

$$v_j = \frac{\sum_{i=1}^{n} \rho_i \, (\mu_{ji})^m \, v_i}{\sum_{i=1}^{n} \rho_i \, (\mu_{ji})^m} \tag{9}$$

and

$$\mu_{ji} = \frac{1}{\sum_{l=1}^{c} \left(\dfrac{\| v_i - v_j \|}{\| v_i - v_l \|} \right)^{2/(m-1)}} \tag{10}$$

Eqs. (9) and (10) constitute an iterative optimization procedure which optimizes the objective function in (7). In the case where all the weights are equal, the weighted fuzzy c-means is equivalent to the classical fuzzy c-means.

To this end, the number of nodes in the RBFNN is equal to the number of clusters c, while the centers of these nodes are the same to the cluster centers v_j $(1 \le j \le c)$. On the other hand, the standard deviations of the RBFNN nodes are calculated as follows,

$$\sigma^j = \frac{2 * d_{max}^j}{3} \qquad (1 \le j \le c) \tag{11}$$

in (11) d_{max} is given in the next equation,

$$d_{max} = \{ \max_k \{ \| x_k - v_j \|^2 \} : x_k \in C_j \text{ such that } u_{jk} \ge \theta \} \tag{12}$$

where C_j is the j-th cluster and u_{jk} is the membership degree of the k-th training data vector to the j-th final fuzzy cluster, and θ a small positive number such that $\theta \in (0,1)$ with $\theta \ll 1$.

2.2 Phase 2

Since the RBFNN's output is linear in the weights w_j, these weights can be estimated by least-squares methods. For each data point x_k, the outputs of the neuron are computed as,

$$s_{kj} = g_j(x_k) \qquad (13)$$

where $g_j(x_k)$ is the Gaussian function. Introducing the weight vector $w = [w_1, w_2, ..., w_c]$ we can write the following matrix equation for the whole data set:

$$\tilde{y} = Aw \qquad (14)$$

where $A = [s_{kj}]$. Finally, the least-squares estimate of the weights w that minimize the network error $e = \tilde{y} - y$ is given next:

$$w = [A^T A]^{-1} A^T \tilde{y} \qquad (15)$$

2.3 Phase 3

In this step, the initial parameter estimations calculated in the previous steps are fed into the back-propagation algorithm. With respect to a certain input-output data pair $(x_k; y_k)$, the task is to minimize the following error function:

$$e = \frac{1}{2}(\tilde{y}_k - y_k)^2 \qquad (16)$$

where y_k and $\tilde{y}_k$ are the real and the predicted network output, respectively.

Learning in an RBF network is accomplished by utilizing the next two stages:
1. Adjusting the parameters of the RBFNN layer, including the RBF centers $v_j \in R^P$ $(1 \leq j \leq c)$ and the variances σ_j $(1 \leq j \leq c)$. Here, p is the dimension of the input space.
2. Calculation of the output weights w_j $(j=1,...,c)$ of the network.

The learning rules for the above stages are:

$$w_j^l(t+1) = w_j^l(t) - e * \frac{\partial e}{\partial w_j^l}\Big|_t \qquad (1 \leq l \leq p) \qquad (17)$$

$$v_j^l(t+1) = v_j^l(t) - e * \frac{\partial e}{\partial v_j^l}\Big|_t \qquad (1 \leq l \leq p) \qquad (18)$$

$$\sigma_j^l(t+1) = \sigma_j^l(t) - e * \frac{\partial e}{\partial \sigma_j^l}\Big|_t \qquad (1 \leq l \leq p) \qquad (19)$$

where e^* is the back-propagation learning parameter. Finally, based on eqs (14) and (16), the partial derivatives in (17), (18) and (19) can easily be calculated.

3 Simulation Experiments

3.1 Mackey–Glass System

In this subsection we use the proposed algorithm in order to predict the Mackey–Glass time series. The Mackey–Glass time series is generated by the following time-delay differential equation,

$$\frac{dx(t)}{dt} = \frac{0.2x(t-\tau)}{1+x^{10}(t-\tau)} - 0.1x(t) \tag{20}$$

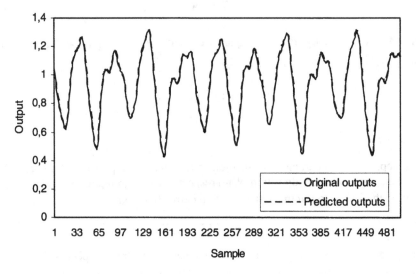

Fig. 2. The original and predicted output respectively for the Mackey Glass example. (Training data).

When the parameter τ is large enough the system appears a chaotic behavior. In our simulations we set $\tau = 17$ and generated a sample of 1000 points. The first 500 points were used as training data, and the last 500 points as test data to validate the RBFNN. The proposed algorithm was applied to build a RBFNN with 4 inputs: $x(t-18)$, $x(t-12)$, $x(t-6)$ and $x(t)$, while the output was the point $x(t+6)$.

The design parameters for the fuzzy c-means and the weighted fuzzy c-means were selected as: $m=2$, $n=20$, $max.\_number\_of\_iteration=30000$ and $min.\_amount\_ of\_ improvement=10^{-5}$. The final number of hidden nodes was equal to $c=14$. For the back-propagation algorithm the learning rate was equal to: $e^{*}=0.001$, and the maximum number of iteration (max_epoch) was equal to 30000. The resulted root mean square error (RMSE) for the training data was equal to 0.00517 and 0.00527 for the test data. The output for the training data is shown in Fig. 2.

Table 1. Comparison results for the Mackey Glass example

Model		Number of hidden nodes	RMSE
Autoregressive model [17]		---	0.19
Cascade correlation NN [17]		---	0.06
Back-propagation NN [17]		---	0.02
Sixth-order polynomial [17]		---	0.04
Linear prediction method [17]		---	0.55
Cho and Wang [15]	(Training)	23	0.0096
	(Test data)	23	0.0114
Kim and Kim, C. [16]		---	0.049
Chen et al. [17]	(Training)	---	0.0069
	(Test data)	---	0.0071
Chen et al. [18]		---	0.0076
Jang et al. [19]		16	0.007
Chen et al. [20]	(Training)	21	0.0107
	(Test data)	21	0.0128
Proposed RBFNN	(Training)	14	0.00517
	(Test data)	14	0.00527

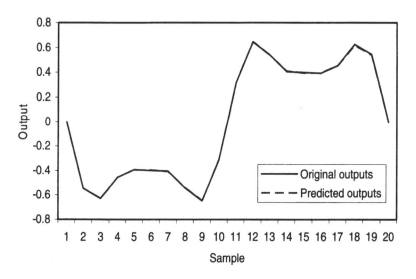

Fig. 3. The original and predicted output respectively for the synthetic one-dimensional data example (test data)

Finally, Table 1 compares different computational paradigms found in the literature to our RBFNN training procedure. According to this table, the proposed algorithm gave the best result with the smallest number of hidden nodes.

3.2 Function Approximation

We consider a single-variable nonlinear function of the following form,

$$y=0.6\sin(\pi x)+0.3\sin(3\pi x)+0.1\sin(5\pi x) \qquad (21)$$

where the input x is defined in the space [-1, 1].

The design parameters for the fuzzy c-means and the weighted fuzzy c-means were selected as: $m=2$, $n=20$, $max.\_number\_of\_iteration=30000$ and $min.\_amount\_ of\_ improvement=10^{-5}$. The implementation of the algorithm gave $c=9$ hidden nodes.

For the back-propagation algorithm the learning rate was equal to: $e^*=0.015$, and the maximum number of iteration (max_epoch) for the training of the RBFNN was equal to 30000. The resulted RMSE for the training data was equal to 0.0036 and for the test data 0.0041. The output for the training data is shown in Fig. 3. The experimental results clearly reveal that the proposed architecture along with the underlying design scheme outperforms the existing networks.

Table 2. Comparison results for the synthetic one-dimensional data example

Model	Number of hidden nodes	RMSE
Classical RBFNN (Test data)	28	0.019
Our model (Training data)	14	0.0036
Out model (Test data)	14	0.0041

Finally, Table 2 compares the performances obtained in this example with the performance of the Classical RBFNN. This table indicates that the performance of the classical RBFNN is inferior when compared to the proposed methodology.

4 Conclusion

In this paper, a new fuzzy logic-based clustering method was developed and evaluated with respect to the parameter estimation of RBFNN. The algorithm utilizes three learning phases to estimate the appropriate values for the network parameters. In the first phase it performs a two-stage fuzzy clustering in such a way that the underlying data structure can be efficiently discovered. In the second phase, the hidden neuron connecting weights are calculated and finally, in the third phase, all the network parameters are fine tuned by applying the back-propagation algorithm. The proposed algorithm offers two important advantages:

1. The training process is controlled by the number of nodes. Thus, in each iteration the number of nodes increases by one until the network performance will be accurate enough. This fact directly implies that the process will stop as soon as the smaller reliable number of nodes is detected maintaining at the same time the smaller number of iterations.

2. The estimation of the hidden node parameters does not depend on any initial random selection of their values but rather it is straightforward result of the clustering process.

Several simulations were performed, which showed that the methodology can be very successful when it is applied to system modeling.

References

1. Luo, W., Karim, M.N., Morris, A., Martin, E.B.: Control Relevant Identification of a pH Wastewater Neutralisation Process using Adaptive Radial Basis Function Networks. Computers and Chemical Engineering, S1017 (1996)
2. Narendra, K.S., Parthasarathy, K.: Identification and Control of Dynamical Systems Using Neural Networks. IEEE Trans. Neural Networks 1, 4 (1990)
3. Leonard, J.A., Kramer, M.A.: Radial Basis Function Networks for Classifying Process Faults. IEEE Control Systems 31 (1991)
4. Lazaro, M., Santamaria, I., Pantaleon, C.: A new EM-based training algorithm for RBF networks. Neural Networks 16(1), 69–77 (2003)
5. Sarimveis, H., Alexandridis, A., Tsekouaras, G., Bafas, G.: A Fast and Efficient Algorithm for Training Radial Basis Function Neural Networks Based on a Fuzzy Partition of the Input Space, Ind. Eng. Chem. Research 41, 751–759 (2002)
6. Lucks, M.B., Oki, N.: A Radial Basis Network for Function Approximation. In: The Proceedings of the 42nd Midwest Symposium on Circuits and Systems, pp. 1099–1101 (1999)
7. Figueiredo, M.A.: On Gaussian Radial Basis Function Approximations: Interpretation, Extensions and Learning Strategies, pp. 618–621 (2000)
8. Karayiannis, N.B.: Reformulated radial basis neural networks trained by gradient descent. IEEE Transactions on Neural Networks 10(3), 657–671 (1999)
9. Gonzalez, J., Rojas, I., Ortega, J., Pomares, H., Fernandez, J., Diaz, A.F.: Multiobjective evolutionary optimization of the size, shape, and position parameters of radial basis function networks for function approximation. IEEE Transactions on Neural Networks 14(6), 1478–1495 (2003)
10. Babuska, R., Verbruggen, H.: Neuro-fuzzy methods for nonlinear system identification. Annual Reviews in Control 27, 73–85 (2004)
11. Pedrycz, W.: Conditional Fuzzy Clustering in the Design of Radial Basis Function Neural Networks. IEEE Trans. Neural Networks 9(4), 601–612 (1998)
12. Staiano, A., Tagliaferri, R., Pedrycz, W.: Improving RBF Networks Performance in Regression Tasks by Means of a Supervised Fuzzy Clustering. Neurocomputing 69, 1570–1581 (2006)
13. Tsekouras, G.: On the use of the weighted fuzzy c-means in fuzzy modeling. Adv. Eng. Softw. 36, 287–300 (2005)
14. Linkens, D.A., Chen, M.: Input selection and partition validation for fuzzy modelling using neural network. Fuzzy Sets Syst. 107, 299–308 (1999)
15. Cho, K.B., Wang, B.H.: Radial basis function based adaptive fuzzy systems their application to system identification and prediction. Fuzzy Sets and Systems 83, 325–339 (1995)
16. Kim, D., Kim, C.: Forecasting time series with genetic fuzzy predictor ensembles. IEEE Trans. Fuzzy Systems 5, 523–535 (1997)
17. Chen, Y., Yang, B., Dong, J., Abraham, A.: Time-series Forecasting using Flexible Neural Tree Model. Information Science 174, 219–235 (2005)

18. Chen, Y., Yang, B., Zhou, J.: Automatic Design of Hierarchical RBF Networks for System Identification, pp. 1191–1199 (2006)
19. Jang, J.-S.R., Sun, C.-T., Mizutani, E.: Neuro-fuzzy and soft computing: a computational approach to learning and machine intelligence. Prentice-Hall, Upper Saddle River (1997)
20. Chen, S., Cowan, C.F.N., Grant, P.M.: Orthogonal least squares learning algorithm for radial basic function network. IEEE Trans. Neural Networks 2, 302–309 (1991)

Speech/Music Discrimination Based on Discrete Wavelet Transform

Stavros Ntalampiras and Nikos Fakotakis

Electrical and Computer Engineering Department, Wire Communication Laboratory,
University of Patras, 26500 Rio - Patras, Greece
sntalampiras@upatras.gr, fakotaki@wcl.ee.upatras.gr

Abstract. In this paper we present an effective approach which addresses the issue of speech/music discrimination. Our architecture focuses on the matter from the scope of improving the performance of a speech recognition system by excluding the processing of information which is not speech. Multiresolution analysis is applied to the input signal while the most significant statistical features are calculated over a predefined texture size. These characteristics are then modeled using a state of the art technique for probability density function estimation, Gaussian mixture models (GMM). A classification scheme consisting of a conventional maximum likelihood decision methodology constitutes the next step of our implementation. Despite the fact that our system is based solely on wavelet signal processing, it demonstrated very good performance achieving 91.8% recognition rate.

Keywords: Computer audition, content-based audio classification, discrete wavelet transform, Gaussian mixture model.

1 Introduction

Over the past decades a lot of work has been conducted in the area of speech processing (SP) and especially in the field of automatic speech recognition. In order to achieve better performance we need the recognizer component to elaborate only on speech data and nothing else. Here enters the idea of speech/music discrimination. Having a system that identifies which part of a sound includes speech or music, we can activate or not the component that processes speech resulting this way to a better accuracy. Afterwards a speech enhancement algorithm may be employed, and in combination with a voice activity detector, have the signal recognized.

Another important scientific domain, audio processing, is getting more and more attention lately. Usually the same techniques as in SP are applied here as well, which shows that there is still much work to be done to attain better performance in such systems. A brief review of the area of speech/music discrimination is following. Slaney and Scheirer [1] utilize signal processing techniques in time and frequency domain to extract 13 features. Their experiments are made in different classification schemes with overall performance of 5.8% error rate. El-Maleh et al present a feature set which combines the line spectral frequencies (LSFs) and zero-crossing-based

J. Darzentas et al. (Eds.): SETN 2008, LNAI 5138, pp. 205–211, 2008.

features for speech/music differentiation [2]. For feature evaluation Bayes error rate with empirical estimation was used while a quadratic Gaussian classifier categorizes the input signal resulting in 95.9% accuracy. In [3] a fusion of two speech/music classification approaches is presented consisting of two different subsystems. The utilization of GMM and spectral features is shown to provide 94% and 90% accuracy for speech and music detection respectively. Tzanetakis et al [4] propose a framework for audio analysis using the Discrete Wavelet transform (DWT) with 12 different subbands while a comparison of Mel Frequency Cepstral Coefficients (MFCC) and Short Time Fourier transform coefficients was made. The application of a Gaussian classifier indicated that features derived from the wavelet transform have similar performance to the other two feature sets. In [5] a multi-layer perceptron forms the basis to classify speech and music and the performance of several features is investigated including mean and variance of DWT. Seven features comprise the final feature set and 96.6% accuracy with ten neurons is obtained. Didiot et al [6] combined energy-based features from the wavelet coefficient of each frequency band and 12 MFCCs in order to train Hidden Markov models. Finally, a class/non-class strategy is employed to distinguish speech and music.

This work is contributing to the field of automatic acoustic analysis for the purpose of "understanding" the surrounding environment by exploiting *only* the perceived auditory information in the way humans exhibit quite effortless. We explore the usage of a feature set based solely on multiresolution processing to achieve efficient speech/music discrimination. The basis of our implementation is the wavelet transform (Fig. 1). Wavelets are usually used in data compression with the well known paradigm of the image compression algorithm, JPEG 2000 proposed by the Joint Photographic Experts Group. We are going to show that the application of the DWT can be of great importance in classification tasks which involve audio processing. The transform has the property of treating with great accuracy the lower frequencies of the signal in contrast to the higher ones. The fundamental property of the Fourier transform is the usage of sinusoids with infinite duration. While sinusoids are smooth and predictable, wavelets tend to be irregular and asymmetric. This is the principal property of the wavelet representation and will be discussed next.

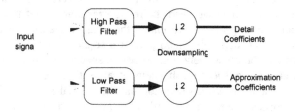

Fig. 1. Discrete Wavelet Transform

2   Wavelet Analysis

During the past years, wavelets techniques have become a common tool in digital signal processing. This kind of analysis has been used in many different researching areas including denoising of signals and applications in geophysics (tropical convention, the dispersion of ocean waves etc) [7]. One can conclude that this emerging type of signal analysis is adequate to provide strong solutions in many and completely different researching areas. The main advantage of the wavelet transform is that it can process time series, which include non stationary power at many different frequencies (Daubechies 1990). Wavelet comprises a dynamic windowing technique which can treat with different precision low and high frequency information. The first step of the DWT is the choice of the original (or mother) wavelet and by utilizing this function, the transformation breaks up the signal into shifted and scaled versions of it. There are many types of mother wavelet functions and in this work the next four are investigated: haar (or db1), db 4, symlet 2 and a biorthogonal function (bior 3.7). A function must have zero mean and be localized in both time and frequency in order to form a mother wavelet (Farge 1992). Several experiments took place before the decision of these functions was made. Although these functions differ a lot (Fig. 2), we will see that the results are pretty much the same. The application of the DWT with these four different original wavelets consists of one-stage filtering of the audio signals as we can see in Fig. 1. Subsequently the data series are downsampled due to the Nyquist theorem in order not to end up having twice as many data comparing to the ones that we started with. In this paper we take under consideration only the *Approximation* coefficients which contain the low frequency information of the input sound, which is considered to contain the most important information as regards to human perception (inspired by the field of image processing).

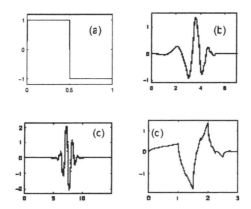

Fig. 2. Shapes of the mother wavelets that were employed (a) Haar (b) Daubechies 4 (c) Biorthogonal 3.7 (d) Symlet 2

2.1 The Feature Set

In our implementation, speech/music discrimination is based on six statistical measurements taken from the low frequency information of the signal. At the primary stage the DWT coefficient is cut into equal chunks of data using a texture size of 480 samples (30 ms), which was determined after extensive experimentations (see Fig. 3). It should be noted that no preemphasis is applied and the analysis is performed onto non overlapping chunks without considering the incomplete ones at the end of each file (in case there is one). For all the experiments the standard wavelet toolbox of the MATLAB^{TM} framework was used.

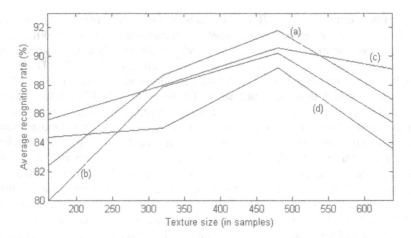

Fig. 3. Average recognition rates against different texture sizes achieved by the following mother wavelet functions: (a) Haar, (b) Daubechies 4, (c) Symlets 2 and (d) Biorthogonal 3.7

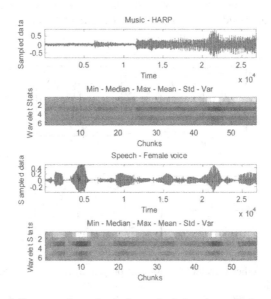

Fig. 4. Feature values of sound samples belonging to both categories

The hierarchical role of the feature extraction process is to capture the characteristics that distinct these two audio classes. Moreover, it is a technique of data reduction and in this case the suppression of the input data is huge having an average proportion of 4.1/100 (feature vector bytes/initial audio signal bytes). We sustain only a small amount of discriminative information of the audio signal using only the following six statistical features measured over the texture size (see also Fig. 4): (i) mean, (ii) variance, (iii) minimum value, (iv) maximum value, (v) standard deviation and (vi) the median. Afterwards we used speech and music data to train probabilistic models

(GMMs), which are described in the next section. Both male and female speech obtained from the TIMIT database, and an EBU music collection [8-9] which incorporates a large variety of musical instruments and compositions were employed to build up speech and music models. All the sounds were sampled at 16 KHz with 16 bit analysis while the average duration was 5.6 seconds.

3   Experimental Setup

The recognition process is consisted of probability computations of two Gaussian mixtures which represent the a priori knowledge that the system includes (Fig. 5). K-means initialization algorithm along with a standard version of the Expectation Maximization (EM) process was used for the training of eight components for each category. Furthermore it should be mentioned that diagonal covariance matrices are utilized for the construction of each mode.

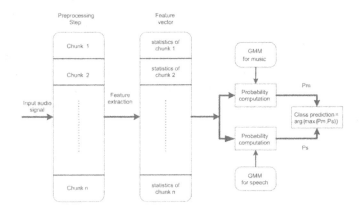

Fig. 5. Overall system architecture

During the testing phase, ten-fold cross validation was employed in order to obtain reliable results. Class decisions are made per frame while care has been taken not to have parts of the same file in both train and test sets simultaneously. As we tabulate in Table 1, the recognition rates achieved by our system are relatively high, concerning the small number of descriptors facilitated by our methodology.

Table 1. Recognition Rates (%)

Mother Wavelet	Music	Speech	Overall
Haar	89.4	94.2	**91.8**
Daubechies 4	89	91.4	**90.2**
Symlets 2	88.5	92.7	**90.6**
Biorthogonal 3.7	88.3	90.1	**89.2**

The four different original wavelets that were examined produced almost equal recognition rates. Despite its simplicity, haar function is proved to have the best performance while the biorthogonal one provides the worst results. Thus, we conclude that *Daubechies 1* mother wavelet function should be used in the task of speech/music discrimination.

4 Conclusions and Future Work

In this paper we explained a wavelet-based architecture for the purpose of efficient and simple speech/music discrimination under the scope of enhancing the performance of a speech recognition system. Our approach utilizes a limited amount of information produced by a well-known multiresolution technique. A comparison between three different wavelet families was conducted and the slight superiority of haar mother wavelet function was made clear.

We conclude that the specific type of analysis provides a strong basis for the implementation of a system with low computational needs as regards the specific classification task. The present contribution proposes a new feature set consisting only by six dimensions, while it reaches very high recognition accuracy. This work completes an initial step towards building a robust system for automatic speech/music discrimination. Our future work includes blind signal separation (BSS) to discriminate overlapping signals, incorporation of a silence detection algorithm and non-redundant fusion of the presented group of descriptors with well known sets, such as MFCC and MPEG-7 low level descriptors.

References

1. Scheirer, E., Slaney, M.: Construction and evaluation of a robust multifeature speech/music discriminator. In: IEEE International Conference on Acoustics, Speech and Signal Processing, vol. 2, pp. 1331–1334 (1997)
2. El-Maleh, K., Klein, M., Petrucci, G., Kabal, P.: Speech/music discrimination for multimedia applications. In: IEEE International Conference on Acoustics, Speech and Signal Processing, vol. 6, pp. 2445–2448 (2000)
3. Pinquier, J., Rouas, J.-L., Andre-Obrecht, R.: A fusion study in speech/music classification. In: International Conference in Multimedia and Expo., vol. 2, pp. 409–412 (2003)
4. Tzanetakis, G., Essl, G., Cook, P.R.: Audio analysis using the wavelet transform. In: International Conference on Acoustics and Music: Theory and Applications (AMTA) (2001)
5. Kashif Saeed Khan, M., Al-Khatib, W.G., Moinuddin, M.: Automatic classification of speech and music using neural networks. In: 2nd ACM international workshop on Multimedia databases, pp. 94–99 (2004)
6. Didiot, E., Illina, I., Mela, O., Haton, J.P., Fohr, D.: A wavelet-based parameterization for speech/music segmentation. In: International Conference on Spoken Language Processing, pp. 653–656 (2006)
7. Torrence, C., Compo, G.P.: A practical guide to wavelet analysis. Bulletin of the American Meteorological society 79, 61–78 (1998)
8. EBU.: SQAM - CD: Sound quality assessment material, Polygram Cat. No 422 204-2, European Broadcasting Union (EBU) (1988)

9. EBU.: Sound quality assessment material, Recordings for subjective tests – Users/ hand-book for the EBU-SQAM Compact Disc, Tech 3253, European Broadcasting Union (EBU) (1988)

10. Mallat, S.G.: A theory for multiresolution signal decomposition: the wavelet representation. IEEE Transactions on Pattern Analysis and Machine Intelligence 2, 674–693 (1989)

11. Foote, J.: An overview of audio information retrieval. ACM Multimedia Systems 7, 2–10 (1999)

12. Nabney, I.: Netlab: Algorithms for Pattern Recognition. Springer, London (2002)

Comparing Datasets Using Frequent Itemsets: Dependency on the Mining Parameters

Irene Ntoutsi and Yannis Theodoridis

Department of Informatics, University of Piraeus, Greece
{ntoutsi,ytheod}@unipi.gr

Abstract. Comparison between sets of frequent itemsets has been traditionally utilized for raw dataset comparison assuming that frequent itemsets inherit the information lying in the original raw datasets. In this work, we revisit this assumption and examine whether dissimilarity between sets of frequent itemsets could serve as a measure of dissimilarity between raw datasets. In particular, we investigate how the dissimilarity between two sets of frequent itemsets is affected by the *minSupport* threshold used for their generation and the adopted compactness level of the itemsets lattice, namely frequent itemsets, closed frequent itemsets or maximal frequent itemsets. Our analysis shows that utilizing frequent itemsets comparison for dataset comparison is not as straightforward as related work has argued, a result which is verified through an experimental study and opens issues for further research in the KDD field.

1   Introduction

Detecting changes between datasets is an important problem nowadays due to the highly dynamic nature of data. A common approach for comparing datasets is to utilize the patterns extracted from these datasets. The intuition behind this approach is that patterns encapsulate (to some degree) the information contained in the original data. In [3], authors measure the deviation between systematically evolving datasets, like the buying habits of customers in a supermarket, in terms of the data mining models they induce. In [5,4], authors measure the dissimilarity between distributed datasets using the corresponding sets of frequent itemsets.

In this work we elaborate on this assumption about the equivalence of dissimilarity in pattern space with the dissimilarity in raw data space, for a very popular pattern type, the frequent itemset patterns. More specifically, we provide a theoretical analysis that shows the dependency of dissimilarity in pattern space on frequent itemsets mining (FIM) settings, namely (a) on the *minSupport* threshold used for the generation of itemsets and (b) on the adopted compactness level for the itemsets lattice (frequent itemsets-FI, closed frequent itemsets-CFI or maximal frequent itemsets-MFI). Regarding the *minSupport* threshold, our analysis shows that the larger this threshold is, the higher the dissimilarity in pattern space is. Regarding the different lattice representations, it turns out that the more compact the representation achieved by the itemset type is, the higher

J. Darzentas et al. (Eds.): SETN 2008, LNAI 5138, pp. 212–225, 2008.

the dissimilarity in pattern space is. Moreover, we describe the different dissimilarity measures proposed so far in the literature ([3,4,5]) through a general common dissimilarity schema and verify the above theoretical results through an experimental study. The results indicate that utilizing pattern comparison for data comparison is not as straightforward as argued by related work and should only be carried out under certain assumptions (e.g., FIM settings).

The rest of the paper is organized as follows: Section 2 describes basic FIM concepts. Section 3 discusses the related work and describes a general formula through which the proposed measures can be described. In Section 4, we present the FIM parameters that affect dissimilarity. In Section 5, we experimentally evaluate the effect of the different parameters on dissimilarity. In Section 6, we present conclusions and outlook.

2   Background on the FIM Problem

Let I be a finite set of distinct items and D be a dataset of transactions where each transaction T contains a set of items, $T \subseteq I$. An *itemset* X is a non-empty lexicographically ordered set of items, $X \subseteq I$. The percentage of transactions in D that contain X, is called the *support* of X in D, $supp_D(X)$. An itemset is *frequent* if $supp_D(X) \geq \sigma$, where σ is the *minSupport* threshold. The set of frequent itemsets (FIs) extracted from D under σ is given by: $F_\sigma(D) = \{X \subseteq I \mid supp_D(X) \geq \sigma\}$. An itemset is frequent iff all of its subsets are frequent (apriori property). This property allows us to enumerate frequent itemsets lattice using more compact representations like closed frequent itemsets (CFIs) and maximal frequent itemsets (MFIs).

A frequent itemset X is called *closed* if there exists no frequent superset $Y \supset X$ with $supp_D(X) = supp_D(Y)$. The set $C_\sigma(D)$ is a subset of $F_\sigma(D)$ since every closed itemset is also frequent.

$$C_\sigma(D) = F_\sigma(D) - \{X \in F_\sigma(D) : Y \supset X \Rightarrow supp_D(X) = supp_D(Y), Y \in F_\sigma(D)\} \tag{1}$$

On the other hand, a frequent itemset is called *maximal* if it is not a subset of any other frequent itemset. The set $M_\sigma(D)$ is also a subset of $F_\sigma(D)$ since every maximal itemset is frequent.

$$M_\sigma(D) = F_\sigma(D) - \{X \in F_\sigma(D) : Y \supset X \Rightarrow Y \in F_\sigma(D)\} \tag{2}$$

By definition, $C_\sigma(D)$ is a subset of $M_\sigma(D)$:

$$M_\sigma(D) = C_\sigma(D) - \{X \in C_\sigma(D) : Y \supset X \Rightarrow Y \in C_\sigma(D)\} \tag{3}$$

$C_\sigma(D)$ is a lossless representation of $F_\sigma(D)$ since both the lattice structure (i.e., frequent itemsets) and the lattice measure (i.e., itemset supports) can be derived from CFIs [7]. Unlike $C_\sigma(D)$, $M_\sigma(D)$ is a lossy representation of $F_\sigma(D)$ since it is only the lattice structure that can be determined from MFIs whereas the measures are lost [7]. By Eq. 1, 2 and 3 it follows that:

$$M_\sigma(D) \subseteq C_\sigma(D) \subseteq F_\sigma(D) \tag{4}$$

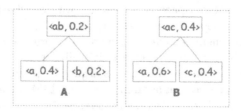

Fig. 1. Two lattices of frequent itemsets to be compared: A (lef), B (right)

Table 1. List of symbols

Symbol	Description
D	a dataset
X	an itemset
$supp_D(X)$	the support of itemset X in dataset D
σ	the *minSupport* threshold
$F_\sigma(D)$	the set of frequent itemsets generated from D under σ
$C_\sigma(D)$	the set of closed frequent itemsets generated from D under σ
$M_\sigma(D)$	the set of maximal frequent itemsets generated from D under σ
$dis(A,B)$	dissimilarity between two set of itemsets A, B

Let us consider for comparison (Fig. 1) two sets of frequent itemsets A, B generated under the same *minSupport* threshold σ from the original datasets D and E, respectively. Each itemset is described as a pair *<structure, measure>* denoting the items forming the itemset (*structure*) and the itemset support (*measure*).

The question is how similar to each other A and B are. There are many cases where the two sets might differ: An itemset, for example, might appear in both A and B sets but with different supports, like the itemset $< a >$ in Fig. 1. Or, an itemset might appear in only one of the two sets, like the itemset $< b >$ which appears in A but not in B. In this case, two things might have occurred: either $< b >$ does not actually exist in the corresponding dataset E or $< b >$ has been pruned due to low support (lower than the *minSupport* threshold σ).

Since the generation of A, B depends on the FIM parameters, namely the *minSupport* threshold σ used for their generation and the adopted lattice representation (FI, CFI or MFI), we argue that the estimated dissimilarity score also depends on these parameters. Furthermore, since dissimilarity in pattern space is often used as a measure of dissimilarity in raw data space we argue that the above mentioned parameters also affect this correspondence.

Table 1 summarizes the symbols introduced in this section.

3   Comparing Frequent Itemset Lattices

Parthasarathy-Ogihara approach: *Parthasarathy and Ogihara* [5] present a method for measuring the dissimilarity between two datasets D and E by using the corresponding sets of frequent itemsets (A and B, respectively). Their metric is defined as follows:

$$dis(A,B) = 1 - \frac{\sum_{X \in A \cap B} \max\{0, 1 - \theta * |supp_D(X) - supp_E(X)|\}}{|A \cup B|} \quad (5)$$

In the above equation, θ is a scaling parameter that reflects how significant are for the user the variations in the support values. This measure works with itemsets of identical structure, i.e., those appearing in $A \cap B$. Itemsets that only partially fit each other like $< ab >$ and $< ac >$ are considered totally dissimilar.

FOCUS approach: The *FOCUS framework* [3] quantifies the deviation between two datasets D and E in terms of the FI sets (A and B, respectively) they induce. A and B are first refined into their union ($A \cup B$) and the support of each itemset is computed with respect to both D and E datasets. Next, the deviation is computed by summing up the deviations of the frequent itemsets in the union:

$$dis(A,B) = \frac{\sum_{X \in A \cup B} |supp_D(X) - supp_E(X)|}{\sum_{X \in A} supp_D(X) + \sum_{X \in B} supp_E(X)} \quad (6)$$

FOCUS measures the dissimilarity between two sets of FIs in terms of their union. Partial similarity is not considered. Indeed, FOCUS tries to find itemsets with identical structures. If an itemset X appears in A with $supp_D(X)$ but not in B, then the corresponding data set E of B is queried so as to retrieve $supp_E(X)$. An upper bound on dissimilarity is provided, which involves only the induced models and avoids the expensive operation of querying the original raw data space. In this case, if an itemset X does not appear in B, it is considered to appear but with zero measure, i.e., $supp_E(X) = 0$.

Li-Ogihara-Zhou approach: *Li Ogihara and Zhou* [4] propose a dissimilarity measure between datasets based on the set of MFIs extracted from these datasets. Let $A = \{X_i, supp_D(X_i)\}$ and $B = \{Y_j, supp_E(Y_j)\}$ where X_i, Y_j are the MFIs in D, E respectively. Then:

$$dis(A,B) = 1 - \frac{2I_3}{I_1 + I_2} \quad (7)$$

$$I_1 = \sum_{X_i, X_j \in A} d(X_i, X_j), \quad I_2 = \sum_{Y_i, Y_j \in B} d(Y_i, Y_j), \quad I_3 = \sum_{X \in A, Y \in B} d(X, Y)$$

$$d(X,Y) = \frac{|X \cap Y|}{|X \cup Y|} * log(1 + \frac{|X \cap Y|}{|X \cup Y|}) * \min(supp_D(X), supp_E(Y))$$

This measure works with the average dissimilarity between pairs of MFIs. Partial similarity is considered; itemsets that have some common part in their structure are compared and their score is aggregated to the total dissimilarity score.

3.1   Common Background of the Three Approaches

All approaches express the dissimilarity between two sets of frequent itemsets as an aggregation of the dissimilarities of their component itemsets:

$$dis(A, B) = \sum_{X \in A, Y \in B} dis(X, Y) \tag{8}$$

where $dis(X, Y)$ is the dissimilarity function between two simple frequent itemsets, defined in terms of their structure and measure components, as follows:

$$dis(X, Y) = f(dis_{struct}(X, Y), dis_{meas}(X, Y)) \tag{9}$$

The function $dis_{struct}()$ evaluates the dissimilarity between the structure components (i.e., frequent itemsets), whereas the function $dis_{meas}()$ evaluates the dissimilarity between their measure components (i.e., supports). The function f aggregates the corresponding structure and measure scores into a total score.

All approaches follow the rationale of Eq. 8 and differentiate only on how $dis_{struct}(X, Y)$, $dis_{meas}(X, Y)$ and f are instantiated. Below, we present how these functions are defined for each of the proposed measures.

In case of the Parthasarathy-Ogihara measure, Eq. 9 can be written as:

$$dis(X, Y) = \max\{0, 1 - dis_{struct}(X, Y) - \theta * dis_{meas}(X, Y)\}$$

$$dis_{struct}(X, Y) = \begin{cases} 0 \text{ , if } X = Y \\ 1 \text{ , otherwise} \end{cases}$$

$$dis_{meas}(X, Y) = \begin{cases} |supp_D(X) - supp_E(Y)| \text{ , if } X = Y \\ 0 \text{ , otherwise} \end{cases}$$

For the FOCUS approach, Eq. 9 is initialized as:

$$dis(X, Y) = (1 - dis_{struct}(X, Y)) * dis_{meas}(X, Y)$$

$$dis_{struct}(X, Y) = \begin{cases} 0 \text{ , if } X = Y \\ 1 \text{ , otherwise} \end{cases}$$

$$dis_{meas}(X, Y) = \begin{cases} |supp_D(X) - supp_E(Y)| \text{ , if } X, Y \in A \cap B \text{ and } X = Y \\ supp_D(X) \text{ , if } X \in A - B \\ supp_E(Y) \text{ , if } Y \in B - A \end{cases}$$

Finally, for the Li-Ogihara-Zhu approach, Eq. 9 becomes:

$$dis(X, Y) = dis_{struct}(X, Y) * \log(1 + dis_{struct}(X, Y)) * dis_{meas}(X, Y)$$

$$dis_{struct}(X, Y) = \frac{|X \cap Y|}{|X \cup Y|}$$

$$dis_{meas}(X, Y) = \min\{supp_D(X), supp_E(Y)\}$$

4   Effect of Mining Parameters on Dissimilarity

In the following subsections, we investigate how the dissimilarity between two sets of frequent itemsets depends on the *minSupport* threshold σ used for their generation and on the adopted lattice representation (FI, CFI or MFI).

4.1   Effect of *minSupport* Threshold on Dissimilarity

Let σ, $\sigma + \delta$ ($0 < \sigma, \delta < \sigma + \delta \leq 1$) be two *minSupport* thresholds applied over a dataset D and let F_σ, $F_{\sigma+\delta}$ be the corresponding sets of frequent itemsets produced by (any) FIM algorithm. The difference set $F_\sigma - F_{\sigma+\delta}$ contains all those itemsets whose support lies between σ and $\sigma + \delta$:

$$Z \equiv F_\sigma - F_{\sigma+\delta} = \{X \subseteq I \mid \sigma < supp(X) \leq \sigma + \delta\} \tag{10}$$

In Fig. 2, an example is depicted which illustrates how the resulting lattice is affected by the increase δ in the *minSupport* threshold. As it is shown in this figure, with the increase of δ, the lattice is reduced. Below, we describe how each of the presented measures is affected by the increase δ in the *minSupport* value.

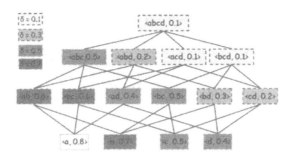

Fig. 2. Effect of δ increase on the lattice structure ($\sigma = 0.1$)

Parthasarathy-Ogihara [5] approach : From Eq. 5, it holds that:

$$dis(F_\sigma, F_{\sigma+\delta}) = 1 - \frac{\sum_{X \in F_\sigma \cap F_{\sigma+\delta}} \max\{0, 1 - \theta * |supp_D(X) - supp_D(X)|\}}{|F_\sigma \cup F_{\sigma+\delta}|}$$

$$= 1 - \frac{\sum_{X \in F_{\sigma+\delta}} \max\{0, 1 - 0\}}{|F_\sigma|}$$

$$\Rightarrow dis(F_\sigma, F_{\sigma+\delta}) = 1 - \frac{|F_{\sigma+\delta}|}{|F_\sigma|} \tag{11}$$

From the above equation, we can conclude that the greater the increase in the *minSupport* threshold value δ is, the smaller the enumerator $|F_{\sigma+\delta}|$ will be (cf. Eq. 10) and thus the greater the distance between the two sets will be.

FOCUS [3] approach: Recalling Eq. 6 and Eq. 10, it holds that[1]:

$$dis(F_\sigma, F_{\sigma+\delta}) = \frac{\sum_{X \in F_\sigma \cup F_{\sigma+\delta}} |supp_D(X) - supp_D(X)|}{\sum_{X \in F_\sigma} supp_D(X) + \sum_{X \in F_{\sigma+\delta}} supp_D(X)}$$

(12)

$$= \frac{\sum_{X:\sigma < supp_D(X) \leq \sigma+\delta} supp_D(X)}{2 * \sum_{X \in F_\sigma} supp_D(X) - \sum_{X:\sigma < supp_D(X) \leq \sigma+\delta} supp_D(X)}$$

For simplicity, let $C = \sum_{X:\sigma < supp_D(X) \leq \sigma+\delta} supp_D(X)$.

$$\Rightarrow dis(F_\sigma, F_{\sigma+\delta}) = \frac{C}{2 * \sum_{X \in F_\sigma} supp_D(X) - C}$$

(13)

In the above equation, if the value of δ increases, the numerator C will also increase whereas the denumerator will decrease (cf. Eq. 10 as well). Thus, as δ increases, the dissimilarity also increases.

Li-Ogihara-Zhou [4]approach: From Eq. 7 and Eq. 10, it holds that:

$$I_1 + I_2 = \sum_{X,Y \in F_\sigma} d(X,Y) + \sum_{X,Y \in F_{\sigma+\delta}} d(X,Y)$$

$$= 2 * \sum_{X,Y \in F_\sigma} d(X,Y) - \sum_{\substack{X:\sigma < supp(X) \leq \sigma+\delta \\ Y:\sigma < supp(Y) \leq \sigma+\delta}} d(X,Y)$$

$$I_3 = \sum_{\substack{X \in F_\sigma \\ Y \in F_{\sigma+\delta}}} d(X,Y) = \sum_{X,Y \in F_\sigma} d(X,Y) - \sum_{\substack{X:\sigma < supp(X) \leq \sigma+\delta \\ Y:\sigma < supp(Y) \leq \sigma+\delta}} d(X,Y)^{'}$$

For simplicity, let $G = \sum_{\substack{X:\sigma < supp(X) \leq \sigma+\delta \\ Y:\sigma < supp(X) \leq \sigma+\delta}} d(X,Y)$.

$$\Rightarrow dis(F_\sigma, F_{\sigma+\delta}) = 1 - \frac{2I_3}{I_1 + I_2} = 1 - \frac{2(I_1 - G)}{2I_1 - G} = \frac{G}{2I_1 - G}$$

(14)

As δ increases, the enumerator G also increases, whereas the denumerator $(2I_1 - G)$ decreases. Thus, dissimilarity increases as δ increases.

To summarize, Eq. 11, 13 and 14 state that, for all approaches, the larger the increase in the $minSupport$ threshold value δ is, the larger the computed dissimilarity score, $dis(F_\sigma, F_{\sigma+\delta})$, will be.

---

[1] Some further explanations on the notation: the term $\sum_{X \in F_\sigma \cup F_{\sigma+\delta}} |supp_D(X) - supp_D(X)|$ corresponds to the sum of the supports of all those itemsets that appear in $(F_\sigma - F_{\sigma+\delta})$. As far, as this set is not empty, this term is > 0.

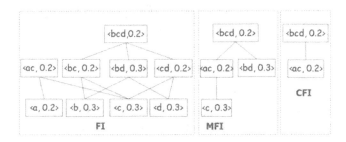

Fig. 3. Effect of representation (FI, CFI, MFI) on the lattice structure ($\sigma = 0.1$)

4.2 Effect of Lattice Representation on Dissimilarity

Let $F_\sigma(D)$, $C_\sigma(D)$, $M_\sigma(D)$ be the sets of FIs, CFIs and MFIs, respectively, extracted from D under (fixed) *minSupport* threshold σ. The example in Fig. 3 illustrates the effect of the different representations (FI, CFI, MFI) in the lattice. These figures affirm Eq. 4 which states that the greater the compactness level is, the "smaller" the resulting lattice will be.

Parthasarathy-Ogihara [5] approach: From Eq. 5, it holds that:

$$
\begin{aligned}
dis(F_\sigma, C_\sigma) = 1 - \frac{\sum_{X \in F_\sigma \cap C_\sigma} \max\{0, 1 - \theta * |supp_D(X) - supp_D(X)|\}}{|F_\sigma \cup C_\sigma|} \\
= 1 - \frac{\sum_{X \in C_\sigma} \max\{0, 1 - 0\}}{|F_\sigma|} = 1 - \frac{|C_\sigma|}{|F_\sigma|} \quad (15)
\end{aligned}
$$

$$
\begin{aligned}
dis(F_\sigma, M_\sigma) = 1 - \frac{\sum_{X \in F_\sigma \cap M_\sigma} \max\{0, 1 - \theta * |supp_D(X) - supp_D(X)|\}}{|F_\sigma \cup M_\sigma|} \\
= 1 - \frac{\sum_{X \in M_\sigma} \max\{0, 1 - 0\}}{|F_\sigma|} = 1 - \frac{|M_\sigma|}{|F_\sigma|} \quad (16)
\end{aligned}
$$

From Eq. 15, 16, it holds that:

$$
dis(F_\sigma, C_\sigma) \le dis(F_\sigma, M_\sigma) \quad (17)
$$

FOCUS [3] approach: Recalling Eq. 6, it holds that:

$$
dis(F_\sigma, C_\sigma) = \frac{\sum_{X \in F_\sigma \cup C_\sigma} |supp_D(X) - supp_D(X)|}{\sum_{X \in F_\sigma} supp_D(X) + \sum_{X \in C_\sigma} supp_D(X)} \quad (18)
$$

$$
(19)
$$

$$
= \frac{\sum_{X \in F_\sigma - C_\sigma} supp_D(X)}{2 * \sum_{X \in F_\sigma} supp_D(X) - \sum_{X \in F_\sigma - C_\sigma} supp_D(X)}
$$

$$dis(F_\sigma, M_\sigma) = \frac{\sum_{X \in F_\sigma \cup M_\sigma} |supp_D(X) - supp_D(X)|}{\sum_{X \in F_\sigma} supp_D(X) + \sum_{X \in M_\sigma} supp_D(X)} \tag{20}$$

$$\tag{21}$$

$$= \frac{\sum_{X \in F_\sigma - M_\sigma} supp_D(X)}{2 * \sum_{X \in F_\sigma} supp_D(X) - \sum_{X \in F_\sigma - M_\sigma} supp_D(X)}$$

where $F_\sigma - C_\sigma$ is given by Eq. 1 and $F_\sigma - M_\sigma$ is given by Eq. 2. From Eq. 18 and Eq. 20, for the FOCUS measure it holds that:

$$dis(F_\sigma, C_\sigma) \le dis(F_\sigma, M_\sigma) \tag{22}$$

Li-Ogihara-Zhou [4]approach: From Eq. 7, it holds that:

$$I_1 + I_2 = \sum_{X,Y \in F_\sigma} d(X,Y) + \sum_{X,Y \in C_\sigma} d(X,Y)$$

$$= 2 * \sum_{X,Y \in F_\sigma} d(X,Y) - \sum_{X,Y \in F_\sigma - C_\sigma} d(X,Y) = 2 * I_1 - \sum_{X,Y \in F_\sigma - C_\sigma} d(X,Y)$$

$$I_3 = \sum_{\substack{X \in F_\sigma \\ Y \in C_\sigma}} d(X,Y) = \sum_{X,Y \in F_\sigma} d(X,Y) - \sum_{X,Y \in F_\sigma - C_\sigma} d(X,Y)$$

$$= I_1 - \sum_{X,Y \in F_\sigma - C_\sigma} d(X,Y)$$

Let $K = \sum_{X,Y \in F_\sigma - C_\sigma} d(X,Y)$. Then:

$$dis(F_\sigma, C_\sigma) = 1 - \frac{2(I_1 - K)}{2I_1 - K} = \frac{K}{2I_1 - K} \tag{23}$$

also it holds that:

$$I_1 + I_2 = \sum_{X,Y \in F_\sigma} d(X,Y) + \sum_{X,Y \in M_\sigma} d(X,Y)$$

$$= 2 * \sum_{X,Y \in F_\sigma} d(X,Y) - \sum_{X,Y \in F_\sigma - M_\sigma} d(X,Y) = 2 * I_1 - \sum_{X,Y \in F_\sigma - M_\sigma} d(X,Y)$$

$$I_3 = \sum_{\substack{X \in F_\sigma \\ Y \in M_\sigma}} d(X,Y) = \sum_{X,Y \in F_\sigma} d(X,Y) - \sum_{X,Y \in F_\sigma - M_\sigma} d(X,Y)$$

$$= I_1 - \sum_{X,Y \in F_\sigma - M_\sigma} d(X,Y)$$

Let $L = \sum_{X,Y \in F_\sigma - M_\sigma} d(X,Y)$. Then:

$$dis(F_\sigma, M_\sigma) = 1 - \frac{2(I_1 - L)}{2I_1 - L} = \frac{L}{2I_1 - L} \tag{24}$$

From Eq. 23 and Eq. 24, for the Li-Ogihara-Zhou measure it holds that:

$$dis(F_\sigma, C_\sigma) \leq dis(F_\sigma, M_\sigma) \qquad (25)$$

Equations 17, 22, 25 state that the more compact the adopted lattice representation (MFIs vs CFIs vs FIs) is, the larger the computed distance becomes.

5   Experimental Evaluation

To evaluate the theoretical results, we experimented with the different dissimilarity measures on datasets from the FIM repository [2]. In particular, we used a real, dense dataset (*Chess*), which consists of 3196 transactions of 76 distinct items and has average transaction length 37. Also, we used a synthetic, sparse dataset (*T10I4D100K*), which consists of 100,000 transactions of 1,000 distinct items and has average transaction length 10. For the extraction of FIs, CFIs and MFIs we used MAFIA [1]. In the case of FOCUS, we used the upper bound of the dissimilarity measure without re-querying the original raw data space. This decision is justified by the fact that we are interested on how patterns capture similarity features contained in the original raw data. For the case of the Parthasarathy-Ogihara measure, we used $\theta = 1$, considering that both structure and measure components contribute equally to the final dissimilarity score.

5.1   Comparing Dissimilarity in Data and Frequent Itemsets Spaces

In this section, we evaluate the argument that dissimilarity in pattern space can be adopted to discuss dissimilarity in data space. In particular, we select a popular pattern representation (FIs) and a specific *minSupport* threshold σ for their generation, while we modify the dataset by adding different proportions of noise. Then, we compare the dissimilarity measured in the FI space with respect to the dissimilarity enforced (by adding noise) in the original raw data space.

Starting with an initial dataset D and for a specific *minSupport* threshold σ, we extracted $F_\sigma(D)$. Then, in every step, we modified an increased number $0\%, 5\%, \ldots, 50\%$ of the transactions of D. The selection of the transactions to be affected was performed in a random way, and for each selected transaction we modified a certain percentage (in particular, 50%) of its items. The modification we made was that the selected item values were reset to 0 (in a preprocessing step both datasets of the experiments where transformed into binary format). As such, the derived pattern sets $F_\sigma(D_{p\%})$ were subsets of the initial set $F_\sigma(D_{0\%})$. Then, we compared the noised pattern sets $F_\sigma(D_{p\%})$ with the initial "un-noised" pattern set $F_\sigma(D_{0\%})$. Regarding the generation of the pattern sets, we used $\sigma = 80\%$ for *Chess* and $\sigma = 0.5\%$ for *T10I4D100K*. The results are illustrated in Fig. 4, where it seems that as the dataset becomes noisier, the distance between the initial (clean) pattern set and the new (noisy) pattern set becomes larger, for all approaches and all datasets.

A comparative study of the two figures shows that the effect of noise is more destructive for the dense dataset (*Chess*), where at is shown in Fig. 4 (right), the

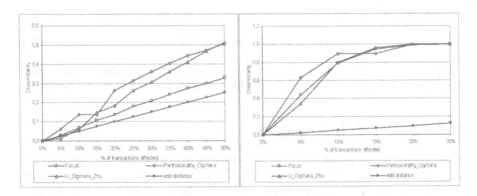

Fig. 4. Impact of dataset noise on FI dissimilarity: $T10I4D100K(\sigma = 0.5\%)$ on the left, $Chess(\sigma = 80\%)$ on the right

dissimilarity increases quickly up to the upper bound 1. This can be explained by the fact that small changes in a dense dataset may cause critical changes in the produced FI lattice. This is not the case for the sparse dataset ($T10I4D100K$) which appears to be more robust in the dataset noise (Fig. 4, left).

5.2   Effect of *minSupport* Threshold

In this section, we evaluate the effect of the *minSupport* threshold on the computed dissimilarity scores. The scenario is as follows: For each dataset D, we fixed initial *minSupport* threshold σ and varied the increase δ in *minSupport* in the range $\delta_0, \delta_1, \ldots, \delta_n(\sigma + \delta_i \leq 1)$. Then, we generate the corresponding FIs for the different *minSupport* values, namely $\sigma + \delta_0, \sigma + \delta_1, \ldots, \sigma + \delta_n$. After that, we compare $FI_{\sigma+\delta_i}$ with the initial $FI_{\sigma+\delta_0}$. We choose different σ, δ parameters for the two datasets based on their cardinality analysis presented in [6]. Since our analysis does not depend on specific support values we choose parameters that yield a reasonable amount of patterns. Thus, in the case of the $D = T10I4D100K$ dataset, we choose for the initial support the value $\sigma = 0, 5\%$ and for *minSupport* increase δ values: $0\%, 0.5\%, \ldots, 4.5\%$, whereas in the case of the $D = Chess$ dataset we choose as initial support value $\sigma = 90\%$ and for *minSupport* increase δ values: $0\%, 1\%, \ldots, 9\%$.

The results are illustrated in Fig. 5. Both charts show that the larger the increase in the minSupport threshold values δ is, the larger the dissimilarity between the corresponding pattern sets is, for all approaches. More specifically, the Parthasarathy-Ogihara approach provides the greatest dissimilarity scores because it only considers for comparison items with identical structure (those belonging to $A \cap B$). On the other hand, FOCUS considers items appearing in the union of the two sets (i.e., $A \cap B$, $A - B$, $B - A$), thus its dissimilarity scores are lower comparing to those computed by the Parthasarathy-Ogihara. As for the Li-Ogihara-Zhu approach, the rationale behind it is slightly different: It considers partial similarity and performs a many-to-many matching between

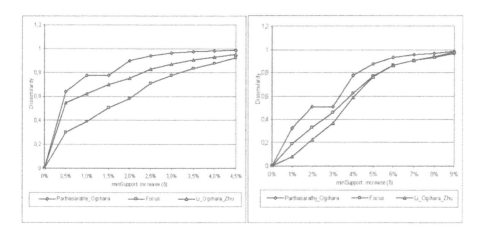

Fig. 5. Impact of *minSupport* increase δ on FI dissimilarity: $T10I4D100K(\sigma = 0.5\%)$ on the left, $Chess(\sigma = 90\%)$ on the right

the itemsets of the two sets. This is in contrast to the other two approaches that perform a one-to-one matching between itemsets that belong to the intersection (Parthasarathy-Ogihara) or to the union (FOCUS) of the two sets.

To summarize, experiments in this subsection have confirmed our theoretical analysis regarding the dependency of the dissimilarity between pattern sets on the *minSupport* threshold that was used for their generation. Indeed, as the *minSupport* becomes more selective, the dissimilarity increases. Generalizing this result, we can state that the more selective the *minSupport* threshold is, the less informative the set of FIs becomes with respect to the raw data space.

5.3   Effect of Lattice Representation

We set the value of the *minSupport* threshold parameter σ to a fixed value and calculate the dissimilarity scores between the different lattice representations under the same noise level, namely $dis(F_\sigma(D_{p\%}), C_\sigma(D_{p\%}))$, $dis(F_\sigma(D_{p\%}), M_\sigma(D_{p\%}))$.

The results for FI - CFI, FI - MFI dissimilarity cases are illustrated in Fig. 6. These charts point out the dependence of the dissimilarity scores on the adopted frequent itemsets lattice compactness level. More specifically, it is clearly shown that CFIs can very well capture the behavior of FIs whereas MFIs cannot; this is true for both datasets.

However, the degree of difference between FI-CFI and FI-MFI dissimilarities scores seems to be subject to the dataset characteristics (sparse vs. dense). More specifically, in case of the sparse dataset $(T10I4D100K)$, CFIs manage to fully capture the behavior of FIs at every noise level while MFIs approximate FIs as the dataset becomes more noisy. On the other hand, in the case of the dense dataset $(Chess)$ we observe that CFIs are closer to FIs than MFIs. In this case

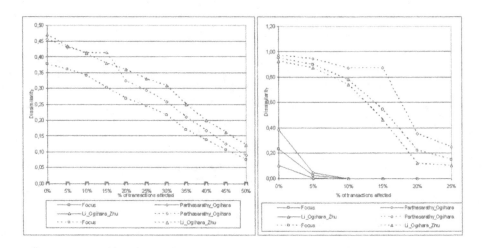

Fig. 6. Impact of noise on dissimilarity for FI-CFI (dotted lines), FI-MFI (solid lines): $T10I4D100K(\sigma = 0.5\%)$ on the left, $Chess(\sigma = 80\%)$ on the right

however the noise effect is more destructive by means that it causes a slower decrease in the differences of CFIs and MFIs from FIs.

To summarize, experiments in this subsection show that the adopted representation for the itemsets lattice affects the derived dissimilarity scores, confirming our theoretical analysis regarding the dependency of dissimilarity on the lattice representations. Also, it seems that CFIs are very good representatives for FIs, whereas this is not the case for MFIs. It turns out that the more compact the representation of the pattern space is, the less informative this space becomes with respect to the initial data space from which patterns were extracted.

6   Conclusions and Future Work

In this work, we investigated whether dissimilarity between sets of frequent itemsets could serve as a measure of dissimilarity between the original datasets. We presented the parameters that affect the problem, namely the *minSupport* threshold used for itemsets generation and the compactness level achieved by the lattice representation (FI, CFI or MFI). Both theoretical and experimental results confirmed that the more "restrictive" the mining parameters are, the larger the dissimilarity between the two sets is. Thus, utilizing pattern comparison for raw data comparison is not as straightforward as related work has argued but it depends on the mining parameters.

As part of our future work, we plan to experiment with more datasets of different characteristics (e.g.,synthetic vs real, dense vs sparse). We also plan to investigate some dissimilarity measure that will better preserve the original raw data space characteristics in the pattern space.

References

1. Burdick, D., Calimlim, M., Gehrke, J.: Mafia: A maximal frequent itemset algorithm for transactional databases. In: International Conference on Data Engineering (ICDE), pp. 443–452. IEEE Computer Society, Los Alamitos (2001)
2. FIMI. Frequent itemsets mining data set repository (valid as of May 2008), http://fimi.cs.helsinki.fi/data/
3. Ganti, V., Gehrke, J., Ramakrishnan, R.: A framework for measuring changes in data characteristics. In: ACM SIGACT-SIGMOD-SIGART Symposium on Principles of Database Systems (PODS), pp. 126–137. ACM Press, New York (1999)
4. Li, T., Ogihara, M., Zhu, S.: Association-based similarity testing and its applications. Intelligent Data Analysis 7, 209–232 (2003)
5. Parthasarathy, S., Ogihara, M.: Clustering distributed homogeneous datasets. In: Zighed, D.A., Komorowski, J., Żytkow, J.M. (eds.) PKDD 2000. LNCS (LNAI), vol. 1910, pp. 566–574. Springer, Heidelberg (2000)
6. Xin, D., Han, J., Yan, X., Cheng, H.: Mining compressed frequent-pattern sets. In: International Conference on Very Large Data Bases (VLDB), pp. 709–720. VLDB Endowment (2005)
7. Zaki, M., Hsiao, C.-J.: Efficient algorithms for mining closed itemsets and their lattice structure. IEEE Transactions on Knowledge and Data Engineering (TKDE) 17(4), 462–478 (2005)

A Theory of Action, Knowledge and Time in the Event Calculus

Theodore Patkos and Dimitris Plexousakis

Institute of Computer Science, FO.R.T.H.
Vassilika Vouton, P.O. Box 1385
GR 71110 Heraklion, Greece
{patkos,dp}@ics.forth.gr

Abstract. To regulate and coordinate the behavior of agents within real-world environments, the participating entities must be able to reason not only about the state of the environment itself, but also about their own knowledge concerning that state, based on information acquired through sensing actions. Considering the dynamic nature of most realistic domains, the study of knowledge evolution over time is a critical aspect. This paper develops a formal account of action, knowledge and time within the context of the Event Calculus and proposes a unified theory that is applicable to diverse scenarios of commonsense reasoning.

Keywords: Event Calculus, Action Theories, Reasoning about Knowledge.

1 Introduction

The formal account of knowledge has been a fascinating active field of research in philosophy, computer science and other disciplines, as well. Still, while formalisms for reasoning about knowledge tell agents how to adjust their beliefs given observations about the changes of an environment, they do not address issues, such as reasoning about the properties of the actual actions that cause the changes, which introduce high-level agent programming and planning requirements. In applications, where agents are situated in complex, uncertain and highly dynamic environments, action theories allow for reasoning about the state and changes of the world, causality and potential effects of actions, qualifications etc. Incorporating knowledge in action theories is essential for modeling real-world applications, where rational agents need to acquire information and decide upon their actions at execution time.

Previous work in the literature has mainly focused on reasoning about knowledge when sensing actions occur and does not cover the temporal aspect of knowledge. Still, to perform effectively in real-world systems, an explicit representation of time and the capability to deal with information about the actual course of events is important [1]. In order to ensure that the behavior of rational agents is properly regulated and also to prove the ability to achieve their objectives, a formal account of knowledge, action and time is required.

J. Darzentas et al. (Eds.): SETN 2008, LNAI 5138, pp. 226–238, 2008.

In this paper we extend the Event Calculus formalism to allow for reasoning not only about the state of affairs in a dynamic domain, but also about the agent's own knowledge concerning that state. The contribution is twofold; first, we provide a unified theory of action, knowledge and time that is able to utilize the solutions for the well-known problems of reasoning about action, such as the frame and ramification problems, non-determinism etc. both for ordinary fluents and for knowledge. Second, although the theory is based on the possible worlds semantics, the treatment of knowledge change given action occurrences does not manipulate this model, but instead handles knowledge as any ordinary fluent, more similar in style to [2], thus resulting in significant computational benefits. Although we restrict the timepoint sort to the integers, the axioms employed here can be trivially extended to a continuous-time axiomatization.

The paper is organized as follows. We start by presenting previous work in the field and discuss how the present study advances the domain. Then, we briefly describe the standard axioms of the version of the Event Calculus we employ and the semantics for our knowledge theory. In the following two sections we describe the foundational axioms of the proposed theory and prove its correctness. Section 7 provides a use case scenario. The paper concludes with remarks on future work.

2 Previous Work

Most previous approaches studied the problems of reasoning about action and knowledge in isolation. A formal account of knowledge in action theories was originally developed within the Situation Calculus. Scherl and Levesque [3,4] investigated a theory for handling knowledge and sensing actions, providing a solution to the frame problem for this type of actions. They introduced the epistemic fluent K and defined its semantics, proving a number of important properties that the specification enjoys. The work was further extended in [5] to accommodate the notion of *ability* to achieve an objective. The fact that this approach did not distinguish between actual effects of actions and what the agent knows about these effects was addressed whitin the Fluent Calculus [6], based on a definition of knowledge over potential states instead of situations. Both frameworks, though, did not explore actions that may cause loss of knowledge. In [7] the action description language $\mathcal{A}_k$ was introduced, which incorporated not only sense actions, but also actions with non-deterministic effects, causing knowledge to be removed from the set of facts known by the agent.

Yet, none of the aforementioned frameworks studied the interaction of sensing and time, which is essential in most real-world applications. The results of sensing occur at a particular point in time, may have a prescribed duration of validity and may even be different depending on other, potentially concurrent, sensing or non-sensing actions. This was acknowledged in the work of Zimmerbaum and Scherl [8], who proposed a unified logical theory of knowledge, sensing, time and concurrency within Situation Calculus. Still, the applicability of the approach is somewhat limited, since the theory adopted an alternative model for the treatment of time, due to the difficulty in combining the different models in this calculus.

The Event Calculus employed in this study, on the other hand, provides a unified framework that handles many aspects of commonsense reasoning, especially those related to time. It uses a linear time representation, where events are considered to be actual, an approach more suitable for real-world implementations as opposed to the branching time representation of the Situation and Fluent Calculi [1]. More importantly, our work also differs to the ones mentioned above in the treatment of knowledge. All previous approaches share a common definition of semantics based on the possible worlds model and perform reasoning by adapting the accessibility relation over possible worlds, in order to determine if a formula is true in each accessible world (epistemic states introduced in $\mathcal{A}_k$ are similar in style). Inevitably, the efficiency of theorem proving of knowledge formulae is disappointing; with n atomic formulae, there are potentially 2^n distinguishable worlds to check truth in. Our approach, though, introduces a set of *knowledge* fluents for each ordinary fluent that are affected by actions the usual way. Therefore, reasoning about knowledge and action is computationally no worse than reasoning about ordinary action occurrences.

Recently, Forth and Shanahan [9] utilized knowledge fluents in the Event Calculus to specify when an agent possesses enough knowledge to execute an action in a partially observable environment. Their approach is significantly different to the one proposed in this study. They have focused on closed and controlled environments, where no exogenous actions with unknown to the agent preconditions could occur. Nor actions with multiple effects, dependent on different sets of preconditions could be supported. The intention was to study ramifications in order to determine which fluents to sense and when, therefore did not provide a general theory of action, knowledge and time within the Event Calculus.

3   Event Calculus Axiomatization

The Event Calculus is a widely adopted formalism for reasoning about action and change that addresses naturally many phenomena of commonsense reasoning. It is a first-order predicate calculus, which uses *events* to indicate changes in the environment and *fluents* to denote any time-varying property. The calculus includes predicates for expressing which fluents hold when (*HoldsAt*), what events happen (*Happens*), which are their effects (*Initiates*, *Terminates*, *Releases*) and whether a fluent is subject to inertia or released from it (*ReleasedAt*). Recall that the commonsense law of inertia captures the property that things tend to stay the same unless affected by some event; when released from inertia, a fluent may have a fluctuant truth value. Furthermore, the Event Calculus uses circumscription (see [10]) to solve the frame problem and to support default reasoning, i.e., reasoning in which a conclusion is reached based on limited information and later retracted when new information is available.

A number of different dialects have been proposed that are summarized in [11] and [12]. Shown below are the fundamental axioms of the version of the Classical Discrete Event Calculus axiomatization we employ, which is thoroughly described in [13]. Variables of the sort event are represented by e, fluent variables by f and variables of the timepoint sort by t, with subscripts where necessary.

Influence of Events on Fluents

$$Happens(e,t) \land Initiates(e,f,t) \Rightarrow HoldsAt(f,t+1) \qquad \text{(DEC1)}$$

$$Happens(e,t) \land Terminates(e,f,t) \Rightarrow \neg HoldsAt(f,t+1) \qquad \text{(DEC2)}$$

$$Happens(e,t) \land Releases(e,f,t) \Rightarrow ReleasedAt(f,t+1) \qquad \text{(DEC3)}$$

$$Happens(e,t) \land (Initiates(e,f,t) \lor Terminates(e,f,t)) \Rightarrow \\ \neg ReleasedAt(f,t+1) \qquad \text{(DEC4)}$$

Inertia of HoldsAt

$$HoldsAt(f,t) \land \neg ReleasedAt(f,t+1) \land \\ \neg \exists e (Happens(e,t) \land Terminates(e,f,t)) \Rightarrow HoldsAt(f,t+1) \qquad \text{(DEC5)}$$

$$\neg HoldsAt(f,t) \land \neg ReleasedAt(f,t+1) \land \\ \neg \exists e (Happens(e,t) \land Initiates(e,f,t)) \Rightarrow \neg HoldsAt(f,t+1) \qquad \text{(DEC6)}$$

Inertia of ReleasedAt

$$ReleasedAt(f,t) \land \neg \exists e (Happens(e,t) \land \\ (Initiates(e,f,t) \lor Terminates(e,f,t))) \Rightarrow ReleasedAt(f,t+1) \qquad \text{(DEC7)}$$

$$\neg ReleasedAt(f,t) \land \neg \exists e (Happens(e,t) \land Releases(e,f,t)) \Rightarrow \\ \neg ReleasedAt(f,t+1) \qquad \text{(DEC8)}$$

4 Semantics of the Epistemic Fluent

Our intention is to express statements that reflect the knowledge agents have about a fluent f at a timepoint t, denoted by $HoldsAt(Knows(f),t)$. We represent knowledge utilizing an epistemic fluent, adapting the possible-worlds semantics and Kripke structures, as done within the Situation Calculus ([4]). Nevertheless, instead of reasoning about knowledge over all accessible worlds, we treat this fluent as any ordinary non-epistemic fluent. The definition of knowledge establishes the correspondence between the semantics based on possible worlds and the treatment of knowledge fluents we present in the next section.

Partial knowledge in the Event Calculus results in different possible models that express the states (set of fluents) that hold at a time instant. We introduce the relation KS to map states to timepoints and write $HoldsAt(KS(z),t)$ to denote that state z is a possible world state at t. We also introduce a binary accessibility relation K over states and write $HoldsAt(K(z',z),t)$, read as "state z' is accessible from state z at time t". We choose to only focus on the temporal aspect of knowledge, avoiding any reference to individual states. This requires that all possible states at a given timepoint be indistinguishable to the agent and accessible from one another, causing K to become an equivalence relation.

We formally define knowledge in the Event Calculus by capturing the notion that a fluent is known at some time instant if it is true in all states that are possible at that instant. This is expressed as an abbreviation shown below, where the notation $f(z')$ specifies that fluent f holds in the given state:

$$HoldsAt(Knows(f), t) \equiv$$
$$\forall z HoldsAt(KS(z), t) \Rightarrow \qquad \text{(KDEF)}$$
$$(\forall z' HoldsAt(K(z', z), t) \Rightarrow HoldsAt(f(z'), t))$$

As is well known, there exists a strong relationship between the properties of the accessibility relation and the characteristic axioms and rules of inference of knowledge. Since our theory restricts K in being reflexive, transitive and symmetric, we can employ an axiomatic system similar to S5 epistemic modal logic system. The knowledge and distribution axioms can be expressed as:

$$HoldsAt(Knows(f_1 \Rightarrow f_2), t) \Rightarrow$$
$$(HoldsAt(Knows(f_1), t) \Rightarrow HoldsAt(Knows(f_2), t)) \qquad \text{(K)}$$

$$HoldsAt(Knows(f), t) \Rightarrow HoldsAt(f, t) \qquad \text{(T)}$$

Finally, for notational convenience, we also define knowledge of the truth value of a fluent ("knows whether"), as:

$$HoldsAt(Kw(f), t) \equiv HoldsAt(Knows(f) \vee Knows(\neg f), t) \qquad \text{(KW)}$$

5 Event Calculus Knowledge Theory

The proposed theory assumes domains involving incomplete knowledge about the initial state, knowledge-producing actions, actions that cause loss of knowledge and actions with context-dependent effects. To simplify the presentation, we assume perfect sensors for the agent, i.e., the result of sensing is always correct.

Typically, the knowledge state of an agent is not only altered through sensing, but also through the occurrence of ordinary events, whose effects are known to the agent. Since these actions affect both the domain and the agent's knowledge base, we can use a knowledge theory to derive changes in what the agent knows when events occur, instead of capturing the effects on separate axioms for the domain and for the agent itself. A general theory of knowledge must consider events that may cause both increase in an agent's knowledge and loss of knowledge. Therefore, in our theory we distinguish two types of events, events with deterministic effects and events with non-deterministic effects, i.e., when two or more alternative effects are possible. In the latter case, loss of knowledge means that the agent does not know either that a fluent is true or false. Deterministic events, whose preconditions are unknown at a particular time instant, are handled as non-deterministic for that time instant. Agents can also perform sensing actions, whose effects, by definition, are limited to changing the agent's knowledge base, with no effect on the domain.

In our theory, the *Knows* fluent is released from inertia at all times, but is always subject to some state constraint while it is released. By releasing the *Knows* fluent from inertia, its truth value varies only with the state constraint that is triggered to specify it. This technique enables us to apply rules, such as the knowledge axiom or the positive introspection axiom, in an elaboration tolerant manner, i.e., without having to write additional rules for each new axiom that we add to the theory.

Nevertheless, there are situations when an ordinary fluent is subject to inertia and, therefore, knowledge about it should persist until some event changes it. In order to "simulate" inertia of knowledge, we introduce an auxiliary fluent, KP (for "knows persistently") that, as long as it holds, forces knowledge to remain true (the KP fluent is always subject to inertia). Finally, knowledge may also hold even when KP does not hold; it can be established indirectly through domain state constraints that result in knowledge ramifications.

5.1   Notational Conventions

We will study how the knowledge state of an agent concerning fluents changes as a consequence of the occurrence of relevant events. We assume that an event e may have positive effect, negative effect and release axioms that are given by the following inference rules, respectively:

$$\bigwedge_{i=1}^{P}[(\neg)HoldsAt(f_i,t)] \Rightarrow Initiates(e,f_{pos},t)$$
$$\bigwedge_{j=1}^{N}[(\neg)HoldsAt(f_j,t)] \Rightarrow Terminates(e,f_{neg},t)$$
$$\bigwedge_{k=1}^{R}[(\neg)HoldsAt(f_k,t)] \Rightarrow Releases(e,f_{rel},t)$$

The expression $\bigwedge_{i=1}^{P}[(\neg)HoldsAt(f_i,t)]$ stands for $(\neg)HoldsAt(f_1,t) \wedge ...$ $\wedge (\neg)HoldsAt(f_P,t)$. A negation symbol in parenthesis denotes that either the fluent or its negation may hold, without any particular significance. The body of each rule denotes the conjunction of preconditions for the given effect. For instance, if all $(\neg)f_1$ to $(\neg)f_P$ hold at t then f_{pos} will be true at $t+1$ after the occurrence of e at t. It should be noted that even when the preconditions are not satisfied the event may still occur, but it will not have the intended effect.

$P, N, R \in \mathbb{N}$. An axiom with no body (i.e., when no fluent conditionals are applicable) is denoted by taking P, N or R equal to zero. For any i, j, k it is possible that $f_i = f_j = f_k$. On the other hand, $f_{pos} \neq f_{neg} \neq f_{rel}$ for a given event e. Free variables in formulas are implicitly universally quantified.

5.2   Foundational Axioms

The theory consists of the following set of axioms[1]:

---

[1] In this study we only concentrate on knowledge about fluent literals. Knowledge about fluent formulas, such as disjunctions, are also supported by a set of additional axioms (KT6) not mentioned in the paper.

Knowledge and the commonsense law of inertia

$$ReleasedAt(Knows((\neg)f), t) \tag{KT1}$$

Knowledge is released from inertia at all times.

Knowledge Persistence

$$HoldsAt(KP(f), t) \Rightarrow HoldsAt(Knows(f), t) \tag{KT2.1}$$

$$HoldsAt(KP(\neg f), t) \Rightarrow HoldsAt(Knows(\neg f), t) \tag{KT2.2}$$

These axioms capture the correlation between the $Knows$ and KP fluents. The latter is always subject to inertia.

Events with known effects

$$\bigwedge_{i=1}^{P}[HoldsAt(Knows((\neg)f_i), t)] \wedge Happens(e, t) \Rightarrow \tag{KT3.1}$$
$$Initiates(e, KP(f_{pos}), t)$$

$$\bigwedge_{i=1}^{P}[HoldsAt(Knows((\neg)f_i), t)] \wedge Happens(e, t) \Rightarrow \tag{KT3.2}$$
$$Terminates(e, KP(\neg f_{pos}), t)$$

$$\bigwedge_{j=1}^{N}[HoldsAt(Knows((\neg)f_j), t)] \wedge Happens(e, t) \Rightarrow \tag{KT3.3}$$
$$Initiates(e, KP(\neg f_{neg}), t)$$

$$\bigwedge_{j=1}^{N}[HoldsAt(Knows((\neg)f_j), t)] \wedge Happens(e, t) \Rightarrow \tag{KT3.4}$$
$$Terminates(e, KP(f_{neg}), t)$$

Positive and negative effect axioms cause deterministic effects. If an agent knows all preconditions of an occurring deterministic action, then it also knows its effect. $KP(f)$ and $KP(\neg f)$ cancel one another to ensure that knowledge remains consistent. Note how these axioms affect knowledge through (KT2).

Knowledge-producing (sense) actions

$$Initiates(sense(f), (KP(f) \vee KP(\neg f)), t) \tag{KT4}$$

From this axiom, $Kw(f)$ is explicitly implied. Only one of the $KP(f)$ and $KP(\neg f)$ fluents will be initiated in each case -the proper one- depending on the result of the sense action (to see why, notice that (KT4) will trigger (KT2.1) or (KT2.2), but not both, due to the Knowledge axiom (T).

Events with non-deterministic effects

$$\bigvee_{i=1}^{P}[\neg HoldsAt(Kw(f_i), t)] \wedge \bigwedge_{i=1}^{P}[\neg HoldsAt(Knows(\neg(\neg)f_i), t)] \wedge \tag{KT5.1}$$
$$\neg HoldsAt(Knows(f_{pos}), t) \wedge Happens(e, t) \Rightarrow$$
$$Terminates(e, (KP(f_{pos}) \vee KP(\neg f_{pos})), t)$$

$$\bigvee_{j=1}^{N} [\neg HoldsAt(Kw(f_j), t)] \wedge \bigwedge_{j=1}^{N} [\neg HoldsAt(Knows(\neg(\neg)f_j), t)] \wedge$$

$$\neg HoldsAt(Knows(\neg f_{neg}), t) \wedge Happens(e, t) \Rightarrow \quad \text{(KT5.2)}$$

$$Terminates(e, (KP(f_{neg}) \vee KP(\neg f_{neg})), t)$$

If an action with *deterministic* effect occurs, which (a) has at least one precondition that the agent does not know if it is true (hence, the agent does not know if the effect axiom is triggered), (b) there is no precondition that the agent knows it does not hold (otherwise, the agent would have been certain that the effect axiom would not be triggered) and (c) the agent does not know that the potential new truth value of the effect fluent is the same as the existing truth value (otherwise the agent is sure that the value will remain the same regardless of whether the action occurs or not), then the agent looses its knowledge about the state of the effect. Note that in this case we must assume that $P, N \neq 0$; if there are no preconditions, then the effect is always deterministic when the event occurs, falling under the scope of (KT3) axioms.

$$\bigwedge_{k=1}^{R} [\neg HoldsAt(Knows(\neg(\neg)f_k), t)] \wedge Happens(e, t) \Rightarrow \quad \text{(KT5.3)}$$

$$Terminates(e, (KP(f_{rel}) \vee KP(\neg f_{rel})), t)$$

If an action that *releases* a fluent occurs, none of whose preconditions is known not to hold (therefore, either the action's preconditions are unambiguously satisfied or the agent does not know if they are satisfied; in either case the result is the same, knowledge about the effect is lost), then the agent cannot infer any knowledge concerning the state of the effect.

State constraints (indirect knowledge)
In contrast to non-deterministic actions that cause loss of knowledge, there may be state constraints that provide indirect knowledge about the state of the environment. For instance, the following formula is a state constraint

$HoldsAt(Holding(user, object), t) \wedge HoldsAt(InRoom(user, room), t) \Rightarrow$

$HoldsAt(InRoom(object, room), t)$,

denoting the indirect effect of an object being in a room if the user holding the object is in that room. One can think of state constraints as a means of increasing the agent's level of intelligence about the world it is situated in.

Let us assume that a domain axiomatization consists of X implication state constraints for a fluent f, Y bi-implication state constraints for f and Z implication state constraints for the knowledge of f, where $X, Y, Z \in \aleph$:
implication sc for f: $\bigwedge_{x=1}^{X}[\bigwedge_l[(\neg)HoldsAt(f_{x_l}, t)] \wedge \Delta_x \Rightarrow (\neg)HoldsAt(f, t)]$
bi-implication sc for f:

$\bigwedge_{y=1}^{Y}[\bigwedge_{m}[(\neg)HoldsAt(f_{y_m},t)] \wedge \Delta_y \Leftrightarrow (\neg)HoldsAt(f,t)]$

implication sc for the knowledge of f:

$\bigwedge_{z=1}^{Z}[\bigwedge_{n}[(\neg)HoldsAt(f_{z_n},t)] \wedge \Delta_z \Rightarrow HoldsAt(Knows((\neg)f),t)],$

where Δ_x, Δ_y, Δ_z are conjunctions of temporal ordering formulas, i.e., conjunctions of timepoint comparisons, event occurrence formulas, i.e., $Happens(e,t)$ or formulas containing predicates $Initiates(e,f,t)$ or $Terminates(e,f,t)$.

$$\bigwedge_{x=1}^{X}[(\bigvee_{l}[HoldsAt(Knows(\neg(\neg)f_{x_l}),t)] \vee \bigvee_{l}[\neg HoldsAt(Kw(f_{x_l}),t)]) \wedge \Delta_x] \wedge$$

$$\bigwedge_{y=1}^{Y}[\bigvee_{m}[\neg HoldsAt(Kw(f_{y_m}),t)] \wedge \Delta_y] \wedge \bigwedge_{z=1}^{Z}[\bigvee_{n}[\neg(\neg)HoldsAt(f_{z_n},t)] \wedge \Delta_z] \Rightarrow$$

$$\neg HoldsAt(Kw(f),t)$$
$$\text{(KT7)}$$

In brief, this axiom states that when knowledge about a fluent f cannot be inferred either directly or indirectly by the agent's KB, then this fluent is unknown to the agent. More specifically, if for each implication state constraint about f either there is at least one precondition which the agent knows that it does not hold or does not know whether it holds and also for each bi-implication state constraint for f the agent does not know whether at least one precondition holds, and finally for each implication state constraint about knowledge of f there is at least one precondition which does not hold, then the agent cannot infer knowledge about the determinant fluent. Note that even (KT2.1) and (KT2.2) are state constraints that fall under the scope of this axiom.

Summarizing

In a nutshell, the procedure of reasoning about knowledge evolves as follows. For any deterministic action with no preconditions or when the agent knows all preconditions, (KT3.1) and (KT3.2) or (KT3.3) and (KT3.4) are triggered and knowledge persists until some other event changes it. If the agent happens not to know at least one precondition, while, at the same time, there is no precondition that it knows it does not hold, then the agent cannot be certain about the action's effects, the action is considered non-deterministic and (KT5.1) or (KT5.2) are triggered. For ordinary non-deterministic actions, (KT5.3) is applied.

When deterministic effects occur, the KP fluent determines the agent's state of knowledge by means of (KT2.1) and (KT2.2) axioms. In addition, knowledge can also be inferred using state constraints. If, however, knowledge about a fluent can neither directly nor indirectly be established, then axiom (KT7) captures the agent's lack of knowledge about f. Whenever a change in the agent's KB causes knowledge to be gained, (KT7) is automatically canceled. As a result, the $Knows$ fluent may hold even when KP does not.

6   Correctness of the Solution

The theory enjoys a number of important properties that justify its correctness.[2] In the following we show that sensing does not affect ordinary fluents and that actions only affect knowledge in the appropriate way, i.e., ensuring that there are no unwanted changes in knowledge, either additions or reductions.

Theorem 1. *Knowledge-producing actions do not change the state of the world.*

Theorem 2. *At any timepoint t, for any fluent f and for any event e, such that (a) e has known effects, (b) f is not (directly or indirectly) affected by e and (c) f is not included in the precondition list of e for any of its effects, then f is known to hold at t + 1 after the occurrence of e at t iff it is known to hold at t.*

Theorem 2 generalizes certain properties of the knowledge theories of previous approaches. First of all, in accordance to [6], we generalize the Situation Calculus approach of [4] by distinguishing between the effect of an action itself and the agent's awareness of it. Moreover, in previous studies once knowledge is gained it is never lost [14,6,4]. The present approach additionally considers non-deterministic actions that cause loss of knowledge; item (a) does not distinguish between deterministic and non-deterministic effects.

Furthermore, there is a number of properties that are automatically generated by the theory. The following are indicative (notice the difference between *knows* and *knows whether* in Propositions 1 and 2). Proposition 3 states that once knowledge of a fluent is established, the theory automatically precludes knowledge of its negation, guarantying freedom from inconsistency.

Proposition 1. *If an agent knows all fluents in the body of an implication state constraint, then it also knows its head.*

Proposition 2. *An agent knows whether fluents on one side of a bi-implication state constraint hold iff it knows whether fluents on the other side of the constraint also hold.*

Proposition 3. *At any timepoint t knowledge of a fluent f implies not knowing the negation of f.*

7   Scenarios and Use Cases

To demonstrate the power of the theory we have designed a number of use-case scenarios focusing mainly on the domain of Ambient Intelligence. The dynamic, social and uncertain context of agent interactions in AmI environments poses the kind of challenges that our theory aims to investigates. Some of the scenarios present adaptations of well-known benchmark problems for automated logic-based commonsense reasoning, while others introduce novel challenges, such as

---

[2] Proofs are omitted due to space limitation.

knowledge preservation for specified durations of time (somewhat similar in nature to the problems studied in [8]). In short, we have confirmed the theory's applicability in use-cases that range from non-deterministic environments (both with released fluents and with *determining* fluents, as described in [11]) to ramifications of knowledge (that are not only accommodated by primitive and derived fluents, but, most importantly, through the use of state constraints), reasoning about only knowing, i.e., what an agent does not know (following the lines of [15]), knowledge of continuously changing fluents, temporal knowledge and other. Many of these issues have been investigated in isolation by other studies in the past, but combining co-existing models for different phenomena has proven to be a very challenging task. An example of temporal knowledge in a non-deterministic environment is the subject of the next example, intending to illustrate the theory's reasoning mechanism in complex domains.

7.1   Opening Safe Example

Imagine a safe that requires the agent to first press a button and then speak the correct combination into a microphone that is located on the safe. The locking mechanism provides a time window of 5 seconds for the user to apply the correct code after the button has been pressed:

$$Happens(Press(B), t) \Rightarrow$$
$$Happens(StartRec(Mic), t) \land Happens(StopRec(Mic), t+5) \tag{6.1}$$

$$Initiates(StartRec(Mic), Recording(Mic), t) \tag{6.2}$$

$$Terminates(StopRec(Mic), Recording(Mic), t) \tag{6.3}$$

If the spoken code is correct the safe's door unlocks automatically. Still, due to noisy conditions, even a correct pronunciation of the combination does not guarantee that the door gets unlocked. In the Event Calculus we can represent this uncertainty by releasing the state of the door from the law of inertia:

$$HoldsAt(Recording(Mic), t) \Rightarrow Releases(Speak("3241"), Open(D), t) \tag{6.4}$$

Thus, after the occurrence of the *Speak* action the truth value of the *Open* fluent will fluctuate in future timepoints, resulting in different models, one where the door is open and one where it is still locked, both of which are possible.

The agent initially knows that the microphone is switched off and the door locked:

$$HoldsAt(KP(\neg Recording(Mic)), 0) \land HoldsAt(KP(\neg Open(D)), 0) \tag{6.5}$$

Notice how we capture the initial state of knowledge of the robot; we use the *KP* instead of the *Knows* fluent, in order to state that this knowledge is subject to inertia and should be kept in the robot's KB, until some event modifies it.

We also assume that all fluent terms are subject to inertia in the initial state (apart from knowledge, of course). Finally, since for all fluents there are no state constraints other than (KT2.1) and (KT2.2), (KT7) is compiled as:

$$\neg HoldsAt(KP(f), t) \wedge \neg HoldsAt(KP(\neg f), t) \Rightarrow \neg HoldsAt(Kw(f), t)\text{(KT7)}$$

A two-action narrative is generated; the agent presses the button at time 0 and speaks into the microphone at time 1.

$$Happens(Press(B), 0) \wedge Happens(Speak("3241"), 1) \tag{6.6}$$

We make a *dynamic* closed world assumption on knowledge, as defined in [16]. We can now prove, using circumscription, that at timepoints >1 the agent does not know whether the door is unlocked and also at timepoints >5 the agent knows that the microphone is not recording any more. In particular, by forming the parallel circumscriptions

$$CIRC[(6.2 - 4) \wedge (KT3.1 - 4) \wedge (KT5.1 - 3); Initiates, Terminates, Releases]$$
$$\text{and } CIRC[(6.1) \wedge (6.6); Happens]$$

in conjunction with (6.5), (6.6), uniqueness-of-names axioms, Event Calculus axioms (DEC1-8), the remaining axioms of our knowledge theory, (K) and (T), it is shown, for instance, that

$$\models \neg HoldsAt(Kw(Open(D)), 2) \wedge HoldsAt(Knows(\neg Recording(Mic)), 6)\text{(6.7)}$$

This result was caused by the triggering of (KT3.1-2) at timepoint 0 and (KT3.3-4) at timepoint 5 for the *Recording* fluent, as well as of (KT5.3) at timepoint 1 for the *Open* fluent. In fact, at timepoints >1 there are two possible models, one in which the door is open and another where it is still locked. In both models, though, (6.7) reflects the agent's state of knowledge.

Imagine now we interleave a sense action at timepoint 2, so that the agent can sense the state of the door:

$$Happens(sense(Open(D)), 2) \tag{6.8}$$

Following the same procedure we will again end up having two different possible models. Now, on the other hand, the agent will have knowledge of the actual state, which is reflected by the next formula, due to (KT4):

$$HoldsAt(Kw(Open(D)), 3) \tag{6.9}$$

8 Conclusion and Future Work

Although previous studies have developed separate solutions for the fundamental problems of commonsense reasoning, combining action, knowledge and time in a unifying framework presents an important step in confronting the challenges posed by reasoning in real-world domains. An appealing extension is to allow for reasoning about the knowledge of other agents and the interaction of communication and knowledge, touching upon subjects, such as group ability, common knowledge and distributed knowledge. Further future plans also involve the study of belief change and belief revision in the Event Calculus.

References

1. Chittaro, L., Montanari, A.: Temporal representation and reasoning in artificial intelligence: Issues and approaches. Annals of Mathematics and Artificial Intelligence 28(1-4), 47–106 (2000)
2. Demolombe, R., del Pilar Pozos Parra, M.: A simple and tractable extension of situation calculus to epistemic logic. In: Ohsuga, S., Raś, Z.W. (eds.) ISMIS 2000. LNCS (LNAI), vol. 1932, pp. 515–524. Springer, Heidelberg (2000)
3. Scherl, R.B., Levesque, H.J.: The frame problem and knowledge-producing actions. In: Proceedings of the 11th National Conference on Artificial Intelligence (AAAI 1993), Washington, DC, USA, pp. 689–697. AAAI Press/MIT Press (1993)
4. Scherl, R.B., Levesque, H.J.: Knowledge, action, and the frame problem. Artificial Intelligence 144(1-2), 1–39 (2003)
5. Lespérance, Y., Levesque, H.J., Lin, F., Scherl, R.B.: Ability and knowing how in the situation calculus. Studia Logica 66(1), 165–186 (2000)
6. Thielscher, M.: Representing the knowledge of a robot. In: Cohn, A., Giunchiglia, F., Selman, B. (eds.) Proc. of the International Conference on Principles of Knowledge Representation and Reasoning (KR), Breckenridge, CO, pp. 109–120 (2000)
7. Lobo, J., Mendez, G., Taylor, S.R.: Knowledge and the action description language A. Theory and Practice of Logic Programming 1(2), 129–184 (2001)
8. Zimmerbaum, S., Scherl, R.B.: Sensing actions, time, and concurrency in the situation calculus. In: Proc. 7th International Workshop on Intelligent Agents VII. Agent Theories Architectures and Languages, London, UK, pp. 31–45 (2001)
9. Forth, J., Shanahan, M.: Indirect and conditional sensing in the event calculus. In: de Mántaras, R.L., Saitta, L. (eds.) ECAI, pp. 900–904. IOS Press, Amsterdam (2004)
10. Lifschitz, V.: Circumscription. Handbook of Logic in Artificial Intelligence and Logic Programming 3, 297–352 (1994)
11. Shanahan, M.: The event calculus explained. In: Veloso, M.M., Wooldridge, M.J. (eds.) Artificial Intelligence Today. LNCS (LNAI), vol. 1600, pp. 409–430. Springer, Heidelberg (1999)
12. Miller, R., Shanahan, M.: Some alternative formulations of the event calculus. In: Computational Logic: Logic Programming and Beyond, Essays in Honour of Robert A. Kowalski, Part II, London, UK, pp. 452–490. Springer, Heidelberg (2002)
13. Mueller, E.: Commonsense Reasoning, 1st edn. Morgan Kaufmann, San Francisco (2006)
14. Levesque, H.J.: What is planning in the presence of sensing? In: Proceedings of 13th National Conference on Artificial Intelligence, vol. 2, pp. 1139–1146 (1996)
15. Lakemeyer, G., Levesque, H.: AOL: A logic of Acting, Sensing, Knowing, and Only Knowing. Principles of Knowledge Representation and Reasoning, 316–329 (1998)
16. Reiter, R.: On Knowledge-Based Programming with Sensing in the Situation Calculus. ACM Transactions on Computational Logic 2(4), 433–457 (2001)

Tensor Space Models for Authorship Identification

Spyridon Plakias and Efstathios Stamatatos

Dept. of Information and Communication Systems Eng.
University of the Aegean
83200 – Karlovassi, Greece
stamatatos@aegean.gr

Abstract. Authorship identification can be viewed as a text categorization task. However, in this task the most frequent features appear to be the most important discriminators, there is usually a shortage of training texts, and the training texts are rarely evenly distributed over the authors. To cope with these problems, we propose tensors of second order for representing the stylistic properties of texts. Our approach requires the calculation of much fewer parameters in comparison to the traditional vector space representation. We examine various methods for building appropriate tensors taking into account that similar features should be placed in the same neighborhood. Based on an existing generalization of SVM able to handle tensors we perform experiments on corpora controlled for genre and topic and show that the proposed approach can effectively handle cases where only limited training texts are available.

Keywords: Authorship identification, Tensor space representation, Text categorization.

1 Introduction

Authorship identification is the task of assigning a text to an author, given a set of candidate authors for whom texts of undisputed authorship are available. Beyond the traditional approach based on human experts, this procedure can be automated by computational tools able to capture and match the stylistic properties of texts and authors [26, 32, 2]. The main idea is that by measuring some textual features we can distinguish between texts written by different authors. Nowadays, such tools are of increasing importance since there are plenty of texts in electronic form in Internet media (e.g., blogs, online forum messages, emails, etc.) indicating many applications of this technology, including forensics (identifying the authors of harassing email messages), intelligence (attributing messages to known terrorists), etc. [1, 23, 8, 33, 35]

One main issue in the research on authorship identification is the definition of appropriate textual features to quantify the stylistic properties of texts [13]. Many different measures have been proposed including simple ones such as word frequencies or character n-gram frequencies and more complex ones requiring some sort of syntactic or semantic analysis [35]. The other main issue is the development of attribution methodologies to assign texts to one candidate author. So far, the proposed attribution models comprise standard discriminative algorithms (e.g., support vector machines) [9] and

J. Darzentas et al. (Eds.): SETN 2008, LNAI 5138, pp. 239–249, 2008.

generative models (e.g., Bayesian methods) [27] as well as models specifically designed for authorship identification tasks [21, 29].

From a machine learning point-of-view, author identification can be viewed as a multi-class single-label text categorization (TC) task [28]. Actually, several studies on TC use this problem as just another testing ground together with other tasks, such as topic identification, language identification, genre detection, etc. [4, 25, 34] However, there are some points that make author identification a special TC task that should be handled carefully, namely:

Feature selection: Author identification is a style-based TC task. In such tasks, usually the most important features are the most frequent ones. On the contrary, in topic-based TC, the most frequent features (e.g., words) are usually excluded since they have little discriminatory power. Note that in case of word-features, the most frequent words carry no semantic information. In style-based tasks, such meaningless (or function) words are used unconsciously by the author, so they offer a way to measure their stylistic properties.

Shortage of training texts: In a typical author identification application only a few (possibly short) texts of undisputed authorship are available for the candidate authors. So, it is crucial for the attribution methodology to be able to effectively handle cases with low amount of training texts.

Class imbalance: It is not unusual to have an unequal distribution of training texts over the candidate authors. Note that beyond the amount of training texts per author, the length of training texts can also produce class imbalance conditions (in case we have short texts for some authors and long texts for other authors). In such cases, the evaluation procedure of an author identification approach should be carefully designed, especially in forensic applications. That is, the fact that there is shortage of training texts for one candidate author in comparison to the other authors does not mean that this person is less likely to be the true author of the text in question.

In this paper, we present an approach that attempts to take into account the aforementioned characteristics of the problem. In particular, we propose second order tensor space models for text representation in contrast to the traditional vector space models. The tensor models are able to handle the same amount of textual features with the vector models but require much fewer parameters to be learnt. Therefore, they are suitable for cases with limited training data. On the other hand, in contrast to vector models, the positioning of each feature into the tensor model plays a crucial role since relevant features should be placed in the same neighborhood. To solve this problem, we propose a frequency-based metric to define the relevance between features and explore several methods for filling the feature matrix. Using an existing generalization of the SVM algorithm able to handle tensors of second order [6] we perform experiments on text corpora controlled for genre and topic under balanced and imbalanced conditions. In the latter case, we pay special attention to the evaluation methodology so that the test corpus distribution over the authors is uncorrelated with the corresponding distribution of the training corpus.

The rest of this paper is organized as follows. Section 2 includes previous work in author identification while Section 3 describes in detail the proposed tensor space models. Section 4 describes the conducted experiments and, finally, Section 5 summarizes the conclusions drawn by this study and proposes future work directions.

2 Previous Work

The majority of the work in authorship identification (or authorship attribution) deals with defining appropriate measures for quantifying the writing style. This line of research is known as *stylometry* [13]. Several hundreds of stylometric features have been proposed attempting to find measures that are reliable and accurate under varying text-length, text types, and availability of text processing tools. The most traditional features are word-based in accordance to work in topic-based text categorization. However, in author identification, the most frequent words have been proved to be the most important ones [5]. Such words, including articles, prepositions, conjunctions, ('stop words' in information retrieval terminology) are usually excluded from topic-based classification since they carry no semantic information. On the other hand, they are closely related to certain syntactic structures. That is why they are also called 'function' words. So, their use indicates the use of certain syntactic structures by the author. Several sets of function words have been defined for English [2, 3, 20]. An alternative way to automatically define the function word set is to extract the most frequent words in a corpus [24, 29]. There are also attempts to use word n-grams to exploit contextual information [27, 7]. However, this process considerably increases the dimensionality of the problem and has not produced encouraging results so far.

Another way to represent text is by using character n-gram frequencies [17, 30]. Again the most frequent character n-grams (n contiguous characters) include the most important information. Although the dimensionality of the problem is increased in comparison to a function word approach, it is much smaller in comparison to a word n-gram approach. Methods based on such features have produced very good results in several author identification experiments and texts in various languages [17, 16, 30, 11]. However, there is still no consensus about the definition of an appropriate n value (the length of character n-grams) for certain natural languages and text types. Another character-based approach makes use of existing text compression tools to estimate the similarity of texts [4, 25].

A more elaborate type of features involves the use of natural language processing tools to extract syntactic [32, 10, 12] or semantic information [3, 10]. In theory, such features should better quantify the stylistic choices of the authors since they are used unconsciously. However, the measurement of such features in raw text is still a difficult procedure and the extracted measures are noisy. As a result, the quantification of writing style is not accurate enough.

Beyond the definition of stylometric features, the research in author identification is dominated by the development of effective attribution methodologies. A significant part of the studies is based on discriminative models utilizing machine learning techniques like SVM [9, 21, 23, 33], neural networks [35], ensemble methods [29] etc. Such powerful machine learning algorithms can effectively cope with high dimensional and sparse data. Another approach is to apply a generative model, like a naïve Bayes model [27]. Yet another approach is to estimate the similarity between two texts [4, 17].

Recently, a number of studies take into account factors such as training set size, imbalanced training data, and the amount of candidate authors in order to build more reliable author identification methods. Marton. et al. [25] and Hirst & Feiguina [12] examine the effectiveness of author identification methods under limited training text conditions. Stamatatos [30] proposes a model for handling limited and imbalanced training texts. In another study, Stamatatos [31] proposes text sampling methods for re-balancing an imbalanced training corpus to improve author identification performance. Finally, Madigan, et al. [24] test various author identification methods in cases with high number of candidate authors.

3 Tensor Space Representation

In a vector space model, a text is considered as a vector in R^n, where n is the number of features. A second order tensor model considers a text as a matrix in $R^x \otimes R^y$, where x and y are the dimensions of the matrix. A vector $x \in R^n$ can be transformed to a second order vector $X \in R^x \otimes R^y$ provided $n \approx x*y$. Notice that the same features are used in both the vector and the tensor model.

A linear classifier in R^n (e.g., SVM) can be represented as $a^T x + b$, that is, there are $n+1$ parameters to be learnt (b, a_i, $i=1,\ldots,n$). Similarly, a linear classifier in $R^x \otimes R^y$ can be represented as $u^T X v + b$, that is, there are $x+y+1$ parameters to be learnt (b, u_i, $i=1,\ldots y$, v_j, $j=1,\ldots x$). Consequently, the number of parameters is minimized when $x=y$ (i.e., square matrix) and this is much lower than n. Therefore, the vector space representation is more suitable in cases with limited training sets since much fewer parameters have to be learnt. Note that in both cases the amount of textual feature is the same (n). However, in the tensor model the position of each feature within the matrix plays a crucial role for the performance of the model since each feature is strongly associated with the features of the same row or column. On the other hand, the position of a feature in the vector model does not affect the performance of the model.

3.1 Support Tensor Machines

To be able to handle tensors instead of vectors, we use a generalization of SVM, called *support tensor machines* (STM) [6]. Initially, this algorithm sets $u=(1,\ldots,1)^T$ and uses it to compute the initial v. Then, it works iteratively by computing in each step a new u and v as follows (given a set of training examples $\{X_i, y_i\}$, $i=1,\ldots,m$):

Computation of v (provided u) solving the following optimization problem:

$$\min_{v,b,\xi} \frac{1}{2}\|u\|^2 v^T v + C\sum_{i=1}^{m}\xi_i \tag{1}$$

$$\text{subject to } y_i(v^T X_i^T u + b) \geq 1 - \xi_i, \xi_i \geq 0, i = 1,\ldots,m$$

Note that this optimization problem is the same as the standard SVM algorithm. So, any computation method used in SVM can also be used here.

Computation of u (provided v) solving the following optimization problem:

$$\min_{v,b,\xi} \frac{1}{2}\|v\|^2 u^T u + C\sum_{i=1}^{m}\xi_i \tag{2}$$

$$\text{subject to } y_i(u^T X_i^T v + b) \geq 1 - \xi_i, \xi_i \geq 0, i = 1,\ldots,m$$

Again, this optimization problem is the same as the standard SVM algorithm and any computation method used in SVM can also be used here. The procedure of calculating new values for u and v is repeated until they tend to converge.

3.2 Feature Relevance

Since the tensor-based model takes into account associations between features (each feature is strongly associated with features in the same row and column) it is crucial to place relevant features in the same neighbourhood. To suitably transform a vector representation to a second order tensor representation, one has to define what features are considered relevant and how relevant features are placed in the same neighbourhood.

As it has been demonstrated by several authorship identification studies, the frequency of features is a crucial factor for their significance [14, 19]. Actually, the frequency information is more important than the discriminatory power of the features when examined individually. Following this evidence, in this paper, we use the frequency of occurrence as the factor that determines relevance among features. Particularly, in a binary classification case, where we want to discriminate author A from author B, the relevance $r(x_i)$ of a feature x_i is:

$$r(x_i) = \frac{f_A(x_i) - f_B(x_i)}{f_A(x_i) + f_B(x_i) + b} \qquad (3)$$

where $f_A(x_i)$ and $f_B(x_i)$ are the relative frequencies of occurrence of feature x_i in the texts of author A and B, respectively, and b a smoothing factor. The higher the $r(x_i)$, the more important the feature x_i for author A. Similarly, the lower the $r(x_i)$, the more important the feature x_i for author B. Note, that the relevance metric is not necessarily associated with the discriminatory power of each feature. However, a feature with high (low) relevance value is likely to be a good discriminator for author A (B) since it is found more times in their texts in comparison to author B (A).

$$\begin{bmatrix} P1 & P2 \\ P3 & P4 \end{bmatrix}$$

P1 comprises features strongly associated with author A
P4 comprises features strongly associated with author B
P2 and P3 comprise neutral features

Fig. 1. A second order tensor divided into four parts according to the feature relevance

3.3 Matrix Filling

Given a ranking of features according to the relevance metric, we need a strategy to fill the matrix of features having in mind that relevant features should be placed in the same neighbourhood. To this end, we consider that each matrix is segmented into four parts of equal size, as it is depicted in figure 1. The upper left part (*P1*) is filled with features strongly associated with author A, the lower right part (*P4*) is filled with features strongly associated with author B, while the two remaining parts (*P2* and *P3*) are filled with relatively neutral features. So, we attempt to create a distance between the features strongly associated with one of the authors in both rows and columns of the feature matrix. Moreover, each row or column of the matrix is strongly associated with one of the authors since it contains some very relevant features for that author and a number of neutral features. That is, the rows and columns of the matrix are composed of a combination of features from *P1* and *P2* or *P3* as well as a combination of features from *P4* and *P2* or *P3*. On the other hand, there are no rows or columns of the matrix that contain features strongly associated with both authors (that is, a combination of features from *P1* and *P4* is not allowed).

To place each feature within each part of the matrix we fill the columns of that part of the matrix from left to right with decreasing relevance values. As a result, the columns of the matrix are filled with features of similar relevance values, while the rows are filled with features of mixed relevance values. This is depicted in figure 2a. We call this matrix filling approach *symmetric cross*.

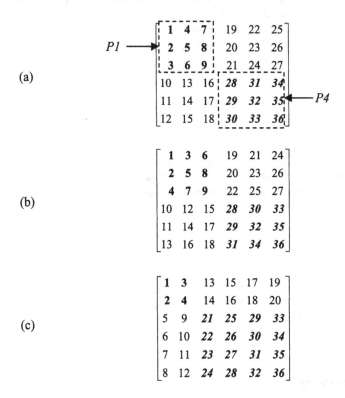

Fig. 2. Examples of matrix filling with the proposed techniques. The feature numbers correspond to the ranking of 36 features according to their relevance (1 correspond to the feature with higher relevance value). Features in boldface (*P1*) are strongly associated with author A, features in boldface italics are strongly associated with author B (*P4*). (a) Symmetric cross: the four parts of the matrix are of equal size and the columns of each part are filled with features of decreasing relevance from the left to the right. (b) Cross-diagonal: the four parts of the matrix are of equal size and each part is filled with features of decreasing relevance from upper left corner to the lower right corner. (c) Asymmetric cross: the four parts of the matrix are of different size and the columns of each part are filled with features of decreasing relevance from the left to the right.

A variation of this technique is to fill each part of the matrix diagonally. In more detail, we start from the upper left corner of each part of the matrix and fill the diagonals with decreasing relevance values, as it is shown in figure 2b. This method distributes the most significant features in a fairer way across the rows and columns of the matrix. We call this matrix filling method *cross-diagonal.*

Yet another variation of the symmetric cross method is to segment the feature matrix into four parts of different size. That is, the upper left part may be smaller than the lower right part of the matrix (see figure 2c). This may correspond to the cases where only a few features are strongly associated with author A while most of the features are associated with author B. The boundaries of the parts of the matrix can be found by examining the relevance values given that positive relevance values indicate features important for author A and negative relevance values indicate features important for author B. We call this matrix filling method *asymmetric cross.*

4 Experiments

4.1 Corpus and Settings

The corpora used for evaluation in this study consist of newswire stories in English taken from the publicly available Reuters Corpus Volume 1 (RCV1) [22]. Although this corpus was not particularly designed for evaluating author identification methods, it offers a large pool of texts of unquestioned authorship that cover a variety of topics and it has already been used by previous studies [24, 18, 31]. We selected the top 10 authors with respect to the amount of texts belonging to the topic class CCAT (about corporate and industrial news) to minimize the topic factor for distinguishing between texts. Given that this corpus is already controlled for genre, we expect the authorship factor to be the most important discriminating factor.

We produced several versions of this corpus by varying the amount of training texts per author. To produce three balanced training corpora, we used 50, 10 or 5 training texts per author, respectively. In all cases, the test corpus comprises 50 texts per author not overlapping with the training texts (see figure 3b).

To produce imbalanced training corpora we applied a Gaussian distribution over the authors. In particular, we set the minimum and maximum amount of training texts per author and an imbalanced Gaussian distribution defines the amount of training texts per author as it is depicted in figure 3a. Three imbalanced training corpora were used, 10:20, 5:10, and 2:10 where the notation $a:b$ means that at least a training texts are available for all the authors and b is the maximum amount of training texts per author. The test corpus comprises 50 texts per author not overlapping with the training texts and it is the same with the test corpus of the balanced training corpora. Note that, by using a balanced test corpus when we know that the training corpus is imbalanced, we attempt to simulate the general authorship identification case where the availability of many training texts for one author should not increase their probability to be the true author of the unknown texts.

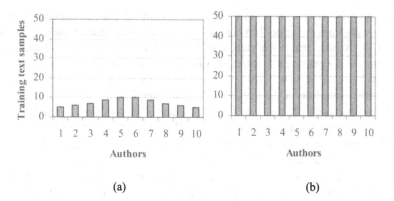

Fig. 3. (a) A 5:10 imbalanced corpus of 10 authors comprising at least 5 texts per author and a maximum of 10 texts for some authors. (b) A balanced corpus of 10 authors comprising 50 texts per author.

4.2 Results

To represent the texts we used a character n-gram approach. Thus, the feature set consists of the 2,500 most frequent 3-grams of the training corpus. This means that a standard linear SVM model [15] can be built using vectors of 2,500 features. Moreover, a second order tensor model can be built based on a 50x50 matrix. Note that since we deal with a multi-class author identification task, we followed a *one vs. one* approach, that is, for each pair of authors a STM model was built and the matrix filling technique was based on the feature relevance for that pair of authors. Based on preliminary experiments, we set the C parameter of linear SVM to 1, the corresponding parameter for STM models to 0.1 and the smoothing parameter b equal to 1.

We tested the STM models with the matrix filling techniques proposed in section 3.3 and compare it to the standard SVM model. The performance results for both the balanced and imbalanced training corpora can be seen in table 1. Recall, that in all the cases the test corpus is the same so these results are directly comparable. The method called STM-Simple is based on a very naïve methodology for filling the feature matrix: the features are ranked in decreasing frequency and, then, the rows of the matrix are filled from the top row to the last raw and from left to right. Therefore, the comparison of the techniques proposed in section 3.3 with this simple method reveals the significance of taking into account the relevance of the features when filling the matrix. As can be seen, the proposed techniques achieve clearly better performance results in comparison to this simple baseline method. So, it is crucial to place similar (relevant) features in the same neighbourhood when filling the feature matrix of an STM model.

Comparing SVM and STM models for the balanced cases reveals that the standard SVM model outperforms STM models when many training texts (50 per author) are available. On the other hand, when the balanced training corpus is limited (10 or 5 texts per author) all the examined STM models are better than SVM. This confirms our hypothesis that the STM can more effectively handle limited training data since much fewer parameters have to be learnt. The imbalanced cases produce more confused results. In two cases the SVM model is the clear winner while in the third case an STM model slightly outperforms the SVM model.

Table 1. Classification accuracy (%) of the SVM and the proposed STM models with various matrix filling techniques

| Model | Training texts per author | | | | | |
| | Balanced | | | Imbalanced | | |
	50	10	5	10:20	5:10	2:10
SVM	**80.8**	64.4	48.2	**64.2**	62.4	**51.0**
STM-Simple	70.4	54.4	44.2	58.2	49.2	39.0
STM-Symmetric cross	76.6	**67.8**	50.4	63.0	59.8	49.2
STM-Cross-diagonal	76.0	64.4	52.4	62.2	**62.6**	49.8
STM-Asymmetric cross	78.0	65.2	**53.4**	61.8	60.6	50.0

5 Conclusions

In this paper, we presented an approach to author identification that is based on a tensor space representation instead of the traditional vector space model. The proposed representation can be used in a classification scheme that requires much fewer parameters to be learnt and it is more suitable in cases where only limited training data are available. Author identification is a representative example of this type of problems where usually extremely limited texts of known authorship are available, especially in forensic applications. A generalization of the SVM algorithm able to handle second order tensors was used. The conducted experiments have shown the effectiveness of the proposed model in cases of shortage of training texts in comparison to a standard SVM model.

A consequence of the second order tensor representation is that the position of each feature within the matrix plays a crucial role since features of the same row or column are strongly associated. To place relevant features in the same neighborhood of the matrix, we first defined a relevance metric that is based on frequency information and then examined various matrix filling techniques. The comparison of the proposed techniques to a baseline method revealed that the information we used is very important for achieving good results. On the other hand, it is not clear which of the proposed techniques is superior. Further experiments should be conducted towards this direction.

The performed experiments were based on both balanced and imbalanced training corpora. However, in all cases the test corpus was balanced to extract more reliable evaluation results since in a typical author identification scenario, the existence of many texts of undisputed authorship for one candidate author should not increase the likelihood of being the true author of the unknown text. Although the proposed models managed to increase the performance (in comparison to a standard vector space SVM model) when dealing with limited training texts, their performance in the imbalanced cases was not encouraging. More elaborated matrix filling techniques and possibly a different definition of feature relevance should be tested for effectively handling imbalanced training corpora in author identification tasks.

References

1. Abbasi, A., Chen, H.: Applying Authorship Analysis to Extremist-group Web Forum Messages. IEEE Intelligent Systems 20(5), 67–75 (2005)
2. Argamon, S., Saric, M., Stein, S.: Style Mining of Electronic Messages for Multiple Authorship Discrimination: First Results. In: 9th ACM SIGKDD, pp. 475–480 (2003)
3. Argamon, S., Whitelaw, C., Chase, P., Hota, S.R., Garg, N., Levitan, S.: Stylistic Text Classification Using Functional Lexical Features. Journal of the American Society for Information Science and Technology 58(6), 802–822 (2007)
4. Benedetto, D., Caglioti, E., Loreto, V.: Language Trees and Zipping. Physical Review Letters 88(4), 048702 (2002)
5. Burrows, J.F.: Word Patterns and Story Shapes: The Statistical Analysis of Narrative Style. Literary and Linguistic Computing 2, 61–70 (1987)
6. Cai, D., He, X., Wen, J.R., Han, J., Ma, W.Y.: Support Tensor Machines for Text Categorization. Technical report, UIUCDCS-R-2006-2714, University of Illinois at Urbana-Champaign (2006)
7. Coyotl-Morales, R.M., Villaseñor-Pineda, L., Montes-y-Gómez, M., Rosso, P.: Authorship Attribution Using Word Sequences. In: 11th Iberoamerican Congress on Pattern Recognition, pp. 844–853. Springer, Heidelberg (2006d)
8. Chaski, C.E.: Who's at the Keyboard? Authorship Attribution in Digital Evidence Investigations. International Journal of Digital Evidence 4(1) (2005)
9. Diederich, J., Kindermann, J., Leopold, E., Paass, G.: Authorship Attribution with Support Vector Machines. Applied Intelligence 19(1/2), 109–123 (2003)
10. Gamon, M.: Linguistic Correlates of Style: Authorship Classification with Deep Linguistic Analysis Features. In: 20th International Conference on Computational Linguistics, pp. 611–617 (2004)
11. Grieve, J.: Quantitative Authorship Attribution: An Evaluation of Techniques. Literary and Linguistic Computing 22(3), 251–270 (2007)
12. Hirst, G., Feiguina, O.: Bigrams of Syntactic Labels for Authorship Discrimination of Short Texts. Literary and Linguistic Computing 22, 405–417 (2007)
13. Holmes, D.I.: The Evolution of Stylometry in Humanities Scholarship. Literary and Linguistic Computing 13(3), 111–117 (1998)
14. Houvardas, J., Stamatatos, E.: N-gram Feature Selection for Authorship Identification. In: 12th International Conference on Artificial Intelligence: Methodology, Systems, Applications, pp. 77–86. Springer, Heidelberg (2006)
15. Joachims, T.: Text Categorization with Support Vector Machines: Learning with Many Relevant Features. In: 10th European Conference on Machine Learning, pp. 137–142 (1998)
16. Juola, P.: Authorship Attribution for Electronic Documents. In: Olivier, M., Shenoi, S. (eds.) Advances in Digital Forensics II, pp. 119–130. Springer, Heidelberg (2006)
17. Keselj, V., Peng, F., Cercone, N., Thomas, C.: N-gram-based Author Profiles for Authorship Attribution. In: Pacific Association for Computational Linguistics, pp. 255–264 (2003)
18. Khmelev, D.V., Teahan, W.J.: A Repetition based Measure for Verification of Text Collections and for Text Categorization. In: 26th ACM SIGIR, pp. 104–110 (2003)
19. Koppel, M., Akiva, N., Dagan, I.: Feature Instability as a Criterion for Selecting Potential Style Markers. Journal of the American Society for Information Science and Technology 57(11), 1519–1525 (2006)

20. Koppel, M., Schler, J.: Exploiting Stylistic Idiosyncrasies for Authorship Attribution. In: IJCAI 2003 Workshop on Computational Approaches to Style Analysis and Synthesis, pp. 69–72 (2003)
21. Koppel, M., Schler, J., Bonchek-Dokow, E.: Measuring Differentiability: Unmasking Pseudonymous Authors. Journal of Machine Learning Research 8, 1261–1276 (2007)
22. Lewis, D., Yang, Y., Rose, T., Li, F.: RCV1: A New Benchmark Collection for Text Categorization Research. Journal of Machine Learning Research 5, 361–397 (2004)
23. Li, J., Zheng, R., Chen, H.: From Fingerprint to Writeprint. Communications of the ACM 49(4), 76–82 (2006)
24. Madigan, D., Genkin, A., Lewis, D., Argamon, S., Fradkin, D., Ye, L.: Author Identification on the Large Scale. In: CSNA 2005 (2005)
25. Marton, Y., Wu, N., Hellerstein, L.: On Compression-based Text Classification. In: European Conference on Information Retrieval, pp. 300–314. Springer, Heidelberg (2005)
26. Mosteller, F., Wallace, D.: Applied Bayesian and Classical Inference: The Case of the Federalist Papers. Addison-Wesley, Reading (1964)
27. Peng, F., Shuurmans, D., Wang, S.: Augmenting Naive Bayes Classifiers with Statistical Language Models. Information Retrieval Journal 7(1), 317–345 (2004)
28. Sebastiani, F.: Machine Learning in Automated Text Categorization. ACM Computing Surveys 34(1) (2002)
29. Stamatatos, E.: Authorship Attribution Based on Feature Set Subspacing Ensembles. International Journal on Artificial Intelligence Tools 15(5), 823–838 (2006)
30. Stamatatos, E.: Author Identification Using Imbalanced and Limited Training Texts. In: 4th International Workshop on Text-based Information Retrieval, pp. 237–241 (2007)
31. Stamatatos, E.: Author Identification: Using Text Sampling to Handle the Class Imbalance Problem. Information Processing and Management 44(2), 790–799 (2008)
32. Stamatatos, E., Fakotakis, N., Kokkinakis, G.: Automatic Text Categorization in Terms of Genre and Author. Computational Linguistics 26(4), 471–495 (2000)
33. de Vel, O., Anderson, A., Corney, M., Mohay, G.: Mining E-mail Content for Author Identification Forensics. SIGMOD Record 30(4), 55–64 (2001)
34. Zhang, D., Lee, W.S.: Extracting Key-substring-group Features for Text Classification. In: 12th Annual SIGKDD International Conference on Knowledge Discovery and Data Mining, pp. 474–483 (2006)
35. Zheng, R., Li, J., Chen, H., Huang, Z.: A Framework for Authorship Identification of Online Messages: Writing Style Features and Classification Techniques. Journal of the American Society of Information Science and Technology 57(3), 378–393 (2006)

Efficient Incremental Model for Learning Context-Free Grammars from Positive Structural Examples

Gend Lal Prajapati[1], Narendra S. Chaudhari[2], and Manohar Chandwani[3]

[1,3] Institute of Engineering & Technology, Department of Computer Engineering
Devi Ahilya University, Indore- 452017, M.P., India
[2] School of Computer Engineering, Nanyang Technological University
Block: N4-2a-32; 50, Nanyang Avenue, Singapore- 639798
gprajapati.iet@dauniv.ac.in, ASNarendra@ntu.edu.sg,
mc.iet@dauniv.ac.in

Abstract. This paper describes a formalization based on tree automata for incremental learning of context-free grammars from positive samples of their structural descriptions. A structural description of a context-free grammar is a derivation tree of the grammar in which labels are removed. The tree automata based learning in this paradigm is early introduced by Sakakibara in 1992, however his scheme assumes that all training examples are available to the learning algorithm at the beginning (i.e., it cannot be employed as an online learning) and also it doesn't optimize the storage requirements as well. Our model has several desirable features that runs in $O(n^3)$ time in the sum of the sizes of the input examples, obtains $O(n)$ storage space saving, achieves good incremental behavior by updating a guess incrementally and infers a grammar from positive-only examples efficiently. Several examples and experimental results are given to illustrate the scheme and its efficient execution.

Keywords: Grammatical inference, incremental learning, machine learning, reversible context-free grammar, structural description, tree automaton.

1 Introduction

We consider the problem of incremental learning of context-free languages from positive-only examples. The problem of learning a correct grammar for the (unknown) target language from a finite set of examples of the language is known as *grammatical inference*. Language learning has been a well-established research area in machine learning originated by Gold [4]. Since his seminal work, there has been a remarkable amount of work to establish a theory of grammatical inference, to find effective and efficient models for learning grammars in polynomial time bound, and to apply those formal learning models to practical problems. For example, grammatical inference has successfully applied to natural language processing [8], pattern recognition [9], designing programming languages [3], biological computations [11], [12], [14], and many others. An excellent survey of challenges and specialized techniques can be found in the literature [5], [6], [15]. A specific survey on grammatical inference has been given in [7], [13].

J. Darzentas et al. (Eds.): SETN 2008, LNAI 5138, pp. 250–262, 2008.
© Springer-Verlag Berlin Heidelberg 2008

The learning of a context-free grammar (CFG) from input examples is an important topic in grammatical inference. This problem is, however, more difficult compared with those for learning of regular grammars, and other more restricted grammars, that can be learned from positive and negative samples. A positive sample presents all and only strings of the unknown language to the learning algorithm, whereas a negative sample contains only those strings that do not belong to the unknown language. Automated learning of CFGs requires additional information, which can be in the form of heuristics, structural descriptions, etc. A string of alphabet (or terminal) symbols in a grammar, with matched parentheses inserted in it (at appropriate places) to indicate the shape of the derivation tree of a CFG, is called a string with grammatical structure. A string with grammatical structure is also called a structured string or a structural description (of string) or simply a skeleton. For example, a structural description for "the old man writes a good story" can be represented as follows: $\langle\,\langle$ the $\langle$ old man $\rangle\,\rangle\,\langle$ writes $\langle$ a $\langle$ good story $\rangle\,\rangle\,\rangle\,\rangle$.

Sakakibara [10] extended the work of Angluin [2], and presented a polynomial time formalization based on tree automata to learn reversible CFGs from positive samples of their structural descriptions. He showed that the reversible CFG is a normal form for CFGs, i.e., reversible CFGs can generate all the context-free languages. In particular, Sakakibara also proposed an algorithm *RC* using his algorithm *RT* to learn reversible CFGs using skeletons. *RT* starts by constructing a *base tree automaton* for the input skeletons and generalizes it to find a particular *reversible tree automaton* by merging its states. Then *RC* constructs and outputs the reversible CFG consistent with the input examples. During merging process *RT* takes a variable LIST to store the pairs of states as merging candidates. However, *RT* also considers those pairs of states as candidates for merging that are rejected at later stage by it during the generalization process, which causes additional operations and storage space as well. In our formulation, we call such pairs as *redundant pairs*. We formally state the storage space required by *RT* for LIST in the following proposition:

Proposition 1. A total of at most $2n(n-1) + (n-1)$ pairs may be placed on LIST by the learning algorithm *RT*, i.e., the storage requirement for LIST in *RT* is $O(n^2)$, where n is the sum of the sizes of the input skeletons.

Furthermore, Sakakibara's scheme is non-incremental because it could not update a guess incrementally whenever some more examples are given to it; instead it restarts from the beginning which is impractical.

For solving above computational problem, we propose an incremental learning model for CFGs from positive structural samples. The main results include:

1. Storage Saving. It avoids redundant pairs of states to place on LIST as candidates for merging and hence reduces the size of LIST by a factor of $O(n)$.

2. Incremental Learning. It uses a simple updating scheme to have good incremental behavior so that it can also be employed as online learning, where the input sample is fed to the learner in an online manner.

3. Computational Complexity. It runs in $O(n^3)$ time. Moreover, it saves a lot of overheads by employing the incremental feature, and computational time to process LIST by rejecting redundant pairs of states to place on it.

We also demonstrate several examples and summary of our experiments to illustrate our model and to prove its theoretical analysis regarding the results.

2 Basic Definitions

Let N^* be the set of finite strings of natural numbers, separated by dots, formed using the catenation as the composition rule and the empty string λ as the identity. For y, $x \in N^*$, write $y \leq x$ iff there is a $z \in N^*$ with $x = y . z$; and $y < x$ iff $y \leq x$ and $y \neq x$. For $x \in N^*$, we define the length of x by $|x|$ as: $|\lambda| = 0$, $|x . n| = |x| + 1$ for $n \in N$.

A *ranked alphabet* V is a finite set of symbols associated with a finite relation, called the rank relation $r_V \subseteq V \times N$. Define $V_n = \{f \in V \mid (f, n) \in r_V \}$. In most of the cases the symbol in V_n are considered as *function symbols*. We say that a function symbol f has an arity n if $f \in V_n$ and a symbol of arity 0 is called a *constant symbol*.

A tree over V is a mapping $t : \mathrm{Dom}(t) \rightarrow V$, where $\mathrm{Dom}(t)$ is a finite subset of N^*. The set $\mathrm{Dom}(t)$ is called the *domain of the tree* t, such that (1) $y \leq x$ where $x \in \mathrm{Dom}(t)$ implies $y \in \mathrm{Dom}(t)$, and (2) $y . i \in \mathrm{Dom}(t)$, $i \in N$ whenever $y . j \in \mathrm{Dom}(t)$ ($1 \leq j \leq i$). An element of $\mathrm{Dom}(t)$ is also called a *node* or *position* of t, where the node λ is the root of the tree. We denote with $t(x)$ the label of a given node x in $\mathrm{Dom}(t)$. Then $t(x) \in V_n$ whenever, for $i \in N$, $x . i \in \mathrm{Dom}(t)$ iff $1 \leq i \leq n$.

V^T denotes the set of all trees over V. Interpreting V as a set of function symbols, V^T can be identified with the well-formed terms over V. This yields a compact string denotation of trees. Hence while declaring "let t be of the form $f(t_1, ...,t_n)...$", we also declare that f is of arity n. The notation $|\mathrm{Dom}(t)|$ denotes the cardinality of $\mathrm{Dom}(t)$, i.e., the size of t. A *frontier* (*terminal*) node in t is a node $x \in \mathrm{Dom}(t)$ such that there is no $y \in \mathrm{Dom}(t)$ with $x < y$. If $x \in \mathrm{Dom}(t)$ is not a frontier node, it is called *interior node*. Let the level of $x \in \mathrm{Dom}(t)$ be $|x|$. Then we define the depth of a tree t as $\mathrm{Depth}(t) = \max \{|x| : x \in \mathrm{Dom}(t)\}$. Let $t \in V^T$ and $x \in \mathrm{Dom}(t)$. The subtree t/x of t at x is a tree such that $\mathrm{Dom}(t/x) = \{y \mid x . y \in \mathrm{Dom}(t)\}$ and $t/x(y) = t(x . y)$, for every $y \in \mathrm{Dom}(t/x)$. Let T be a set of trees. We define the set $\mathrm{Sub}(T)$ of subtrees of elements of T by: $\mathrm{Sub}(T) = \{t/x \mid t \in T \text{ and } x \in \mathrm{Dom}(t)\}$.

A *partition* of some set S is a set of pairwise disjoint nonempty subsets of S whose union is S. If π is a partition of S, then for any element $s \in S$ there is a unique element of π containing s, which is denoted by $B(s, \pi)$ and we call the block of π containing s. The *trivial partition* of a set S is the class of all singleton sets $\{s\}$ such that $s \in S$.

Let π be a partition of some set S. Let $s_1, s_2 \in S$. We say that the pair (s_1, s_2) is *redundant* iff $B(s_1, \pi) = B(s_2, \pi)$ and it is *nonredundant* iff $B(s_1, \pi) \neq B(s_2, \pi)$.

Note that, if (s_1, s_2) is a redundant pair, then the pair (s_2, s_1) will also be redundant.

Example 1: As an example, consider the set $S = \{q_1, q_2, q_3, q_4, q_5, q_6, q_7, q_8, q_9, q_{10}, q_{11}\}$ and $\pi = \{ \{q_1, q_4\}, \{q_2\}, \{q_3\}, \{q_5, q_8, q_9\}, \{q_6\}, \{q_7\}, \{q_{10}\}, \{q_{11}\} \}$, then the pairs (q_1, q_4), (q_5, q_8), (q_5, q_9) and (q_8, q_9) are redundant, and (q_5, q_{10}) is one of the pairs which are nonredundant for the given instance.

Let V be a ranked alphabet and m be the maximum rank of the symbols in V. A tree *automaton* over V is a quadruple $A = (Q, V, \delta, F)$ such that Q is a finite set ($Q \cap V_0 = \varnothing$), F is a subset of Q, and $\delta = (\delta_0, \delta_1, ..., \delta_m)$ consists of the following maps:

$\delta_k : V_k \times (Q \cup V_0)^k \to 2^Q$ $(k = 1, 2, ..., m)$, and $\delta_0(a) = a$ for $a \in V_0$.

Q is the set of states, F is the set of final states of A and δ is the *state transition function* of A. In this definition, the symbols of 0 ranks on the frontier nodes are taken as *initial* states. The transition relation δ can be recursively defined to V^T by letting

$$\delta(f\ (t_1, ..., t_k)) = \begin{cases} \bigcup_{q_i \in \delta(t_i), i = 1, ..., k} \delta_k\ (f, q_1, ..., q_k) & \text{if } k > 0, \\ \{f\} & \text{if } k = 0. \end{cases}$$

The tree t is *accepted* by A iff $\delta(t) \cap F \neq \varnothing$. The *language* of A, i.e., the set of trees accepted by A, denoted $T(A)$, is defined as $T(A) = \{t \in V^T \mid \delta(t) \cap F \neq \varnothing\}$. It is clear that the tree automaton A cannot accept any tree of depth 0.

A tree automaton is *deterministic* iff for each k-tuple $q_1, ..., q_k \in Q \cup V_0$ and each symbol $f \in V_k$, there is at most one element in $\delta_k(f, q_1, ..., q_k)$. Deterministic tree automata can be viewed as algorithms for labelling the nodes of a tree with states.

Let $A = (Q, V, \delta, F)$ be any tree automaton. If π is any partition of Q, then we define another tree automaton $A/\pi = (Q', V, \delta', F')$ induced by π as follows: Q' is the set of blocks of π (i.e., $Q' = \pi$). F' is the set of all blocks of π that contain an element of F (i.e., $F' = \{B \in \pi \mid B \cap F \neq \varnothing\}$). δ' is a mapping from $V_k \times (\pi \cup V_0)^k$ to 2^π and for B_1, ..., $B_k \in Q' \cup V_0$ and $f \in V_k$, the block B is in $\delta'_k (f, B_1, ..., B_k)$ whenever there exist $q \in B$ and $q_i \in B_i \in \pi$ or $q_i = B_i \in V_0$ for $i = 1, ..., k$ such that $q = \delta_k(f, q_1, ..., q_k)$.

Let S_+ be a finite set of trees of V^T. We denote by $Bs(S_+) = (Q, V, \delta, F)$, the base tree automaton for S_+, where: $Q = \text{Sub}(S_+) - V_0$, $F = S_+$, $\delta_k(f, u_1, ..., u_k) = f(u_1, ..., u_k)$ whenever $u_1, ..., u_k \in Q \cup V_0$ and $f(u_1, ..., u_k) \in Q$, and $\delta_0(a) = a$ for $a \in V_0$.

An *alphabet* Σ is a finite nonempty set of symbols. Σ^* denotes the set of all finite strings over Σ. We denote the empty string by ε. The length of a string w is denoted $|w|$, so $|\varepsilon| = 0$. A language L over Σ is a subset of Σ^*.

A grammar is a quadruple $G = (N, \Sigma, P, S)$, where N and Σ are alphabets of non-terminals and terminals, respectively, such that $N \cap \Sigma = \varnothing$. P is a finite set of productions and $S \in N$ is a special nonterminal called the *start symbol*. A grammar $G = (N, \Sigma, P, S)$ is called *context-free* if all production rules are of the form $A \to \alpha$, where $A \in N$ and $\alpha \in (N \cup \Sigma)^*$. If $A \to \beta$ is a production of P and α and γ are any strings in $(N \cup \Sigma)^*$, then $\alpha A \gamma \Rightarrow \alpha \beta \gamma$. $\Rightarrow^*$ is the reflexive and transitive closure of $\Rightarrow$. The *language generated* by the grammar G, denoted $L(G)$, is $\{w \in \Sigma^* \mid S \Rightarrow^* w\}$.

Let $G = (N, \Sigma, P, S)$ be a CFG. For $A \in N \cup \Sigma$, we denote by $D_A(G)$ the set of trees over $N \cup \Sigma$ and is defined by

$$D_A(G) = \begin{cases} \{a\} & \text{if } A = a \in \Sigma, \\ \{A(t_1, ..., t_k) \mid A \to B_1, ..., B_k, \ t_1 \in D_{B_1}(G), ..., t_k \in D_{B_k}(G)\} & \text{if } A \in N. \end{cases}$$

An element of $D_A(G)$ is called a *derivation tree* of G from A. For the set $D_S(G)$ of derivation trees of G from the start symbol S, the S-subscript will be omitted.

A *skeletal alphabet* Sk is a ranked alphabet consisting of exactly one special symbol σ with the rank relation $r_{Sk} \subseteq \{\sigma\} \times \{1, 2, 3, ..., m\}$, where m is the maximum rank of the symbols in *Sk*. A *skeleton* is a tree defined over $Sk \cup V_0$. The *skeletal* (or *structural*) *description* of $t \in V^T$, denoted $s(t)$, is a skeleton with $\mathrm{Dom}(s(t)) = \mathrm{Dom}(t)$:

$$s(t)(x) = \begin{cases} t(x) & \text{if } x \text{ is a frontier node,} \\ \sigma & \text{if } x \text{ is an interior node.} \end{cases}$$

Thus a skeleton is a tree in which all interior nodes are labelled with a special label σ. The skeletal description of a tree conserves the structure of the tree, but not the label names describing that structure. The *skeletal set* of the set of trees T, denoted $K(T)$, is $\{s(t) \mid t \in T\}$. A skeleton in $K(D(G))$ is called a *structural description* of G. A tree automaton over $Sk \cup \Sigma$ is called a *skeletal tree automaton*.

3 Reversible CFGs

A skeletal tree automaton $A = (Q, Sk \cup \Sigma, \delta, F)$ is *reversible* iff A has only one final state, A is deterministic, and for no two distinct states q_1 and q_2 in Q do there exist a $\sigma \in Sk_k$, a state $q_3 \in Q$, an integer $i \in N$ ($1 \leq i \leq k$), and $k-1$ -tuple $u_1, ..., u_{k-1} \in Q \cup \Sigma$ such that $\delta_k(\sigma, u_1, ..., u_{i-1}, q_1, u_i, ..., u_{k-1}) = q_3 = \delta_k(\sigma, u_1, ..., u_{i-1}, q_2, u_i, ..., u_{k-1})$.

Let $A = (Q, Sk \cup \Sigma, \delta, \{q_f\})$ be a reversible skeletal tree automaton for a skeletal set. The corresponding CFG $G'(A) = (N, \Sigma, P, S)$ is defined as: $N = Q, S = q_f$, and $P = \{\delta_k(\sigma, x_1, ..., x_k) \to x_1, ..., x_k \mid \sigma \in Sk_k, x_1, ..., x_k \in Q \cup \Sigma$ and $\delta_k(\sigma, x_1, ..., x_k)$ is defined $\}$.

A CFG $G = (N, \Sigma, P, S)$ is reversible, iff: (1) there do not exist productions $A \to \alpha$ and $B \to \alpha$ in P for any two distinct nonterminals $A, B \in N$, where $\alpha \in (N \cup \Sigma)^*$, and (2) there do not exist productions $A \to \alpha B \beta$ and $A \to \alpha C \beta$ in P for any two distinct nonterminals $B, C \in N$, where $A \in N$, and $\alpha, \beta \in (N \cup \Sigma)^*$.

Remark 1. (Sakakibara [10]). For any context-free language L, there is a reversible CFG G such that $L(G) = L$, and if all strings in L are of length ≥ 2, then there is an ε-free reversible CFG G such that $L(G) = L$.

4 Learning Algorithms

In this section, we first describe and analyze our algorithm *IRT* for incremental learning of reversible skeletal tree automata from positive samples. Next we apply this algorithm to learn CFGs from positive samples of their structural descriptions. We restrict our consideration to only ε-free CFGs, because ε-free grammars are important in practical applications such as efficient parsing.

A *positive sample* of a tree automaton A is a finite subset of $T(A)$. A positive sample S_+ of a reversible skeletal tree automaton A is a *characteristic sample* for A iff for any reversible skeletal tree automaton A', $T(A') \supseteq S_+$ implies $T(A) \subseteq T(A')$. The use of a characteristic sample is to avoid the problem of guessing a language that is a strict superset of the target language and for identification from positive data.

4.1   The Incremental Learning Algorithm *IRT* for Tree Automata

The inputs to *IRT* are the reversible skeletal tree automaton A_+ (if any) computed by *IRT* on input a nonempty positive sample S_+, and a finite nonempty set of skeletons S_{++} given in the subsequent stage of the learning. That is *IRT* may start with the zero experience or it might have some intermediate results based on the examples in S_+ it has received earlier for the target language. The output is a particular reversible skeletal tree automaton $A_{++} = IRT(A_+, S_{++})$ whose characteristic sample is precisely the input sample, *i.e.* $S_+ \cup S_{++}$. Depending on inputs to *IRT*, it performs as follows:

(1) A_+ *is empty.* If S_{++} is a finite nonempty set of skeletons given to the algorithm at the initial stage of the learning process, then on input S_{++}, *IRT* first constructs the base tree automaton $A_{++} = Bs(S_{++})$. It then begins with the trivial partition π_0 of the set Q of states of A_{++} and by employing merging passes it finds the final partition π_f of the set Q with the properties that the tree automaton A_{++}/π_f is reversible, and $S_{++} \subseteq T(A_{++}/\pi_f)$.

(2) A_+ *is not empty.* In this case *IRT* correctly sets the initial partition (i.e., the partition from which merging process starts), say π_1 so that the merging passes required to find A_+ is reduced. Thus *IRT* updates a guess using previous result and based on the example(s) in S_{++} without starting from the trivial partition of the set of states of $Bs(S_+ \cup S_{++})$. It then makes necessary initializations to find the base tree automaton A_{++} for $S_+ \cup S_{++}$, and goes on merging passes to obtain the final partition π_f of the set Q of states of A_{++} such that A_{++}/π_f is reversible and $S_+ \cup S_{++} \subseteq T(A_{++}/\pi_f)$.

So, for constructing π_f, *IRT* begins with the initial partition π_0 or π_1 (as the case may be) and repeatedly merges any two distinct blocks that violate any one of the criteria in the definition of reversible tree automaton. When no longer remains any such pair of blocks, the resulting partition is π_f. *IRT* outputs A_{++}/π_f and halts.

The learning algorithm *IRT* is illustrated in Fig. 1. This completes the description of *IRT*, and we next determine its computational time.

4.2   Time Complexity of *IRT*

Theorem 2. The execution time of the learning algorithm *IRT* in worst case is bounded by $O(n^3)$, where parameter n is the sum of the sizes of the input skeletons and size of a skeleton (or tree) t is the number of nodes in t, i.e., $|Dom(t)|$.

Proof. Let $Sa = S_+ \cup S_{++}$ be the set of input skeletons, n be the sum of the sizes of the skeletons in Sa, and d be the maximum rank of the symbol σ in Sk. The automaton A_{++} can be found in $O(n)$ time and contains at most n states. Similarly, the time to output the final reversible tree automaton is $O(n)$. The partition π_f of the states of A_{++} may be queried and updated using the simple MERGE and FIND operations described in [1]. Processing each pair of states from LIST entails two FIND operations to determine the blocks containing the two states. Then they are merged with a MERGE operation. At most $2(d+1)n(n-1)$ and $2dn(n-1)$ FIND operations are required by *p*-UPDATE and *s*-UPDATE, respectively excluding the number of FIND operations required by the function *n*-REDUNDANT. At most $n-1$ new pairs may be requested to place on LIST, so the maximum of $2(n-1) + 2(n-1) = 4(n-1)$ FIND operations are processed by the function

Input: the reversible skeletal tree automaton $A_+ = Bs(S_+)/\pi'_f$ (may be empty) computed by *IRT* on input a nonempty positive sample S_+ and a finite nonempty set of skeletons S_{++}, where π'_f is the final partition of the set of states of $Bs(S_+)$ constructed by *IRT*;
Output: a reversible skeletal tree automaton A_{++};
Procedure:
/* *Initialization, if A_+ is not empty, machine has experience* */
Let $A' = (Q', V, \delta', F')$ be $Bs(S_{++})$;
Let π_0 be the trivial partition of Q';
Let $\pi_1 = \pi_0 \cup \pi'_f$;
Let $B_f \in \pi'_f$ be the block such that it is the final state of A_+;
Let π_t be π_1 with B_f and all blocks $B(q, \pi_0)$ such that $q \in F'$ merged;
Choose some $q_f \in B_f$; Let LIST contain all pairs (q_f, q') such that $q' \in F'$;
Let $i = 1$;
Let $A_{++} = (Q, V, \delta, F)$ such that Q = the set of states of $Bs(S_+) \cup Q'$,
  δ = the set of state transitions of $Bs(S_+) \cup \delta'$, F = the set of final states of $Bs(S_+) \cup F$,
  V_0 = the set of constant symbols in $S_+ \cup$ the set of constant symbols in S_{++};
/* *Initialization, if A_+ is empty, machine has no experience* */
Let $A_{++} = (Q, V, \delta, F)$ be $Bs(S_{++})$; Let π_0 be the trivial partition of Q;
Let π_t be π_0 with all blocks $B(q, \pi_0)$ such that $q \in F$ merged;
Choose some $q \in F$; Let LIST contain all pairs (q, q') such that $q' \in F - \{q\}$;
Let $i = 0$;
While LIST $\neq \varnothing$ **do**   /* *Merging* */
  Begin
    Remove first element (q_1, q_2) from LIST;
    Let $B_1 = B(q_1, \pi_i)$ and $B_2 = B(q_2, \pi_i)$;
    Let π_{i+1} be π_i with B_1 and B_2 merged;
    p-UPDATE(π_{i+1}) and s-UPDATE(π_{i+1}, B_1, B_2);
    Increase i by 1;
  End
Let $f = i$ and output the tree automaton A_{++}/π_f.
where
  p-UPDATE(π_{i+1}) is:
    For all pairs of states $q'_1 = \sigma(u_1, ..., u_k)$ and $q'_2 = \sigma(u'_1, ..., u'_k)$ in Q with $B(u_j,$
      $\pi_{i+1}) = B(u'_j, \pi_{i+1})$ or $u_j = u'_j \in \Sigma$ for $1 \leq j \leq k$ and $B(q'_1, \pi_{i+1}) \neq B(q'_2, \pi_{i+1})$
    Do if n-REDUNDANT(π_t, q'_1, q'_2) then add the pair (q'_1, q'_2) to LIST;
  s-UPDATE(π_{i+1}, B_1, B_2) is:
    For all pairs of states $\sigma(u_1, ..., u_k) \in B_1$ and $\sigma(u'_1, ..., u'_k) \in B_2$ with $u_l, u'_l \in Q$
      and $B(u_l, \pi_{i+1}) \neq B(u'_l, \pi_{i+1})$ for some l $(1 \leq l \leq k)$ and $B(u_j, \pi_{i+1}) = B(u'_j, \pi_{i+1})$
      or $u_j = u'_j \in \Sigma$ for $1 \leq j \leq k$ and $j \neq l$
    Do if n-REDUNDANT$(\pi_t, q'_1 = u_l, q'_2 = u'_l)$ then add the pair (u_l, u'_l) to LIST.
  n-REDUNDANT(π_t, q'_1, q'_2) is:
    Let $B'_1 = B(q'_1, \pi_t)$ and $B'_2 = B(q'_2, \pi_t)$;
    If $B'_1 \neq B'_2$ **then**
      Begin Let π'_t be π_t with B'_1 and B'_2 merged; Let $\pi_t = \pi'_t$; return TRUE; **End**
    Else return FALSE.

Fig. 1. The incremental learning algorithm *IRT* for reversible tree automata

n-REDUNDANT to check nonredundant pairs. Therefore for processing each pair of states from LIST requires a maximum of $2(n-1)$ $((2d+1)$ $n + 2) + 2$ FIND operations. LIST can contain maximum $n-1$ pairs. Thus a total of at most $(2(n-1)$ $((2d+1)$ $n + 2) + 2)$ $(n - 1)$ FIND operations and a total of at most 2 $(n - 1)$ MERGE operations are required during merging passes. The operation MERGE takes $O(n)$ time and the operation FIND takes constant time, so the learning algorithm IRT requires a total time of $O(n^3)$.   □

4.3 The Learning Algorithm IRC for CFGs

A positive structural sample of a CFG G is a finite subset of $K(D(G))$. A positive structural sample S_+ of a reversible CFG G is a characteristic structural sample for G iff for any reversible CFG G', $K(D(G')) \supseteq S_+$ implies $K(D(G)) \subseteq K(D(G'))$.

The algorithm IRC to learn reversible CFGs from positive samples of their structural descriptions is sketched in Fig. 2. IRC outputs a particular reversible CFG $G = IRC(A_+, S_{++})$ whose characteristic sample is precisely the input sample. We state running time of IRC in Theorem 3, which can be easily shown by using Theorem 2.

> **Input**: the reversible skeletal tree automaton $A_+ = Bs(S_+)/\pi'_f$ (may be empty) computed by IRT on input a nonempty positive sample S_+ and a finite nonempty set of positive structural examples S_{++}, where π'_f is the final partition of the set of states of $Bs(S_+)$ constructed by IRT ;
> **Output**: a reversible CFG G;
> *Procedure*:
> Run IRT on the tree automaton A_+ and the set S_{++};
> Let $G = G'(IRT(A_+, S_{++}))$ and output the grammar G.

Fig. 2. The incremental learning algorithm IRC for reversible grammars

Theorem 3. The execution time of the learning algorithm IRC is bounded by $O(n^3)$ in the sum of the sizes of the input skeletons, where size of a skeleton (or tree) t is the number of nodes in t, i.e., $|Dom(t)|$.

5 An Example

In the following, we demonstrate an example to show the incremental learning process of the learning algorithm IRC.

First suppose that the learning algorithm IRC is given the following sample:

Sample 1: $\{\langle\langle\langle$ the $\rangle\langle\langle$ girl $\rangle\rangle\rangle\langle\langle$ likes $\rangle\langle\langle$ a $\rangle\langle\langle$ cat $\rangle\rangle\rangle\rangle\rangle$,
$\langle\langle\langle$ the $\rangle\langle\langle$ girl $\rangle\rangle\rangle\langle\langle$ likes $\rangle\langle\langle$ a $\rangle\langle\langle$ dog $\rangle\rangle\rangle\rangle\rangle\}$.

Clearly $S_+ = \varnothing$ and A_+ is empty (the algorithm starts with zero experience). The base tree automaton $A_{++} = Bs(Sample\ 1) = (Q, V, \delta, F)$ can be constructed as follows. $V_0 = \{$the, girl, likes, a, cat, dog$\}$. All states of A_{++} are of the form $q = \sigma(u_1, ..., u_k)$, where $\sigma \in Sk$, and $u_1, ..., u_k \in Q \cup V_0$, and the trivial partition π_0 of the set Q are:

$$q_1 = \sigma(\text{the})$$
$$q_2 = \sigma(\text{girl})$$
$$q_3 = \sigma(q_2)$$
$$q_4 = \sigma(q_1, q_3)$$
$$q_5 = \sigma(\text{likes})$$
$$q_6 = \sigma(\text{a})$$
$$q_7 = \sigma(\text{cat})$$
$$q_8 = \sigma(q_7)$$

$$q_9 = \sigma(q_6, q_8)$$
$$q_{10} = \sigma(q_5, q_9)$$
$$q_{11} = \sigma(q_4, q_{10}) \in F$$
$$q_{12} = \sigma(\text{the})$$
$$q_{13} = \sigma(\text{girl})$$
$$q_{14} = \sigma(q_{13})$$
$$q_{15} = \sigma(q_{12}, q_{14})$$
$$q_{16} = \sigma(\text{likes})$$

$$q_{17} = \sigma(\text{a})$$
$$q_{18} = \sigma(\text{dog})$$
$$q_{19} = \sigma(q_{18})$$
$$q_{20} = \sigma(q_{17}, q_{19})$$
$$q_{21} = \sigma(q_{16}, q_{20})$$
$$q_{22} = \sigma(q_{15}, q_{21}) \in F.$$

$\pi_0 = \{\{q_1\}, \{q_2\}, \{q_3\}, \{q_4\}, \{q_5\}, \{q_6\}, \{q_7\}, \{q_8\}, \{q_9\}, \{q_{10}\}, \{q_{11}\}, \{q_{12}\}, \{q_{13}\},$
$\{q_{14}\}, \{q_{15}\}, \{q_{16}\}, \{q_{17}\}, \{q_{18}\}, \{q_{19}\}, \{q_{20}\}, \{q_{21}\}, \{q_{22}\}\}.$

IRT begins with π_0 and finds the following final partition π_f, and outputs A_{++}/π_f :

$\pi_f = \{\{q_1, q_{12}\}, \{q_2, q_{13}\}, \{q_3, q_{14}\}, \{q_4, q_{15}\}, \{q_5, q_{16}\}, \{q_6, q_{17}\}, \{q_7\}, \{q_8\}, \{q_9\},$
$\{q_{10}\}, \{q_{11}, q_{22}\}, \{q_{18}\}, \{q_{19}\}, \{q_{20}\}, \{q_{21}\}\}.$

If we call the blocks of the above partition, π_f as $NT1 = \{q_1, q_{12}\}$, $NT2 = \{q_2, q_{13}\}$, $NT3 = \{q_3, q_{14}\}$, $NT4 = \{q_4, q_{15}\}$, $NT5 = \{q_5, q_{16}\}$, $NT6 = \{q_6, q_{17}\}$, $NT7 = \{q_7\}$, $NT8 = \{q_8\}$, $NT9 = \{q_9\}$, $NT10 = \{q_{10}\}$, $NT11 = \{q_{18}\}$, $NT12 = \{q_{19}\}$, $NT13 = \{q_{20}\}$, $NT14 = \{q_{21}\}$, $S = \{q_{11}, q_{22}\}$, then CFG corresponding to A_{++}/π_f is defined by:

$S \rightarrow NT4\ NT10$
$S \rightarrow NT4\ NT14$
$NT1 \rightarrow \text{the}$
$NT2 \rightarrow \text{girl}$
$NT3 \rightarrow NT2$
$NT4 \rightarrow NT1\ NT3$

$NT5 \rightarrow \text{likes}$
$NT6 \rightarrow \text{a}$
$NT7 \rightarrow \text{cat}$
$NT8 \rightarrow NT7$
$NT9 \rightarrow NT6\ NT8$
$NT10 \rightarrow NT5\ NT9$

$NT11 \rightarrow \text{dog}$
$NT12 \rightarrow NT11$
$NT13 \rightarrow NT6\ NT12$
$NT14 \rightarrow NT5\ NT13.$

However, it is not reversible (for example, the productions $S \rightarrow NT4\ NT10$, and $S \rightarrow NT4\ NT14$ do not satisfy the condition (2) in the definition of the reversible CFG). So *IRC* merges distinct nonterminals repeatedly and outputs the reversible CFG:

$S \rightarrow NT1\ NT2$
$NT1 \rightarrow NT3\ NT4$
$NT2 \rightarrow NT6\ NT7$
$NT3 \rightarrow \text{the}$

$NT4 \rightarrow NT5$
$NT5 \rightarrow \text{girl}$
$NT6 \rightarrow \text{likes}$
$NT7 \rightarrow NT8\ NT9$

$NT8 \rightarrow \text{a}$
$NT9 \rightarrow NT10$
$NT10 \rightarrow \text{cat}$
$NT10 \rightarrow \text{dog}.$

Suppose that in the next stage the following examples are added to the sample:

Sample 2: $\{\langle\langle\langle a\rangle\langle\langle dog\rangle\rangle\rangle\langle\langle chases\rangle\langle\langle the\rangle\langle\langle girl\rangle\rangle\rangle\rangle\rangle,$
$\langle\langle\langle a\rangle\langle\langle dog\rangle\rangle\rangle\langle\langle chases\rangle\langle\langle a\rangle\langle\langle cat\rangle\rangle\rangle\rangle\rangle\}.$

Now $S_{++} = $ *Sample 2* and A_+ is the automaton found in the previous stage on input *Sample 1*. The states of the base tree automaton $A' = Bs(S_{++}) = (Q', V, \delta', F')$ are of the form $q = \sigma(u_1, ..., u_k)$ and the trivial partition π_0 of the set Q' are:

$$q_{23} = \sigma(a)$$ $$q_{31} = \sigma(q_{28}, q_{30})$$ $$q_{39} = \sigma(a)$$

$$q_{24} = \sigma(dog)$$ $$q_{32} = \sigma(q_{27}, q_{31})$$ $$q_{40} = \sigma(cat)$$

$$q_{25} = \sigma(q_{24})$$ $$q_{33} = \sigma(q_{26}, q_{32}) \in F'$$ $$q_{41} = \sigma(q_{40})$$

$$q_{26} = \sigma(q_{23}, q_{25})$$ $$q_{34} = \sigma(a)$$ $$q_{42} = \sigma(q_{39}, q_{41})$$

$$q_{27} = \sigma(chases)$$ $$q_{35} = \sigma(dog)$$ $$q_{43} = \sigma(q_{38}, q_{42})$$

$$q_{28} = \sigma(the)$$ $$q_{36} = \sigma(q_{35})$$ $$q_{44} = \sigma(q_{37}, q_{43}) \in F'.$$

$$q_{29} = \sigma(girl)$$ $$q_{37} = \sigma(q_{34}, q_{36})$$

$$q_{30} = \sigma(q_{29})$$ $$q_{38} = \sigma(chases)$$

$$\pi_0 = \{\{q_{23}\}, \{q_{24}\}, \{q_{25}\}, \{q_{26}\}, \{q_{27}\}, \{q_{28}\}, \{q_{29}\}, \{q_{30}\}, \{q_{31}\}, \{q_{32}\}, \{q_{33}\}, \{q_{34}\}, \{q_{35}\}, \{q_{36}\}, \{q_{37}\}, \{q_{38}\}, \{q_{39}\}, \{q_{40}\}, \{q_{41}\}, \{q_{42}\}, \{q_{43}\}, \{q_{44}\}\}.$$

Initialization for $A_{++} = (Q, V, \delta, F)$:

$V_0 = \{a, dog, chases, the, girl, cat, likes\}$. $F = \{q_{11}, q_{22}\} \cup \{q_{33}, q_{44}\}$.

$$Q = \{q_1, q_2, q_3, q_4, q_5, q_6, q_7, q_8, q_9, q_{10}, q_{11}, q_{12}, q_{13}, q_{14}, q_{15}, q_{16}, q_{17}, q_{18}, q_{19}, q_{20}, q_{21}, q_{22}\} \cup \{q_{23}, q_{24}, q_{25}, q_{26}, q_{27}, q_{28}, q_{29}, q_{30}, q_{31}, q_{32}, q_{33}, q_{34}, q_{35}, q_{36}, q_{37}, q_{38}, q_{39}, q_{40}, q_{41}, q_{42}, q_{43}, q_{44}\}.$$

Now *IRT* sets the following initial partition π_1 to start merging:

$$\pi_1 = \{\{q_1, q_{12}\}, \{q_2, q_{13}\}, \{q_3, q_{14}\}, \{q_4, q_{15}\}, \{q_5, q_{16}\}, \{q_6, q_{17}\}, \{q_7\}, \{q_8\}, \{q_9\}, \{q_{10}\}, \{q_{11}, q_{22}\}, \{q_{18}\}, \{q_{19}\}, \{q_{20}\}, \{q_{21}\}, \{q_{23}\}, \{q_{24}\}, \{q_{25}\}, \{q_{26}\}, \{q_{27}\}, \{q_{28}\}, \{q_{29}\}, \{q_{30}\}, \{q_{31}\}, \{q_{32}\}, \{q_{33}\}, \{q_{34}\}, \{q_{35}\}, \{q_{36}\}, \{q_{37}\}, \{q_{38}\}, \{q_{39}\}, \{q_{40}\}, \{q_{41}\}, \{q_{42}\}, \{q_{43}\}, \{q_{44}\}\}.$$

Beginning with the partition π_1 the algorithm *IRT* finds the final partition π_f:

$$\pi_f = \{\{q_1, q_{12}, q_{28}\}, \{q_2, q_{13}, q_{29}\}, \{q_3, q_{14}, q_{30}\}, \{q_4, q_{15}, q_{31}\}, \{q_5, q_{16}\}, \{q_6, q_{17}, q_{23}, q_{34}, q_{39}\}, \{q_7, q_{40}\}, \{q_8, q_{41}\}, \{q_9, q_{42}\}, \{q_{10}\}, \{q_{11}, q_{22}, q_{33}, q_{44}\}, \{q_{18}, q_{24}, q_{35}\}, \{q_{19}, q_{25}, q_{36}\}, \{q_{20}, q_{26}, q_{37}\}, \{q_{21}\}, \{q_{27}, q_{38}\}, \{q_{32}\}, \{q_{43}\}\}.$$

Finally *IRC* outputs the following reversible CFG:

$S \rightarrow NT1\ NT2$	$NT4 \rightarrow NT5$	$NT8 \rightarrow dog$
$NT1 \rightarrow NT3\ NT4$	$NT5 \rightarrow girl$	$NT9 \rightarrow likes$
$NT1 \rightarrow NT6\ NT7$	$NT6 \rightarrow a$	$NT9 \rightarrow chases.$
$NT2 \rightarrow NT9\ NT1$	$NT7 \rightarrow NT8$	
$NT3 \rightarrow the$	$NT8 \rightarrow cat$	

Next suppose that the following examples are further added to the sample:

Sample 3: $\{\langle\langle\langle a \rangle\langle\langle dog \rangle\rangle\rangle\langle\langle chases \rangle\langle\langle a \rangle\langle\langle girl \rangle\rangle\rangle\rangle\rangle,$
$\langle\langle\langle the \rangle\langle\langle dog \rangle\rangle\rangle\langle\langle chases \rangle\langle\langle a \rangle\langle\langle young \rangle\langle\langle girl \rangle\rangle\rangle\rangle\rangle\rangle\}.$

Then by utilizing the incremental feature as described above, *IRC* outputs the following reversible CFG:

$$S \rightarrow NT1\ NT2$$
$$NT1 \rightarrow NT3\ NT4$$
$$NT4 \rightarrow NT5$$

$$NT\,4 \rightarrow NT\,6\;NT\,4$$
$$NT\,2 \rightarrow NT\,7\;NT\,1$$
$$NT\,3 \rightarrow \text{the}$$
$$NT\,3 \rightarrow \text{a}$$
$$NT\,5 \rightarrow \text{girl}$$
$$NT\,5 \rightarrow \text{cat}$$
$$NT\,5 \rightarrow \text{dog}$$
$$NT\,6 \rightarrow \text{young}$$
$$NT\,7 \rightarrow \text{likes}$$
$$NT\,7 \rightarrow \text{chases.}$$

6 Experimental Results

The experiments that we have done are to see how our results compare with those of Sakakibara [10]. We take the following primary structural examples of grammars:

Sample 4: Sample 1 ∪ Sample 2.

Sample 5: Sample 1 ∪ Sample 2 ∪ Sample 3.

Sample 6: { ⟨ v := ⟨⟨⟨ v ⟩⟩ + ⟨⟨⟨ v ⟩⟩⟩⟩ ⟩, ⟨ v := ⟨⟨⟨ v ⟩ × ⟨⟨ v ⟩⟩⟩⟩ ⟩,

         ⟨ v := ⟨⟨⟨ v ⟩⟩ + ⟨⟨⟨ v ⟩ × ⟨⟨ v ⟩⟩⟩⟩⟩ ⟩,

         ⟨ v := ⟨⟨⟨ '(' ⟨⟨⟨ v ⟩⟩ + ⟨⟨⟨ v ⟩⟩⟩⟩ ')' ⟩ × ⟨⟨ v ⟩⟩⟩⟩ ⟩ }.

Sample 7:

   { ⟨ while ⟨⟨⟨⟨ v ⟩⟩⟩ > ⟨⟨⟨ v ⟩ × ⟨⟨ v ⟩⟩⟩⟩⟩ do ⟨ v:= ⟨⟨⟨ v ⟩⟩ + ⟨⟨⟨ v ⟩⟩⟩⟩ ⟩ ⟩,

   ⟨ if ⟨⟨⟨⟨ v ⟩⟩⟩ > ⟨⟨⟨ v ⟩ × ⟨⟨ v ⟩⟩⟩⟩⟩ then ⟨ v:= ⟨⟨⟨ v ⟩⟩ + ⟨⟨⟨ v ⟩⟩⟩⟩ ⟩ ⟩,

   ⟨ begin ⟨⟨ v:= ⟨⟨⟨ v ⟩⟩ + ⟨⟨⟨ v ⟩⟩⟩⟩ ⟩ ; ⟨⟨ v:= ⟨⟨⟨ v ⟩ × ⟨⟨ v ⟩⟩⟩⟩ ⟩⟩⟩ end ⟩,

   ⟨ begin ⟨⟨ v:= ⟨⟨⟨ v ⟩ × ⟨⟨ v ⟩⟩⟩⟩ ⟩⟩⟩ end ⟩ }.

Sample 8: Sample 6 ∪ Sample 7.

Sample 9: { ⟨⟨ ab ⟩⟨ cd ⟩⟩, ⟨⟨ a ⟨ ab ⟩ b ⟩⟨ c ⟨ cd ⟩ d ⟩⟩, ⟨⟨ ab ⟩⟨ c ⟨ cd ⟩ d ⟩⟩ }.

Sample 10: { ⟨ a ⟨⟨⟨ bc ⟩⟩⟩ d ⟩, ⟨ a ⟨⟨ a ⟨⟨ b ⟨ bc ⟩ c ⟩⟩ d ⟩⟩ d ⟩,

        ⟨ a ⟨⟨⟨ b ⟨ bc ⟩ c ⟩⟩⟩ d ⟩ }.

Sample 11: Sample 9 ∪ Sample 10.

The first set of experiments investigates how by rejecting redundant pairs of states to add to LIST reduces the storage space and improves the efficiency of the learning algorithm. Our *IRT*, and *RT* [10] are executed for the above training examples. The results are summarized in Table 1. These results verify the theoretical analysis regarding the space savings by a factor of $O(n)$ introduced due to our scheme over *RT*.

The experimental results of the second set of experiments demonstrate the need of incremental behavior in the learning algorithm over the non-incremental one. We have executed the algorithms *RC* [10] and ours *IRC* on the respective samples for learning three different kinds of grammars. In all these cases the training examples are supplied stage by stage to the algorithms. The summary of our experiments is shown in Table 2, Table 3 and Table 4. In these tables *#IRC* and *#RC* refer the number of merging passes required by *IRC* and *RC* respectively, "*Empty*" is used to denote that $S_+ = \varnothing$ (i.e., A_+ is empty), and examples in the S_{++} are given in the subsequent stages.

Table 1. LIST Behavior in *RT* and *IRT*

Structural Examples (S_{++})	Number of States in $Bs(S_{++})$	Size of LIST in RT N_a	Size of LIST in IRT N_b	$\#P_m$
Sample 5	68	1447	43	97.03%
Sample 8	92	13372	86	99.36%
Sample 11	27	94	19	79.79%

$\#P_m$: $((N_a - N_b) / N_a) \times 100$, the percentage of memory saving in IRT.

Table 2. Experimental results to learn a CFG for a simple natural language

Sample S_+	The Set S_{++}	$S_{++} \cup S_+$	#IRC	#RC
Empty	*Sample 1*	*Sample 1*	07	07
Sample 1	*Sample 2*	*Sample 4*	19	26
Sample 4	*Sample 3*	*Sample 5*	17	43
	Total		43	76

Table 3. Experimental results to learn a CFG for a subset of the syntax for a programming language Pascal

Sample S_+	The Set S_{++}	$S_{++} \cup S_+$	#IRC	#RC
Empty	*Sample 6*	*Sample 6*	30	30
Sample 6	*Sample 7*	*Sample 8*	56	86
	Total		86	116

Table 4. Experimental results to learn a CFG for the language $\{a^m b^m c^n d^n \mid m \geq 1, n \geq 1\} \cup \{a^m b^n c^n d^m \mid m \geq 1, n \geq 1\}$

Sample S_+	The Set S_{++}	$S_{++} \cup S_+$	#IRC	#RC
Empty	*Sample 9*	*Sample 9*	07	07
Sample 9	*Sample 10*	*Sample 11*	12	19
	Total		19	26

It is clear that computational time of the learning algorithms is proportional to the number of merging passes required by them. Therefore an efficient learning aims to learn CFGs with less number of passes. The results in the above tables confirm that our *IRC* requires less number of passes for learning the target grammars compared to *RC*.

7 Concluding Remarks

In this paper, we have considered the problem of incremental learning of CFGs from positive samples of their structural descriptions and investigated the effects of updating a guess incrementally and rejecting redundant pairs of states to add to LIST. We have shown that our scheme conserves the storage space requirements by a factor of $O(n)$ by managing LIST efficiently and runs in $O(n^3)$ time in the worst case. We have presented the summary of our experiments using representative selected examples of grammars. Our experiments confirm the theoretical analysis regarding the space savings introduced due to our scheme, and demonstrate the need of incremental behavior in the learning algorithm.

References

1. Aho, A.V., Hopcroft, J.E., Ullman, J.D.: Data Structures and Algorithms. Addison-Wesley, Reading (1983)
2. Angluin, D.: Inference of Reversible Languages. J. ACM 29, 741–765 (1982)
3. Crespi-Reghizzi, S., Melkanoff, M.A., Lichten, L.: The use of Grammatical Inference for Designing Programming Languages. Comm. ACM 16, 83–90 (1973)
4. Gold, E.M.: Language Identification in the Limit. Inform. and Control 10, 447–474 (1967)
5. de la Higuera, C.: Current Trends in Grammatical Inference. In: Amin, A., Pudil, P., Ferri, F.J., Iñesta, J.M. (eds.) SPR 2000 and SSPR 2000. LNCS, vol. 1876, pp. 28–31. Springer, Heidelberg (2000)
6. Higuera, C., de la Higuera, C.: A Bibliographical Study of Grammatical Inference. Pattern Recognition 38, 1332–1348 (2005)
7. Lee, S.: Learning of Context-Free Languages: A Survey of the Literatures. Technical Report TR-12-96, Centre for Research in Computing Technology, Harvard University Cambridge, Massachusetts (1996)
8. Pereira, F., Schabes, Y.: Inside-Outside Reestimation for Partially Bracketed Corpora. In: Proc. 30th Ann. Meeting of the Assoc. for the Comput. Linguistics, pp. 128–135 (1992)
9. Richetin, M., Vernadat, F.: Efficient Regular Grammatical Inference for Pattern Recognition. Pattern Recognition 17(2), 245–250 (1984)
10. Sakakibara, Y.: Efficient Learning of Context-Free Grammars from Positive Structural Examples. Inform. and Comput. 97, 23–60 (1992)
11. Sakakibara, Y.: Grammatical Inference in Bioinformatics. IEEE Transactions on Pattern Analysis and Machine Intelligence 27, 1051–1062 (2005)
12. Sakakibara, Y., Brown, M., Underwood, R.C., Mian, I.S., Haussler, D.: Stochastic CFGs for Modeling RNA. In: Proc. of the 27th Hawaii Int. Conf. on Syst. Sciences, pp. 284–293 (1994)
13. Sakakibara, Y.: Recent Advances of Grammatical Inference. Theoret. Comput. Sci. 185, 15–45 (1997)
14. Searls, D.: The Language of Genes. Nature 420, 211–217 (2002)
15. Valiant, L.G.: A Theory of the Learnable. Comm. ACM 27, 1134–1142 (1984)

Enhancing NetLogo to Simulate BDI Communicating Agents

Ilias Sakellariou[1], Petros Kefalas[2], and Ioanna Stamatopoulou[2]

[1] Department of Applied Informatics, University of Macedonia, Thessaloniki, Greece
iliass@uom.gr
[2] Department of Computer Science, CITY College, Thessaloniki, Greece
kefalas@city.academic.gr, istamatopoulou@seerc.org

Abstract. The implementation process of complex agent and multi-agent systems (AMAS) can benefit significantly from a simulation platform that would allow rapid prototyping and testing of initial design ideas and choices. Such a platform, should ideally have a small learning curve, easy implementation and visualisation of the AMAS under development, while preserving agent oriented programming characteristics that would allow to easily port the design choices to a fully-fledged agent development environment. However, these requirements make such a simulation platform an ideal learning tool as well. We argue that NetLogo meets most of the requirements that suit our criteria. In addition, we describe two extra NetLogo libraries, one for BDI-like agents and one for ACL-like communication that allow effortless development of goal-oriented agents, that communicate using FIPA-ACL messages. We present one simulation scenario that employs these libraries to provide an implementation in which agents cooperate under a Contract Net protocol.

Keywords: Multi-Agent Systems, Simulation Platforms.

1 Introduction

Development of Agents and Multi-Agent Systems (AMAS) is a challenging and complex task. Due to the complexity that AMAS exhibit, simulation of AMAS models becomes an important step that can facilitate understanding of how the intended system will perform when it will be actually implemented, since it allows rapid prototyping and testing of initial design ideas and choices. It is crucial that simulation output should be meaningful enough for the developers to draw conclusions and drive the actual implementation. For instance, in multi-agent systems with spatial reasoning and behaviour, a visual output which displays agents moving in a two or three dimensional space is necessary.

On the other hand, future developers and researchers must be educated in AMAS theory and trained in practice, which presents an equally challenging task. Firstly, the topic is too broad to fit within specific time constraints. Especially at the University level, a course on AMAS can hardly fit itself among a plethora of other mainstream/popular topics, despite the fact that AMAS is listed in the ACM/IEEE Computing Curricula [1] as part of Intelligent Systems

J. Darzentas et al. (Eds.): SETN 2008, LNAI 5138, pp. 263–275, 2008.

area. Secondly, due to the wide foundations and applicability of AMAS, it is only natural that there is a lot of diversity with respect to learning outcomes and content, teaching and assessment, theory and practice etc. But definitely, educators' common aim is to introduce AMAS as a useful new software paradigm, and for this appropriate educational tools are necessary.

Such educational tools for AMAS should ideally have a small learning curve, easy implementation and visualisation of the system under development, while preserving agent oriented programming characteristics that would allow to easily port the design choices to a fully-fledged agent development environment.

The requirements for choosing a simulation environment that can also act as an educational tool (or vice-versa) are complemented by the following criteria:

- have a simple environment that presents the minimum installation problems,
- provide easy visualization for a better view of the agent behaviour,
- easy to learn and use language thus keeping a small learning curve,
- clearly demonstrate the difficulties in AMAS programming,
- support the basic agent architectures (reactive, BDI, hybrid)
- provide means for communication, message exchange and interaction.

A number of languages and environments are available as options to use in practice for training future AMAS developers and researchers. Apparently, deciding which one to choose is not easy, since all the above criteria are not fully met by existing development or simulation environments. For the work presented in this paper, we chose NetLogo [2]. NetLogo has a number of important features as an educational and simulation tool for AMAS, but lacks support for goal-oriented and communicating agents. Our aim is to present how these issues are dealt with, by enhancing NetLogo with libraries that support BDI architecture and FIPA-ACL message exchange. The libraries have been developed for educational purposes but can also be used for more advanced AMAS simulation.

The structure of the paper is as follows: In section 2, we discuss in more detail the requirements and the choices available for using a simulation tool in education. NetLogo is briefly presented in Section 3 and the extensions we suggest and implemented are listed in Section 4. In the following section 5, we present a case study of forest fires simulation, which demonstrates the applicability of our proposal. Finally, we conclude with discussion and future work.

2   Simulation and Education in AMAS

It is commonly accepted that simulation tools and environments are not only useful for AMAS research and development but also useful for training future AMAS researchers and developers. We will generally refer to the latter as learners; they might be students, researchers or professionals who undergo further training. In order to understand the theoretical concepts and complexity of AMAS, some kind of practical program development is needed. There are several choices educators may follow, depending on the emphasis and focus they give to certain aspects of AMAS, e.g. architectures, communication and interaction protocols, applications etc. Various tools and environments for AMAS have been reported to assist the educational process, like RoboCup, NetLogo, TAC, FIPA-OS, JADE, JadeX, Jason, Protégé etc. [3,4,5,6,7,8]. All aim to improve active learning in the context of AMAS, some by engaging learners in writing code (e.g. Java), others by allowing development of peripheral to agents structures (e.g. Ontologies).

Many success stories in integrating such environments are reported, but admittedly, most of the times there are concerns about the tool's complexity, the time spent by some educators to build such a tool and the time spent by the learners to reach the required level of skills in order to produce something useful to their eyes. Given the time restrictions for training, which normally last a few weeks, it is rather difficult for them to develop something simple but also meaningful, easy to implement but also challenging, artificial but also realistic enough, at some level of abstraction but also practical.

It would be preferable if learners see a more realistic view of their work in a simulation. If this is the process of a competition like game (e.g. Trading Agent Competition) or a visualisation of the agents' environment (e.g. RoboCup simulation), the understanding and satisfaction seems to be increasing. One could also argue that a simple robotic platform (e.g. Lego Mindstorms, RoboSapiens, i-SOFT etc.) [9,10] could also serve the purpose, since in the average mind perception a robot (from a science fiction perspective) matches with that of an agent. Although, such an approach is feasible and sometimes desirable in more engineering-oriented courses, it still suffers from complexity issues due to the fact that learners need to take into account non-symbolic percepts and additional hardware devices and protocols.

In principle, learners should receive instruction and training on:

- basic AI theory and techniques;
- basic notions of intelligent agents;
- types of intelligent agents, their architecture, strengths and limitations;
- issues involved in AMAS communication and interaction;
- possible application areas of the AMAS technology;
- demonstration on how AMAS revolutionise human-computer interaction;
- advantages of the agent-based approach to engineering complex software systems.

More particularly, the topics covered may include, among others, agent architectures (logic, reactive, BDI, hybrid), communication and interaction protocols (speech acts theory, agent communication languages, knowledge communication, Contract Net protocol, auctions, negotiation), biology-inspired agents (Ant Colony Optimization, Bio-Networking, Artificial Life), planning, learning and mobile agents, agent theories (intentional notions: information, motivation and social attitudes), AMAS software engineering methodologies (AAII methodology, Cassiopeia, Agent UML), Semantic Web basics etc. [11,12,13]. It is important, however, that some hands-on experience is also provided.

In the current work, we argue that NetLogo can act as the basis of an educational tool as well as an environment for AMAS research, provided that it is enriched with certain features. NetLogo meets several of the requirements listed in Section 1. It is easy to install and provides with a plethora of interesting examples ready to run. The language is functional and the learning curve seems to be small, since someone is ready to produce a descent program in a short period of time. Additionally, it allows visualisation of the developed systems, which is extremely valuable to AMAS with spatial reasoning and behaviour as it helps them both to gain a better understanding as well as get immediate feedback from the simulation. However, NetLogo suffers from not being able to provide ready-made constructs for goal-oriented agents, communication and coordination. In order to compensate for the features that NetLogo lacks, we decided to work towards the implementation of an appropriate set of of extensions that aim to facilitate the development of BDI-like communicating agents.

3   NetLogo as a Modelling Tool

NetLogo is a modelling environment targeted to the simulation of multi-agent systems that involve a large number of agents. The platform aims to provide "a cross-platform multi-agent programmable modelling environment" [2].

The main entities of NetLogo are the patches, the turtles and the observer.[1] The *observer* simply controls the experiment, in which turtles and patches actively participate. *Patches* are stationary "agents", i.e. components of a grid on which *turtles* exist, i.e. agents that are able to move, "live" and interact. Both patches and turtles can inspect the environment around them, for example detect the existence of other agents, view the state of their surrounding patches/turtles, and modify the environment. Agents can be organised in groups under a user specified name, and thus agents of different *breeds* can exist in the simulation world. Probably the feature that most greatly enhances the modelling expressiveness of the platform is the fact that each patch and turtle can have its own user-defined variables: in the case of patches this allows modelling complex environments by including an adequate number of variables that describe it sufficiently and in the case of turtles it simply means that each agent can carry its own state, stored again in a number of user defined variables.

The NetLogo programming language allows the specification of the behaviour of each patch and turtle, and of the control of execution. The language is simple and expressive and has a rather functional flavour. There is an extensive set of programming primitives, good support for floating point mathematics, random numbers and plotting capabilities. Programming primitives include for example, commands for "moving" the turtles on the grid, commands for environment inspection (i.e. the state of other turtles and patches), classic programming constructs (branching, conditionals, repetition) etc. The main data structure is *lists* (following the functional Lisp approach), and the language supports both functions, called *reporters*, as well as procedures. Monitoring and execution of the agents is controlled by the *observer* entity that "asks" each agent to perform a

---

[1] The recent version (4.0) also offers links, not involved in this work.

specific computational task. The programming environment also offers GUI creation facilities, through which custom visualizations of the studied multi-agent systems can be created with particular ease.

As it has been argued elsewhere [14][2], we found NetLogo to be suitable for modelling agent systems that we are dealing with, since each NetLogo agent:

- perceives its environment and acts upon it,
- carries its own thread of control, and
- is autonomous,

i.e. it falls under the classic definition of agency found in [11]. NetLogo is an excellent tool for rapid prototyping and initial testing of multi-agent systems, particularly suited to systems with agents situated and operating in a restricted space, as well as an excellent animation tool of the modelled system. However, it lacks build-in support for implementing communicating agents with intentions and beliefs, making the task of modelling/testing more complex architectures and protocols rather hard.

4   Extending NetLogo with Libraries for Intelligent Agents

Being a platform that is primarily targeted to modelling social and natural phenomena, NetLogo fully supports the creation and study of reactive agent systems. Indeed, one of the first models that we studied was the Luc Steels Mars Explorer experiment [15] that presents many similarities to the ant colony foraging behaviour experiment included in the library models (ants can be modelled and studied as reactive agents [16,17]). However, the study of BDI agents that are able to communicate with explicit symbolic message exchange was not supported. Thus, taken into consideration the fact that NetLogo fulfilled the majority of our requirements we decided to extend the platform by providing one library for building simple BDI-like agents and one for FIPA-ACL-like message exchange, that are described in the sections that follow.

4.1   BDI-Like Agents in NetLogo

Although, we could have adopted as an option to link through the JAVA interface of NetLogo an already existing BDI development platform, as for example JAM [18], this would have made the installation of the complete platform a lot more difficult and would have significantly increased the learning curve, due to the complexity of such a fully fledged development environment. Thus, we decided to provide a simpler alternative, i.e. to develop limited BDI agents by providing the necessary primitives through a NetLogo library.[3]

---

[2] Actually, the excellent page that Jose Vidal (http://jmvidal.cse.sc.edu/) maintains for NetLogo models, was a starting point for our work.

[3] NetLogo being a simple platform, does not support libraries in the classic sense found in other programming environments; by the term library we refer to a set of procedures and functions that the user is given and includes in its own code.

Intentions in NetLogo. The simple BDI architecture that we have followed, follows a PRS-like [19] model, i.e. there is a set of intentions (goals) that are pushed into a stack; of course the implementation is far from delivering all the features of systems like JAM, but can still be effectively used in implementing simple BDI agents in the NetLogo simulation platform.

An intention (*I*) consists of two parts: the *intention name (I-name)* and a condition that we call *intention-done (I-done)*. The former maps to a NetLogo procedure (usually user defined), while the latter maps to boolean NetLogo reporter (function) (again usually user defined). The semantics are the standard followed by other architectures: an agent must pursue an intention until the condition described in the intention-done part evaluates to true. For example consider the following intention of an example ground unit agent working at a fire site (see Section 5):

```
["move-towards-dest [23 20]" "at-dest [23 20]"]
```

The above simply states that the agent is currently committed to moving towards the point with coordinates (23, 20) and it will retain the intention until the reporter **at-dest** [23 20] evaluates to true. Note that the user in this case has to specify both the procedure and the reporter, that map to the two parts of the intention, since they are not part of the built-in NetLogo primitives.

As mentioned, the main concept behind the present implementation is the *intention stack* where all the intentions of the agent are stored. Agents execute intentions popping them from the stack as shown below:

```
IF the intention stack is not empty THEN do:
  Get intention I from the top of the stack;
  Execute I-name;
  IF I-done evaluates to true THEN pop I from stack;
ELSE do nothing
```

The NetLogo implementation is rather straightforward: each intention is represented by a list of two elements, one for each part. The intention stack is a list stored in a specific "turtle-own" variable, i.e. each agent carries its own stack. The execution model is encoded in the procedure **execute-intentions**, that is called to invoke the agent's proactive behaviour. Notice that running a NetLogo simulation involves multiple execution cycles, invoking in each a procedure for each participating agent (or group of agents), thus ensuring their "parallel" execution. In consequence, procedure **execute-intentions** concerns only the execution of one intention in each such cycle.

The library also provides a set of reporters and procedures to the user for adding and removing intentions on the stack, inspecting the current intentions, set time-outs as *intention-done* conditions etc. For example, the following line:

```
add-intention "move-towards-dest [23 20]" "at-dest [23 20]"
```

will add the corresponding intention to the stack of the calling agent. A slightly more complicated example is shown below, that allows an agent to participate as a manager in a Contract Net protocol:

```
add-intention "select-best-and-reply" "true"
add-intention "wait-for-proposals" timeout_expired 15
add-intention "send-cfp" "true"
```

Since `add-intention` simply pushes intentions in the stack, such "plans" as the above have to be encoded in reverse order. Intention `send-cfp` concerns the agent broadcasting a call-for-proposals message to all agents and is executed only once, since it has as *I-done* condition the NetLogo reserved word `true`. `wait-for-proposals` remains as the agent's intention for the next 15 execution cycles (*clicks* according to NetLogo jargon), indicated by the provided `timeout_expired` BDI-library reporter. Finally, `select-best-and-reply` will be executed once, selecting the best contractor among the received bids. Note that `wait-for-proposals`, `select-best-and-reply` and `send-cfp` are user defined procedures.

Managing Beliefs. To further support the BDI architecture, facilities for managing *beliefs* were also created. Although, the latter was not really necessary, since it is rather simple to store any information on a related turtle variable, we have designed a set of procedures and reporters that would form an abstraction layer that facilitates the user to manage agent beliefs, without getting into too many details about how to program in the NetLogo language.

A belief consists of two elements: the *type* and the *content*. The former declares the type of the belief, i.e. indicates a "class" that the belief belongs to. Examples could include any string, e.g. "position" "agent" etc. Types facilitate belief management, since they allow to check for example whether a belief of a specific type exists or the removal of multiple beliefs at once. The content on the other hand, is the specific information stored in the belief. It can be any NetLogo structure (integer, string, list etc.). Obviously, there might be multiple beliefs of the same type with a different content, however two beliefs of the same type and content cannot be added. For instance, `["fire-at" [23 15]]` and `["location" [3 7]]` are examples of beliefs that the agent can have.

Belief management is done through a set of reporters and procedures that allow the creation, removal, checking of the current agent's beliefs. For example, the following line:

```
add-belief create-belief "fire-at" [23 15]
```

will include a belief of type "fire-at" with content "[23 15]" in the agent's beliefs. In the current implementation, all agent beliefs are stored in an agent own variable named `beliefs`.

4.2   FIPA-Like Message Passing

The ability to exchange symbolic messages is rather important in order to model agent communication and interaction protocols, such as the Contract Net protocol. Thus, it was necessary to somehow enhance NetLogo with explicit message communication primitives. Messages closely follow the FIPA ACL message format, i.e. are lists of the form:

```
[<performative> sender:<sender> receiver:<receiver>
    content: <content>..]
```

For example, the following message was sent by agent (turtle) 5 to agent 3, its content is "fire-at [23 15]" and the message performative (FIPA) is "inform".

```
["inform" "sender:5" "receiver:3" "content:" "fire-at [23 15]"]
```

A message may include the above fields (performative, sender, receiver, content), omitting others such as the ontology field specified by FIPA, assuming that all agents use the same ontology. The library, however, allows the creation and addition of any custom field that may be considered necessary.

Participating agents are uniquely characterized by an ID, which is in fact an integer automatically assigned by NetLogo at the time of their creation ("who" NetLogo variable). This naming scheme was adopted since it greatly facilitates the development of the message passing facilities. Message passing is asynchronous and a set of library reporters and procedures allow easily creating, sending, receiving, and processing messages between NetLogo agents. For example the code below (assuming that the calling agent is 8):

```
let   somemsg create-message "inform"
set   somemsg add-receiver 5 somemsg
set   somemsg add-content "fire-at [23 15]" somemsg
send somemsg
```

will send to agent 5 the following message:

```
["inform" "sender:8" "receiver:5" "content:" "fire-at [23 15]"]
```

Of course, both libraries (BDI and FIPA) take advantage of functional features of the NetLogo programming language. The exact message can be send by simply issuing the following one-line command:

```
send add-content "fire-at [23 15]" add-receiver 5
    create-message "inform"
```

It should be noted that there are also primitives that allow broadcasting a message to a group (breed) of NetLogo agents. For instance the following line of code sends a cfp message containing the location of a fire (see Section 5) to all agents of the breed "units":

```
broadcast-to units add-content (list "fire-at" list pxcor pycor)
    create-message "cfp"
```

Incoming messages for each agent are stored in the incoming-queue variable, which is in fact a list. This is a "user-defined" variable that each agent must have in order to be able to communicate. Sending a message to an agent simply means adding the message to its incoming-queue list; it does not require an explicit receive command to be invoked on the receivers side. At any time the agent has the ability to obtain and process the messages from its queue using the reporters and procedures provided.

Both libraries, FIPA-ACL and BDI, were fully implemented in the NetLogo language and thus, their "installation" is trivial, since users have only to include the given library code in their models. The library offers limited debugging facilities, by allowing the user to inspect the list of messages exchanged and the current intention stack of the agents in the NetLogo programming environment. Limited error checking and debugging facilities was a design choice, in order to avoid having efficiency issues.

We have used both libraries for a number of years in a Intelligent Agents course: students have found them easy to use and did manage to implement multi agent systems that involve communicating BDI agents under an interaction protocol. The interested reader may find the libraries, brief manuals and examples at [20].

5   Forest Fires Scenario: A Case Study

Using the programming facilities described in the previous sections, we have designed and implemented a simulation of a multi-agent system in a fictional scenario that involves forest fires detection and suppression. The aim of the

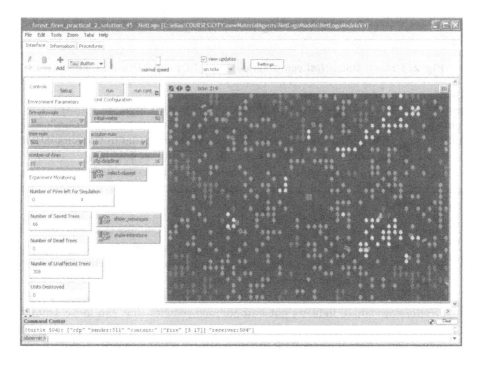

Fig. 1. NetLogo ScreenShot of the Forest Fires Scenario

```
to collect-msg-update-intentions
 let msg []
 let performative []
 while [not empty? incoming-queue]
 [set msg get-message
  set performative get-performative msg
  if performative = "cfp" [reply-to-cfp msg]
  if performative = "accept-proposal" [reply-to-accept-proposal msg]
  if performative = "reject-proposal" [do-nothing]            ]
end

to reply-to-cfp [msg]
 ifelse current-intention = "find-target-fire"
 [let dist distancexy get-content msg
   send add-content dist create-reply "propose" msg          ]
 [send add-content "busy" create-reply "refuse" msg           ]
end

to reply-to-accept-proposal [msg]
 ifelse current-intention = "find-target-fire"
 [let crds get-content msg
   add-belief msg
   add-intention "send-confirmation"  "true"
   add-intention "put-out-fire" "fire-out"
   add-intention (word "move-towards-dest " crds) (word "at-dest " crds)]
 [send add-content get-content msg create-reply "failure" msg    ]
end
```

Fig. 2. Ground Units NetLogo Code using the BDI and FIPA-ACL libraries

specific AMAS was the constant monitoring of the forest area, as well as taking immediate action in the case of a fire spot, so that fire spreading is disallowed and creation of fire fronts is avoided. We have tried to model in the simulation many real-world features so that it is as close to reality as possible.

The environment (shown in Fig. 1) is inhabited by two types of communicating agents with different capabilities and constraints, that cooperate to perform the above task:

- *Ground units* are fire extinguishing autonomous vehicles that are able to travel around the forest environment and put out any fires they detect. Their speed is rather low and depends on the water supplies carried at each moment. Their sensors have a limited range, and while moving they have to avoid obstacles such as other ground units and forest areas already on fire.
- *Scouters* are small, light weight autonomous vehicles that can move around the environment relatively fast. Although they do not have the ability to extinguish a fire, they have close-range fire detection sensors. The only obstacle for scouters are trees on fire; they are small enough to co-exist in the same area with ground units or other scouters.

In the model, fire spots (i) appear randomly during the execution of the experiment, and (ii) are spreading in adjacent trees over time, if they are not put out, i.e. there is a close to reality fire spread model against which the agents were competing. Various parameters can be set as, for example, the number of fire spots that randomly appear (number-of-fires drop-down menu in Fig. 1), the density of the forest (tree-num drop-down menu in Fig. 1), the number of scouter and ground units, the initial water supplies of each unit etc.

Cooperation in this simple MAS is rather straightforward: scouters patrol the forest area, detect fire spots and inform ground units that move to the specific location and extinguish the fire. Cooperation takes place under the Contract-Net protocol where scouters assume the role of managers and ground units that of contractors: for each announced contract (cfp-message) by a scouter, bids (proposals) are submitted by ground units and one is awarded the contract. There is a single evaluation criterion of the bids, the distance of the ground units from the fire location. Although it could have been possible to directly implement the agents' behaviour in NetLogo, the BDI and FIPA-ACL libraries presented, greatly facilitated the development process. For instance, the code in Fig. 2 shows part of the ground units code, that involves message handling, proposal creation and commitment to an "accept-proposal" message that has as content the coordinates of the forest fire, i.e. a large part of the agent's reasoning. It should be noted that the partial implementation of this simulation was given as a coursework in the context of an Intelligent Agents course: details may be found at [20].

6   Discussion and Conclusions

We started using NetLogo as an educational tool five years ago, in the context of coursework of an undergraduate course in Intelligent Agents. Soon we started to realise the necessity of being able to simulate more complex AMAS than those composed of simple reactive agents. It was also encouraging to see a growing interest of the academic as well as research community towards this simulation environment. Both acted as motivations to create the BDI and FIPA-ACL NetLogo libraries.

In education, we have already seen the benefits that NetLogo provides. We use the same assignment template each year but with a different scenario (case study) such as agents rescuing victims in a disaster area, taxis transporting passengers in a city, vacuum cleaning a floor, luggage carriers solving logistics problems in airports, space satellite aligning to provide telecommunication services etc. [20]. The overall impression was that students enjoyed the practical aspect of the course. It is also clear to us that the overall student performance is increased due to better and deeper understanding of the issues around AMAS. Most of the students attributed much of the success of their solutions to the libraries provided. Learners really enjoyed the sense of seeing and experimenting with the virtual world they developed with minimum programming effort, thus having more time to design and prototype the "real thing".

In research, NetLogo provided us a platform in which we could immediately prototype and test our theoretical findings. Our interest in formal modelling of swarm intelligence and biology inspired AMAS exactly matched the initial Net-Logo requirements [16,17]. We are currently investigating how formal models for goal-oriented agents exhibiting dynamic re-organisation can be directly mapped to NetLogo constructs, so that a semi or fully automatic translation would be possible. It is important that the libraries needed for such experimentation have been already implemented.

Through the two libraries described in this paper, we managed to have a platform that facilitates development of something more than a reactive agent. The libraries are by no means complete or fully fledged to meet the complete BDI and FIPA-ACL requirements. Future work is directed towards such extensions. For example, NetLogo offers a JAVA interface, through which a link to complete BDI packages, such as JAM, might be possible. Although we consider it unsuitable for educational purposes, as argued above, the idea could offer new opportunities for using the platform in research. In addition, we are considering a rather ambitious development of an extendible/customisable game platform in NetLogo, in which educators will be able to setup game environments and rules, such as tank battles, RoboCup etc., that would allow an even simpler use of the platform in the context of a course. Towards this direction, we have already implemented a prototype, but the work is not fully completed yet.

References

1. Joint ACM/IEEE Task Force on Computing Curricula: Computing curricula 2001. ACM Journal of Educational Resources in Computing 1 (2001)
2. Wilensky, U.: Netlogo. Center for Connected Learning and Computer-based Modelling, Northwestern University, Evanston, IL (1999), http://ccl.northwestern.edu/netlogo
3. Beer, M.D., Hill, R.: Teaching multi-agent systems in a UK new university. In: Proceedings of 1st AAMAS Workshop on Teaching Multi-AgentSystems (2004)
4. Hara, H., Sugawara, K., Kinoshita, T.: Design of TAF for training agent-based framework. In: Proceedings of 1st AAMAS Workshop on Teaching Multi-AgentSystems (2004)
5. Williams, A.B.: Teaching multi-agent systems using AI and software technology. In: Proceedings of the 1st AAMAS Workshop on Teaching Multi-AgentSystems (2004)
6. Beer, M.D., Hill, R.: Multi-agent systems and the wider artificial intelligence computing curriculum. In: Proceedings of the 1st UK Workshop on Artificial Intelligence in Education (2005)
7. Bordini, R.H.: A recent experience in teaching multi-agent systems using Jason. In: Proceedings of the 2nd AAMAS Workshop on Teaching Multi-Agent Systems (2005)
8. Fasli, M., Michalakopoulos, M.: Designing and implementing e-market games. In: Proceedings of the IEEE Symposium on Computational Intelligence in Games, pp. 44–50. IEEE Press, Los Alamitos (2005)

9. Behnke, S., Müller, J., Schreiber, M.: Playing Soccer with RoboSapien. In: Breden-feld, A., Jacoff, A., Noda, I., Takahashi, Y. (eds.) RoboCup 2005. LNCS (LNAI), vol. 4020, pp. 36–48. Springer, Heidelberg (2006)

10. Ferme, E., Gaspar, L.: RCX+PROLOG: A platform to use Lego Mindstorms^{TM} robots in artificial intelligence courses. In: Proceedings of the 3rd UK Workshop on AI in Education (2007)

11. Wooldridge, M.: An Introduction to MultiAgent Systems. J. Wiley & Sons, Chichester (2002)

12. Weiss, G. (ed.): Multiagent Systems: A Modern Approach to Distributed Artificial Intelligence. MIT Press, Cambridge (1999)

13. Russell, S., Norvig, P.: Artificial Intelligence: A Modern Approach. Prentice Hall, Englewood Cliffs (2002)

14. Vidal, J.M., Buhler, P., Goradia, H.: The past and future of multiagent systems. In: Proceedings of 1st AAMAS Workshop on Teaching Multi-AgentSystems (2004)

15. Steels, L.: Cooperation between distributed agents through self-organisation. In: Towards a New Frontier of Applications, Proceedings of the IEEE International Workshop on Intelligent Robots and Systems (IROS 1990), pp. 8–14 (1990)

16. Gheorghe, M., Stamatopoulou, I., Holcombe, M., Kefalas, P.: Modelling dynami-cally organised colonies of bio-entities. In: Banâtre, J.P., Fradet, P., Giavitto, J.L., Michel, O. (eds.) UPP 2004. LNCS, vol. 3566, pp. 207–224. Springer, Heidelberg (2005)

17. Stamatopoulou, I., Sakellariou, I., Kefalas, P., Eleftherakis, G.: Formal mod-elling for in-silico experiments with social insect colonies. In: Papatheodorou, T., Christodoulakis, D., Karanikolas, N. (eds.) Current Trends in Informatics, Patras, Greece, May 18-20. Proceedings of the 11th Panhellenic Conference in Informatics (PCI 2007), vol. B, pp. 79–89 (2007)

18. Huber, M.J.: JAM: a BDI-theoretic mobile agent architecture. In: Proceedings of the 3rd Annual Conference on Autonomous Agents, pp. 236–243. ACM, New York (1999)

19. Georgeff, M.P., Lansky, A.L.: Reactive reasoning and planning. In: Proceedings of the AAAI Conference on Artificial Intelligence (AAAI 1987), pp. 677–682 (1987)

20. Sakellariou, I.: Extending NetLogo to Support BDI-like Architecture and FIPA ACL-like Message Passing: Libraries, Manuals and Examples (2008), http://eos.uom.gr/~iliass/projects/NetLogo

Integration of Computational Intelligence Applications in Engineering Design

Kostas M. Saridakis and Argiris J. Dentsoras

Machine Design Lab, Dept. of Mech. Eng. & Aeronautics,
University of Patras, 26500, Patras, Greece
{saridak,dentsora}@mech.upatras.gr

Abstract. Several techniques have been proposed recently as a result of the intensive research done in the field of computational intelligence. These techniques seem to act beneficially in a variety of scientific domains by performing better than the conventional methodologies. The current paper focuses on the domain of engineering design, whose demanding nature motivates the research for domain-independent and efficient tools and methodologies. In this context, the application of computational intelligence techniques in engineering design is reviewed and the enhancement of basic design processes such as knowledge representation, optimal solution search and design knowledge re-utilization by soft-computing techniques is described. The purpose of the paper is to present ways that Fuzzy Logic (FL), Genetic Algorithms (GA) and Artificial Neural Networks (ANN) can be beneficially used in engineering design as stand-alone tools or within a stepwise generic methodological approach for parametric engineering design is described, which is compliant with the existing design theory and practice. E

1 Introduction

Engineering design is a human activity aiming to the development and/or evolution of products, machines, devices, etc. The effort for a systematic approach of design has guided researchers towards establishing a design theory [1,2] by developing several methodologies. The existing design models may be classified into three general categories: *descriptive*, *prescriptive* and *computer-based*. On the basis of these three general design models, several design methods have been proposed with the characterization 'design-for-X', each of them viewing a different aspect of design, e.g. design for assembly, design for manufacture, design for maintenance, design for cost, design for reliability, design for service ability, etc. The applicability of a design model or a design methodology depends on the type of the problem under consideration. Ueda [3] provides an overview of the methodology of emergent synthesis, while discussing analysis, synthesis and emergence in the realm of artifacts. In this theory, the difficulties in synthesis are categorized into three classes. Through the review of the existing literature, the following issues regarding the design process become distinguishable [4]: a) the design knowledge representation (modeling), b) the search for optimal solutions, c) the retrieval of pre-existing design knowledge and the learning of new knowledge. Recent research activity in engineering design is deployed by taking into account one or more of these issues, which, in the context of the present paper, are

J. Darzentas et al. (Eds.): SETN 2008, LNAI 5138, pp. 276–287, 2008.

used as a basis in order to evaluate the proposed approaches as they arise from the relevant literature.

The scientific community was convinced that well-established artificial intelligence techniques could be used as tools in problems where conventional approaches fail or perform poorly. In the early 1990s, L. A. Zadeh proposed the concept of Soft Computing (SC). Soft Computing is an evolving collection of artificial intelligence methodologies that can: a) exploit the tolerance for imprecision and uncertainty inherent in human thinking and in real life problems b) deliver robust, efficient and optimal solutions to problems and c) explore further and capture the available design knowledge. Fuzzy Logic (FL) [5,6], Artificial Neural Networks (ANN) [5,6], and Genetic Algorithms (GAs) [7] are the core methodologies of soft computing. SC yields rich knowledge representation (symbol and pattern), flexible knowledge acquisition (machine learning), and flexible knowledge processing (inference by structuring interfaces between symbolic and pattern knowledge). Additionally, SC techniques can either be deployed as separate tools or be integrated in unified and hybrid architectures [8,9].

The fusion of SC techniques causes a paradigm shift (breakthrough) in engineering and science fields (including engineering design) by solving problems, which could not be solved with conventional or stand-alone computational tools. In the present paper, the conjunction of engineering design and soft computing is studied. Fuzzy logic is utilized for modeling the uncertainty in the design problem in two ways. Firstly, the relations among the design entities are stated in an approximated and qualitative manner and a fuzzy reasoning scheme is deployed for delivering a digraph that captures the relations among the design entities. Moreover, fuzzy logic is used for stating design objectives in terms of fuzzy preferences on specific values (or value intervals) of the design entities. The aggregation of these fuzzy preferences results to an overall design preference, which is used as optimization criterion for the genetic algorithm. Finally, artificial neural networks are used as retrieving mechanisms in a case-based reasoning framework, which is deployed for reusing past design solutions. Case-based reasoning (CBR) – that could be called as case-based design (CBD) within the context of the present paper - is a technique that resembles the human analogical thinking and considering of past solutions when solving a new problem [10].

2 Formulating the Design Problem

The current methodology has been developed with the consideration of maintaining the domain independency in the design knowledge representation, thus providing the possibility of utilizing the methodology in every parametric design problem. Initially, some terms that are used for the design problem's representation are briefly defined (extended definitions and descriptions of the proposed representation model can be found in [11,12,13].

Design parameters (DPs) DP_1, DP_2,.., DP_n are design entities that formally represent physical, functional or behavioral attributes of the designed product, system, component, etc. The values of the DPs distinguish the alternative designs in the Design Parameter

Space (DPS) formed by the set of all alternative design vectors $\overrightarrow{DP_1}$, $\overrightarrow{DP_2}$,..., $\overrightarrow{DP_n}$ with each of these vectors containing all the DPs. Each design problem is stated in the context of the relative DPs and their associative relationships. These relationships can be expressed either quantitatively or qualitatively depending on the interrelated DPs and the explicitness of the available design knowledge. Three types of associative relations are identified: a) computational formulas, b) empirical rules and c) fuzzy rules. A performance variable (PV) is defined as a design entity that models a performance characteristic. This characteristic may refer to the physical, functional or behavioral domain of the product or system under design. A performance variable may represent a design parameter that is critical for the efficiency of the final outcome and its variation has to be closely examined. Some of the aforementioned design entities (DPs and PVs) are more important than others. In order to model this relative importance, weighting factors are used, denoted by the symbol w. The weighting factors are used in order to determine quantitatively and subjectively the most important and critical performance variables according to the designer's beliefs.

The design problem formulation proceeds by an initial listing of, all design parameters characterizing the design problem. The enlisted DPs are interrelated with diverse formalisms (rules, computational relations, etc.), which in the current step are not considered. Alternatively, the designer may approximate the strength of these interdependencies with a qualitative linguistic way (e.g. weak, strong, very strong, etc.) through fuzzy sets. These fuzzy relations are used for constructing a fuzzy design structure matrix [14], as shown in figure 2.

After performing standard DSM operations (partitioning and tearing) on the basis of a fuzzy inference mechanism [14], a directed graph that contains all the design parameters may be constructed by using the partitioned fuzzy DSM (figure 3). Case-DeSC supports three types of relations: computational formulas, rules and fuzzy rules. The utilization of a FL approach in order to structure a DSM and a hierarchical tree of DPs provides two basic benefits: a) qualitative modelling of relationships strengths, b) collaborative structuring of DSM by multiple designers.

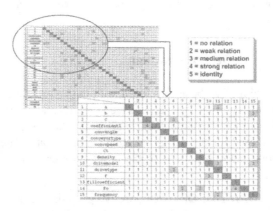

Fig. 2. Formulation of the fuzzy Design Structure Matrix (DSM)

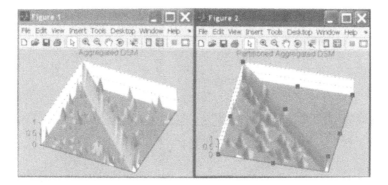

Fig. 3. Partitioning the fuzzy Design Structure Matrix

3   Stating the Criterion for Solutions Optimality

The design objectives for the design problem are modeled through fuzzy preferences on performance variables, denoted as μ (PV). These preferences must be aggregated by using a specific aggregation function that reflects a strategy about how competing attributes should be traded-off. The function P attempts to formalize the balance of conflicting design goals and constraints by preserving the design rationality. The evaluation of the design solutions is performed after the various individual preferences have been combined or aggregated. The overall preference $\mu_o(\overrightarrow{PV})$ is a function that combines all the preferences on the performance variables for a particular design case:

$$\mu_o(\overrightarrow{PV}) = P\left(\mu(PV_1), \mu(PV_2), ..., \mu(PV_j)\right) \qquad (1)$$

The aggregation functions should be extracted from a list of available aggregation functions, judged as suitable for use in design problems [15].

The selection of the trade-off strategy requires adequate knowledge of the design problem under consideration and specific knowledge on the relation between the aggregated parameters. Two contradictory trade-off strategies bound the family of the available aggregation functions appropriate for design. Assume the case of applying preferences on the PVs of a machine assembly. A high preference corresponding to a demand for high strength for one component cannot compensate against a low preference corresponding to low strength for another component because a possible failure of the latter may result to failure of the whole system. This example projects the non-compensating trade-off strategy (P_{min}) that takes into consideration the lowest preference during the stage of aggregation:

$$P_{min}((\mu_1, w_1), ...(\mu_j, w_j)) = \min(\mu_1, ..., \mu_j) \qquad (2)$$

Next consider the case of a battery and two of its performance variables, namely, the unit cost and its energy capacity. As it is obvious in this case, low unit cost can compensate for low energy capacity and high-energy capacity can compensate for high unit cost. This example demonstrates how a more acceptable attribute partially compensates for a less

a. fuzzy preferences for PV_1, PV_2

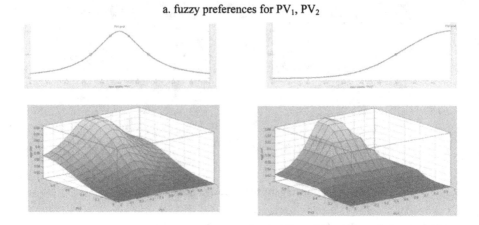

Fig. 4. Aggregation of the fuzzy preferences (a) assigned to performance variables PV_1, PV_2 dimensions for compensating (b) and non-compensating (c) preference aggregation)

acceptable one and is modeled as a fully compensating strategy (P_{II}) for which the aggregation function is the geometric weighted mean:

$$P_{II}((\mu_1, w_1), \ldots (\mu_n, w_n)) = \left(\prod_{k=1}^{n} \mu_\kappa^{w_\kappa} \right)^{\frac{1}{w}}, w = w_1 + \ldots + w_n \qquad (3)$$

Figures 4-a, 4-b and 4-c depict the aggregation of the preferences on intervals of values of the equally weighted ($w_1 = w_2 = 0.5$) performance variables PV_1, PV_2 using the two contradicting trade-off strategies. The designer should define an aggregation strategy between the limiting cases of the functions P_{min} and P_{II}.

4 Searching for Optimal Solutions

In the context of the current methodology an optimization technique is needed for finding an optimal solution. The genetic algorithm has been selected as an optimization technique in Case-DeSC for four main reasons: 1) it doesn't require any kind of information about the design problem and the underlying PV relations, 2) it provides fast and robust solutions, 3) it avoids trapping in local extremes and 4) it is performs population-based optimization, which is used for populating the case base with solutions.

It is important to remark that the performance of the genetic algorithm (accuracy and speed) is depended on some selections related with the genetic evolution. So, in the context of the oscillating conveyors' design, several genetic optimizations were deployed before these selections are set. Another characteristic of superiority for the genetic algorithms when compared to other optimization methods is that their operation is based upon populations of solutions instead of point-to-point optimization. This contributes towards avoiding local extremes and provides the ability to take advantage of the generated solutions. Additionally, as optimization technique could be any non-gradient-based heuristic technique, while several hybrid architectures can be used combining GAs with other optimization techniques [12,13].

5   Utilization and Learning of Design Knowledge

The present approach relies on a case-based reasoning framework that facilitates the retrieving and reusing of past design cases. The case-based reasoning process is performed through submitting an artificial neural network to unsupervised learning of the designer's preferences. The fuzzy preferences assigned to the variable PVs must be compiled to single scalar values before they can be used. The definition of these scalar values must be done under the consideration that they must capture adequately the information modeled through a fuzzy preference. It is reasonable for this compiled scalar value to represent a value of the PV with high levels of preference in the initial fuzzy set representation. Moreover, in case of continuous spaces of values of PVs, these high levels of preferences may occur for multiple values. Case-DeSC provides the 'medium-of-maximums' as the default defuzzification method for compiling the fuzzy preferences into scalar values. This method searches for the value of the considered PV for which the preference is maximum and if there are more than one values then the medium of them is considered. Optionally the designer may select the 'centroid' defuzzification method [3,6], if it is estimated that it models better the specific fuzzy preference. Case-DeSC provides the possibility of selecting one of the available defuzzification methods for each variable PV independently.

The determination of the retrieval pattern and the deffuzification method for each fuzzy preference makes possible the retrieval process of similar past solutions. The general idea of this retrieval process is summarized on the statement 'Search for past cases (solutions) characterized by PV values that are similar to the PV values defined by the designer'. Case-DeSC utilizes competing neural networks [3,6,8] and k-means clustering algorithm [8] for accomplishing the retrieval tasks. This algorithm revises the centroids of the clusters incorporating this movement. A competing neural network with a number of input nodes corresponding to the PVs is also considered. This neural network is submitted to unsupervised learning [3,6,8] using as training examples all the cases stored in the case base. The neural network is designated to classify the training example cases by using the k-means algorithm [8,9].. The selection of a function for defining the global weighted distance measures depends on the characteristics of the considered design problem and on the designer's preferences. By the completion of training, the neural network is able to classify the input vector of scalar designer's preferences to a specific cluster. Then all design cases contained in this cluster will be retrieved and proposed as similar solutions to the designer. The ANN serve as an efficient tool to retrieve past solutions from the case base. The retrieved solution can then be utilized with different ways in order to enhance final design performance [12,13].

6   Integrated SCAD Approach and Example Case

The present paper aims at introducing a domain-independent design methodology that integrates fuzzy logic (FL), genetic algorithm (GA) and artificial neural networks (ANN) for representing and solving design problems in a stepwise manner. There are four basic steps that are depicted in figure 5.

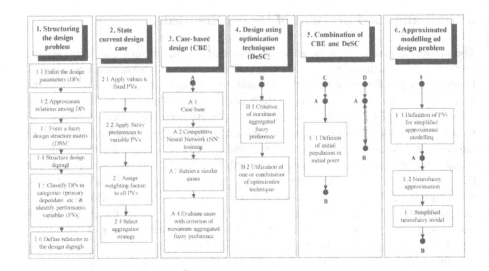

Fig. 5. A methodological approach with six steps for designing with soft-computing techniques

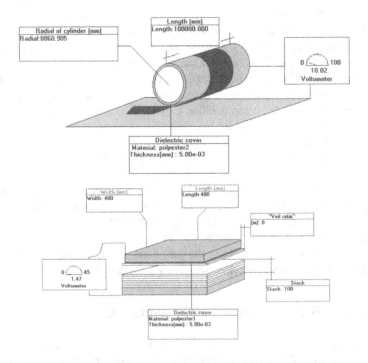

Fig. 6. Two types of electrostatic grippers for fabrics: roller and plate

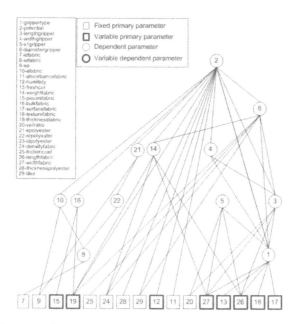

Fig. 7. Hierarchical tree of design parameters and performance variables for the problem of robotic gripper

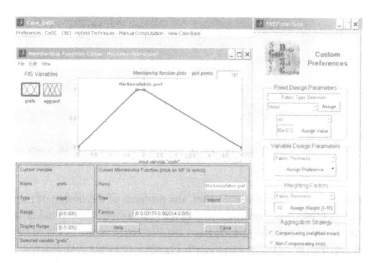

Fig. 8. Two types of electrostatic grippers for fabrics: roller and plate

Figure 5 summarizes the content of the previous paragraphs in the present paper and provides a roadmap on how the soft-computing techniques can enhance different design processes resulting to formulating an integrated framework for engineering design parametric design. Additionally, in the same figure an optional sixth step is presented related to the simplification of the initial design problem through neuro-fuzzy approximation [11,12].

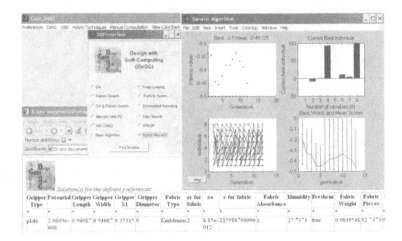

Fig. 9. Optimal solution extraction in Case-DeSC using one of the available optimization techniques or a hybrid technique

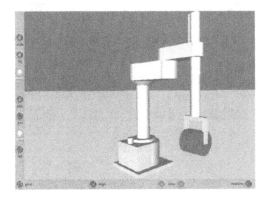

Fig. 10. VRML visualization using the design parameters' values of the extracted optimal solution

As an example case in the current paper, the problem of the design of robotic grippers for fabric is presented. The specific problem starts from conceptual design phase during where a type of robotic gripper must be selected. For the needs of the specific problem the electrostatic gripper is selected. Figure 6 presents two different types for the electrostatic gripper. The problem is analyzed by using the DSM matrix and finally a hierarchical tree for the problem design parameters is formulated as depicted in figure 7.

The design preferences for the performance variables are assigned in Case-DeSC (figure 8) together with other required design knowledge (aggregation strategy, etc.) and the system extracts a design solution by using an available stand-alone or hybrid optimization technique (figure 9). The extracted solution is visualized in 3-d VRML environment. Through this example problem, Case-DeSC manages to extract optimal design solution with fuzzy preferences and to visualize this optimal solution. Other components such CBD procedures could be also utilized to retrieve past solutions [12].

7 Discussion

Design is a process that may integrate diverse activities (i.e. exploratory, creative, rational, interactive, decision-making, iterative, etc.) during the three main design steps of design knowledge representation, optimal solution search and design knowledge learning. The present article has been authored with the intention of focusing on the opportunities provided by the implementation of soft computing techniques in engineering design. The combination of these fields leads to more complex architectures able to address multiple - and more complex - design tasks such as representation, solution search, learning, etc. For example, fuzzy logic-based approaches provide the capability of modeling uncertainty in design and managing this uncertainty under individual or collaborative frameworks on the basis of both qualitative and quantitative design entities. The optimal solution could be searched through a set-based optimization method such as the genetic algorithm, which provides the possibility of avoiding local extremes, resulting in both optimal and robust 'elite' solutions, which can be further used in various ways. Furthermore, the neural networks may act as powerful learning or approximation tools and be implemented at specific design phases. An important research opportunity is to use the SC techniques with the same aforementioned success to learn and develop descriptive processes, related with designers' best practices and strategies. Furthermore, it could be possible to use SC techniques to imitate/simulate/emulate processes and systems from other scientific or social domains, which could be applied in engineering design under specific aspects. The fusion of SC techniques provides more robust design frameworks, which address the set of the identified design issues more efficiently. The combination of SC techniques with case-based design approaches provides even more efficient results regarding these issues, resembling the way that human designers provide solutions (through analogical reasoning).

A significant research activity has been reported towards addressing the collaboration, communication and coordination in engineering design, but according to the author's beliefs there are a lot of opportunities in the context of integrating the three aforementioned issues with the support of SC techniques. Although collaborative discourses may enhance the design approaches in detailed parametric design by communicating the knowledge and reasoning about the individual objectives, the creative and conceptual design problems require a different perspective. This perspective should preserve the rationality and consistency of the design knowledge representation, whereas the designers' intuition and experience should be utilized in a systematic way. Identifying this difference of perspectives on developing approaches for design problems, different development paths should be followed, according to the degree of the encapsulated design knowledge. Although, SC techniques are applied to both problem types, there is no unified SC-based approach that addresses both conceptual and detailed designs. Furthermore, it becomes evident that the SC approaches are not capable of addressing specific design issues that commonly underlie in industries design activities even in the case that the SC techniques are integrated and combined in different ways of hybridism. However, it seems that -until now- the soft computing techniques are mainly utilized to perform specific design tasks (e.g. representation, optimization, etc.) and the approaches that deploy soft computing methods in an integrated manner are rare (if not non-existent). Additionally, although many of the existing SC approaches perform efficiently in domain-specific design problems, they miss applicability or robustness in case of problems with multiple compensating characteristics (e.g. co-existence of quantitative and qualitative design parameters, common models for the conceptual and detailed design phases). Therefore, the expansion of SC systems in industrial activities render the deployment of supportive technologies such as agent-based, web-based and AI systems imperative in order to address the identified obstructive design issues. In order to have these design issues addressed, there is a need for holistic approaches using information and AI technologies as integrated frameworks capable of accommodating the SC modules as autonomous and independent agents. Although the ideal SC system should provide the maximum level of automation and/or support to the designer, its underlying architecture should preserve a human-centric character that takes advantage of human intuition, creativity and experience. The abovementioned remarks may be considered as future opportunities for the SC research community, with the perspective of developing integrated multi-tasking and multi-purpose Soft-Computing-Aided Design (SCAD) methodologies/systems in applied design centers and in the industry. The present paper makes a projection of SC techniques towards the field of engineering design in order to optimize design solutions and confront the common difficulties inherent in parametric engineering design.

Acknowledgments. University of Patras is partner and University of Aegean is associate partner of the EU-funded FP6 Innovative Production Machines and Systems (I*PROMS) Network of Excellence.

References

1. Finger, S., Dixon, J.R.: A review of research in mechanical engineering design. Part I: descriptive, prescriptive, and computer-based models of design processes. Part II: representations, analysis and design for the life cycle, Research in Engineering Design (1), 51-67 & 121-137 (1989)
2. Evbuowman, N.F.O., Sivaloganathan, S., Jebb, A.: A survey of design philosophies, models, methods and systems. Proc. Insts. Mech. Engrs. Part B: Journal of Engineering Manufacture 210, 301–320 (1996)
3. Ueda, K.: Synthesis and emergence: research overview. Advanced Engineering Informatics 15, 321–327 (2001)
4. Saridakis, K.M., Dentsoras, A.J.: Soft Computing in Engineering Design-A Review. The journal of Advanced Engineering Informatics 22(2), 202–221 (accepcted, 2008)
5. Kosko, B.: Neural networks and fuzzy systems: a dynamical systems approach to machine intelligence. Prentice-Hall, Englewood Cliffs (1992)

6. Kasabov, K.N.: Foundation of neural networks, fuzzy systems and knowledge engineering. MIT Press, Cambridge (1996)
7. Koza, R.J.: Genetic programming, on the programming of computers by means of natural selection. The MIT Press, Cambridge (2000)
8. Gallant, S.I.: Neural Network Learning and Expert Systems. The MIT Press, Cambridge (1993)
9. Pal, K.S., Dillon, S.T., Yeung, S.D.: Soft Computing in Case Based Reasoning. Springer, London (2001)
10. Maher, L.M., Balachandran, B.M., Zhang, M.D.: Case-based reasoning in design. Lawrence Erlbaum Associates, Mahwah (1995)
11. Saridakis, K.M., Dentsoras, A.J.: Integrating fuzzy logic, genetic algorithm and neural network in a collaborative parametric design framework. Advanced Engineering Informatics 20, 379–399 (2006)
12. Saridakis, K.M., Dentsoras, A.J.: Case-DeSC: Case-based design with soft computing techniques. Expert Systems and Applications 32, 641–657 (2007)
13. Saridakis, K.M., Dentsoras, A.J.: Using case-based reasoning and soft-computing techniques for the initialization of engineering design optimization. In: 8th Biennial Conference on Engineering Systems Design and Analysis, Torino, Italy (2006)
14. Saridakis, K.M., Dentsoras, A.J.: A fuzzy rule-based approach for the collaborative formation of design structure matrices, Applications and Innovations in Intelligent Systems XIII, pp. 81–94. Springer, Heidelberg (2005)
15. Law, S.: Evaluating imprecision in engineering design, Ph.D thesis, California Institute of Technology, Pasadena, CA (1996)
16. Saridakis, K.M., Dentsoras, A.J.: Evolutionary neuro-fuzzy modeling of parametric design. In: I*PROMS Virtual International Conference on Intelligent Production Machines and Systems (2005)

A Prolog Based System That Assists Experts to Construct and Simulate Fuzzy Cognitive Maps

Athanasios Tsadiras

Department of Information Technology
Alexander Technological Educational Institute of Thessaloniki
P.O.BOX 141, 57400, Thessaloniki, Greece
tsadiras@it.teithe.gr

Abstract. The method of Fuzzy Cognitive Map (FCM) is a combination of Fuzzy Logic and Artificial Neural Networks that is heavily used by experts and scientists of a diversity of disciplines, for strategic planning, decision making and predictions. A system that would assist decision makers to represent and simulate their own developed Fuzzy Cognitive Maps would be highly appreciated by them, especially from those that do not possess adequate computer skills. In this paper, a Prolog based system is designed and implemented to assist experts to both construct and simulate of their own FCMs. The representation capabilities of the system and the design choices are discussed and a variety of examples are given to demonstrate the use of the system.

Keywords: Fuzzy Cognitive Maps; Prolog; Simulation; Decision Making; Predictions.

1 Introduction

Fuzzy Cognitive Map (FCM) is a formal method for making predictions and taking decisions, that is used by scientists from various disciplines such as economy & management [1-3], industry [4], medicine [5], political science [6,7], and ecology [8,9]. The success of the construction of an FCM is heavily depended on the degree of the expertise of the domain experts involved in the FCM construction. Making decision through FCM, requires the simulation of the FCM model, which is a difficult task especially from those scientists that do not possess the necessary computer skills. The need of a system that would assist scientists to construct and simulate their own FCMs led us to the creation of a Prolog-based such system, which we will discuss in the following sections.. After a short introduction to FCMs in section 2, the representation capabilities of the system are discussed (section 3). The design of the system is presented in section 4, followed by a section concerning the implementation and the demonstration of the system's capabilities. Finally, in section 6, a summary is presented, accompanied by a number of conclusions.

2 Fuzzy Cognitive Maps

Based on Axelord's work on Cognitive Maps [10], Kosko introduced in 1986, Fuzzy Cognitive Maps (FCMs) [11,12]. FCMs are considered a combination of fuzzy logic

J. Darzentas et al. (Eds.): SETN 2008, LNAI 5138, pp. 288–300, 2008.

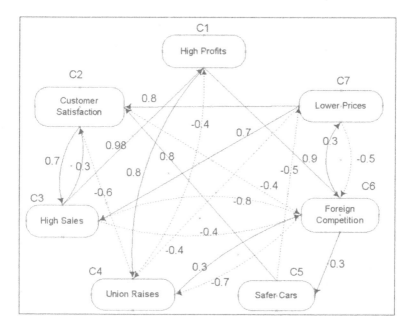

Fig. 1. An FCM concerning a car industry (modified version of original taken from [18])

and artificial neural networks and many researchers have made extensive studies on their capabilities (see for example [13-17]). An example of an FCM, concerning a car industry [18], is given in Figure 1.

FCMs create models as collections of concepts and the various causal relations that exist between these concepts. So in figure 1, the nodes represent the concepts that are involved in the model and the directed arcs between the nodes represent causal relationships between the corresponding concepts. The type of the causal relation between the two nodes is determined by the weight that each arc is accompanied. A positive (negative) causal relation between two concepts C_i and C_j means that an increase of the activation level of concept C_i will increase (decrease) C_j and also a decrease of concept C_i will decrease (increase) C_j.

The level of activation of each concept C_i at time step t is determined by number A_i^t, i=1,...,n with n to be the number of concepts of the FCM. We define w_{ij} as the weight of the arc that connects C_i and C_j. There is synchronous updating to the FCM system which means that A_i^{t+1}, i=1,...,n is calculated by one of the following formulas:

i) $A_i^{t+1} = f_M(A_i^t, S_i^t) - d A_i^t$ (using Certainty Neurons [15]) or $\qquad$ (1)

ii) $A_i^{t+1} = f_M(w_{1i} A_1^t f_M(w_{2i} A_2^t f_M(....f_M(w_{n-1,i} A_{n-1}^t, w_{n,i} A_n^t)))) - dA_i^t$ $\qquad$ (2)
(using Recursive Certainty Neurons [19])

where,

- $S_i^t = \sum_j w_{ji} A_j^t$ is the sum of the weight influences that concept C_i receives at time step t from all other concepts,

- d is a decay factor and

$$f_M(A_i^t, S_i^t) = \begin{cases} A_i^t + S_i^t(1 - A_i^t) = A_i^t + S_i^t - S_i^t A_i^t & \text{if } A_i^t \geq 0, S_i^t \geq 0 \\ A_i^t + S_i^t(1 + A_i^t) = A_i^t + S_i^t + S_i^t A_i^t & \text{if } A_i^t < 0, S_i^t < 0 \quad |A_i^t|, |S_i^t| \leq 1 \\ (A_i^t + S_i^t)/(1 - \min(|A_i^t|, |S_i^t|)) & \text{if } A_i^t S_i^t < 0 \end{cases} \quad (3)$$

is the function that was used for the aggregation of certainty factors at the MYCIN expert system [20].

3 Representation Capabilities

As it is discussed above, an FCM contains a number of arcs that represent causal relationships between the concepts of the FCM. For example, in figure 1, the arc between concept C_2:"Customers Satisfaction" and C_3:"High Sales" with weight 0.7 can be interpreted as the following statement

customers satisfaction influences sales with weight 0.7.

This means that in order to completely represent the FCM we will need as many such statements as many arcs/causal relationships exist in the FCM. Each of these statements would be in the following format:

Concept1 "influences" Concept2 "with weight" W.

Statements that are also useful for the simulation of the FCM are such as the ones below:

foreign competition is set to 0.8.   or
foreign competition is set to high.

Statements like those would mean that the activation level of the specific concept should be kept steady to that certain level through the whole simulation process. Such statements can have the following form:

Concept "is set to" Value.

Another parameter that must be specified is that of the decay factor d, described in section 2. That parameter can be set to a value by a statement such as the following:

"decay is set to" Value.
(e.g. *decay is set to 0.1.*)

The type of transfer function that should be used in the simulation can be either a) that described in equation (1) or b) that of equation (2). This can be stated in the following way.

"use mycin."	for equation (1) or
"use recursive."	for equation (2)

From the all the above, we define a grammar, written in valid Prolog syntax [21], for assisting the representation of FCMs, in figure 2.

```
statement-->concept,[influences], concept,[with,weight],weight.
statement-->concept,[influence], concept,[with,weight],weight.
statement -->[decay, is,set, to], number05.
statement-->concept,[is,set,to],value.
statement-->concept,[are,set,to],value.
statement-->[use, recursive].
statement-->[use,mycin].

concept-->[X].
concept-->[X],concept.

weight-->[X],{number(X)},{X=<1}, {-1=<X}.

value-->weight.
value--> [X], {member(X,[low,average, high])}.
value-->[very,high].
value-->[very,low].

number05--> [X],{number(X)},{X=<0.5}, {0=<X}.
```

Fig. 2. Grammar for representing FCMs

In the above grammar, special treatment exists for allowing the use of plural in statements. So

a) "influences" and "influence" are equivalent
b) "is set to" and "are set to" are equivalent

Also "concept" is defined as a series of one or more words and "weights" are only numbers in the interval [-1,1]. Correct "value" is considered not only any number in the interval [-1,1], but also a linguistic value such as "very low", "low", "average", "high" and "very high" in a manner closer to fuzzy logic. These linguistic values during simulation are translated to the values 0.1, 0.3, 0.5, 0.7, 0.9, respectively. Finally, the decay factor is allowed to take only a value of a number in the interval [0, 0.5]. The above syntax contains all the necessary facilities for describing easily and clearly an FCM structure but also contains limitations in order to prevent the user from presenting ambiguities or misunderstandings.

Using the above syntax, the FCM of figure 1 can be represented, as it is shown in figure 3.

The statements of figure 3 are not close to a computer programming language, but closer to natural language, requiring limited computer skills from its author. This means that the author can concentrate on the FCM itself and not to its computer representation. The simplicity of the grammar used leads the domain experts themselves to describe and execute their own FCMs.

profits influences strikes with weight 0.8.
profits influences foreign competition with weight 0.9.
customers satisfaction influences sales with weight 0.7.
customers satisfaction influences foreign competition with weight -0.4.
sales influence profits with weight 0.98.
sales influence customers satisfaction with weight 0.6.
sales influence foreign competition with weight -0.4.
strikes influence profits with weight -0.4.
strikes influence customers satisfaction with weight -0.3.
strikes influence foreign competition with weight 0.7.
safe cars influence customers satisfaction with weight 0.8.
safe cars influence foreign competition with weight 0.3.
safe cars influence low prices with weight -0.5.
foreign competition influences sales with weight -0.8.
foreign competition influences strikes with weight -0.7.
foreign competition influences safe cars with weight 0.3.
foreign competition influences low prices with weight 0.3.
low prices influence customers satisfaction with weight 0.8.
low prices influence profits with weight 0.7.
low prices influence strikes with weight -0.4.
low prices influence foreign competition with weight -0.5.

foreign competition is set to 0.5.
decay is set to 0.1.
use recursive.

Fig. 3. Statements representing the FCM of figure 1

4 Designing the FCM Assisting System

Looking the statements in figure 3 we can see that they resemble facts of the Prolog language. The type of these statements suits to the declarative nature of Prolog. Furthermore, the grammar of figure 2 can be directly handled by Prolog which has been used successfully in Natural Language Processing [22]. These are the main reasons for choosing the Prolog programming language for implementing the FCM Assistance System. The design of the system is shown in figure 4.

The system has the following three components/stages:

Stage 1 - Lexical Analysis. Statements such as those of figure 3, concerning the structure of the FCMs are written in a simple text file having ".fcm" extension. This file is read character by character and valid words, numbers and punctuation marks, such as the full stop, are identified. If this stage succeeds, the system continues with stage 2, otherwise the system ends its process and returns a corresponding error message.

In this stage, regular expressions [23] as the following are used:

- Natural number = [0-9]+
- Integer=(+|-)?{natural}
- Decimal= {integer}(".".{natural})?(E {natural})?

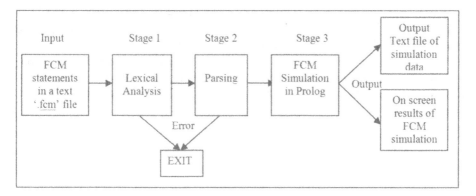

Fig. 4. The design of the FCM Assistance System

<u>Stage 2 - Parsing:</u> In this stage a parser is required. After the identification of words, numbers and punctuation marks, the parser checks whether or not the statements follow the syntactical rules defined in section 3. If the parser succeeds then the system continues with stage 3, otherwise it ends its process and returns a relevant error message in order to help the user to identify the syntax error and correct it.

<u>Stage 3 - FCM Simulation in Prolog:</u> After passing Stage 2, the system has all the necessary information for constructing and simulating the FCM. Prolog code is written in order to simulate the FCM imposed to the system. The results of the simulation are either a) shown on the computer screen or b) stored in a data file. FCM can reach a) an equilibrium point, b) a limit circle periodical behaviour or c) a chaotic behaviour [24]. If an equilibrium point is reached, this point is shown on the computer screen and the activation levels of the concepts of the FCM model through the transition phase towards this equilibrium point are saved in a ".txt" data file for further examination. If the FCM model reaches a limit circle behaviour or a chaotic behaviour then the system simulates 5000 time steps of interaction, and saves the activation levels of the concepts of the FCM model through these 5000 time steps to a ".txt" file for further examination to a graph drawing software tool.

5   Implementation and Demonstration of the FCM Assistance System

The FCM assistance system is implemented in Prolog, using SWI-Prolog [25], following the design described in section 4. To demonstrate its implementation, the FCM of figure 1 will be used. To do that, the statements of figure 3 are saved in a file called 'car_industry.fcm' and inserted to the system. The output is shown in figure 5, where seven (7) different concepts are identified and one of them, concept "Foreign Competition", is set constant to 0.5. This means that the imposed scenario #1, attempts to predict the consequences of having a moderate increase in foreign competition. After 70 time steps, FCM reaches an equilibrium point to the state that is presented in figure 5. From that, we can conclude that according to the FCM predictions, the reaction of the car industry to the increase of foreign competition is to create safer cars (0.6). This leads to increase in prices (low prices=-0.638) so sales are significantly decreased (-0.880) and the same applies to profits (-0.898) and movements toward strikes (-0.883). All the above, cause customers satisfaction to decrease (-0.798). The data of the interactions during the 70 time steps are saved to an output text file. This file can be processed by a graph drawing software tool and the transition phase towards equilibrium can be exhibited. This transition phase for the case above is shown in figure 6.

Fig. 5. Simulation of FCM concerning a Car Industry. Scenario #1: Foreign Competition is set to 0.5.

We can conclude the car industry is vulnerable to foreign competition. Other "what-if" scenarios can be imposed to the system, in order the decision maker to examine the predicted outcomes according to the FCM. For example, a decision maker would like to see the prediction of the system in the case, the direct way that foreign competition affect movements towards safer cars is lowered. This means that the initial statement

foreign competition influences safe cars with <u>weight 0.3</u>.
should be changed to

foreign competition influences safe cars with <u>weight 0.1</u>.

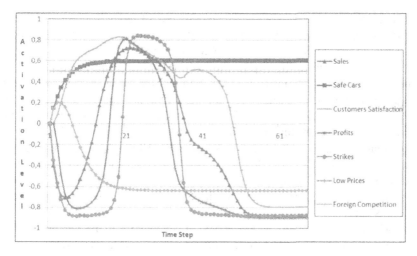

Fig. 6. Transition phase towards equilibrium for scenario #1

```
  SWI-Prolog -- c:/FCM/fcm.pl                                    _ □ X
File  Edit  Settings  Run  Debug  Help
%  lexical.pl compiled 0.00 sec, 8,212 bytes
%  syntax.pl compiled 0.00 sec, 9,212 bytes
% c:/FCM/fcm.pl compiled 0.00 sec, 39,372 bytes
Welcome to SWI-Prolog (Multi-threaded, Version 5.6.27)
Copyright (c) 1990-2007 University of Amsterdam
SWI-Prolog comes with ABSOLUTELY NO WARRANTY. This is free software,
and you are welcome to redistribute it under certain conditions.
Please visit http://www.swi-prolog.org for details.

For help, use ?- help(Topic). or ?- apropos(Word).

1 ?- fcm('car_industry.fcm').

Process of FCM based on file <car_industry.fcm> started

Concept 1 corresponds to: [sales]
Concept 2 corresponds to: [safe, cars]
Concept 3 corresponds to: [customers, satisfaction]
Concept 4 corresponds to: [profits]
Concept 5 corresponds to: [strikes]
Concept 6 corresponds to: [low, prices]
Concept 7 corresponds to: [foreign, competition]
Concept [foreign, competition] is set stable to: 0.5

**** Equilibrium found at time step 99 ****
Values at Equilibrium are the following
[sales]=0.70715
[safe, cars]=0.3333
[customers, satisfaction]=0.77839
[profits]=0.82721
[strikes]=0.83706
[low, prices]=-0.16376
[foreign, competition]=0.5

Transition Phase is saved at file <profucoma_output.txt>

Yes
2 ?- ▪
```

Fig. 7. Simulation of FCM concerning scenario #2

Imposing this new scenario #2, the outcome is that of figure 7. Once again seven (7) different concepts are identified, and one of them, concept "Foreign Competition", is set constant to 0.5. After 99 time steps, FCM reached an equilibrium point, with the concepts to be activated to the degree shown in figure 7. It can be concluded that according to the FCM predictions, now the reaction of the car industry to the increase of foreign competition is to create safer cars (0.333) but to a smaller degree than that of scenario #1 where safer cars=0.6. This leads to smaller increases in prices (low prices=-0.163) so sales are increased (0.707), profits are also increased (0.827) and the same applies to movements toward strikes (0.837). All the above, cause customers satisfaction to increase (0.778). This is a much better scenario for the car industry than that of scenario #1. An output text file contains the data of the interactions during the 99 time steps. Figure 8 exhibits the transition phase towards equilibrium, based on this data.

It should also be mentioned that any lexical or syntax error in the statements of the ".fcm" file will lead to the end of the compilation. A corresponding error message is produced that will help the user to identify the lexical or syntax error. For example if in the 'car_industry.fcm' file the statement

customers satisfaction <u>influences</u> sales with weight 0.7.

was mistakenly written as

customers satisfaction <u>inffluences</u> sales with weight 0.7.

then the error will be identified and an error message will be shown. Such an error message for the case above is shown in figure 9. The user can see where the error is and an example of a correct similar statement is given.

Foreign Competition is set to 0.5. Now "foreign competition" influences construction of "safer cars" with weight 0.1. Equilibrium found after 99 time steps.

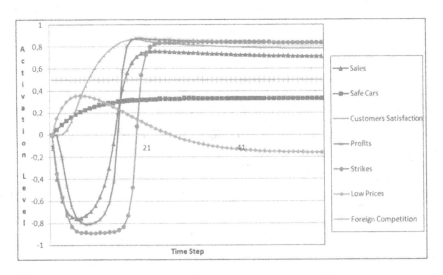

Fig. 8. Transition phase towards equilibrium for scenario #2

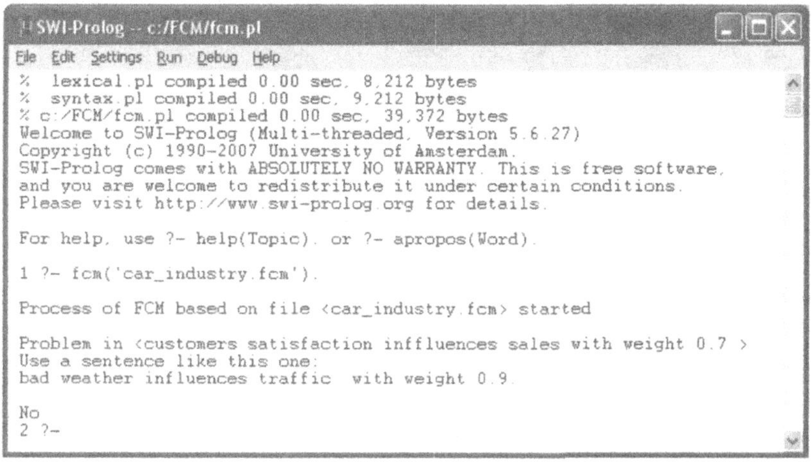

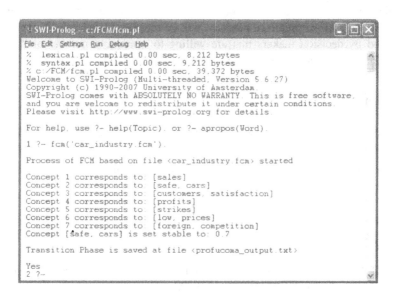

Fig. 9. Syntax error found. User is guided by a similar correct statement.

Another scenario (scenario #3) is imposed to the system, to examine the consequences of a decision to manufacture much safer cars ("safer cars" is set stable to 0.7). Imposing this scenario to the system, the outcome is that of figure 10.

Fig. 10. Simulation of FCM concerning a Car Industry for scenario#3. "Safer cars" is set to 0.7. No equilibrium point found.

No equilibrium point is found even after making 5000 time steps. Conclusions can only be drawn by studying the transition phase, which is saved in the output text file. The transition phase according to the data of this output file is shown in figure 11. It

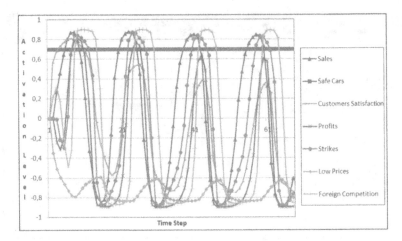

Fig. 11. Transition phase of an FCM concerning a car industry for scenario #3. A limit cycle - periodical behaviour is reached

is apparent that FCM enters a limit cycle, periodical behaviour where all concepts are getting high and low periodically. We can conclude for the car industry that these concepts will constantly increase and decrease, meaning that there will be strong interactions in the car industry and no equilibrium for this scenario. More scenarios can be inserted and the predicted outcome will be received in seconds. This is highly appreciated by decision makers that are willing to examine a number of "what-if" scenarios and check the predicted consequences of these imposed scenarios.

6 Summary – Conclusions

After a short introduction to Fuzzy Cognitive Maps, a Prolog based system is designed and implemented in order to:

a) read statements that represent causal relationships among concepts of the FCM
b) simulate the scenario written in the form of the above statements
c) present the equilibrium point of the imposed scenario (if such a point exists)
d) produce an output file that contain the data of the transition phase to visualize it if the user is interested in it.

The use of the system above is illustrated using a number of examples, to prove that it is useful for decision makers working with FCMs. The main reasons for that are:

a) the representation of FCMs is simple using close to natural language statements.
b) no programming skills are required to create and simulate FCMs,
c) a number of different scenarios can be tested easily and quickly, having the decision maker focused on the FCM itself and not to computer technicalities.

References

1. Billman, B., Courtney, J.F.: Automated Discovery in Managerial Problem Formulation: Formation of Causal Hypotheses for Cognitive Mapping. Decision Sciences 24, 23–41 (1993)
2. Bougon, M.G.: Congregate Cognitive Maps: A Unified Dynamic Theory of Organization and Strategy. Journal of Management Studies 29, 369–389 (1992)
3. Çoban, O., Seçme, G.: Prediction of socio-economical consequences of privatization at the firm level with fuzzy cognitive mapping. Information Sciences: an International Journal 169(1-2), 131–154 (2005)
4. Papageorgiou, E., Groumpos, P.: A Weight Adaptation Method for Fuzzy Cognitive Maps to a Process Control Problem. In: Bubak, M., van Albada, G.D., Sloot, P.M.A., Dongarra, J. (eds.) ICCS 2004. LNCS, vol. 3037, pp. 515–522. Springer, Heidelberg (2004)
5. John, R.I., Innocent, P.R.: Modeling uncertainty in clinical diagnosis using fuzzy logic. IEEE Transactions on Systems, Man and Cybernetics 35(6), 1340–1350 (2005)
6. Andreou, A.S., Mateou, N.H., Zombanakis, G.A.: The Cyprus Puzzle And The Greek - Turkish Arms Race: Forecasting Developments Using Genetically Evolved Fuzzy Cognitive Maps. Defence and Peace Economics 14(4), 293–310 (2003)
7. Tsadiras, A.K., Kouskouvelis, I., Margaritis, K.G.: Using Fuzzy Cognitive Maps as a Decision Support System for Political Decisions. In: Manolopoulos, Y., Evripidou, S., Kakas, A.C. (eds.) PCI 2001. LNCS, vol. 2563, pp. 172–182. Springer, Heidelberg (2003)
8. Ramsey, D., Veltman, C.: Predicting the effects of perturbations on ecological communities: what can qualitative models offer? Journal of Animal Ecology 74, 905–916 (2005)
9. Hobbs, B.F., Ludsin, S., Knight, R.L., Ryan, P.A., Biberhofer, J.: Fuzzy Cognitive Mapping as a Tool to Define Management Objectives for Complex Ecosystems. Ecological Applications 12(5) (October 2002)
10. Axelrod, R.: Structure of Design. Princeton University Press, Princeton (1976)
11. Kosko, B.: Fuzzy Cognitive Maps. Inter. Jour. of Man-Machine Studies 24, 65–75 (1986)
12. Kosko, B.: Neural Networks and Fuzzy Systems. Prentice-Hall, Englewood Cliffs (1992)
13. Khan, M.S., Chong, A., Gedeon, T.: A Methodology for Developing Adaptive Fuzzy Cognitive Maps for Decision Support. Journal of Advanced Computational Intelligence 4(6), 403–407 (2000)
14. Khan, M.S., Quaddus, M.: Group Decision Support Using Fuzzy Cognitive Maps for Causal Reasoning. Group Decision and Negotiation 13, 463–480 (2004)
15. Tsadiras, A.K., Margaritis, K.G.: Cognitive Mapping and Certainty Neuron Fuzzy Cognitive Maps. Information Sciences 101, 109–130 (1997)
16. Taber, R., Yager, R.R., Helgason, C.M.: Quantization effects on the equilibrium behavior of combined fuzzy cognitive maps. International Journal of Intelligent Systems 22(2), 181–202 (2006)
17. Stach, W., Kurgan, L., Pedrycz, W., Reformat, M.: Genetic learning of fuzzy cognitive maps. Fuzzy Sets and Systems 153(3), 371–401 (2005)
18. Eberhart, R.C., Dobbins, R.W.: Neural Network PC Tools. Academic Press, London (1990)
19. Tsadiras, A.K., Margaritis, K.G.: Recursive Certainty Neurons and an Experimental Study of their Dynamical Behaviour. In: Proceedings of the European Congress on Intelligent Techniques and Soft Computing (EUFIT 1997), Aachen, Germany, September 1997, pp. 510–515 (1997)
20. Buchanan, B.G., Shortliffe, E.H.: Rule-Based Expert Systems. The MYCIN Experiments of the Stanford Heuristic Programming Project. Addison-Wesley, Reading (1984)

21. Bratko, I.: Prolog–Programming for Artificial Intelligence, 3rd edn. Addison Wesley, Reading (2000)
22. Sharp, R.: CAT2: An Experimental Eurotra Alternative. Machine Translation 6, 215–228 (1991)
23. Aho, V., Sethi, R., Ullman, J.D.: Compilers: Principles, Techniques and Tools. Addison-Wesley, Reading (1986)
24. Tsadiras, A.K., Margaritis, K.G.: An Experimental Study of Dynamics of the Certainty Neuron Fuzzy Cognitive Maps. NeuroComputing 24, 95–116 (1999)
25. SWI-Prolog, Free Software Prolog environment, licensed under the Lesser GNU Public License, http://www.swi-prolog.org

Incremental Relevance Vector Machine with Kernel Learning

Dimitris Tzikas[1], Aristidis Likas[1], and Nikolaos Galatsanos[2]

[1] Department of Computer Science, University of Ioannina, Ioannina, Greece, 45110
{tzikas,arly}@cs.uoi.gr
[2] Department of Electrical Engineering, University of Patras, Rio, Greece, 26500
ngalatsanos@upatras.gr

Abstract. Recently, sparse kernel methods such as the Relevance Vector Machine (RVM) have become very popular for solving regression problems. The sparsity and performance of these methods depend on selecting an appropriate kernel function, which is typically achieved using a cross-validation procedure. In this paper we propose a modification to the incremental RVM learning method, that also learns the location and scale parameters of Gaussian kernels during model training. More specifically, in order to effectively model signals with different characteristics at various locations, we learn different parameter values for each kernel, resulting in a very flexible model. In order to avoid overfitting we use a sparsity enforcing prior that controls the effective number of parameters of the model. Finally, we apply the proposed method to one-dimensional and two-dimensional artificial signals, and evaluate its performance on two real-world datasets.

1 Introduction

In regression problems we are given a training set $\{x_n, t_n\}_{n=1}^{N}$, with noisy measurements $t_n = y(x_n)$ of the output of a function y when its input is a d-dimensional vector x_n and we want to make predictions for the output $y(x)$ of the function at any arbitrary test point x. Usually, the outputs t_n are assumed to contain additive noise ϵ_n:

$$t_n = y(x_n) + \epsilon_n. \tag{1}$$

In order to make predictions we typically assume a parametric form for the function y, for example the linear model is commonly used:

$$y(x|w) = \sum_{i=1}^{M} w_i \phi_i(x), \tag{2}$$

where $w = (w_1, \ldots, w_M)^T$ are the weights of the linear model and $\{\phi_i(x)\}_{i=1}^{M}$ is the basis function set. A common assumption for the noise ϵ_n is to be independent, identically distributed (i.i.d.), following a Gaussian distribution with zero mean and variance β^{-1}:

$$p(\epsilon_n|\beta) = N(\epsilon_n|0, \beta^{-1}), \tag{3}$$

J. Darzentas et al. (Eds.): SETN 2008, LNAI 5138, pp. 301–312, 2008.

where $N(\boldsymbol{x}|\boldsymbol{\mu}, \boldsymbol{\Sigma}) = (2\pi)^{-\frac{M}{2}}|\boldsymbol{\Sigma}|^{-\frac{1}{2}}\exp[-\frac{1}{2}(\boldsymbol{x}-\boldsymbol{\mu})^T\boldsymbol{\Sigma}^{-1}(\boldsymbol{x}-\boldsymbol{\mu})]$ denotes the multivariate M-dimensional Gaussian distribution over $\boldsymbol{x}$ with mean $\boldsymbol{\mu}$ and covariance matrix $\boldsymbol{\Sigma}$. Then, the likelihood of this model is obtained from (1) and (3):

$$p(\boldsymbol{t}|\boldsymbol{w}, \beta) = \prod_{n=1}^{N} N(t_n|y(\boldsymbol{x}_n|\boldsymbol{w}), \beta^{-1}), \qquad (4)$$

where $\boldsymbol{t} = (t_1, \ldots, t_N)^T$.

The linear model of (2) is a very popular model, that allows efficient estimation of its parameters. However, 'designing' a linear model (i.e. selecting an appropriate basis functions set) is very difficult, but essential in order to achieve good performance. Many recent approaches use a very large set of basis functions, but make estimations that are *sparse*, meaning that they use only few of the available basis functions. Sparsity has attracted a lot of attention lately since it is an efficient way to control the complexity of the model and avoid overfitting. Furthermore, making predictions with sparse models is very computationally efficient.

The relevance vector machine (RVM) [1] [2] is a sparse Bayesian treatment of the linear model, which assumes that the basis function set consists of kernel functions 'centered' at each training point:

$$y(\boldsymbol{x}|\boldsymbol{w}) = \sum_{n=1}^{N} w_n K(\boldsymbol{x}, \boldsymbol{x}_n), \qquad (5)$$

where $K(\boldsymbol{x}_1, \boldsymbol{x}_2)$ is some kernel function. This model has as many parameters as training points, however overfitting can be avoided by assuming a suitable prior distribution for the weights $\boldsymbol{w}$. The simplest choice is to use a Gaussian distribution on the weights, $p(\boldsymbol{w}) = \prod_{i=1}^{M} N(w_i|0, \alpha^{-1})$, which provides good regularization and allows inference at small computational cost, but it does not support sparse estimations. Instead, in order to obtain sparse estimations, the RVM assumes a Gaussian prior with *different* precision α_i for each weight w_i:

$$p(\boldsymbol{w}|\boldsymbol{\alpha}) = \prod_{i=1}^{M} N(w_i|0, \alpha_i^{-1}), \qquad (6)$$

where $\boldsymbol{\alpha} = (\alpha_1, \ldots, \alpha_M)^T$. Under the Bayesian framework, the parameters α_i can be estimated by maximizing the marginal likelihood $p(\boldsymbol{t}|\boldsymbol{\alpha}, \beta)$, obtained by integrating out the weights $\boldsymbol{w}$. Sparsity is achieved because most of the parameters α_i are estimated to very large values, thus pruning the corresponding basis functions by forcing their weights to become zero.

Selection of the kernel function is typically achieved using a cross-validation technique. However, recently there have been several attempts to learn the kernel function simultaneously with parameters of the model. For example, in [3, 4, 5] the kernel has been modeled as a linear combination of other basis functions. Furthermore, in [6] the width parameter of Gaussian kernels is learned by maximizing the marginal likelihood of the model and in [7] feature selection has been achieved by learning the variances of anisotropic Gaussian kernel functions after applying to them a sparsity enforcing prior. All these methods attempt to learn parameters of kernels that are centered at many different locations, however they assume that all these kernels have the same parameters. This might be a significant limitation if the function that we attempt to model has different characteristics at different locations, such as a signal with varying frequency.

In this paper we consider sparse linear models whose basis functions are isotropic and anisotropic Gaussian kernels, but unlike the existing literature we assign different values to the parameters of each basis function. More specifi-caly, anisotropic Gaussian kernels use a separate variance parameter h^j for each input dimension and are given by:

$$\phi(\boldsymbol{x}; \boldsymbol{m}, \boldsymbol{h}) = \exp\left[-\sum_{j=1}^{d}(h^j)^{-2}(x^j - m^j)^2\right],$$ (7)

while isotropic Gaussian kernels have the same form, but use the same variance prameter $h = h^1 = \cdots = h^d$ for all input dimensions. The main contribution of this paper is a methodology to automatically learn the mean $\boldsymbol{m}$ and variance $\boldsymbol{h}$ parameters of these basis functions, simultaneously with learning the parameters of the sparse linear model. Because in the proposed methodology we assume that the parameters of each basis function are given different values, the proposed model is very flexible.

For this reason we use a sparsity enforcing prior that directly controls the number of effective parameters of the model. This prior, has previously been used for orthogonal wavelet basis function sets [8], but here we extend it for arbitrary basis function sets. Learning in the proposed model is achieved using an algorithm that is similar to the incremental RVM algorithm [9]. It starts with an empty model and at each iteration it adds to the model an appropriate basis function, in order to maximize the marginal likelihood of the model. However, unlike the typical RVM where selection of which basis function to add to the model is a discrete optimization procedure in the finite space of basis functions, in the proposed method we use a continuous optimization procedure in the infinite space defined by the parameters of the basis functions.

There are several advantages of the proposed methodology as compared to traditional RVM [1]. First, there is no need to select the parameters of the kernel via cross validation, since they are selected automatically and therefore the corresponding computational cost is avoided. Furthermore, because each kernel may have different parameter values, the model is very flexible and it can accurately estimate a wide variety of functions. Finally, because of the sparsity enforcing prior the obtained models are typically much sparser compared to the typical RVM.

2   Sparse Bayesian Linear Models

2.1   Sparse Bayesian Regression

In this section we will review sparse Bayesian learning of linear models [1]. The posterior distribution of the weights can be computed using Bayes's law:

$$p(w|t, \alpha, \beta) = p(t|w, \beta)p(w|\alpha)p(\alpha)/p(t), \qquad (8)$$

where $p(t|w, \beta)$ is the likelihood function given by (4), $p(w|\alpha)$ is the prior distribution of the weights given by (6) and $p(\alpha)$ is a prior distribution on the weight precisions α. Defining the 'design' matrix $\Phi = (\phi(x_1), \ldots, \phi(x_M))^T$, with $\phi(x) = (\phi_1(x), \ldots, \phi_M(x))^T$, it can be shown that the weight posterior distribution is given by [1]:

$$p(w|t, \alpha, \beta) = N(w|\mu, \Sigma), \qquad (9)$$

where

$$\mu = \beta\Sigma\Phi^T t, \qquad\qquad \Sigma = (\beta\Phi^T\Phi + A)^{-1}, \qquad (10)$$

and $A = \text{diag}(\alpha)$. The weight precisions α can be ubdated by maximizing the logarithm of the marginal likelihood $p(t|\alpha, \beta) = \int p(t|w, \beta)p(w|\alpha)p(\alpha)\mathrm{d}\,w$. Assuming an uninformative prior for α ($p(\alpha) = const$), this is given by [1]:

$$L = \log p(t|\alpha, \beta) = -\frac{1}{2}\left(N\log 2\pi + |C| + t^T C^{-1} t\right), \qquad (11)$$

where $C = \beta^{-1}I + \Phi A^{-1}\Phi^T$. Setting the derivative of the marginal likelihood with respect to $\log\alpha_i$ to zero, gives the following update formula for α_i [1]:

$$\alpha_i = \frac{\gamma_i}{\mu_i^2}, \qquad\qquad \gamma_i = 1 - \alpha_i\Sigma_{ii}. \qquad (12)$$

Furthermore, setting the derivative of the marginal likelihood with respect to $\log\beta$ to zero, we obtain the following update formula for β [1]:

$$\beta = \frac{N - \sum_{i=1}^{N}\gamma_i}{\|t - \Phi\mu\|^2}. \qquad (13)$$

2.2   Incremental Optimization for Sparse Bayesian Learning

The computational cost of the sparse Bayesian learning algorithm is high for large datasets, because the computation of Σ in (10) requires $O(N^3)$ operations. A more computationally efficient incremental algorithm has been proposed in [9]. It initially assumes that $\alpha_i = \infty$, for all $i = 1, \ldots, M$, which corresponds to assuming that all basis functions have been pruned because of the sparsity constraint. Then, at each iteration one basis function may be either added to the model or re-estimated or removed from the current model. When adding a basis functions to the model, the corresponding parameter α_i is set to the value that maximizes the marginal likelihood.

More specifically, the terms of the marginal likelihood (11) that depend on a single parameter α_i are [9]:

$$l(\alpha_i) = \frac{1}{2}\left(\log\alpha_i - \log(\alpha_i + s_i) + \frac{q_i^2}{\alpha_i + s_i}\right), \qquad (14)$$

where

$$s_i = \phi_i^T C_{-i}^{-1} \phi_i, \qquad\qquad q_i = \phi_i^T C_{-i}^{-1} t, \qquad\qquad (15)$$

and $C_{-i} = \beta I + \sum_{j \neq i} \alpha_j \phi_j \phi_j^T$.

It has been shown in [10] that $l(\alpha_i)$ has a single maximum at:

$$\alpha_i = \frac{s_i^2}{q_i^2 - s_i} \qquad\qquad \text{if } q_i^2 > s_i,$$

$$\alpha_i = \infty \qquad\qquad \text{if } q_i^2 \leq s_i. \qquad\qquad (16)$$

Based on this result the incremental algorithm proceeds iteratively, adding each time a basis function ϕ_i if $q_i^2 > s_i$ and removing it otherwise.

An important question that arises in the incremental RVM algorithm is which basis function to update at each iteration. There are several possibilities, for example we could choose a basis function at random or with some additional computational cost, we could test several and select the one whose addition will cause the largest increase to the marginal likelihood.

2.3   Adjusting Sparsity

In order to control the amount of sparsity, we define a prior on α that directly penalizes models with large number of effective parameters [8]. Notice, that the output of the model at the training points $y = (y(x_1), \ldots, y(x_N)))^T$ can be evaluated as $y = St$, where $S = \beta \Phi \Sigma \Phi^T$ is the so called smoothing matrix. The 'degrees of freedom' of S, given by the trace of the smoothing matrix trace(S), measure the effective number of parameters of the model. This motivates the following sparsity prior [8]:

$$p(\alpha) \propto \exp(-c \, \text{trace}(S)), \qquad\qquad (17)$$

where the sparsity parameter c provides a mechanism to control the amount of desired sparsity. When using specific values of the sparsity parameter c, some known model selection criteria are obtained [11]:

$$c = \begin{cases} 0 & \text{None (typical RVM)}, \\ 1 & \text{AIC (Akaike information criterion)}, \\ \log(N)/2 & \text{BIC (Baysian information criterion)}, \\ \log(N) & \text{RIC (Risk inflation criterion)}. \end{cases} \qquad (18)$$

Learning using this prior is achieved by maximizing the posterior $p(\alpha, \beta | t) \propto p(t | \alpha, \beta) p(\alpha) p(\beta)$. If the basis function set is orthogonal ($\Phi^T \Phi = I$) this prior reduces to:

$$p(\alpha_i) \propto \exp(-\frac{c}{1 + \alpha_i/\beta}). \qquad\qquad (19)$$

An incremental algorithm that maximizes the marginal likelihood has been proposed in [8], which is similar to the typical incremental RVM algorithm [9].

However, because of the sparsity prior analytical updates for the weight precisions α_i (such as (16)) cannot be obtained, but instead a numerical solution is used to update them.

In this work, we consider the general case of non-orthogonal basis functions. Since $\text{trace}(\beta\boldsymbol{\Phi}\boldsymbol{\Sigma}\boldsymbol{\Phi}^T) = M - \sum_{i=1}^{M} \alpha_i \Sigma_{ii}$, we can write the proposed sparsity prior as:

$$p(\alpha_i) \propto \exp\left(-c\left(M - \sum_{i=1}^{M} \alpha_i \Sigma_{ii}\right)\right). \tag{20}$$

Maximizing the posterior $p(\boldsymbol{\alpha}, \beta|\boldsymbol{t}) \propto p(\boldsymbol{t}|\boldsymbol{\alpha}, \beta)p(\boldsymbol{\alpha})p(\beta)$ leads to adding to the marginal log-likelihood of (11) an additional term that is obtained from (20):

$$L^s = L - c(M - \sum_{i=1}^{M} \alpha_i \Sigma_{ii}). \tag{21}$$

Setting, the derivative of L^s with respect to $\log \alpha_i$ to zero,

$$\frac{\partial L^s}{\partial \log \alpha_i} = \frac{1}{2}(1 - \alpha_i \Sigma_{ii} - \alpha_i \mu_i^2) + c(1 - \alpha_i \Sigma_{ii})\alpha_i \Sigma_{ii} = 0, \tag{22}$$

we obtain the following update formula for α_i:

$$\alpha_i = \frac{\gamma_i}{\mu_i^2 - 2c\gamma_i \Sigma_{ii}}. \tag{23}$$

Similarly, we update β by setting the derivative of L^s with respect to $\log \beta$ to zero:

$$\frac{\partial L^s}{\partial \log \beta} = \frac{1}{2}\left[\frac{N}{\beta} - \|\boldsymbol{t} - \boldsymbol{\Phi}\boldsymbol{\mu}\|^2 - \text{trace}(\boldsymbol{\Sigma}\boldsymbol{\Phi}^T\boldsymbol{\Phi})\right] - \beta c\,\text{trace}(\boldsymbol{\Phi}\boldsymbol{\Sigma}\boldsymbol{\Phi}) = 0. \tag{24}$$

Because of the sparsity prior, we cannot solve this equation analytically. However, we can easily obtain a numerical solution that we use to update β.

Regarding the incremental algorithm, keeping only the terms of L that depend on a single parameter α_i and because $\Sigma_{ii} = 1/(\alpha_i + s_i)$ [9], we obtain:

$$l^s(\alpha_i) = l(\alpha_i) - c(1 - \frac{\alpha_i}{\alpha_i + s_i}), \tag{25}$$

whose gradient is given by:

$$\frac{\partial l^s(\alpha_i)}{\partial \alpha_i} = \frac{1}{2}\left[\frac{1}{\alpha_i} - \frac{1}{\alpha_i + s_i} - \frac{q_i^2 - 2cs_i}{(\alpha_i + s_i)^2}\right]. \tag{26}$$

Setting this gradient to zero, we find that $l^s(\alpha_i)$ is maximized at

$$\alpha_i = \frac{s_i^2}{q_i^2 - (2c+1)s_i} \qquad \text{if } q_i^2 > (2c+1)s_i,$$

$$\alpha_i = \infty \qquad \qquad \text{if } q_i^2 \leq (2c+1)s_i. \tag{27}$$

It may be surprising that we can analytically maximize $l^s(\alpha_i)$, while in [8] where in the prior term $p(\boldsymbol{\alpha})$ the basis functions are approximated as orthogonal ($\boldsymbol{\Phi}^T\boldsymbol{\Phi} \approx \boldsymbol{I}$), analytical maximization is infeasible.

3 Kernel Learning

In this section we propose an algorithm for learning the parameters for the proposed model. Notice that the typical RVM algorithm cannot be applied here, since it is based on the assumption that θ_i are fixed in advance. Instead, the proposed algorithm is based on the incremental RVM algorithm and therefore it works with only a subset of the basis functions, which are named *active basis functions*. In order to explore the basis function space, there are mechanisms to convert inactive basis functions to active and vice versa.

Specifically, at each iteration we select the most appropriate basis function to add to the model as measured by the increment of the marginal likelihood. Therefore, in order to select a basis function for addition to the model we perform an optimization of the marginal likelihood with respect to the parameters of the basis function. In a typical RVM, where the basis functions are kernels, this optimization is performed with respect to the locations of the kernels. Furthermore, because the kernels are assumed to be located at the training points, this optimization is discrete. In contrast, the proposed model assumes continuous parameters for the basis functions, and therefore continuous optimization must be employed, which uses the derivatives of the marginal likelihood with respect to the parameters of the basis functions.

Furthermore, in contrast to the incremental RVM algorithm, which at each iteration selects a single basis function and it either adds it to the model or reestimates its parameters or removes it from the model, the proposed algorithm performs at each iteration all these three operations; it first attempts to add a basis function to the model, then updates all parameters of active basis functions and finally removes any active basis functions that no longer contribute to the model. The additional operations speed up convergence without introducing significant computational cost, since there are only few active basis functions.

Hereafter, we discuss the steps of the proposed algorithm in detail.

1) Select an inactive basis function to add to the model. Addition of a basis functions to the model is always performed in a way that increases the marginal likelihood. This is achieved using a continous numerical optimization method (such as BFGS), which requires the derivatives of the marginal likelihood. The derivative for α_i is given by (26) and for θ_{ik} it is given by:

$$\frac{\partial l^s(\alpha_i)}{\partial \theta_{ik}} = -\left(\frac{1}{\alpha_i + s_i} + \frac{q_i^2 + c\alpha_i}{(\alpha_i + s_i)^2}\right) r_i + \frac{q_i}{\alpha_i + s_i} w_i, \tag{28}$$

where

$$r_i \equiv \frac{1}{2}\frac{\partial s_i}{\partial \theta_{ik}} = \phi_i(\theta_i)^T C_{-i}^{-1}\frac{\partial \phi_i(\theta_i)}{\partial \theta_{ik}}, \qquad w_i \equiv \frac{\partial q_i}{\partial \theta_{ik}} = t^T C_{-i}^{-1}\frac{\partial \phi_i(\theta_i)}{\partial \theta_{ik}}. \tag{29}$$

Notice that since we use a local optimization method (in our case the quasi-Newton BFGS method), we can only attain a local maximum of the marginal likelihood, which depends on the initialization. For this reason, we perform this maximization several times, each time with different initialization and then we use the parameters that correspond to the best solution.

2) Optimize active basis functions. Although we optimize the parameters of each basis function at the time that we add it to the model, it is possible that the optimal values for the already existing parameters will change, because of the addition of the new basis function. For this reason, after the addition of a basis function, we optimize the parameters α_i and θ_i of all the active basis functions of the current model. Specifically, the weight precision parameters α_i are updated using (16), while the basis function parameters θ_i are updated using an optimization algorithm. Instead of computing separately the derivative for each θ_i from (28), we use the following formula [1]:

$$\frac{\partial L^s}{\partial \theta_{ik}} = \sum_{n=1}^{N} \frac{\partial L^s}{\partial \phi_i(\boldsymbol{x}_n; \boldsymbol{\theta}_i)} \frac{\partial \phi_i(\boldsymbol{x}_n; \boldsymbol{\theta}_i)}{\partial \theta_{ik}}, \tag{30}$$

where $\frac{\partial L^s}{\partial \phi_i(\boldsymbol{x}_n; \boldsymbol{\theta}_i)} \equiv D_{ni}$ is given by

$$\boldsymbol{D} = \beta \left[(\boldsymbol{t} - \boldsymbol{\Phi}\boldsymbol{\mu})\boldsymbol{\mu}^T - \boldsymbol{\Phi}\boldsymbol{\Sigma} \right] + 2c\beta\boldsymbol{\Phi}\boldsymbol{\Sigma}\boldsymbol{A}\boldsymbol{\Sigma}.$$

3) Optimize hyperparameters and noise precision. The hyperparameters $\boldsymbol{\alpha}$ of the active basis functions are updated at each iteration using (23) and the noise precision β is updated by numerically solving (24).

4) Remove basis functions. After updating the hyperparameters $\boldsymbol{\alpha}$ of the model it is possible that some of the active basis functions will no longer have any contribution to the model. This happens because of the sparsity property, which allows only few of the basis functions to be used in the estimated model. For this reason, we remove from the model those basis functions that no longer contribute to the estimate, specifically those with $\alpha_i > 10^{12}$. Removing these basis functions is important, not only because we avoid the additional computational cost of updating their parameters, but also because we avoid possible singularities of the covariance matrices due to numerical errors in the updates.

5) Repeat until convergence. We assume that the algorithm has converged when the increment of the marginal likelihood is negligible ($\Delta L^s < 10^{-6}$). Because at each iteration we consider only a subset of the basis functions for addition to the model, we assume that convergence has occured only when the above criterion is met for ten successive iterations.

4   Numerical Experiments

In this section we present results from the application of the proposed method to artificial and real regression problems. Specifically, we compare the proposed method with isotropic and anisotropic kernels (denoted with aRVM and aRVM^d repsectively) and the typical RVM with Gaussian kernels [1] (denoted with RVM).

In the first experiment we generated $N = 128$ points from the "Doppler" function

$$g(x) = \sqrt{x(1-x)} \sin\left(\frac{2\pi(1+\delta)}{(x+\delta)}\right), \tag{31}$$

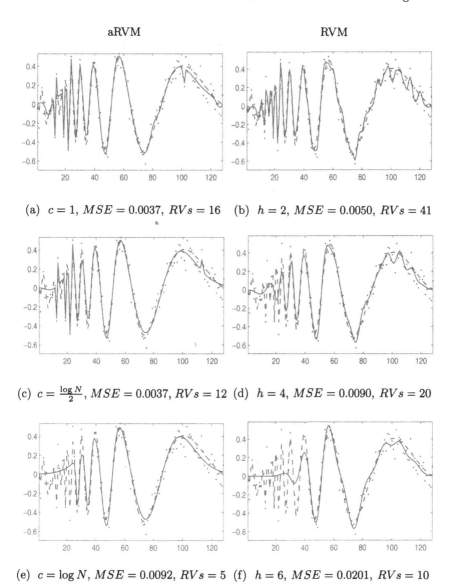

Fig. 1. Regression example with Doppler signal. Estimates obtained with (a),(d),(g) aRVM, (b),(e),(h) RVM and (c),(f),(i) sRVM. The dashed line shows the true signal, the dots are the noisy observations and the solid line shows the estimate. Under each image the values of the kernel width h or sparseness parameter c, the mean square error (MSE) of the estimate and the number of relevance vectors are shown.

with $\delta = 0.01$ and added white Gaussian noise of variance $\sigma^2 = 1$. We then applied the proposed aRVM and the typical RVM regression methods and evaluated the obtained estimates by computing the mean square error $MSE = \sum_{i=1}^{N}(g(x_i) - \hat{y}_i)/N$. Notice that in the one-dimensional case, there is no difference between isotropic and anisotropic kernels. The results are shown in Fig. 1. In the above experiments the smoothness of the estimates of the traditional RVM method depends on the width of the kernel h. Here, we have selected the value of the width that results to the lowest MSE and two other values for illustration purposes. On the other hand, the smoothness of the estimates of aRVM depends on the sparseness parameter c. Here, we have used the values $c = 1$, $c = \log(N)/2$ and $c = \log(N)$ that correspond to the AIC, BIC and RIC model selection criteria.

Notice that the optimal estimates in terms of MSE were obtained using the aRVM method. Also, the optimal solutions of aRVM are consistently sparser as compared to the optimal solutions of traditional RVM. Furthermore, we observe that traditional RVM fails to estimate signals that contain both high and low frequencies. Specifically, using a large kernel fails to estimate the high frequencies while using a small kernel results in noisy estimates of the low frequencies. The herein proposed aRVM method overcomes this problem by assigning to each kernel a separate width parameter.

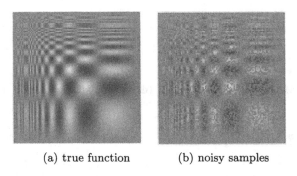

(a) true function   (b) noisy samples

Fig. 2. (a) True signal and (b) samples of the two-dimensional 'Doppler' signal that was used for training of the methods

aRVM   aRVM^d RVM

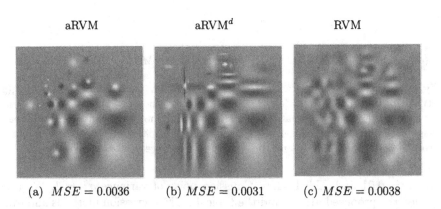

(a) $MSE = 0.0036$ (b) $MSE = 0.0031$ (c) $MSE = 0.0038$

Fig. 3. Estimation of the aRVM method with (a)isotropic and (b)anisotropic Gaussian kernel functions

Table 1. Comparison of aRVM and RVM on regression

Dataset	N	d	MSE	RVs	MSE	RVs	MSE	RVs
			— aRVM^d —		— aRVM —		— RVM —	
computer hardware	209	6	21677	5.1	22379	5.0	30004	140.5
concrete	1030	8	26.24	22	34.515	9.10	44.204	140.2

In the next experiment we used a two-dimensional generalization of the 'Doppler' function [8]:

$$g(x_1, x_2) = g(x_1)g(x_2), \tag{32}$$

where $g(x)$ is given by 31. We then set $\delta = 0.01$ and generated a 128×128 image by sampling this function on a grid. We trained the compared methods using a subset of these samples, containing a proportion of $r = 0.5$ randomly selected samples. Furthermore, we added to the observations white Gaussian noise of variance $\sigma^2 = 0.1$. We then applied (i) the proposed aRVM method with anisotropic Gaussian kernels (ii) aRVM with isotropic kernels and (iii) the typical RVM method with a fixed Gausian kernel that was selected using cross-validation. The result of each method was evaluated by measuring the mean square error with respect to the true function (without noise) on the whole 128×128 image. For aRVM, we set the sparsity parameter to $c = \log(N)/2$ and for the kernel width of RVM we test several values and select to illustrate the case $h = 2$, which is the value that produced the smallest MSE among all tested values of h. The samples of the training set are shown in Fig. 2 and the estimations of the algorithms are shown in Fig. 3. Observing these results, it is obvious that using anisotropic kernels improves the accuracy of the estimation.

Finally, in a last experiment we test the performance of the proposed method aRVM on two real datasets obtained from the UCI Machine learning repository, using either isotropic or anisotropic Gaussian kernels. The results, which are shown in Table 1 have been obtained by performing ten-fold cross-validation and demonstrate the potential of the proposed method for real regression problems.

5 Conclusions

We have presented a learning methodology according to which the parameters of the basis functions of sparse linear models can be determined automatically. More specifically, we assume that the basis functions of this model are kernels and, unlike most kernel methods, for each kernel we learn distinct values for a set of parameters (i.e. location, scale). Because many parameters are adjusted, the proposed model is very flexible. Therefore, to avoid overfitting we use a sparsity prior that controls the effective number of parameters of the model, in order to encourage very sparse solutions. The proposed approach has several avantages; it automatically learns the parameters of the kernel without the need to perform cross-validation, the model is very flexible and it can approximate a wide variety of functions efficiently, and it produces very sparse models. Finally, it is straightforward to extend the proposed methodology for learning the parameters of any type of basis functions, as long as their derivatives can be computed.

References

1. Tipping, M.E.: Sparse Bayesian learning and the relevance vector machine. Journal of Machine Learning Research 1, 211–244 (2001)
2. Bishop, C.M.: Pattern Recognition and Machine Learning. Springer, Heidelberg (2006)
3. Lanckriet, G.R.G., Cristianini, N., Bartlett, P., Ghaoui, L.E., Jordan, M.I.: Learning the kernel matrix with semidefinite programming. J. Mach. Learn. Res. 5, 27–72 (2004)
4. Girolami, M., Rogers, S.: Hierarchic Bayesian models for kernel learning. In: ICML 2005: Proceedings of the 22nd international conference on machine learning, pp. 241–248. ACM, New York (2005)
5. Sonnenburg, S., Rätsch, G., Schäfer, C., Schölkopf, B.: Large scale multiple kernel learning. J. Mach. Learn. Res. 7, 1531–1565 (2006)
6. Quiñonero-Candela, J., Hansen, L.K.: Time series prediction based on the relevance vector machine with adaptive kernels. In: Proceedings of the International Conference on Acoustics, Speech, and Signal Processing, Piscataway, New Jersey, vol. 1, pp. 985–988. IEEE, Los Alamitos (2002)
7. Krishnapuram, B., Hartemink, A.J., Figueiredo, M.A.T.: A Bayesian approach to joint feature selection and classifier design. IEEE Trans. Pattern Anal. Mach. Intell. 26(9), 1105–1111 (2004)
8. Schmolck, A., Everson, R.: Smooth relevance vector machine: a smoothness prior extension of the RVM. Machine Learning 68(2), 107–135 (2007)
9. Tipping, M., Faul, A.: Fast marginal likelihood maximisation for sparse Bayesian models. In: Proc. of the Ninth International Workshop on Artificial Intelligence and Statistics (2003)
10. Faul, A.C., Tipping, M.E.: Analysis of sparse Bayesian learning. In: Advances in Neural Information Processing Systems, pp. 383–389. MIT Press, Cambridge (2001)
11. Holmes, C.C., Denison, D.G.T.: Bayesian wavelet analysis with a model complexity prior. In: Bernardo, J.M., Berger, J.O., Dawid, A.P., Smith, A.F.M. (eds.) Bayesian Statistics 6: Proceedings of the Sixth Valencia International Meeting. Oxford University Press, Oxford (1999)

Histogram-Based Visual Object Recognition for the 2007 Four-Legged RoboCup League*

Souzana Volioti and Michail G. Lagoudakis

Intelligent Systems Laboratory
Department of Electronic and Computer Engineering
Technical University of Crete
Chania 73100, Crete, Greece
{sbolioti,lagoudakis}@intelligence.tuc.gr

Abstract. RoboCup is an annual international robotic soccer competition. One of its most popular divisions is the Four-Legged League, whereby each team consists of four fully autonomous Sony AIBO robots and researchers focus solely on software development over a standard robotic platform. To compete successfully each robot needs to address a variety of problems: visual object recognition, legged locomotion, localization, and team coordination. This paper focuses on the problem of visual object recognition and suggests a uniform approach for recognizing the key objects in the RoboCup 2007 field: the two goals, the two beacons (colored landmarks), and the ball. The proposed method processes the color-segmented camera image and delivers the type, as well as an estimate of the distance and the angle with respect to the robot, of each recognized object in the current field of view. Our approach is based on a number of procedures which are used uniformly for all three types of objects: horizontal and vertical scanning of the image, identification of large colored areas through a finite state machine, clustering of colored areas through histograms, formation of a bounding box indicating possible presence of an object, and customized filtering for removing implausible indications. Our approach is compared against the approaches of two RoboCup teams (German Team 2004 and SPQR-Legged 2006) and is shown to be equally good or better in many cases. The proposed approach has been used successfully by Team Kouretes of the Technical University of Crete, Greece during various RoboCup events.

1 Introduction

In its short history, the RoboCup competition [1] has grown to a well-established annual event bringing together the best robotics researchers from all over the world. The initial conception by Hiroaki Kitano in 1993 led to the formation of the RoboCup Federation with a bold vision: *"By the year 2050, to develop a team of fully autonomous humanoid robots that can win against the human world soccer champions"*. The uniqueness of RoboCup stems from the real-world challenge it poses, whereby the core problems of robotics (perception,

---

* This project was partially supported by the European Marie-Curie International Reintegration Grant MCIRG-CT-2006-044980 awarded to Michail G. Lagoudakis.

J. Darzentas et al. (Eds.): SETN 2008, LNAI 5138, pp. 313–326, 2008.

Fig. 1. Four-legged league shot from RoboCup German Open 2007

cognition, action, coordination) must be addressed simultaneously under real-time constraints. The proposed solutions are tested on a common benchmark environment through soccer games in various leagues, thus setting the stage for demonstrating and promoting the best research approaches, and ultimately advancing the state-of-the-art in the area. Beyond soccer, RoboCup now includes also competitions in search-and-rescue missions (RoboRescue), homekeeping tasks (RoboCup@Home), robotic performances (RoboDance), and simplified soccer leagues for K-12 students (RoboCup Junior).

The Four-Legged League of the RoboCup competition (Figure 1) is among the most popular leagues, featuring four robot players (SONY AIBO robots) per team. The AIBO ERS-7 robot resembles a small pet dog and is equipped with a 576MHz 64-bit RISC processor, 64 Mb of RAM, a color CCD camera, three infrared distance sensors, touch sensors, wifi communication, and a number of servos on its legs, head, tail, and ears, providing 20 degrees of freedom in total. Games take place in a $4m \times 6m$ field marked with white lines, colored goals, and two colored beacons which serve as landmarks (Figure 2). Each game consists of two 10-minute halves and teams switch colors and side at halftime. There are several rules enforced by human referees during the game. For example, a player is punished with a 30-seconds removal from the field if he performs an illegal action, such as pushing an opponent for more than three seconds, grabbing the ball between his legs and head for more than three seconds, leaving the field, or entering his own goal area as a defender.

The main characteristic of the Four-Legged League is that no hardware changes are allowed; all teams use the exact same robotic hardware and differ only in terms of their software. This convention results to the league's characterization by a unique combination of features: autonomous vision-based player operation, legged locomotion and action, uniform robotic platform. Given that the underlying robotic hardware is common for all competing teams, research efforts have focused on developing more efficient algorithms and techniques for visual perception, active localization, omnidirectional motion, skill learning, and coordination strategies. During the course of the years an independent observer could easily notice a clear progress in all research directions.

Fig. 2. The 2007 Four-Legged RoboCup League field [2]

This paper focuses solely on the problem of object recognition within visual perception. Through its color CCD camera mounted on the head, each robot must be able to extract all information needed for successful play. The key objects in the RoboCup 2007 field are the two goals, the two beacons (colored landmarks), and the ball. Each object has its own appearance characteristics (color, shape, size) which discriminate it from the other objects. The goal is to identify such objects in the current field of view, if visible, and to estimate their distance and angle with respect to the robot's current pose. This information can subsequently be used for localization and decision making.

Briefly, the proposed object recognition method processes the color-segmented camera image; the image is scanned pixel by pixel, both horizontally and vertically along specific scan lines. Large areas of the same color are then determined through a finite state machine, which permits a specific tolerance of non-colored pixels within the colored areas. These areas are likely to be part of an object of interest. In order to figure this out, the colored areas are clustered through histograms, which expose the density of the colored areas and give an indication of their extent. Such indications help in finding out the dimensions of an object within the image. Then, after a series of transformations using the robot head pose, the object dimensions within the image, the camera parameters, and the real world dimensions of each object, an estimate of the distance and the orientation to each object with respect to the robot is computed. Some checks are also performed to filter out implausible indications. The proposed approach is compared against the solutions given by two RoboCup teams (German Team 2004 and SPQR-Legged 2006) by means of statistics in successful object recognition and processing time. The results indicate that our approach is equally good or even better in many cases.

The remainder of the paper is structured as follows: Section 2 describes the target field objects and states the problem. Section 3 describes the proposed histogram-based approach, while Section 4 discusses related work by other RoboCup teams. Finally, Section 5 summarizes the results of comparing our approach to other known approaches.

2   Problem Statement

This section states the problem we address in this paper and places our work in the broader context of the robot architecture in terms of input and output.

2.1   Raw Image Correction and Color Segmentation

Our object recognition method relies on visual input. Each AIBO robot is equipped with a color CCD camera which delivers YUV images at 30 frames per second with a resolution of 160 × 208 pixels. The viewing angle is 56.9 deg wide and 45.2 deg high. The camera is mounted on the robot head which can move independently of the robot body using its three degrees of freedom (head pan, head tilt, neck tilt). This freedom of motion offers an effective viewing angle which is approximately 240 deg wide and 175 deg high with respect to the robot body. Several hardware tests have shown that the raw image from the ERS-7 camera is radiometrically distorted (becoming blue away from the center and close to the corners), therefore we pass the image through an inverse transformation to cancel this distortion [3].

The corrected image is then segmented according to color. Given that there is only a handful of colors used for the objects in the field (green, orange, sky-blue, yellow, white), it is essential to map each pixel of the image to its correct color class and then use the color-segmented image for object recognition. Color segmentation has proved to be quite sensitive to lighting conditions, therefore RoboCup organizers arrange for ceiling lights that deliver approximately uniform lighting at about 1000 lux. Our color segmentation is based on a method that first transforms the image into the HSI format and then uses mainly the H-dimension to separate the colors, continuously adapting the separation thresholds every 20 frames for best separation [4]. This method is more effective when the raw image is saturated, therefore we typically set the camera parameters to high gain and medium shutter speed.

These preprocessing steps are essential for obtaining a color-segmented image, however they are not unique; other alternative choices could be used to produce the desired color-segmented image. A detailed coverage of image correction and color segmentation methods is outside the scope of this paper.

2.2   RoboCup 2007 Field Objects

The problem studied in this paper is the problem of object recognition using color-segmented images from the robot camera. The main objects of interest are the two goals, the two beacons, and the ball. The robot needs to recognize at least these objects in order to play soccer successfully; the ball is the central object around which the game revolves, the goals are the areas where the ball must go to or stay away from, and the beacons are essential landmarks in determining the current robot position and orientation in the field. These five objects are depicted in Figure 3. There are also other objects of secondary importance in the field; these are the lines marked on the fields and the partner/opponent robot

Fig. 3. Objects of interest in the 2007 field: goals, beacons, and ball

players. Recognition of lines is not really helpful due to high perceptual aliasing (there are several areas in the field that yield the same line perception), and recognition of other players is a rather complicated task to be performed in real-time, given the multitude of possible robot poses and angles. The displayed goal structure was first introduced in 2007. In previous years, a completely different goal structure with compact colored areas was used, as well as four beacons instead of two marked with combinations of pink, yellow, and skyblue.

Each goal is either skyblue or yellow in color and is composed of four main pieces. Two cylindrical pieces of 30cm in length and 10cm in diameter stand for the vertical goalposts and support the upper horizontal goalpost which has a length of 80cm and a diameter of 5cm. There is also a rectangular part (10cm high) attached to the lower part of the vertical goalposts and delimits the goal region in the field. The beacons are cylindrical pieces 40cm in height and 10cm in diameter, which are painted white, skyblue, and yellow in two different combinations. Starting from the bottom, each beacon consists of a white part (20cm high) and then, depending on its type, two colored parts (10cm high). The beacons are placed at fixed positions on either side of the field middle line with the intention of facilitating the localization process, if recognized correctly. The ball is an orange plastic sphere 10cm in diameter. Its smooth surface typically leads to light reflections visible as yellow or white spots on the top part making its recognition even harder. Note that any of the above objects may be only partially visible through the camera. This is especially true for the ball when the robot is very close to it, a situation that occurs frequently. Even in such cases, it is imperative to obtain the correct type, distance, and angle of the object.

In order to address the recognition problem fully, the color-segmented camera images are combined with the pose of the robot at each moment, whereby by robot pose we mean the complete joint configuration of the robot. In particular, the pose of the robot's head, where the camera is placed, must be taken into account, so as to conclude the correct relationship between the camera image, the real world, and the orientation of the image with respect to the robot body.

3   Histogram-Based Object Recognition

This section covers the recognition process in detail. Recognition of the various objects relies on a few basic operations which are used in common. These basic operations are described first, followed by a description of the specialized recognition procedure for each object. Note that the parameter values given below are indicative only and may be changed at will. These numbers represent the choices that worked best for our specific recognition problem.

3.1   Basic Operations

The basic operations on the color-segmented image that are used uniformly for all objects are: construction of the horizon line, construction of scan lines, identification of large parts, construction of density histograms, and estimation of distance and angle. Note that these operations may be applied selectively on demand in each case; for example, vertical scanning may not be called, if horizontal scanning reveals no object indication.

Construction of the horizon line. Since all objects in the field are aligned horizontally or vertically with respect to the horizon line, it is essential to construct first the horizon of the image. The horizon line is the intersection of the projection plane (the plane where the real world's objects are projected by the camera lens and is perpendicular to the optical axis of the robot camera) and a plane parallel to the ground at the height of the camera. To this end, the rotation of the camera relatively to the robot's body is necessary information; the rotation can be extracted accurately from the current robot pose and the forward kinematics of the robot.

Construction of scan lines. Each image is scanned horizontally and if needed vertically to identify segments of uniform color. The horizontal scan lines are constructed to be parallel to the horizon line and to cover the entire image, whereas the vertical scan lines are constructed to be perpendicular to the horizon line and may or may not cover the entire image depending on the target object. Each scan line is eventually transformed into a pixelated line over the image pixels that approximates the true geometric line.

Identification of large parts. Scanning along any scan line, whether horizontal or vertical, is conducted using a finite state machine. This machine alternates between two states depending on whether the target color has been found (TargetColor state) or not (NoTargetColor state) along the line, under some transition tolerance to account for noise in the image. Persistence of the machine in the TargetColor state for more than a threshold number of pixels indicates the existence of a large part within the scan line that carries the target color. Through the state machine, the beginning and end of each such large part can be easily marked.

Construction of density histograms. All large parts (if found) are grouped into a single structure that reveals their density along the scanning direction. Each histogram is a discretization of the entire viewing angle along the scanning direction and the value of each bin indicates the number of overlapping large parts at that particular angle. The peak of the histogram indicates the highest density of large parts along the scanning direction.

Estimation of distance and angle. If an object has been successfully identified on the image and its size in pixels has been determined, then using the

Fig. 4. Basic operations: raw image, horizon line, scan lines, large parts, indications

known real dimensions of the object, the camera parameters and characteristics, and common geometry one can easily estimate the distance of the object from the camera, as well as its horizontal angle with respect to the camera focal point. Furthermore, utilizing knowledge of the current robot pose, the angle with respect to the robot body can also be estimated.

Figure 4 demonstrates some of these basic operations in the context of a horizontal scanning for a goal. Note that all operations are applied to the color-segmented image.

3.2   Goal Recognition

The recognition procedure for goals aims at recognizing the vertical and the horizontal goal posts separately. To this end, the image is first scanned horizontally in its entirety using 40 uniformly-distant horizontal scan lines and the large parts found are collected into the horizontal histogram. A peak in this histogram (if any) above a prespecified threshold (10% of the total number of scan lines) indicates the presence of a vertical post. If two peaks are found, then both vertical posts are visible and the entire goal is contained within the image (horizontally, at least); in this case, the horizontal goal indication extends between the two posts and receives high confidence on both sides. If only one peak is found then only one vertical post is visible and therefore the goal is only partially visible in the image; in this case, the horizontal goal indication extends from that post to the appropriate end of the image as indicated by the density of large parts and receives high confidence only on the side of the vertical post. If no peaks are found, then either no goal is currently visible in the image, or the robot is so close to the goal that only sees the central goal part in the image; in these two cases, either no horizontal goal indication is formed at all, or the one formed comes with a low confidence.

The procedure continues with a vertical scanning using 50 uniformly-distant vertical scan lines that extend only above and below the horizon to a fixed viewing angle (±30 deg). This is dictated by the fact that the goals lie around the horizon line and cannot be far away from it. Note that if the current pose of the robot is such that the horizon line is not visible, vertical scanning may not take place at all, if no vertical scan lines lie within the image. Vertical scanning proceeds exactly as the horizontal one, accumulating large parts into

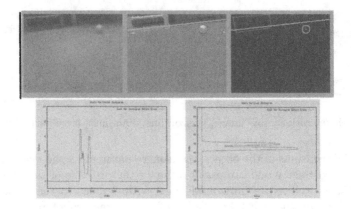

Fig. 5. Goal recognition: raw image, color-segmented image, recognized objects

the vertical histogram and identifying two, one, or no vertical goal posts as before. A vertical goal indication is similarly derived from the vertical scanning with the appropriate confidence level.

During the final goal recognition phase, it is decided whether the horizontal and vertical indications can be combined to offer a single goal indication (a bounding box). Three filters are used to reject implausible indications: (a) the confidence filter rejects indications that come with low horizontal and vertical confidence, that is, no goal post has been identified; (b) the ratio filter rejects indications that come with high confidence, but their horizontal vs. vertical size ratio fails to comply (within an interval) to the true size ratio of the real goal; and, (c) the distance filter rejects indications that yield an invalid distance estimation (negative or larger than the field size). If a set of indications passes successfully through these filters, the combined indication is used to determine the distance and the angle of the recognized object with respect to the robot body and the resulting information is passed to the higher level of cognition. The entire procedure is repeated in a interleaved manner for both the skyblue and the yellow colors. Figure 5 shows an instance of successful goal recognition along with the corresponding horizontal and vertical histograms. The two peaks in each histogram indicate the corresponding goal posts.

3.3 Beacon Recognition

The recognition procedure for beacons aims at recognizing areas where the two colors are aligned on top of each other. To this end, the image is first scanned horizontally in its entirety using 60 uniformly-distant horizontal scan lines and the large parts found are collected into horizontal histograms, one for each color (yellow and skyblue). Then, we seek to find a peak in the product of the two histograms; the idea is that only areas where the two colors coincide will yield large product values. Before multiplication, each entry in the two histograms is also checked for having a value above a prespecified threshold (10% of the

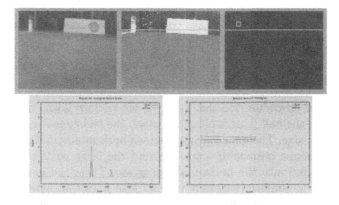

Fig. 6. Beacon recognition: raw image, color-segmented image, recognized objects

total number of scan lines) and whether the values in the two histograms are not too far from each other (each beacon colored area must yield approximately the same number of large parts). If such a peak is found, it clearly indicates two aligned colored areas of approximately the same size, likely a beacon indication. The horizontal extent of the mode corresponding to the peak in the product of histograms is calculated to mark the horizontal limits of this indication.

If a horizontal indication is found, the procedure proceeds with a vertical scanning using only 7 uniformly-distant vertical scan lines that extend horizontally only within the limits of the horizontal indication and vertically only above and below the horizon to a fixed viewing angle (±30 deg). Again, this is dictated by the fact that the beacons, like the goals, lie around the horizon line and cannot be far away from it. Note that if the current pose of the robot is such that the horizon line is not visible, vertical scanning may not take place at all, if no vertical scan lines lie within the image. Vertical scanning proceeds exactly as the horizontal one, accumulating large parts into two vertical histograms and identifying one peak in each, one for each color. The relative positioning of the picks indicates which color is above and which one is below.

During the final beacon recognition phase, it is decided whether the horizontal and vertical indications found can be combined to offer a single beacon indication. Three filters are used to reject implausible indications: (a) the proximity filter rejects indications where the vertical peaks of the two colors are far apart and not adjacent to each other; (b) the ratio filter rejects indications whose horizontal vs. vertical size ratio fails to comply (within an interval) to the true size ratio of the real beacon; and, (c) the distance filter rejects indications that yield an invalid distance estimation (negative or larger than the field size). If a set of indications passes successfully through these filters, the combined indication is used to determine the distance and the angle of the recognized beacon with respect to the robot body and the resulting information is passed to the higher level of cognition. Figure 6 shows an instance of successful beacon recognition along with the corresponding horizontal and vertical histograms. The two peaks in each histogram indicate the corresponding areas for the skyblue and the yellow colors. Notice that the two peaks coincide in the horizontal histogram, but are adjacent in the vertical one.

3.4   Ball Recognition

The recognition procedure for the ball is a bit trickier given that, when the robot is close to the ball (this is often the case), the ball cannot be viewed in its entirety in the image. The image is first scanned horizontally in its entirety using 40 uniformly-distant horizontal scan lines and the large parts of orange color found are collected into the horizontal histogram. Then, the image is scanned vertically using 50 uniformly-distant vertical scan lines that extend only below the horizon line (the ball cannot lie above the horizon line—this is also a way of avoiding false positives caused by spectators around the field wearing orange clothes) and the large parts of orange color found are collected into the vertical histogram. Then, we seek to find the peaks in both the horizontal and the vertical histograms. The resulting point (if found) indicates a central point of an orange blob in the image which might correspond to a ball.

In the next phase, a ray scanning begins from the central point along eight directions in the image (the four orthonormal and the four diagonal) aiming at identifying at least three points on the periphery of the ball. If a ray scan reaches the end of the image before the end of the orange color, it is rejected as it corresponds to a direction where the ball is hidden. However, if a transition to another color (with appropriate tolerance) is observed, then the last orange point is kept as belonging to the periphery of the ball. From three such points the center and the radius of the circle marking the ball can be calculated using the middle perpendiculars. In particular, the center of the circle is the intersection point of the middle perpendiculars of the circle chords formed by the three points on the periphery. The radius is calculated as the distance of the center to any of the three points. Note that the center of the ball may lie outside the image.

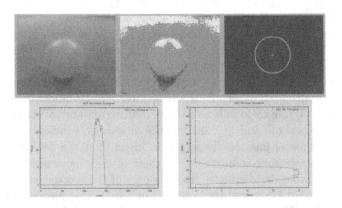

Fig. 7. Ball recognition: raw image, color-segmented image, recognized objects

The resulting circle indication goes through two filters: (a) the horizon filter rejects indications which place the center of the ball above the horizon; and, (b) the distance filter rejects indications that yield an invalid distance estimation for the ball (negative or larger than the field size). If the circular indication passes successfully through these filters, it is used to determine the distance of the recognized ball and its angle with respect to the robot body; this information is then passed to the higher level of cognition. Figure 7 shows an instance of successful ball recognition along with the corresponding horizontal and vertical histograms. This is the simple case where the ball lies entirely within the image.

4   Related Work

This section reviews the work of four representative RoboCup Four-Legged teams in object recognition. Given that the type of goals used before 2007 were different, we only review beacon and ball recognition, which remained largely unchanged.

The German Team is a collaboration of the Humboldt University of Berlin, the University of Bremen, the Technical University of Darmstadt, and the University of Dortmund and has been participating in RoboCup competitions since 2001. In German team's object recognition [3], the horizon line is calculated first and the image is scanned using non-uniform horizontal and vertical scan lines. These scan lines are denser near the horizon and sparser away from it. For beacon detection (2004 type), the scan lines look for pink pixels, which are clustered to form a horizontal indication of pink color. From that indication, 4 scan lines are constructed perpendicular to it and scan for the other color (yellow or skyblue) of the beacon. After that, different sanity and comparison checks are performed to verify the existence of a beacon and, if they come true, the center of the pink area is calculated. From this center point, the image is scanned in four directions to detect the beacons' borders. Regarding ball detection, the scan lines search for the largest orange part of the image and, when found, its center is calculated. The center is the starting point of eight scan lines that scan horizontally, vertically, and diagonally in order to find edges of the ball. Some further color checks are performed to avoid false ball perceptions and, the ending points of the scan are saved in a list. The center and radius of the ball are found using the Levenberg-Marquardt method taking into account points from this list.

The rUNSWift team is based at the University of New South Wales in Sydney, Australia, and first participated in the RoboCup competition in 1999. Object recognition for rUNSWift [5] begins with feature detection. During this phase, the image is scanned using grids of scan lines constructed around the horizon line and each pixel's value is compared with its previous one in the scan line. If a rapid value transition occurs an object feature is detected. For example, changing from green to orange indicates a ball feature. The features detected are grouped together to create the complete object. Beacon recognition (2004 type) relies on detecting pink features. An eight-point asterisk scan is performed around each beacon feature to figure out the beacon type. Then, the height and centroid of the pink features are calculated and from them the height of the whole beacon. Ball recognition is based on detecting ball edge features to form the outline of the ball, according to a generalization of Siegel's repeated median line fitting algorithm to circles. These features are combined in all possible triplets in order to form a circle using the middle perpendiculars.

Cerberus is a four-legged league team based at Bogazici University in Istanbul, Turkey and has been participating in the RoboCup competition since 2001. Regarding object recognition, first the regions of interest that may contain an object are determined. The Cerberus team applies a region growing algorithm on the raw image [6]. Regions of the same color are merged, if they are adjacent to each other, and are blobbed to form areas likely to contain a specific object (candidate regions). For accurate processing, each candidate region is rotated according to the robot's head pose. The team has introduced the bounding octagon of a region, which intends to reduce the rotation cost, as octagons are more easily rotated than rectangular boxes. To detect an object, a number of checks is performed on each candidate region. First, the neighboring regions should meet the specifications of the field colors. For instance, the bottom neighboring region of the ball should be green. Furthermore, the regions' size is taken into account and compared with the actual size and dimensions' ratio of the real objects.

The SPQR-Legged team is based at the University of Rome "La Sapienza" in Rome, Italy and has been participating to RoboCup competitions since 1998. Object recognition begins by scanning the whole color-segmented image trying to find regions of the same color. This process consists of two phases: first, the image is scanned for different particles of the same color and then the particles of the same color are blobbed in order to create a region where most pixels belong to a single color class. So, for recognition of particular objects, one needs to scan only the relevant colored regions. Each region is a rectangular box within the image frame. Once the blobbing of candidate regions is finished, each region is scanned for an object according to the color that the region contains. Concerning recognition of the beacons and the ball, SPQR-Legged has been using the approaches of the German Team applied separately to each candidate region and not to the whole image as the German Team does.

5   Results

This section attempts an evaluation of the proposed approach and a comparison to related approaches of two other RoboCup teams: GermanTeam 2004 and SPQR-Legged 2006. It was not possible to compare goal recognition, since the present goals have been introduced in 2007 and are significantly different than those used in previous RoboCup competitions. The statistics of object recognition using all approaches is shown in Table 1. Evaluation is based on three log files[1] recorded during the RoboCup German Open 2007 competition totalling several hundreds of frames.

---

[1] Log files contain continuous camera frames along with the robot pose at each frame.

Table 1. Recognition performance of the proposed and other approaches

Object	Approach	True Positives	False Negatives	True Negatives	False Positives
Goals	Proposed	**61%**	**39%**	**93%**	**7%**
Beacons	Proposed	**34%**	**66%**	**99%**	**1%**
	GT2004	28%	72%	99%	1%
	SPQRL2006	31%	69%	99%	1%
Ball	Proposed	**79%**	**21%**	**100%**	**0%**
	GT2004	67%	33%	99%	1%
	SPQRL2006	51%	49%	100%	0%

GT2004 approach exhibits better performance than SPQRL2006 in recognizing the ball, whereas SPQRL2006 performs better beacon recognition. Our proposal compares favorably to both of them; it improves on both beacon and ball recognition. The average processing time of our approach stands at 51 ms per frame, where GT2004 stands at 42 ms and SPQRL2006 at 52 ms. GT2004 achieves the lowest processing time, since the grids of scan lines focus only on regions that are most likely to contain objects of interest, whereas SPQRL2006 needs more time per frame to determine the candidate regions and process separately each one of them. Our approach exhibits a processing time similar to that of SPQRL2006, however with a significant increase in performance. The increase compared to GT2004 is due to the larger and uniform grids of scan lines we use.

False negatives are mostly due to the rapid movement of the robot's head, which can degrade the original appearance and shape of the object within the image. In addition, bad color segmentation, shadows, and reflections within the image can lead to misunderstanding of the object shape, which may lead to a false negative. Regarding the relatively low percentage of true positives for the beacons, this is likely due to their relatively small colored area. On the other hand, ball perception reaches excellent performance perhaps due to the fact that the ball is the only orange object in the field, in contrast to the skyblue and yellow colors which appear in multiple places.

6 Conclusion

The robots operate autonomously in the four-legged league, therefore all processes (object recognition, motion control, behavior control, localization, etc.) must run in real time on the onboard processor. This limitation prevents the use of sophisticated machine vision techniques for object recognition. In this paper, we proposed a simple approach for efficient recognition of the objects in the 2007 field that yields good performance while observing the real-time requirements. The proposed approach has been used successfully by Team Kouretes of the Technical University of Crete, Greece during various RoboCup events.

References

1. Kitano, H., Asada, M., Kuniyoshi, Y., Noda, I., Osawa, E., Matsubara, H.: Robocup: A challenge problem for AI. AI Magazine 18(1), 73–85 (1997)
2. Four-Legged Technical Committee: Robocup four-legged league rule book (2007)
3. Rofer, T., Laue, T., Burkhard, H.D., Hoffmann, J., Jungel, M., Gohring, D., Lotzsch, M., Duffert, U., Spranger, M., Altmeyer, B., Goetzke, V., von Stryk, O., Brunn, R., Dassler, M., Kunz, M., Risler, M., Stelzer, M., Thomas, D., Uhrig, S., Schwiegelshohn, U., Dahm, I., Hebbel, M., Nistico, W., Schumann, C., Wachter, M.: GermanTeam 2004. Technical report, Humboldt-Universität zu Berlin, Universität Bremen, Technische Universität Darmstadt, and Universität Dortmund (2004)
4. Iocchi, L.: Robust color segmentation through adaptive color distribution transformation. In: Lakemeyer, G., Sklar, E., Sorrenti, D.G., Takahashi, T. (eds.) RoboCup 2006: Robot Soccer World Cup X. LNCS (LNAI), vol. 4434, pp. 287–295. Springer, Heidelberg (2007)
5. North, A.: Object recognition from subsampled image processing. Honor's thesis, University of New South Wales, Sydney, Australia (2005)
6. Akin, H.L., Mericli, C., Gokce, B., Geleri, F., Tasdemir, N., Celik, B., Celik, M.: Cerberus 2006 team report. Technical report, Bogazici University, Turkey (2006)

Learning Ontologies of Appropriate Size

Elias Zavitsanos[1,2], Sergios Petridis[1], Georgios Paliouras[1],
and George A. Vouros[2]

[1] Institute of Informatics and Telecommunications, NCSR "Demokritos", Greece
[2] University of Aegean, Department of Information and Communication Systems
Engineering, Artificial Intelligence Laboratory, Samos, Greece
{izavits,petridis,paliourg}@iit.demokritos.gr, georgev@aegean.gr

Abstract. Determining the size of an ontology that is automatically learned from text corpora is an open issue. In this paper, we study the similarity between ontology concepts at different levels of a taxonomy, quantifying in a natural manner the quality of the ontology attained. Our approach is integrated in a recently proposed method for language-neutral learning of ontologies of thematic topics from text corpora. Evaluation results over the Genia and the Lonely Planet corpora demonstrate the significance of our approach.

1 Introduction

Ontology learning is a relatively new field of research, aiming to support the continuous and low-cost development and maintenance of ontologies, especially in fast evolving domains of knowledge. These tasks, when performed manually, require significant human effort. Thus, automated methods for ontology construction are very much needed.

Ontology learning is commonly viewed [1, 5, 13, 14] as the task of *extending* or *enriching* a seed ontology with new ontology elements mined from text corpora. Depending on the ontology elements being discovered, existing approaches deal with the identification of concepts, subsumption relations among concepts, instances of concepts, or concept properties/relations. Furthermore, we may classify existing ontology learning approaches to be either of the linguistic, statistical, or machine learning type, depending on the specific techniques employed.

While much work concentrates on enriching existing ontologies, few approaches deal with the construction of an ontology without prior knowledge. Among the difficulties of such an endeavor, is the determination of the appropriate depth of the subsumption hierarchy, given the text collection at hand. The benefit of being able to determine the appropriate depth of a taxonomy is that the hierarchy captures accurately the domain knowledge provided by the texts, reducing the extent of overlap among concepts and providing a coherent representation of the domain. The determination of the appropriate hierarchy depth prohibits both over-engineered representations and generic ones, since it constitutes a criterion for a well-structured hierarchy. However, there is a strong dependence of such a method to the corpus, since an imbalanced corpus could lead to a misleading decision for the appropriate depth.

J. Darzentas et al. (Eds.): SETN 2008, LNAI 5138, pp. 327–338, 2008.
© Springer-Verlag Berlin Heidelberg 2008

In this paper, we propose an automated statistical approach to ontology learning, without presupposing the existence of a seed ontology, or any other type of external resource, except the corpus of text documents. The proposed method tackles the tasks of concept identification and subsumption hierarchy construction. Moreover, it tries to optimize the size of the learned ontology for the given text collection.

In the proposed method, concepts are identified and represented as multinomial distributions over terms in documents[1]. Towards this objective, the Markov Chain Monte Carlo (MCMC) process of Gibbs sampling [9] is used, following the Latent Dirichlet Allocation (LDA) [4] model. To discover the subsumption relations between the identified concepts, conditional independence tests among these concepts are performed. Finally, statistical measures between the discovered concepts at different levels of the hierarchy are used to optimize the size of the ontology. The statistical nature of the approach guarantees language independence.

In what follows, section 2 states the problem, refers to existing approaches that are related to the proposed method, and motivates our approach. In section 3, we present the new method, while section 4 describes the derivation of a criterion for determining the appropriate depth of the hierarchy according to the corpus. Section 5 presents experiments and evaluation results, and finally, section 6 concludes the paper sketching plans for future work.

2   Problem Definition and Related Work

2.1   Problem Definition

In this paper we address three major problems related to the ontology learning task:

1. The discovery of the concepts in a corpus.
2. The ordering of the discovered concepts by means of the subsumption relation.
3. The determination of the depth of the subsumption hierarchy.

In other words, assuming only the existence of a text collection, we aim to (a) discover the concepts that express the content of documents in the corpus, independently of the terms' surface appearance, i.e. without taking into account simple TF/IDF values or the order of words in the texts, (b) form the ontology subsumption hierarchy backbone, using only statistical information concerning the discovered concepts, and (c) explore how deep in the subsumption hierarchy the text collection allows us to go, by measuring the similarity between the discovered concepts.

---

[1] "Terms" does not necessarily denote domain terms, but words that will constitute the vocabulary over which concepts will be specified. In the following, we use "terms" and "words" interchangeably.

2.2   Related Work

Towards the automated learning of ontologies, much work concerns concept identification and taxonomy construction. In this paper we are interested in statistical techniques, and thus, we discuss here related approaches.

On the task of concept identification with statistical techniques, the authors in [2] extend an ontology with new concepts considering words that co-occur with each of the existing concepts. The method requires that there are several occurrences of the concepts to be classified, so that there is sufficient contextual information to generate topic signatures. The work reported in [1] follows similar research directions. In [5], the authors apply statistical analysis on Web pages in order to identify word clusters that are proposed as potential concepts to the knowledge engineer. In this case, the ontology enrichment task is based on statistical information of word usage in the corpus and the structure of the original ontology.

More sophisticated schemes include the use of TF/IDF weighting in conjunction with Latent Semantic Indexing (LSI) [6], towards revealing latent topics in a corpus of documents. A classification task assigns words to topics, making each topic a distribution over words. Probabilistic Latent Semantic Indexing (PLSI) [10] extends LSI assuming that each document is a probability distribution over topics and each topic is a probability distribution over words. Although PLSI provides more accurate modelling than LSI, it must be pointed out that this model is prone to overfitting (being corpus specific), involving a large number of parameters that need to be estimated [4]. Latent Dirichlet Allocation (LDA) [4] improves on PLSI, providing a model that samples topics for each word that appears in each document.

Hierarchical extensions have also been proposed to the above models. Hierarchical Probabilistic Latent Semantic Analysis (HPLSA) has been proposed in [7], in order to acquire a hierarchy of topics, by enabling data to be hierarchically organized based on common characteristics. Hierarchical Latent Semantic Analysis (HLSA) has been introduced in [12] to identify hierarchical dependencies among concepts by exploiting word occurrences among concepts (latent topics). This approach actually computes relations among topics, based on the words that they contain. Different topics might share common words, and therefore these words are collected at a higher level. Both of these methods, inherit known problems of PLSI, such as overfitting.

Moreover, the method of hLDA [3], a hierarchical extention of LDA, has been proposed to deal with the problem of the hierarchical organization of topics. However, this method assumes that each document is a mixture of topics along a path from the root topic to a leaf, making this way a document to comprise only one specific topic and its abstractions. Finally, the model of hPAM [11] deals with some limitations of hLDA. It allows multiple inheritance between topics, but on the other hand the fixed-depth hierarchy that produces and the need for predefining the number of topics are its basic limitations. In general, determining the appropriate depth of the hierarchy still remains to our knowledge an open issue.

In this paper we address the problems of concept identification and taxonomy construction using statistical and machine learning techniques. The statistical nature of the proposed method assures that the method is not dependent on the language of the corpus, but only on the statistical information that the corpus provides, i.e., the word frequencies. In addition, having no prior knowledge, we aim to determine statistically the depth of the hierarchy.

3   The Method

As depicted in figure 1, given a corpus of documents, the method first extracts the terms. The extracted terms constitute the term space, over which the latent topics are defined. In the second step, feature vectors are constructed for each document, based on term frequency. Next, the latent topics are generated as distributions over vocabulary terms according to the documents in the corpus and the terms observed. Through an iterative process, latent topics are discovered and organized into hierarchical layers until the criterion for appropriate depth is satisfied.

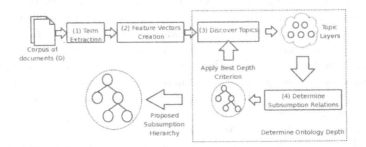

Fig. 1. The proposed ontology learning method

More specifically, the stages followed by the proposed method are as follows:

(1) *Term Extraction* - From the initial corpus of documents, treating each document as a bag of words, we remove stop-words using statistical techniques. The remaining words constitute the vocabulary, forming the term space for the application of the topic generation model.

(2) *Feature Vector Creation* - This step creates a Document - Term matrix, each entry of which records the frequency of each term in each document. This matrix is used as input to the topic generation model.

(3) *Discover Topics* - In this step, the iterative task of the learning method is initiated. To generate the topics we follow the Latent Dirichlet Allocation (LDA) [4] approach. LDA belongs in the family of Probabilistic Topic Models (PTMs). These models are based on the idea that documents are mixtures of thematic topics, which are represented by means of probability distributions over terms. PTMs are based on the bag-of-word assumption, assuming that words are independently and identically distributed in the texts, given the thematic

topics of each text. PTMs are generative models for documents: they specify a probabilistic procedure by which documents are generated as combinations of latent variables, i.e. topics. Generally, this procedure states that topics are probability distributions over a predefined vocabulary of words and according to the probability that a topic participates in the content of each document, words are sampled from the corresponding topic in order to generate the documents.

The LDA model specifies a generative process, according to which, topics are sampled repeatedly in each document. Specifically, given a predefined number of topics K, for each document:

1. Choose $N \sim \text{Poisson}(\xi)$.
2. Choose $\theta \sim \text{Dirichlet}(\alpha)$.
3. For each of the N words w_n:
   - Choose a topic $z_n \sim \text{Multinomial}(\theta)$
   - Choose a word w_n from $\text{p}(w_n \mid z_n, \beta)$, a multinomial probability distribution conditioned on the topic z_n.

$p(z_n = i)$ stands for the probability that the ith topic was sampled for the nth word and indicates which topics are important, i.e., reflect the content of a particular document. $p(w_n \mid z_n = i)$ stands for the probability of the occurrence of word w_n given the topic i and indicates the significance of each word for each topic.

In this paper, we are not interested in the generative process per se, but rather in the inverse process. Documents are known and words are observations towards assessing the topics of documents, as combinations of words. Thus, we aim to infer the topics that generated the documents and then organize these topics hierarchically. In order to infer the latent topics, the proposed method uses the Markov Chain Monte Carlo (MCMC) process of Gibbs sampling [8].

At each iteration of this step, sets of topics, that we call layers, are generated by the iterative application of LDA. Starting with one topic and by incrementing the number of topics in each iteration, layers with more topics are generated. A layer comprising few topics attempts to capture all the knowledge of the corpus through generic concepts. As the number of topics increases, the topics become more focused, capturing more detailed domain knowledge. Thus, the method starts from "general" topics, iterates, and converges to more "specific" ones.

(4) *Determine Subsumption Relations* - In each iteration, the method identifies the subsumption relations that hold between topics of different layers. The discovered topics are arranged in a hierarchical manner according to their conditional independencies, determined by the following condition:

$$|\hat{P}(A \cap B \mid C) - \hat{P}(A \mid C)\hat{P}(B \mid C)| \leq th. \tag{1}$$

Equation (1) is best explained through an information theoretic framework. Specifically, since the generated topics are random variables, e.g. A and B, by measuring their mutual information we obtain an estimate of their mutual dependence. Therefore, given a third variable C that makes A and B (almost) conditionally independent, the mutual information of topics A and B is reduced and C contains a large part of the common information of A and B, i.e., C is a broader topic than the others. In this case we may safely assume that C subsumes both A and B and the corresponding relations are added to the ontology. We should also point out, that C has been generated before A and B. Thus, it belongs in a layer that contains topics that are broader in meaning than the ones in the layer of A and B.

In addition, we search for a topic C that makes topics A and B as much conditionally independent as possible. Therefore, between two possible parent topics C_1 and C_2 we will choose the one that maximizes the difference of the following mutual informations:

$$\Delta = I(A, B) - I(A, B \mid C) \qquad (2)$$

Therefore, equation (1) is not used as an absolute measure to judge the subsumption relations between concepts, but as a relative way of finding the best concept C that can be considered as the father of A and B.

In order to calculate the conditional independencies between topics, we use the document-topic matrix generated by the LDA model. Each entry of this matrix expresses the probability of a specific topic given a document. The estimation of the probabilities of equation (1) is explained in [15]. Moreover, the threshold th has been introduced to avoid small rounding problems at the calculations. Therefore, it has a very small value near zero.

(5) *Determine Ontology Depth* - A significant contribution of the proposed method is the determination of the appropriate depth of the hierarchy from the given corpus of documents. As already mentioned, the topics are probability distributions over the term space. We use a criterion based on the similarity of these distributions that indicates the convergence towards the appropriate depth. We thus improve on our recently proposed work [15] by proposing algorithm 1. The way in which the appropriate depth of the taxonomy is determined is explained in the following section.

4   Measuring Similarity between Concepts

In the proposed method, concepts are represented as multinomial distributions over terms in documents. In order to determine the depth of the subsumption hierarchy we define a criterion based on the symmetric KL divergence between concepts of different levels that participate in subsumption relations. The intuition behind this is that the symmetric KL divergence between concepts that belong in the top levels of the hierarchy should be higher than the KL divergence between concepts that belong in the lower levels of the hierarchy. This is due to the fact that the top concepts are broader in scope than lower ones and the "semantic distance" between them and their children is expected to be higher than this of more specific concepts and their children.

In order to validate this assumption, we have experimented with two golden standard ontologies and the corresponding corpora:

Data: Document - Term Matrix
Result: Subsumption hierarchy of topics
initialization;
start with number of topics K=1;
while *Stop Criterion not achieved* **do**
 | Generate a pair of topic layers in parallel (for current value of K and for
 | K+1);
 | **for** *every topic* i *in 1st topic layer of pair* **do**
 | | **for** *every pair of topics (j, k) in 2nd topic layer* **do**
 | | | **if** *(conditional independence of j and k given i is the maximum*
 | | | *among other pairs) AND (satisfies the threshold* th*)* **then**
 | | | | *i* is parent of *j* and *k*
 | | | **end**
 | | **end**
 | **end**
 | **if** *Stop Criterion achieved* **then**
 | | **end**;
 | **else**
 | | increase number of topics;
 | **end**
end

Algorithm 1. Constructing a subsumption hierarchy of appropriate depth

1. The Genia[2] ontology comprises 43 concepts connected with 41 subsumption relations, which is the only type of relation among the concepts. The corresponding corpus consists of 2000 documents from the domain of molecular biology.
2. The Lonely Planet ontology contains 60 concepts and 60 subsumption relations among them. The Lonely Planet corpus is a collection of about 300 Web pages from the Lonely Planet Web site[3], providing touristic information.

In order to measure the similarity of the concepts in the ontologies, we represented the concepts of each gold standard ontology as probability distributions over the term space of the corresponding corpus, as shown in figure 2. This representation allows the application of statistical measures concerning the similarity between concepts.

To represent each concept as distribution over terms we have to measure the frequency of the terms that appear in the context of each concept. In both corpora, the concept instances are annotated in the texts, providing direct population of the concepts in the golden standard ontologies with their instances. Since we have populated each concept with its instances, it is possible to associate each document to the concept(s) that it refers to, by counting the concept instances that appear in the document. Thus, we are able to create feature vectors based on the document in which each concept appears. These feature vectors actually form a two-dimensional matrix that records the frequency of each term in the context

---

[2] The GENIA project, http://www-tsujii.is.s.u-tokyo.ac.jp/GENIA
[3] The Lonely Planet travel advise and information, http://www.lonelyplanet.com/

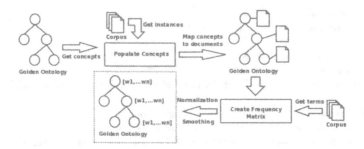

Fig. 2. The process of representing the golden ontology concepts as probability distributions over the term space

of each concept. That is, we have a "concept - term" matrix that represents each concept as a distribution over the term space of the text collection.

For each concept, frequencies are normalized giving a probability distribution over the term space. Since the goal is to measure the symmetric KL divergence between concepts that participate in subsumption relations, we also performed smoothing of the probability distributions to eliminate possible zero values of unseen terms. For this purpose, we applied the Laplace law (3) on the probability distribution of each concept.

$$\hat{P}_L(w_i) \doteq \frac{\hat{P}(w_i) + 1}{N + 1}, \forall i, \tag{3}$$

where N is the vocabulary size.

In order to measure the symmetric KL divergence between two concepts p and q that are related with a subsumption relation, we used the following formula:

$$D_{KL} = \frac{1}{2}[\sum_i P(w_i) log \frac{P(w_i)}{Q(w_i)} + \sum_i Q(w_i) log \frac{Q(w_i)}{P(w_i)}], \tag{4}$$

where $P(\cdot)$ and $Q(\cdot)$ are the distributions corresponding to concepts p and q. Small values of KL divergence indicate high similarity between concepts. Figure 3 depicts the results obtained by measuring the similarity between concepts that participate in subsumption relations, in the case of the Genia and the Lonely Planet gold standard ontologies.

Figure 3 confirms our assumption that concepts at the lower levels of the hierarchy are more similar to their children than concepts at higher levels of the hierarchy. For both corpora, KL divergence is minimized at the leaf level of the ontologies.

Based on this approach of measuring the KL divergence of subsumed concepts, we define a relative criterion that indicates how deep the hierarchy should be according to the information provided by the corpus of documents. This criterion, which corresponds to the stop criterion of Algorithm 1, is defined as follows:

$$1 - \frac{KL_{bottom}}{KL_{top}} < \varepsilon. \tag{5}$$

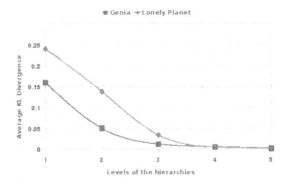

Fig. 3. Average KL Divergence of subsumed concepts in the Genia and the Lonely Planet gold standard ontology

In equation (5), KL_{top} corresponds to the average symmetric KL divergence between the concepts of level l and the concepts of level $l + 1$. KL_{bottom} is the average symmetric KL divergence between the concepts at level $l + 1$ and the concepts of level $l + 2$. Values close to 0 indicate that the new level of concepts added does not differ much from the parent concepts. Thus we are reaching maximum "specificity" and therefore optimal depth. Values near 1 indicate that the hierarchy can go deeper. Actually, the parameter ε, does not depend on the application. It has a very small value very close to zero to avoid small rounding errors during the computations.

5 Evaluation

We have evaluated the proposed method on both corpora introduced in section 4. The parameters that have been introduced in this paper are the parameter ε in the stop criterion (5) and the threshold th for the significance of subsumption relations. Both parameters are introduced in order to provide control over the process, although the method is robust to the values of the parameters. Typically, one would choose very small values for these parameters, independent of the particular application.

The evaluation procedure that we followed uses the representation of the golden standard concepts as probability distributions over the term space of the documents, as explained in section 4. In addition, the concepts of the produced hierarchy have exactly the same representation. They are probability distributions over the same term space. We can, thus, perform a one-to-one comparison of the golden concepts and the produced topics. More specifically, a topic is matched to a concept if their corresponding distributions were the "closest" compared to all the other and their KL divergence (4) was below a fixed threshold th_{KL}. Obviously, small values of KL divergence indicate high similarity between golden concepts and discovered topics.

The quantitative results have been produced using the metrics of *Precision* and *Recall*. Regarding the concept identification, we define *Precision* as the ratio of the number of concepts correctly detected to the total number of concepts detected, and *Recall* as the ratio of the number of concepts correctly detected to the number of concepts in the gold standard. Accordingly, for the subsumption relations (SRs): *Precision* is the ratio of the number of SRs correctly detected to the total number of SRs detected, and *Recall* is the ratio of the number of SRs correctly detected to the number of SRs in the gold standard.

The choice of threshold th_{KL} affects the quantitative results, since a strict choice would force few topics to be matched with golden concepts, while a loose choice would cause many topics to be matched with golden concepts. We have chosen a value of $th_{KL} = 0.2$ for the purposes of our evaluation, as we observed relative insensitivity of the result for values between 0.2 and 0.4 and we opted for the more conservative value in this plateau. Tables 1 and 2 depict the experimental results in the case of the Genia and Lonely Planet corpora respectively.

Table 1. Evaluation results for the Genia corpus

Concept Identification		
Precision	Recall	F-measure
94%	76%	84%
Subsumption Hierarchy Construction		
Precision	Recall	F-measure
93%	75%	83%

Table 2. Evaluation results for the Lonely Planet corpus

Concept Identification		
Precision	Recall	F-measure
62%	36%	44%
Subsumption Hierarchy Construction		
Precision	Recall	F-measure
53%	35%	42%

In order to obtain a more detailed picture of the performance of the method, we replaced the stopping criterion of Algorithm 1 with predefined depths for the learned hierarchy and we experimented in both corpora. Figures 4 and 5 present the evaluation results in terms of the F-measure for various depths of the hierarchy, using the same configuration ($th_{KL} = 0.2$) for the evaluation method.

Figure 4 depicts that for a predefined depth of 8 levels of the produced hierarchy, the F-measure is maximized compared to the Genia gold standard. Respectively, in the case of the Lonely Planet corpus, the F-measure is maximized for a predefined depth of 10 levels of the produced hierarchy (figure 5). Tables 1 and 2

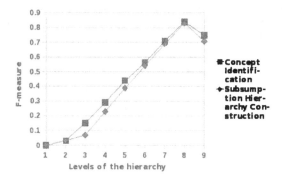

Fig. 4. F-measures for Concept Identification and Subsumption Hierarchy Construction for the Genia corpus

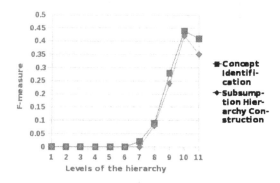

Fig. 5. F-measures for Concepts Identification and Subsumption Hierarchy Construction for the Lonely Planet corpus

confirm that the proposed method, using the stop criterion that we derived in section 4, managed to achieve the best results in both corpora. Therefore, the method determined correctly the appropriate depth in both corpora.

Concerning the quantitative results, in the case of the Genia corpus, where the golden concepts were instantiated sufficiently in the documents, i.e. the texts contain many concepts instances, the numerical results were higher than the ones in the case of the Lonely Planet corpus, where half of the golden concepts had only one instance and generally most of the concepts were insufficiently instantiated. The difficulty of the model to retrieve some very specific concepts in the Lonely Planet corpus is due to this fact.

6 Conclusions

In this paper, we have presented a method for concept identification and taxonomy construction that determines automatically the appropriate size of the subsumption hierarchy. We improved our recently proposed method that relies on conditional independence tests between thematic topics, by incorporating a statistical criterion that determines the appropriate depth of the produced hierarchy. We have also experimented with two corpora, where we have showed that the presented method managed to determine the most appropriate size for the subsumption hierarchy, producing the best quantitative results.

Future work includes further experiments to validate the proposed method on new ontologies and corpora.

References

1. Agirre, E., Ansa, O., Hovy, E., Martinez, D.: Enriching very large ontologies using the www. In: ECAI 2000 Workshop on Ontology Construction (2000)
2. Alfonseca, E., Manandhar, S.: An unsupervised method for general named entity recognition and automated concept discovery. In: International Conference on General WordNet (2002)
3. Blei, D., Griffiths, T., Jordan, M., Tenenbaum, J.: Hierarchical topic models and the nested chinese restaurant process. In: NIPS (2004)
4. Blei, D.M., Ng, A.Y., Jordan, M.I.: Latent dirichlet allocation. In: Journal of Machine Learning Research (2003)
5. Faatz, A., Steinmetz, R.: Ontology enrichment with texts from the www. In: Semantic Web Mining Workshop ECML/PKDD (2002)
6. Fortuna, B., Mladevic, D., Grobelnik, M.: Visualization of Text Document Corpus. In: ACAI (2005)
7. Gaussier, E., Goutte, C., Popat, K., Chen, F.: A hierarchical model for clustering and categorising documents. In: BCS-IRSG (2002)
8. Griffiths, T., Steyvers, M.: A probabilistic approach to semantic representation. In: Conference of the Cognitive Science Society (2002)
9. Griffiths, T.L., Steyvers, M.: Finding scientific topics. In: National Academy of Science (2004)
10. Hofmann, T.: Probabilistic latent semantic indexing. In: SIGIR (1999)
11. Mimno, D., Li, W., McCallum, A.: Mixtures of hierarchical topics with pachinko allocation. In: Proceedings of the 24th International Conference on Machine Learning (2007)
12. Paaß, G., Kindermann, J., Leopold, E.: Learning prototype ontologies by hierarchical latent semantic analysis. In: Knowledge Discovery and Ontologies (2004)
13. Roux, C., Proux, D., Rechermann, F., Julliard, L.: An ontology enrichment method for a pragmatic information extraction system gathering data on genetic interactions. In: ECAI Workshop on Ontology Learning (2000)
14. Wagner, A.: Enriching a lexical semantic net with selectional preferences by means of statistical corpus analysis. In: ECAI Workshop on Ontology Learning (2000)
15. Zavitsanos, E., Paliouras, G., Vouros, G.A., Petridis, S.: Discovering subsumption hierarchies of ontology concepts from text corpora. In: Proceedings of the IEEE/WIC/ACM International Conference on Web Intelligence - WI 2007. Springer, Heidelberg (2007)

Bayesian Model of Recognition on a Finite Set of Events

Vladimir Berikov and Gennady Lbov*

Sobolev Institute of mathematics,
Koptyg av. 4, 630090 Novosibirsk, Russia
berikov@math.nsc.ru,
lbov@math.nsc.ru
http://www.math.nsc.ru

Abstract. In the paper, we consider the Bayesian model of recognition on a finite set of events. The model is not oriented on the most "unfavorable" distribution and on the asymptotic case. We consider the questions of the model applicability in problems of decision tree construction and rare events recognition.

1 Introduction

One of the main problems in pattern recognition is finding "good" decision function from the given family of functions. Under the "good" decision function is usually understood the one for which the risk (the expected losses of wrong forecasting for a new object) is small. In real life problems the probability distribution is unknown. One should make a decision by learning sample of limited size. It is known that the complexity of decision function class is an important factor influencing the quality of decisions [8]. For the best quality, one should reach certain compromise between the complexity of class and the accuracy of decisions on learning sample.

Two main approaches to the solution of the problem are available: empirical approach (one-hold-out, bootstrap, cross-validation methods) and probabilistic approach (based on a preliminary model of data generating process).

Simple and intuitively clear, empirical approach however suffers from lack of theoretical grounds, large variance of risk estimates and computational expensiveness.

Probabilistic approach is more substantiated, nevertheless it has a number of drawbacks: need to formulate distribution model for patterns in variable space, tendency to get worst-case estimations and asymptotic results.

In [1,2,5] the Bayesian model of recognition on a finite set of events was introduced. The model is based on probabilistic approach (in fact, the Bayesian learning theory), however it is not oriented on the most "unfavorable" distribution and asymptotic case. In the given work, we aim to discuss the applicability of the model in real pattern recognition problems.

* The authors were supported by the Russian Foundation of Basic Research, grants 07-01-00331a, 08-07-00136a, and by Human Capital Foundation, grant 144.

J. Darzentas et al. (Eds.): SETN 2008, LNAI 5138, pp. 339–344, 2008.

The rest of paper is organized as follows. In the next section we give formal definition of the model and describe its base properties. The applicability of the model in problems of decision tree construction and rare events recognition is considered in section 3. In the summary we make the concluding remarks.

2   Bayesian Model

Let each object under investigation be described by a collection of heterogeneous (qualitative or quantitative) features. The Bayesian model of recognition on a finite set of events is introduced by formulating some propositions, the sense of which consists in abstracting from local metric properties of feature space and local characteristics of learning method:

1. Instead of points of feature space we shall consider the "events". Under the "event" we understand some sub-region of feature space. Let us form a system of events taking into account relations between them (neighborhood or common ancestor).
2. We shall use the notion of learning method (mapping from the set of all possible samples into the set of decision functions).
3. The Bayesian learning theory gives us the possibility to formalize expert knowledge about forecasting problem, not demanding "hard" definition of probability distribution model in variable space.

Let us consider the following discrete random variables: input variable X with set of unordered values $D_X = \{c_1, \ldots, c_M\}$, where c_j is j-th value ("event"), and output (predicted) variable Y with the set of unordered values $D_Y = \{\omega^{(1)}, \ldots, \omega^{(K)}\}$, where $\omega^{(l)}$ is l-th value, called l-th class, $K \geq 2$ is a number of classes. Let us encode the values of variable X by event numbers, and classes – by class numbers. Under the complexity of model we shall understand the value M.

Let $p_j^{(l)}$ be a probability of joint event "$X = j, Y = l$", $p_j^{(l)} \geq 0$, $\sum_{j,l} p_j^{(l)} = 1$, $j = 1, \ldots, M, l = 1, \ldots, K$. Denote $\theta = (p_1^{(1)}, \ldots, p_j^{(l)}, \ldots, p_M^{(K)})$. We also shall call θ as "strategy of nature". To solve the recognition problem it is required to find decision function $f : D_X \to D_Y$.

Suppose we are given loss function $L_{r,l}$. The losses appear if the decision is $Y = r$, but the true class is l. We shall call loss function an indicator loss function, if $L_{l,l} = 0$ and $L_{r,l} = 1$ for $r \neq l$, $r, l = 1, \ldots, K$.

Let certain set Φ of decision functions be given. For each decision function f from Φ, the expected losses (risk) of wrong recognition for an arbitrary object equals $R_f(\theta) = \sum_{j=1}^{M} \sum_{l=1}^{K} L_{f(j),l} p_j^{(l)}$. In case of indicator loss function, the risk coincides with misclassification probability: $P_f(\theta) = 1 - \sum_j p_j^{(f(j))}$. For optimal Bayes decision function f_B we have $P_{f_B}(\theta) = 1 - \sum_j \max\{p_j^{(1)}, p_j^{(2)}, \ldots, p_j^{(K)}\}$.

In applied recognition problems vector θ is usually unknown. Decision function is chosen from Φ according to random sample of observations of X and Y

(learning sample) by means of certain given method μ (learning algorithm). Let N be sample size, $n_j^{(l)}$ – number of objects of l-th class that correspond to j-th event; $\sum_{l,j} n_j^{(l)} = N$. Denote $s = (n_1^{(1)}, \ldots, n_j^{(l)}, \ldots, n_M^{(K)})$. The empirical risk for decision function f equals $R_f^e(s) = \frac{1}{N} \sum_{j=1}^{M} \sum_{l=1}^{K} L_{f(j),l} n_j^{(l)}$.

Learning method μ can be considered as a function defining a mapping from finite set of all samples $\mathbf{S}=\{s\}$ of the given size into the set of decision functions Φ: $f = \mu(s)$. An example of learning method is empirical risk minimization method.

Let $S = (N_1^{(1)}, \ldots, N_M^{(K)})$ be random frequency vector having multinomial distribution: $p(s|\theta) = \dfrac{N!}{\prod_{l,j} n_j^{(l)}!} \prod_{l,j} \left(p_j^{(l)}\right)^{n_j^{(l)}}$. Consider a family of multinomial distribution models of frequency vector with set of parameters $\Theta = \{\theta\}$.

Let us consider random vector Θ taking values from Θ with certain known distribution density $p(\theta)$. We shall suppose that Θ has Dirichlet distribution (conjugate with multinomial distribution): $p(\theta) = \dfrac{1}{Z} \prod_{l,j} (p_j^{(l)})^{d_j^{(l)}-1}$, where $d_j^{(l)} > 0$ are some given real numbers expressing a priori knowledge on distribution of Θ, $l = 1, \ldots, K$, $j = 1, \ldots M$; Z is normalizing constant, $Z = \dfrac{\Gamma(D)}{\prod\limits_{i,j} \Gamma(d_j^{(i)})}$, where $\Gamma(\cdot)$ is gamma function, $D = \sum\limits_{j,l} d_j^{(l)}$. For $d_j^{(l)} \equiv 1$ we shall have uniform a priori distribution that can be used in case of uncertainty at the specification of model parameters. Thus the risk is a function $R_f(\Theta)$ depending from random strategy of nature.

In [1,2,5] the properties of the model have been studied. The investigations follow three main directions. Firstly, we established some properties that facilitate the specification of model. The expected probability of error $\mathbf{E}P_{f_B}(\Theta)$ for optimal Bayes decision function f_B was found. This value corresponds to the expected degree of intersection between classes. The averaging was done over all possible strategies of nature θ with distribution density $p(\theta) = \dfrac{1}{Z} \prod\limits_{l,j} (p_j^{(l)})^{d-1}$.

Theorem 1. *[1] Let $K = 2$, $d_j^{(l)} \equiv d$, ($l = 1, 2$, $j = 1, \ldots, M$), where $d > 0$ is some parameter. Then $\mathbf{E}P_{f_B}(\Theta) = I_{0.5}(d + 1, d)$, where $I_x(p, q)$ is beta distribution function with parameters x, p, q (Figure 1).*

The remarkable property is that the expected risk does not depend on the complexity M.

Parameter d can be used for the definition of a priori distribution on recognition tasks: when this parameter decreases, the classes become less intersected in average. On the contrary, one can evaluate this parameter by available expert knowledge on the degree of intersection. In real recognition problems, feature space never comes accidently. It means that the error probability is not likely to be too large for f_B. For example, if an expert affirms that the expected

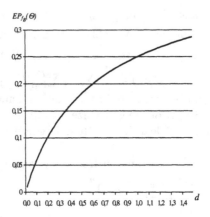

Fig. 1. Dependence between parameter d and the expected misclassification probability for the Bayes decision function

misclassification probability for ideal decision is about 0.15, then we should set $d = 0.38$.

Secondly, we received the expressions for the Bayes estimates of risk $R_\mu = ER_{\mu(S)}(\Theta)$ for arbitrary learning method μ. The averaging was carrying out over all strategies of nature and all learning samples of the given size N; some variants of additional information about strategies of nature were considered. The estimates allow us to choose the optimal complexity of decision function class depending on sample size and available expert knowledge.

Thirdly, we obtained the Bayes estimates of risk for arbitrary decision function f. Here the averaging was carrying out over all possible strategies of nature. The estimates can be used for the comparison of different decision functions in the class and choosing the best one.

Theorem 2. *[1] A posteriori mathematical expectation of risk $R_f(\Theta)$ for given sample s and decision function f equals*

$$R_{f,s} = \frac{1}{N+D} \sum_{j,l} L_{f(j),l}(n_j^{(l)} + d_j^{(l)}).$$

It follows from the estimation theory [6] that a posteriori mathematical expectation of risk is optimum Bayes estimate of $R_f(\Theta)$ under quadratic loss function.

It is easy to show that for $d_j^{(l)} \equiv 1$ the Bayes estimate of misclassification probability equals $P_{f,s} = \dfrac{N_f^{er} + (K-1)M}{N + KM}$, where N_f^{er} is a number of misclassified objects. For small N_f^{er} and M the following approximate expression holds: $P_{f,s} \approx \dfrac{N_f^{er}}{N} + (K-1)\dfrac{M}{N}$. For $K = 2$ and $d_j^{(l)} \equiv d$ we have $P_{f,s} \approx \dfrac{N_f^{er}}{N} + d\dfrac{M}{N}$. We use these expressions as quality criteria for decision tree with M leaves (see next section).

3   Decision Tree Construction and Rare Events Recognition

Consider the problem of decision tree construction. For optimal tree finding, two variants of using the Bayesian model can be suggested.

1. Let us form the "initial" partition tree dividing variable space on sufficiently large number of subregions. It is possible to consider the initial set of subregions, as well as any amalgamation of them in accordance with tree structure, as a finite set of "events". Note that learning sample and tree structure are independent. The Bayes estimates of risk are used as a quality criterion for decision functions defined on each partition. The directed search algorithms can be applied for examining different variants of partition in order to choose the best one.

2. Let us use another way of forming the initial tree. Divide learning sample into two random parts of equal size. The first part is used for the construction of classification tree by some algorithm (CART [3], C4.5 [7], LRP, R-method [5] etc). Parameters of algorithm should be assigned to ensure minimum empirical risk. Thus, the resulted tree is overtrained usually. The second part is used for pruning the tree in order to minimize the Bayes estimate of risk. An advantage of this method, in comparison with the previous one, is that the initial tree has not "empty" leaves (i.e., without any object). So the number of leaves is relatively small.

For each variant, the class of decision functions has finite number of elements. So the convergence of the Bayes estimate of risk to true risk is uniform over the class. Thus, the minimization of the Bayes estimate allows to find a decision function, close to optimum in the family. Because of the optimality property of the Bayes estimate, the rate of convergence is maximal. Note that the principle of uniform convergence in pattern recognition was suggested by Vapnik and Chervonenkis [8], however from non-Bayes point of view and for the empirical risk estimate.

Let us give an example. Let $M = 3$, $K = 2$, $f(1) = 1$, $f(2) = 1$, $f(3) = 2$, $N = 5$, $d = 1$. By Monte Carlo method we estimate the probability of an event $|R_{f,S} - R_f(\Theta)| \geq 0.1$. Simulations show that the estimate equals 0.49 (for 10^6 MC evaluations). The standard empirical risk estimate give larger probability of the deviation: 0.65.

A number of computer simulations on artificial data sets show that our algorithms often demonstrate better generalization performance in comparison with CART and C4.5, especially in case of complex dependencies between features (for example, having chessboard structure with many irrelevant features [5]).

As another application of the model, we consider a problem of rare events recognition. It should be mentioned that the data sets for rare events analysis are very unbalanced and may have small size. To obtain good prediction quality, one should use as much as possible relevant features (as a rule, of heterogeneous type). All available additional expert information (for example, estimates of a priori probabilities of events) should be used as well.

Let us give an example of numerical experiment. Two random Boolean and one random numerical sequences were generated independently. The length of sequences equals 1000. It was set that for the defined combination of previous values of these sequences the "undesirable" event should take place with probability 0.75. The event does not occur for other combinations. Thus, Boolean sequence denoting the presence or absence of the undesirable event was formed (the frequency of this event was about 0,09). The sequences were analyzed by pruning algorithm described above. The recursive algorithm ("R-method" [5]) was used for tree growing. For quality estimation, 10-fold cross-validation technique has been applied. As a result, the test sequence having 100 situations describing undesirable events was recognized with one error. Cross-validation risk estimate was 0.086. At the same time, Reduced-Error Pruning (REP) algorithm [4] exceeded the estimate by 7%.

A number of applied problems in the field of bioinformatics, analysis of anthropological and medical data, low water-level prediction, etc were successfully solved by the proposed algorithms.

4   Summary

The Bayesian model of recognition on a finite set of events, unlike other models, is not oriented on the most "unfavorable" distribution and on the asymptotic case. The model allows to substantiate decision quality criteria depending on the available expert knowledge. The application of the model to decision tree construction and rare events recognition was suggested. In our opinion, the results show that the model can be applied in real recognition problems, especially in case of heterogeneous, categorical or unbalanced data sets of small size.

References

1. Berikov, V.B.: Bayes estimates for recognition quality on a finite set of events. Pattern Recognition and Image Analysis 16(3), 329–343 (2006)
2. Berikov, V.B., Litvinenko, A.G.: The influence of prior knowledge on the expected performance of a classifier. Pattern Recognition Letters 24(15), 2537–2548 (2003)
3. Breiman, L., Friedman, J., Olshen, R., Stone, C.: Classification and Regression Trees. Wadsworth International, California (1984)
4. Esposito, F., Malerba, D., Semerato, G.: A comparative analysis of methods for pruning decision trees. IEEE Trans. Pattern Anal. and Machine Intelligence 19(5), 476–491 (1997)
5. Lbov, G.S., Berikov, V.B.: Stability of decision functions in problems of pattern recognition and analysis of heterogeneous information. Novosibirsk, Sobolev Institute of mathematics (in Russian) (2005)
6. Lehmann, E.L., Casella, G.: Theory of point estimation. Springer, Heidelberg (1998)
7. Quinlan, J.R.: C4.5: Programs forMachine Learning. Morgan Kaufmann, San Francisco (1993)
8. Vapnik, V.: Estimation of dependencies based on empirical data. Springer, Heidelberg (1982)

The DR-Prolog Tool Suite
for Defeasible Reasoning and Proof Explanation
in the Semantic Web

Antonis Bikakis[1,2], Constantinos Papatheodorou[2], and Grigoris Antoniou[1,2]

[1] Institute of Computer Science, FO.R.T.H., Heraklion, Greece
[2] Computer Science Department of University of Crete, Greece
bikakis@ics.forth.gr, cpapath@csd.uoc.gr, antoniou@ics.forth.gr

Abstract. In this work we present the design and general architecture of DR-Prolog, a system for defeasible reasoning and proof explanation in the Semantic Web, and the implementation of three different tools that constitute the DR-Prolog Tool Suite: (a) the DR-Prolog API; (b) the DR-Prolog Web application; and (c) the DR-Prolog desktop application. DR-Prolog supports reasoning with Defeasible Logic theories and ontological knowledge in RDF(S) and OWL, is compatible with RuleML, and enables extracting meaningful proof explanations for the answers it computes.

1 Introduction

The development of the Semantic Web proceeds in steps, each step building a layer on top of another. At present, the highest layer that has reached sufficient maturity is the ontology layer in the form of the description logic-based language OWL [1]. The next step will be the logic and proof layers. The implementation of these two layers will allow the user to state any logical principles, and permit the computer to infer new knowledge by applying these principles on the existing data. Rule systems appear to lie in the mainstream of such activities.

Most studies on the integration of rules and ontologies in the Semantic Web have been based on monotonic logics. Some prominent approaches are: (a) the Description Logic Programs proposed in [2]; (b) the integration of Description Logics and Datalog rules, followed in [3,4]; (c) the F-logic based rule language TRIPLE [5]; and (d) the Semantic Web Rules Language (SWRL [6]), which extends OWL-DL with Horn-style rules.

Approaches based on non-monotonic logics, on the other hand, constitute another interesting solution, as they offer more expressive capabilities and are closer to commonsense reasoning. Four recently developed non-monotonic rule systems are: (a) DR-Prolog [7], which stands in the core of the tools that we describe in this paper; (b) DR-DEVICE [8], a defeasible reasoning system for the Web, which is implemented in Jess and integrates well with RuleML and RDF; (c) SweetJess [9], which implements defeasible reasoning through the use of situated courteous logic programs; and (d) dlvhex [10], which integrates rules and ontologies using answer-set semantics.

J. Darzentas et al. (Eds.): SETN 2008, LNAI 5138, pp. 345–351, 2008.

This paper describes the integration of the DR-Prolog defeasible reasoning engine described in [7] with the Prolog-based proof explanation service, presented in [11], in three different implementations that constitute the DR-Prolog Tool Suite: (a) the DR-Prolog API; (b) the DR-Prolog web application; and (c) the DR-Prolog desktop application. The main features of the system are:

1. It is based on Prolog. The core of the system consists of a well-studied translation of defeasible knowledge into logic programs under Well-Founded Well-Founded, Kunen and Answer-Set Semantics [12].
2. It accepts both strict (monotonic) and defeasible (non-monotonic) rules and uses priorities between competing rules to resolve potential conflicts.
3. It is compatible with RuleML [13], the main standardization effort for rules on the Semantic Web.
4. It can reason with RDF, RDF Schema and (parts) of OWL ontologies.
5. It automatically generates an explanation for every answer it computes in a formal and meaningful representation.

2   DR-Prolog Architecture

The DR-Prolog architecture consists of seven main modules, which exchange information with each other or with the user in the way depicted in Figure 1.

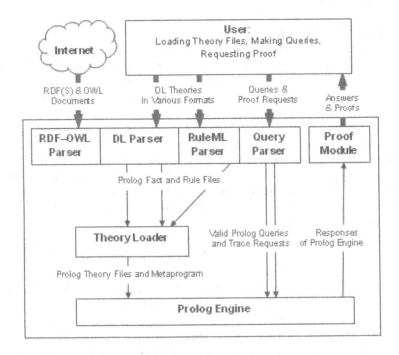

Fig. 1. DR-Prolog Architecture

RDF-OWL Parser. The role of this module is to download ontology (RDF(S), OWL) documents from a given URL, parse the documents, and translate the ontology triples into logical facts of the form: *Predicate(Subject, Object)*. The module also contains a set of Prolog rules that capture the semantics of RDFS constructs and of some of OWL constructs. A detailed description of these rules is available at [7]. For the implementation of the parser, we used the SWI-Prolog RDF library, which we enriched with a couple of functions that aim at properly handling the namespaces and forming the triples in the desirable format.

DL Parser. The DL Parser is responsible for parsing the rule theories imported by the user and encoded in the syntax of Defeasible Logic, checking for their validity, notifying the user about possible errors, and translating the DL theories into the native DR-Prolog syntax, as this is described in [7].

RuleML Parser. Its rule is to parse RuleML DL theories, check their validity and notify the user about possible errors, and translate them into DR-Prolog syntax. RuleML theories may contain factual knowledge, rules and queries. The module is implemented using XSL Transformation along with a second processing level for tasks such as the extraction of queries from the RuleML document.

Query Parser. The Query Parser undertakes the user query and proof requests. It identifies the type of request, checks if the query is valid, notifies the user about possible errors, and translates it into a valid Prolog query.

Theory Loader. This module imports the Prolog files created by the RDF-OWL Parser, the DL-Parser, and the RuleML Parser, and dispatches them to the Prolog engine, along with a Prolog metaprogram that simulates the proof theory of Defeasible Logic. The user can select between two variants of DL - *ambiguity blocking* and *ambiguity propagating*. Both variants and the corresponding metaprograms are described in detail in [7].

Prolog Engine. In the core of the system stands a Prolog engine that compiles the Prolog files dispatched by the Theory Loader, and computes answers to the user queries, returning the answer to the user and its trace to the Proof Module. For the implementation of the engine, we used the XSB logic programming system, as it supports well-founded semantics of logic programs through the use of tabled predicates and its negation (*sk_not*) operator.

Proof Module. This module is responsible for generating meaningful and well-formed proof explanations to the user. It processes the trace of the answer computed by the Prolog engine by cutting out redundant information, and returns a tree-like sequence of rules that explains how the computed answer derives from the imported theory and data. [11] contains a more detailed description of the design and functionality of this module.

3 Implementation

3.1 DR-Prolog API

DR-Prolog API is a Java Library, which enables integrating the DR-Prolog functionality in defeasible reasoning based AI applications. It provides two types of functions: (a) functions that handle the underlying communication with the central Prolog engine; and (b) utility functions that enable writing DR Prolog programs. For the back-end communication mentioned above, DR-Prolog API in turn uses the Interprolog API (www.declarativa.com/Interprolog). This is an open-source Java front-end that supports various Prolog engines including SWI and XSB. It provides access to Prolog engines over TCP/IP sockets and launches Prolog processes in the background, outside the Java Virtual Machine. In the DR-Prolog API, Interprolog is used as follows:

- In the RDF-OWL Parser it is used to launch SWI in order to translate the ontology data into Prolog clauses.
- In the Theory Loader, it initiates the Prolog engine and consults the Prolog theory and data files along with the appropriate DL metaprogram.
- It fully implements the XSB Prolog Engine functionality.

The utility functions implement the functionality of the four parsers of the system, the Proof module, and some extra functionality (e.g. initializing various DR-Prolog parameters, downloading files from given URLs etc.).

The following java code demonstrates the use of the API through a very simple scenario. In this scenario, we initiate DR-Prolog, define a DL theory, which contains a single defeasible rule ($r: a \Rightarrow b$) and issue a query about b.

```
import drprolog.*; public class HelloDrPrologWorld {
    public static void main(String[] args){
        prologApi DREngine =
            new prologApi("/XSB/bin","amb_metaprogram.P", true);
        DREngine.quickLoadDLTheory("r: a => b. ");
        String answer = DREngine.executePrologQuery("b", false);
        DREngine.shutdownPrologEngine(); } }
```

DR-Prolog API, as well as detailed documentation describing the use of the API and the available functions, and some sample programs are available at the DR-Prolog web site (http://www.csd.uoc.gr/~bikakis/DR-Prolog).

3.2 DR-Prolog Web Application

The DR-Prolog Web interface is a server-client application based on the Model View Controller (MVC) architecture (slightly modified, though, in order to suit the needs of its multiple-layered nature). It enables different web users to simultaneously interact with the DR-Prolog reasoning engine.

The Web Application is implemented on top of the DR-Prolog API. The interface is a JSP page, which provides access to the DR-Prolog functionality.

Users start sessions, which are uniquely identified by the server, to support simultaneous accesses. The server keeps track of the user-uploaded files by saving them to individually created folders for each session and maintaining the required references in vectors. All user requests are routed through the jsp interface to the *controller* servlet. After validating the request and depending on the contact code passed on to the controller, a decision is made and control is dispatched to the appropriate servlet implementing the requested function. The system responses are then placed in Java Bean objects (Model), and finally forwarded to the appropriate Jsp page for presentation (View). In case an error occurs at any level, control flow stops and user is redirected to a jsp error page, which provides the user with an error description, the point of occurrence and possibly a solution to the problem.

The remaining servlets implement the functionality of DR-Prolog. Most of them are wrapper functions of their respective DR-Prolog API functions, which are called with parameters that are acquired through user submitted forms.

- The *uploadHandler* servlet unifies the functionality of all parsers. It uploads the theory/data files submitted by the user through forms, and depending on their format, it calls the appropriate parser of the DR-Prolog API. If no errors are encountered, it updates the appropriate user session objects and vectors containing the newly created parsed file paths.
- The *drTheory* servlet is responsible for: (a) session creation and deletion; (b) file management; (c) implementing the functionality of the Query Parser. When a query or proof request is made, after validating the query, *drTheory* forwards execution control to *drpMachine*
- The *drpMachine* servlet implements the Theory Loader, the Prolog Engine and the Proof module. Specifically, it calls the appropriate DR-Prolog API functions in order to: (a) load the user theory and data files and queries to the Prolog Engine; (b) construct the proof explanation; and (c) return the answers to the user.
- The *urlFileDownloader* servlet undertakes downloading files from given URLs.

3.3   DR-Prolog Desktop Application

The DR-Prolog Desktop Application is a single-user application, which implements the functionality of DR-Prolog. Comparing to the DR-Prolog Web Application, it merely provides some extra text-editing type of operations, such as (quick) creating, saving and loading Theory files. It has been implemented around the DR-Prolog API. In fact, there is a direct, one-to-one mapping of DR-Prolog API functions to application buttons. To be able to use this application, a user must download: (a) The DR-Prolog.zip file containing a java executable (DR-Prolog.jar) and some library files, from the DR-Prolog web site; (b) the XSB Prolog engine from http://xsb.sourceforge.net/; and (c) the SWI-Prolog engine from http://www.swi-prolog.org.

4   Conclusion

In this paper, we presented the DR-Prolog Tool Suite; a suite of tools that implement the functionality of DR-Prolog. All tools have been built according to the DR-Prolog architecture, presented in Section 2, and provide defeasible reasoning capabilities on DL theories and ontology data encoded in various formats including a RuleML representation for DL theories, and the RDF syntax for ontology data. In the future, we plan to (a) test the usability of the tools through user evaluation; (b) integrate the modal and deontic extensions of Defeasible Logic, described in [14]; (c) extend the tools with verbal and visual proof explanations; and (d) integrate the DR-Prolog API in defeasible reasoning applications for brokering, bargaining, automated agent negotiation and mobile computing.

References

1. McGuinness, D.L., van Harmelen, F.: OWL Web Ontology Language Overview W3C Recommendation (2004), http://www.w3.org/TR/owl-features/
2. Grosof, B.N., Horrocks, I., Volz, R., Decker, S.: Description logic programs: combining logic programs with description logic. In: WWW, pp. 48–57 (2003)
3. Levy, A.Y., Rousset, M.C.: Combining Horn rules and description logics in CARIN. Artificial Intelligence 104(1-2), 165–209 (1998)
4. Rosati, R.: On the decidability and complexity of integrating ontologies and rules. WSJ 3(1), 41–60 (2005)
5. Sintek, M., Decker, S.: TRIPLE - A Query, Inference, and Transformation Language for the Semantic Web. In: Horrocks, I., Hendler, J. (eds.) ISWC 2002. LNCS, vol. 2342, pp. 364–378. Springer, Heidelberg (2002)
6. Horrocks, I., Patel-Schneider, P.F.: A proposal for an OWL Rules Language. In: WWW 2004: Proceedings of the 13th international conference on World Wide Web, pp. 723–731. ACM Press, New York (2004)
7. Antoniou, G., Bikakis, A.: DR-Prolog: A System for Defeasible Reasoning with Rules and Ontologies on the Semantic Web. IEEE Transactions on Knowledge and Data Engineering 19(2), 233–245 (2006)
8. Bassiliades, N., Antoniou, G., Vlahavas, I.P.: DR-DEVICE: A Defeasible Logic System for the Semantic Web. In: Ohlbach, H.J., Schaffert, S. (eds.) PPSWR 2004. LNCS, vol. 3208, pp. 134–148. Springer, Heidelberg (2004)
9. Grosof, B.N., Gandhe, M.D., Finin, T.W.: SweetJess: Translating DAMLRuleML to JESS. In: RuleML (2002)
10. Eiter, T., Ianni, G., Schindlauer, R., Tompits, H.: dlvhex: A System for Integrating Multiple Semantics in an Answer-Set Programming Framework. In: WLP, pp. 206–210 (2006)
11. Antoniou, G., Bikakis, A., Dimaresis, N., Genetzakis, M., Georgalis, G., Governatori, G., Karouzaki, E., Kazepis, N., Kosmadakis, D., Kritsotakis, M., Lilis, G., Papadogiannakis, A., Pediaditis, P., Terzakis, C., Theodosaki, R., Zeginis, D.: Proof explanation for a nonmonotonic semantic web rules language. Data and Knowledge Engineering 64(3), 662–687 (2008)

12. Antoniou, G., Billington, D., Governatori, G., Maher, M.J.: Embedding defeasible logic into logic programming. Theory Pract. Log. Program 6(6), 703–735 (2006)
13. RuleML: The RuleML Initiative website (2006), http://www.ruleml.org/
14. Antoniou, G., Dimaresis, N., Governatori, G.: A system for modal and deontic defeasible reasoning. In: Orgun, M.A., Thornton, J. (eds.) AI 2007. LNCS (LNAI), vol. 4830, pp. 609–613. Springer, Heidelberg (2007)

Modeling Stroke Diagnosis with the Use of Intelligent Techniques

S. Lalas[1], N. Ampazis[1], A. Tsakonas[1], G. Dounias[1], and K. Vemmos[2]

[1] Department of Financial and Management Engineering, University of the Aegean,
Chios, Greece
[2] Unit of Acute Stroke, Therapeutic Clinic, "Alexandra" General Hospital,
Athens, Greece

Abstract. The purpose of this work is to test the efficiency of specific intelligent classification algorithms when dealing with the domain of stroke medical diagnosis. The dataset consists of patient records of the "Acute Stroke Unit", Alexandra Hospital, Athens, Greece, describing patients suffering one of 5 different stroke types diagnosed by 127 diagnostic attributes / symptoms collected during the first hours of the emergency stroke situation as well as during the hospitalization and recovery phase of the patients. Prior to the application of the intelligent classifier the dimensionality of the dataset is further reduced using a variety of classic and state of the art dimensionality reductions techniques so as to capture the intrinsic dimensionality of the data. The results obtained indicate that the proposed methodology achieves prediction accuracy levels that are comparable to those obtained by intelligent classifiers trained on the original feature space.

1 Introduction

The importance of timesaving and accurate methods for stroke diagnosis makes the domain a suitable candidate in applying modern approaches of intelligent computer-aided diagnosis. The handling of medical decisions concerning stroke type diagnosis by using intelligent techniques is not a new approach since a decade ago the problem had been primarily faced using inductive machine learning algorithms [1], [2], [3], [4]. The entire stroke database which we use in the present study consists of 243 diagnostic characteristics which describe 10 different types of stroke. In our experiments we used 1000 patient records of the "Acute Stroke Unit", Alexandra Hospital, Athens, Greece, describing patients suffering one of 5 different stroke types, diagnosed by 127 diagnostic attributes / symptoms. Reduction of the size of the database took place with the aid of medical experts. Attributes are nominal or numerical (in certain cases, part of the attributes are blank, or represent unknown or missing data, or even dont care values). The diagnoses took place in two phases, as there was a Primary Diagnosis (PD) which occurred in emergency conditions and a Final Diagnosis (FD) which came out later with the aid of laboratory examinations, within the acute stroke unit.

J. Darzentas et al. (Eds.): SETN 2008, LNAI 5138, pp. 352–358, 2008.

2   Stroke Registry Variables

The dataset consists of 127 decision variables which are classified into 12 diagnostic categories which can be abstracted to five (5) or even further to two (2) categories. In fact, in medical practice there exist two major categories, namely ischemic and hemorrhagic stroke. Considering their major subdivisions, 5 stroke types of interest arise. These are: *large vessel atherosclerosis, cardioembolic stroke, lacune, infarcts of unknown cause*, and *intracerebral hemorrhage*. As already mentioned above, our experimental data describe a real world problem, that is, the diagnostic characteristics of 1000 patients of the "Acute Stroke Unit", Alexandra Hospital, Athens, Greece. The most frequent class is Cardioembolic Stroke (33% of the total) against the other 4 classes that reach up a 15-18% each. Additional data have also been stored in the stroke database which are related to the hospitalization and recovery phase of the patients.

In order to process the two datasets (PD and FD) with the intelligent methodology used in this paper, all features that contained unknown values were removed and the resulted datasets were further processed. Namely, all discrete (e.g. multi-class) features were decomposed into 1-to-n binary features (where n is the number of each features classes). This resulted in retaining 58 diagnostic characteristics for the final PD dataset and 25 characteristics for the final FD dataset.

3   Intelligent Methodologies for Stroke Diagnosis

3.1   Genetic Programming

One powerful search methodology of the evolutionary computation (EC) is Genetic Programming (GP) [5]. GP is widely applied in a large number of real-world problems. Extending the inherited characteristics of the EC, GP adopts a flexible variable-length solution representation and the elimination of premature convergence of the solution population. The primary GP allows for the automatic creation of expressions in mathematical, logical or algorithmic forms. As with most EC algorithms, a population of candidate solutions is commonly maintained, and successive generations are expected to enhance the solution pool (i.e. the algorithms population), enabling search into large and discontinued spaces. The unique feature of GP is a tree-like solution representation that may correspond to mathematical expressions, offering the ability to GP to perform the so-called symbolic regression. In this classification problem, the function set is comprised of logical operators, arithmetic operators and conditional IFs, adopting a GP-classification tree model [6]. As fitness measure, the total number of correct classifications was used. We have applied 10-fold cross-validation, with further use of a validation set during training, to reduce data overfitting.

3.2   Techniques for Dimensionality Reduction

Dimensionality reduction is the process of transforming high-dimensional data into a meaningful representation of reduced dimensionality. Dimensionality reduction is important in many domains, since it facilitates classification, visualization, and compression of high-dimensional data, by reducing the effects of the curse of dimensionality and other undesired properties of high-dimensional spaces [7]. In this study the following linear and non-linear dimensionality reduction techniques were employed: *Principal Components Analysis (PCA)* [8], *Probabilistic PCA (probPCA)* [9], *Kernel PCA* [10], *Stochastic Proximity Embedding (SPE)* [11], *Diffusion Maps (DM)* [12], [13], *Restricted Boltzmann Machines (RBM) multilayer autoencoder (AutoRBM)* [14], *Evolutionary Algorithm multilayer autoencoder (AutoEA)* [15], and *Manifold charting* [16].

Ideally, the target dimensionality should be set equal to the *intrinsic dimensionality* of the dataset which is the minimum number of parameters needed to account for the observed properties of the data [17]. In order to estimate the intrinsic dimensionality of the PD and FD datasets, the following intrinsic dimensionality estimators were employed which are based on local and global properties of the data: *Correlation dimension estimator* [18], *Nearest neighbor estimator* [18], *Maximum likelihood estimator* [19], *Eigenvalue-based estimator* [20], *Packing numbers estimator* [21], and *Geodesic Minimum Spanning Tree (GMST) estimator* [18].

4 Results and Discussion

4.1 Previously Obtained Results

Results obtained using approaches based on inductive decision trees [4], [1], [3] achieved an accuracy ranging from 83% for primary diagnosis to 86.3% for final diagnosis using 10-fold cross validation on the entire data set (850 cases were used for training and 150 cases for testing at each fold). When fuzzy modeling was attempted a slightly higher accuracy was obtained for final diagnosis, reaching at 88.7% (a smaller subset of 38 decision variables was used for the final diagnosis experiments in this case). In both, crisp and fuzzy modeling of the problem, the highest misclassifications were generally obtained for classes 2 and 4. All the above results refer to the five-class problem. Large decision trees equivalent to more than 200 decision rules, were obtained in most experiments.

4.2 GP-Results

We applied the GP algorithm to the derived PD and FD datasets and then we proceeded with the dimensionality reduction methods. All the intrinsic dimensionality estimation and dimensionality reduction task were carried out in MATLAB using the "Matlab Toolbox for Dimensionality Reduction" [18]. For the PD dataset the dimensionality was reduced from 58 to 12 which is the intrinsic dimensionality estimate provided by the majority of all the intrinsic dimensionality estimators. For the FD diagnosis dataset the dimensionality was reduced from 25 to 5 which is again the intrinsic dimensionality estimate provided by the majority of all the intrinsic dimensionality estimators.

Tables 1 and 2 summarize the feature dimensionality of the abovementioned data configurations. Before every GP run, we normalized the input data sets into the [-1, 1] range (using the *min-max* criterion), to facilitate the search.

Table 1. Dimensionality Reduction (PD)

PD	Features without missing values	Resulted Features	Reduced Feature Set
Binary	20	20	0
Discrete	9	35	0
Continuous	3	3	12
Total	32	58	12

Table 2. Dimensionality Reduction (FD)

FD	Features without missing values	Resulted Features	Reduced Feature Set
Binary	10	10	0
Discrete	5	15	0
Continuous	0	0	5
Total	15	25	5

Table 3 summarizes the results of the GP search, for the PD dataset. For this task, each of the data sets derived by the various dimensionality reduction methods produced lower accuracy results than the original data GP search (i.e. having available 58 features). However, it is interesting to note that this decrease in the results was relatively small for at least two cases, namely PCA (91.33%) and ProbPCA (87.86%). Fold #7 produced a simple and comprehensible classification tree which classifies correctly 61.29% of the cases in the test set.

Table 3. PD (Primary Diagnosis)

Feature set	10-Fold Cross Validation	Std. Dev.	Best Solution	Relative Success to Original
Original	0.5807	0.0565	0.6559	
ProbPCA	0.5102	0.0529	0.6022	87.86%
AutoRBM	0.3560	0.1156	0.5161	61.30%
AutoEA	0.3540	0.0685	0.4731	60.96%
DM	0.4631	0.0584	0.5699	79.75%
KernelPCA	0.3690	0.0791	0.5269	63.54%
SPE	0.3744	0.0590	0.5054	64.47%
PCA	0.5304	0.0463	0.6064	91.33%

Table 4. FD (Final Diagnosis)

Feature set	10-Fold Cross Validation	Std. Dev.	Best Solution	Relative Success to Original
Original	0.7701	0.0447	0.8387	
PCA	0.3851	0.0851	0.5914	50.00%
SPE	0.3743	0.0777	0.5161	48.60%
KernelPCA	0.3872	0.0877	0.5484	50.28%
DM	0.3723	0.0716	0.5054	48.34%
Manifold	0.3648	0.0733	0.4839	47.37%
AutoEA	0.3455	0.0626	0.4516	44.86%
AutoRBM	0.3669	0.0863	0.5484	47.64%
ProbPCA	0.3732	0.0726	0.4946	48.46%

In the FD task, the reduced data sets derived by the various dimensionality reduction methods also resulted in lower success rates for the GP. This time the loss of accuracy was higher (the best reduced-data model achieved only 38.72 % accuracy in the test set whereas the original data set enabled the GP to produce a 77.01% cross-validation result in the test set). Table 4 summarizes the GP results for the final diagnosis problem. A simple, easily interpretable, classification tree was derived during Fold #9, and carries 76.34% accuracy in the test set.

Overall, the GP managed to produce competitive results, deriving in some cases small and comprehensible solutions. We believe that further investigation should be performed, at least, for the PCA and ProbPCA methods in the primary diagnosis problem, since they seem promising in that task due to their ability to maintain data information in a high degree and of course due to the lower computational complexity that they induce to the GP classifier. Medical experts could apply further investigation into the resulted GP trees, in order to examine potential knowledge extraction.

Another issue that is of interest and in need of further investigation is the determinaton of the reasons why all the dimensionality reduction techniques failed to provide good results in the FD case, as compared to the PD case. A possible explanation is that usually dimensionality reduction is more meaningful when dealing with projections from very high dimensional spaces to just a few dimensions whereas the FD does not have a significantly large feature space to start with. The original feature space of only 25 attributes for the FD problem was derived, however, due to constraints imposed by the large number of missing values in the dataset. It would be therefore interesting to investigate the performance of other dimensionality reduction techniques that are able to deal with missing data such as the one described in [22].

5   Conclusion and Further Research

The field of sroke medical diagnosis is very complicated. The definition of the patients' condition must be very accurate and therefore close collaboration with the expert is required. The early evaluation performed by the expert has a success rate of about 70% or even less. This fact makes it clear that a computer-based evaluation greater than 80% would be a great asset on the physician's side. Future research of the team in this area includes the study and modeling of the error of the expert MDs between primary and final diagnosis of stroke, the effective handling of missing or "don't care" values existing among specific decision variables of the stroke database, and the evaluation of the performance of more advanced classifiers.

References

1. Alexopoulos, E., Dounias, G., Vemmos, K.: Medical diagnosis of stroke using inductive machine learning. In: Proceedings of ACAI 1999: Advanced Course on Artificial Intelligence (W13) Workshop on Machine Learning in Medical Applications (1999)
2. Nomikos, I., Dounias, G., Vemmos, K.: Comparison of alternative criteria for the evaluation of machine learning in the medical diagnosis of stroke. In: Proceedings of 3rd International Data Analysis Symposium, pp. 63–66 (1999)
3. Alexopoulos, E., Dounias, G., Vemmos, K., Nomikos, I.: Knowledge discovery & machine learning for medical diagnosis of stroke. In: 21st Annual Meeting of the Medical Decision Making Society (1999)
4. Nomikos, I., Dounias, G., Tselentis, G., Vemmos, K.: Conventional vs. fuzzy modeling of diagnostic attributes for classifying acute stroke cases. In: ESIT-2000, European Symposium on Intelligent Techniques, pp. 192–197 (2000)
5. Koza, J.: Genetic Programming: On the Programming of Computers by Means of Natural Selection. MIT Press, Cambridge (1992)
6. Tsakonas, A., Dounias, G.: Hierarchical classification trees using type-constrained genetic programming. In: Proc. of 1st Intl. IEEE Symposium in Intelligent Systems (2002)
7. Jimenez, L., Landgrebe, D.: Supervised classification in high-dimensional space: geometrical,statistical, and asymptotical properties of multivariate data. IEEE Transactions on Systems, Man and Cybernetics 28(1), 39–54 (1997)
8. Hotelling, H.: Analysis of a complex of statistical variables into principal components. Journal of Educational Psychology 24, 417–441 (1933)
9. Tipping, M., Bishop, C.: Probabilistic prinicipal component analysis. Technical Report NCRG/97/010, Neural Computing Research Group, Aston University (1997)
10. Scholkopf, B., Smola, A., Muller, K.R.: Nonlinear component analysis as a kernel eigenvalue problem. Neural Computation 10(5), 299–319 (1998)
11. Agrafiotis, D.: Stochastic proximity embedding. Journal of Computational Chemistry 24(10), 1215–1221 (2003)
12. Lafon, S., Lee, A.: Diffusion maps and coarse-graining: A unified framework for dimensionality reduction, graph partitioning, and data set parameterization. IEEE Transactions on Pattern Analysis and Machine Intelligence 28(9), 1393–1403 (2006)
13. Nadler, B., Lafon, S., Coifman, R., Kevrekidis, I.: Diffusion maps, spectral clustering and the reaction coordinates of dynamical systems. Applied and Computational Harmonic Analysis: Special Issue on Diffusion Maps and Wavelets 21, 113–127 (2006)
14. Hinton, G., Salakhutdinov, R.: Reducing the dimensionality of data with neural networks. Science 313(5786), 504–507 (2006)

15. Raymer, M., Punch, W., Goodman, E., Kuhn, L., Jain, A.: Dimensionality reduction using genetic algorithms. IEEE Transactions on Evolutionary Computation 4, 164–171 (2000)
16. Brand, M.: Charting a manifold. In: Advances in Neural Information Processing Systems, vol. 15, pp. 985–992. The MIT Press, Cambridge (2002)
17. Fukunaga, K.: Introduction to Statistical Pattern Recognition. Academic Press Professional, Inc., San Diego (1990)
18. van der Maaten, L.: An introduction to dimensionality reduction using matlab. Technical Report MICC 07-07, Maastricht University, Maastricht, The Netherlands (2007)
19. Levina, E., Bickel, P.: Maximum likelihood estimation of intrinsic dimension. In: Advances in Neural Information Processing Systems, vol. 17. The MIT Press, Cambridge (2004)
20. Fukunaga, K., Olsen, D.: An algorithm for finding intrinsic dimensionality of data. IEEE Transactions on Computers 20, 176–183 (1971)
21. Kegl, B.: Intrinsic dimension estimation based on packing numbers. In: Advances in Neural Information Processing Systems, vol. 15, pp. 833–840. The MIT Press, Volume (2002)
22. Kurucz, M., Benczur, A.A., Csalogany, K.: Methods for large scale svd with missing values. In: Proc. KDD-Cup and Workshop at the 13th ACM SIGKDD International Conference on Knowledge Discovery and Data Mining (2007)

Introducing Parallel Computations to a PTTP-Based First-Order Reasoning Process in the Oz Language

Adam Meissner

Poznań University of Technology, Institute of Control and Information Engineering,
pl. M. Skłodowskiej-Curie 5, 60-965 Poznań, Poland
Adam.Meissner@put.poznan.pl

Abstract. We present a method of adding parallel computations to a reasoning process based on the Prolog Technology Theorem Proving approach. For this purpose, the input set of first-order logic formulas is translated into a logically equivalent program in the Oz language.

Keywords: parallel reasoning, first-order logic, PTTP, the Oz language.

1 Introduction

This work concerns the problem of adding parallel computations to first-order logic (FOL) reasoning systems. A comprehensive survey of the considered issue can be found, e.g., in [1]. In programs of this type a method by which the computations are parallelized and distributed is usually a hardwired part of the execution strategy. It complicates the whole construction increasing the probability of error and brings up possible scalability difficulties. In this paper we take a different approach based on the Prolog Technology Theorem Proving (PTTP) technique [5]. The input set of formulas (given by a user) is translated into a logically equivalent *relational program* in the Oz language [3], whose declarative semantics is similar to that one of Prolog [2]. However, the operational semantics of relational programs, unlike in Prolog, is not fixed in the runtime environment. The execution strategy for programs of this type is implemented as a *search engine* [3] – a special object which runs the given program; some of these objects are provided in libraries of the Mozart system [3] being a runtime environment for Oz. Hence, the user can select the execution strategy, which makes possible to perform the computations in various ways, also in parallel on distributed machines. In this paper we describe the crucial elements of FOL to Oz translation and we present results of experiments for estimating a speedup and a work granularity of parallel processing. It should be remarked that the PTTP technique (or similar methods) has already been taken into account in parallel reasoning systems with the hardwired execution strategy (e.g. METEOR, PARTHENON, PARTHEO, see [1]).

J. Darzentas et al. (Eds.): SETN 2008, LNAI 5138, pp. 359–364, 2008.

2   Translating First-Order Logic Formulas into Oz

The general idea is based on concepts applied in PTTP. A set of input FOL formulas (comprising axioms and a given hypothesis) is translated into a program in the Oz language. Besides the representation of formulas, the program contains the procedures extending the semantics of Oz in order to obtain a sound and complete inference system for first-order logic. Furthermore, three elements are added ([5]):

- the sound unification procedure with the *occur-check*,
- the *model elimination* technique enabling complete inferences for FOL,
- the *consecutively bounded depth-first search* method (CB-DFS) which modifies the search strategy to handle derivation trees with infinite branches.

The set of input formulas is initially transformed to the set of implications. We will only outline this process for the sake of brevity. The hypothesis Q is converted to the form *query* $\leftarrow Q'$ where *query* is a reserved predicate name and Q' is the formula Q in disjunctive normal form regarding the reversal of quantifiers before the skolemization. For example, the hypothesis $\forall x(p(x) \land q(x))$ is converted to the formula $p(g) \land q(g)$ where g is a Skolem function (see [5] for details). The set of axioms in turn is firstly translated to a set of clauses of the form $L_1 \lor \ldots \lor L_k$, where L_i for $i = 1, \ldots, k$ are *literals* [2] and all variables are assumed to be universally quantified over the whole formula. Then, every clause is converted to k implications $L_i \leftarrow \neg L_1 \land \ldots \land \neg L_{i-1} \land \neg L_{i+1} \land \ldots \land \neg L_k$ for $i = 1, \ldots, k$. Let n denote the number of implications containing the given predicate p in the consequent literal. Also, let T_i stand for the sequence of arguments of such a literal in the i-th implication of this kind and let B_i represent the antecedent of that implication consisting of $m(i)$ literals containing the respective $p_{i,1} \ldots p_{i,m(i)}$ predicate symbols, for $i = 1, \ldots, n$. Then, all the implications are transformed to formulas of the form F given below, where X is a sequence of new variables implicitly universally quantified over the whole formula.

$$p(X) \leftarrow X = T_1 \land B_1 \lor \ldots \lor X = T_n \land B_n$$

Next, every formula F is translated into the Oz procedure p whose schematic definition is given below. Expressions of the Oz language are set in typewriter font while metaexpressions are in italics; slightly abusing the notation, we also use FOL expressions as metaexpressions. A set of procedures of the form p is an example of the relational program whose computational model is closely relevant to SLD-resolution based reasoning for *logic programs* [2]. During a procedure call its actual arguments are unified with the formal arguments of the procedure definition. Any finite execution of the procedure can either terminate successfully or it can result in failure. A failure, among other cases, occurs as an effect of an unsuccessful unification or it may be generated by the `fail` statement. The statement `choice` contained in the procedure corresponds to the alternative of its all parts, i.e., each part is executed independently from the others and the final result is a set of all results computed in every part. The statement fails if its all parts return a failure. The program execution is modeled as a *search tree* whose every node represents a sequence of statements. In particular, the choice statement causes a branching of the tree into all statement parts.

```
proc {p X PosAnc NegAnc DepIn DepOut}                                    % 1
    Declaration_of_variables_not_included_in_X                           % 2
in                                                                       % 3
    if {IdMember p(X) PosAnc} then fail                                  % 4
    elseif {IdMember ¬p(X) NegAnc} orelse                               % 5
           {UniMember ¬p(X) NegAnc} then DepOut = DepIn                  % 6
    else                                                                 % 7
        choice                                                           % 8
            [X] = [T₁]                                                   % 9
        ...
        [] [X] = [Tᵢ]                                                    % 10
           {Unify uᵢ,₁ wᵢ,₁}                                            % 11
           ...
           {Unify uᵢ,ᵥ(ᵢ) wᵢ,ᵥ(ᵢ)}                                     % 12
           (DepIn >= m(i)) = true                                        % 13
           DepIn1 = DepIn - m(i)                                         % 14
           Anc1 = p(X)|PosAnc                                            % 15
           {pᵢ,₁ Tᵢ,₁ Anc1 NegAnc DepIn1 D1}                           % 16
           {pᵢ,₂ Tᵢ,₂ Anc1 NegAnc D1 D2}                               % 17
           ...
           {pᵢ,ₘ(ᵢ) Tᵢ,ₘ(ᵢ) Anc1 NegAnc Dm DepOut}                    % 18
        ...
        [] [X] = [Tₙ]                                                    % 19
        ...
        end                                                              % 20
    end                                                                  % 21
end                                                                      % 22
```

When the procedure p is called, its arguments X are unified with the sequence T_i for $i = 1, \ldots, n$ in every part of the statement choice. The unification should respect the occur-check test to avoid the creation of cyclic terms. On the other hand this test, due to its cost, should be performed only when necessary. For this reason the *linearization* method [5] is implemented, which indicates variables that should be unified with the occur-check. This kind of unification is performed in lines 11–12 (the numeration of lines is discontinuous); $u_{i,j}$ and $w_{i,j}$ denote the j-th pair of considered variables in the i-th part of the statement choice, for $i = 1, \ldots, n$ and $j = 1, \ldots, v(i)$ where $v(i)$ is the number of the variable pairs in the given statement part. The procedure p has four additional arguments (except X). Moreover, the arguments PosAnc and NegAnc are introduced with regard to the implementation of the model elimination technique in order to extend the computational model into a complete reasoning method for FOL. The arguments represent the positive (PosAnc) and negative (NegAnc) literals, which precede the current literal (here, represented by the procedure p) on the branch of the search tree. The current literal is regarded as proved if it is identical to (line 5) or unifiable with (line 6) any of its negated ancestors. The two remaining arguments, namely DepIn and DepOut, are added for the implementation of the CB-DFS method, which controlls the exploration of infinite branches in the search tree. The tree can be explored only up to the level restricted by the input depth (DepIn). Furthermore, the sequence of procedure calls representing literals in the current tree node can be executed only if their number is less or equal to the input depth (line 13). Otherwise, the computations result in failure and the search can be repeated with an increased value of the input depth. Infinite calls can also be detected by a simple test employing the ancestor list (line 4).

Below, we give an example of the procedure p being a representation of the formula $\neg q(x_1, x_2) \leftarrow (x_1, x_2) = (a, f(y)) \vee (x_1, x_2) = (f(y), g(y)) \wedge p(y, z) \wedge \neg r(z)$. The unique prefixes Not and not_ mark negative literals in procedure names and, respectively, on the ancestor lists (predicates in positive literals on the lists start from lowercase letters due to syntactic demands of Oz).

```
proc {NotQ X1 X2 PosAnc NegAnc DepIn DepOut}
   DepIn1 Anc1 D1 Y Y1 Z
in
   if {IdMember not_Q(X1 X2) NegAnc} then fail
   elseif {IdMember  q(X1 X2) PosAnc} orelse
          {UniMember q(X1 X2) PosAnc} then DepOut = DepIn
   else
      choice
         [X1 X2] = [a f(Y)]
         DepOut = DepIn
      [] [X1 X2] = [f(Y) g(Y1)]
         {Unify Y Y1}
         (DepIn >= 2) = true
         DepIn1 = DepIn - 2
         Anc1 = not_Q(X1 X2)|NegAnc
         {P Y Z PosAnc Anc1 DepIn1 D1}
         {NotR Z PosAnc Anc1 D1 DepOut}
      end
end
```

Following from the semantics of the statement choice, every branch of the search tree represents independent computations and therefore the tree can be explored by various strategies. For this purpose we use a parallel search engine [4] provided by the Mozart system. When the engine is created, it initiates a number of processes, namely a *manager* and a group of *workers*; the number of workers as well as the list of machines forming the computational environment are the parameters of the engine creation. The manager controls the computations by finding a work for idle workers and collecting the results whereas the workers construct fragments of the search tree. In order to start the reasoning process, the engine is told to run the procedure Query (corresponding to the literal *query*) with the given input depth and initially empty ancestor lists. In the subsequent steps, the input depth is incremented according to the CB-DFS strategy. It has to be remarked that this strategy should be rather implemented directly in the engine (since it cannot handle infinite search trees) and not in the program representing input formulas. However, due to some technical obstacles this task is left for future work.

3 Experimental Results

In the experiments, the inference process was executed in various variants of the computational environment in order to prove exemplary hypotheses. We considered two evaluation criteria, namely a *speedup* caused by parallelization of the computational process and a *work granularity* ([4]), that is a degree of the dispersion of the search tree on particular workers. The computational environment in general consisted of six machines powered by the Mozart system 1.3.2; their

Table 1. Parameters of machines forming the computational environment

Name	Processor	RAM	Ethernet	Operating system
M	Pentium P4D, 3.4 GHz	1 GB	1 GBit	Win. 2000 Prof. 5.00
W1	Pentium-M 760, 2.0 GHz	1 GB	1 GBit	Win. XP HE 2002
W2–W4	Pentium P4D, 3.4 GHz	1 GB	1 GBit	Win. 2000 Prof. 5.00
W5	Pentium(R) 4, 1.8 GHz	256 MB	100 MBit	Win. 2000 Prof. 5.00

parameters are given in Table 1. The machine marked by the symbol M was designated for running the manager while each of the other computers processed one worker. In tests we considered various configurations of the environment comprising from 1 up to 5 workers of distinct computational power. Machines processing the workers were introduced to the environment according to their order given in Table 1. For example, a variant of the configuration with three workers was compounded of the machines M, W1, W2 and W3. All the testing hypotheses come form the TPTP library [6]. The formulas were chosen under the general criterion that the time of computations performed by one worker should range from 30 to 120 sec. Tests were run with the initial input depth equaled 1 which was successively increased by 1.

The results of the tests are presented in Figure 1. The bar chart on the left side depicts a speedup of computations. The speedup is understood as a quotient of the time of computations performed in the environment consisting of the manager and one worker, and in the environment with subsequently growing number of workers. Every cluster of bars represents a speedup obtained for the particular input problem whose name is given on the left; a shading of a bar indicates the number of workers in the computational environment as it is given in the legend on the right side of the chart. All given measurements are approximations due to the heterogeneity of the computational environment. Distinct speedup characteristics obtained for the input problems confirm rather an expected effect that the speedup depends on the structure of the constructed search tree. The increment of the speedup is always positive, however it fluctuates with the increasing number of workers. The lowest speedup is achieved for the worker W5, whose computational power is essentially lower than the power of the others. However, different results are obtained for the problem HEN008-3 where the speedup systematically grows. The explanation of this behavior requires more tests concerning more workers.

The pie chart on the right-hand side of Figure 1 illustrates the work granularity measured by the number of the search tree nodes explored by the respective worker in relation to the total size of the tree. The tests show that the work granularity almost exclusively depends on the power of workers and it is nearly not influenced by the input data. For the machine W1, the standard deviation of the work granularity taken for each problem equals 3.4% and for the other workers it never exceeds 1.6%. This indicates that the task decomposition strategy implemented in the parallel search engine can efficiently adapt a load of the individual worker to its computational capabilities. It is not clear, however, why

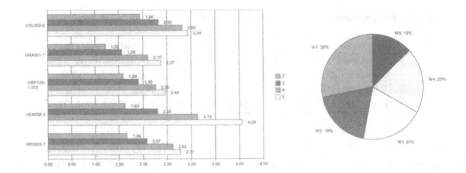

Fig. 1. Speedup and work granularity

the strategy "prefers" the worker W1 to W2, W3 and W4 since its computational power seems to be lower. This issue remains to be explained in future.

4   Final Remarks

We present a method by which parallel computations can be introduced to a first-order reasoning process based on the PTTP methodology. The experiments show a reasonable speedup, however the comprehensive analysis of the system efficiency requires more tests. The tests should consider a greater number of workers as well as more exemplary problems selected with regard to the structure of the search tree. Also, the behavior of the task decomposition strategy requires more thorough observations. The complete search strategy should be implemented directly in the parallel search engine rather than in the program representing input formulas. These works are planned for future.

References

1. Bonacina, M.P.: A taxonomy of parallel strategies for deduction. Annals of Mathematics and Artificial Intelligence 29(1-4), 223–257 (2000)
2. Nilsson, U., Małuszyński, J.: Logic, Programming and Prolog, 2nd edn. John Wiley & Sons Ltd., Chichester (1995)
3. Van Roy, P., Haridi, S.: Concepts, Techniques, and Models of Computer Programming. The MIT Press, Cambridge (2004)
4. Schulte, C.: Programming Constraint Services, Ph.D thesis. University of Saarlandes (2000)
5. Stickel, M.: A Prolog Technology Theorem Prover: A New Exposition and Implementation in Prolog, Technical Note No. 464, SRI Int., Menlo Park (1989)
6. Sutcliffe, G., Suttner, C.B., Yemenis, T.: The TPTP Problem Library. In: Bundy, A. (ed.) CADE 1994. LNCS, vol. 814, pp. 252–266. Springer, Heidelberg (1994)

A Dense Stereo Correspondence Algorithm for Hardware Implementation with Enhanced Disparity Selection

Lazaros Nalpantidis[1], Georgios Ch. Sirakoulis[2], and Antonios Gasteratos[1]

[1] Democritus University of Thrace, Department of Production and Management Engineering,
GR-67 100 Xanthi, Greece
[2] Democritus University of Thrace, Department of Electrical and Computer Engineering, GR-67
100 Xanthi, Greece
lanalpa@pme.duth.gr, gsirak@ee.duth.gr, agaster@pme.duth.gr

Abstract. In this paper an effective, hardware oriented stereo correspondence algorithm, able to produce dense disparity maps of improved fidelity is presented. The proposed algorithm combines rapid execution, simple and straightforward structure as well as comparably high quality of results. These features render it as an ideal candidate for hardware implementation and for real-time applications. The proposed algorithm utilizes the Absolute Differences (AD) as matching cost and aggregates the results inside support windows, assigning Gaussian distributed weights to the support pixels, based on their Euclidean distance. The resulting Disparity Space Image (DSI) is furthered refined by Cellular Automata (CA) acting in all of the three dimensions of the DSI. The algorithm is applied to typical as well as to self-recorded real-life image sets. The disparity maps obtained are presented and quantitatively examined.

Keywords: stereoscopic vision, stereo correspondence, cellular automata.

1 Introduction

Stereo correspondence remains in the focus of the machine vision community for a few decades [1]. It is implied by the biological finding that two, slightly moved from each other, images of the same scene are enough in order to perceive the depth of the objects depicted. Thus, the importance of stereo correspondence is obvious in the fields of machine vision, virtual reality, robot navigation, Simultaneous Localization and Mapping (SLAM) [2], depth measurements and 3D environment reconstruction [3].

The advances in the field of stereo vision during the recent years are, to a large extent, dominated and guided by the test-bench of Scharstein and Szeliski [4]. Very accurate results have been reported on the corresponding web site [5]. However, the high quality of these results comes most of the times at the expense of computation power and thus processing time. The stereo correspondence methods utilized in order to obtain such accuracy and coverage typically involve some kind of global disparity consideration. This direction, although impressive in terms of results, disregards the need for simple, real-time solutions that would be ideally hardware implementable. Hardware implementations provide high execution speeds and often eliminate the

J. Darzentas et al. (Eds.): SETN 2008, LNAI 5138, pp. 365–370, 2008.

need of a dedicated computer. These features are highly appreciated in the case of autonomous robots, where the payload, power consumption, resources management and timing constraints are very strict. However, hardware implementation to be feasible and efficient the utilized algorithm should be simple and parallel in structure avoiding repetitive calculations.

This paper presents a stereo correspondence algorithm, able to produce dense disparity maps with satisfactory output fidelity, for most of the practical autonomous behavior cases. The main merit of the proposed algorithm is its simplicity, rendering it as an ideal choice for real-time operations and hardware implementation. AD is utilized as matching cost function since it is the simplest one. The aggregation step is a 2D process performed inside fix-sized square support windows upon a slice of the Disparity Space Image (DSI). The pixels inside each support window are assigned to a Gaussian distributed weight during aggregation. The weight of each pixel is a Gaussian function of its Euclidean distance towards the central pixel of the current window. The resulting aggregated values of the DSI are furthered refined by applying Cellular Automata (CA). Finally, the best disparity value for each pixel is decided by a Winner-Take-All (WTA) selection step.

Space, time and even the dynamical variables are discrete in CA [6]. CA comprise a very effective computational tool in simulating physical systems and solving scientific problems, because they can capture the essential features of systems where global behavior arises from the collective effect of simple components which interact locally through simple rules. Moreover, CA can easily be implemented in hardware and deal with image processing problems due to the parallel nature of their structure.

2 Proposed Algorithm

The algorithm proposed in this paper follows the structure presented in Fig. 1. The matching cost aggregation step of the proposed algorithm consists of two sub-steps rather than one. In addition, the disparity selection process is a non-iterative one.

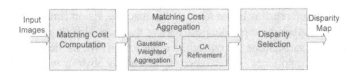

Fig. 1. Block diagram of the proposed stereo correspondence algorithm

The results are refined during the aggregation procedure, rather than during an additional final step. This point of view coincides with the authors' motivation. The key idea is that instead of refining the resulting 2D disparity map, refinement should be performed inside the 3D DSI. Thus, all the available information could be taken into consideration. After all, WTA is a rigid information rejection method, often rejecting useful information as well. This alteration preserves the quality of the produced results while removing any iterative stage from the algorithm's flow.

The matching cost function utilized is the AD. It is inherently the simplest metric of all, involving only summations and finding absolute values of intensity values.

$$AD(i, j, d) = |I_l(i, j) - I_r(i, j - d)|$$ (1)

where I_l and I_r denote the left and right image's pixel intensity value, d is the value of the disparity under examination and i,j are the coordinates of the pixel. The main merit of AD is the speed of calculations and its potential to be easily hardware implemented.

The AD calculated in the previous step comprise the DSI. These results are aggregated inside fix-sized square windows for constant value of disparity. The width of the window plays an important role on the final result. Small windows generally preserve details but suffer from noise, whereas big windows have the inverse behavior. After extensive testing to perform best, the width of the square window is selected to be 11 pixels.

However, the AD summation is weighted. Each pixel is assigned a weight $w(i,j,d)$, the value of which results from the 2D Gaussian function of the pixel's Euclidean distance from the central pixel. The center of the function coincides with the central pixel and has a standard deviation equal to the one third of the distance from the central pixel to the nearest window-border. The Gaussian weight function remains the same for fixed width of the support window. Thus, it can be considered as a fixed mask that can be computed once, and then applied to all the windows.

The weighted SAD comprises the DSI:

$$DSI(i, j, d) = \sum_{\mu=-11}^{11} \sum_{v=-11}^{11} w(i + \mu, j + v, d) \cdot AD(i + \mu, j + v, d)$$ (2)

The resulting aggregated values of the DSI are furthered refined by applying CA. Two CA transition rules are applied to the DSI. The values of parameters used by them were determined after extensive testing to perform best. The first rule attempts to resolve disparity ambiguities. It checks for excessive consistency of results along the disparity (d) axis and, if necessary, corrects on the perpendicular (i,j) plane. It can be expressed as follows:

- if at least one of the two pixels lying from either sides of a pixel across the disparity axis (d) differs from the central pixel less than half of its value, then its value is further aggregated within its 3x3 pixel, constant-disparity neighborhood,

$$if \begin{bmatrix} |DSI(i, j, d) - DSI(i, j, d - 1)| < \frac{1}{2} DSI(i, j, d) \\ or \ |DSI(i, j, d) - DSI(i, j, d + 1)| < \frac{1}{2} DSI(i, j, d) \end{bmatrix}$$

$$then \ DSI(i, j, d) = \frac{1}{9} \sum_{\mu=-1}^{1} \sum_{v=-1}^{1} DSI(i, j, d)$$ (3)

The second rule is placed in order to smoothen the results and at the same time to preserve the details. It checks and acts on constant-disparity planes and can be expressed as follows:

- if there are at least 7 pixels in the 3x3 pixel neighborhood which differ from the central pixel less than half of the central pixel's value, then the central pixel's value is scaled down by the factor 1.3 as dictated by exhaustive testing.

$$if \quad |DSI(i+\mu, j+\nu, d) - DSI(i,j,d)| < \tfrac{1}{2} DSI(i,j,d)$$

$$\forall (\mu, \nu \in \{-1,1\} \ while \ (\nu \neq 0 \wedge \mu \neq 0))$$

$$for \ at \ least \ 7 \ pixels \ in \ the \ 3x3 \ pixel \ neighborhood \qquad (4)$$

$$then \quad DSI(i,j,d) = \tfrac{1}{1.3} DSI(i,j,d) \ .$$

The two rules are applied once. Their outcome comprises the enhanced DSI that will be used in order the optimum disparity map to be produced.

In the last stage the best disparity value for each pixel is decided by a WTA selection procedure. For each image pixel coordinates (i,j) the smaller value is searched for on the d axis and its position is declared to be the pixel's disparity value.

$$D(i,j) = \ \arg\min DSI(i,j,d) \ . \qquad (5)$$

3 Experimental Results

The algorithm was applied to standard image sets [4, 7] as well as to self-recorded real-life ones, in order to be evaluated. Results are presented in terms of calculated images in Fig. 2 and quantitative metrics in Table 1.

Table 1. Percentage of pixels whose absolute disparity error is greater than 1 in various regions of the images

Data Sets	Non-occluded (%)	All (%)	Discontinuities (%)
Tsukuba	10.3	12.3	23.5
Venus	8.86	10.2	35.8
Teddy	24.5	31.5	35.2
Cones	20.6	28.8	31.1

Table 2, on the other hand, presents the Normalized Mean Square Error (NMSE) for the calculated disparity maps of the four image sets, for a simplified version of the proposed algorithm, which makes no use of CA, as well as for the complete version of the algorithm. The addition of CA substantially improves the quality as shown from the last column.

Table 2. Calculated NMSE for various versions of the algorithm

Data Sets	Normalized Mean Square Error (NMSE)		Improvement (%)
	Proposed without CA	Proposed with CA	
Tsukuba	0.0627	0.0593	5.42
Venus	0.0545	0.0447	17.98
Teddy	0.1149	0.1108	3.57
Cones	0.0809	0.0768	5.07

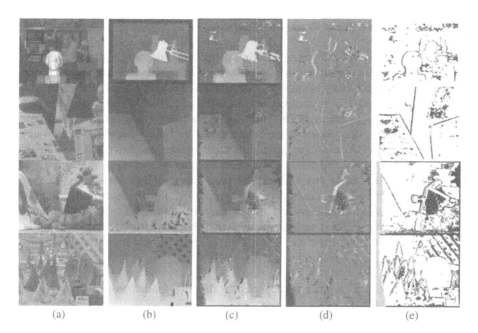

(a)   (b)   (c)   (d)   (e)

Fig. 2. Results for the Middlebury data sets. From top to bottom: the Tsukuba, Venus, Teddy and Cones images From left to right: the reference (left) images (a), the provided ground truth disparity maps (b), the disparity maps calculated by the proposed method (c), maps of signed disparity error (d), and maps of pixels with absolute computed disparity error bigger than 1 (e).

The proposed algorithm was also applied to two self-recorded non synthetic stereo pairs, as well. The stereo pairs and the calculated disparity maps are presented in Fig. 3.

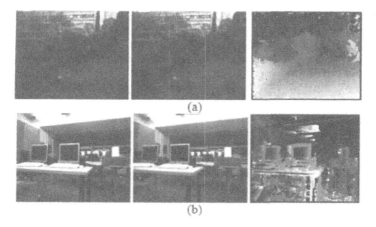

Fig. 3. Self-recorded scenes. (a) outdoor scene, (b) indoor scene. From left to right: left image, right image, calculated disparity map.

4   Conclusions

The proposed algorithm exhibits satisfactory performance despite its simple structure. Gaussian weighted aggregation and CA refinement inside the DSI have been proven to comprise an effective computational combination. Disparity maps of standard image sets, as well as of self-recorded ones are calculated. The data show that the proposed algorithm is in the right direction for a hardware implementable, real-time solution. However, the quality of the results could be further improved by refining further the applied CA rules. The possibilities concerning the nature and the number of the applied CA rules are practically endless and the chosen ones, although effective, are only one of those possibilities. The proposed algorithm's ability to calculate disparity maps of real-life scenes is highly appreciated. Finally, it can be concluded that the algorithm's serial flow and low complexity combined with the presented satisfactory results render it as an appealing candidate for hardware implementation. Thus, depth calculation could be performed efficiently in real-time by autonomous robotic systems.

Acknowledgments. This work is supported by the E.U. funded project View-Finder, FP6-IST-2005-045541.

References

1. Marr, D., Poggio, T.A.: Cooperative Computation of Stereo Disparity. Science 194, 283–287 (1976)
2. Murray, D., Little, J.J.: Using Real-Time Stereo Vision for Mobile Robot Navigation. Autonomous Robots 8, 161–171 (2000)
3. Jain, R., Kasturi, R., Schunck, B.G.: Machine vision. McGraw-Hill, Inc., New York (1995)
4. Scharstein, D., Szeliski, R.: A Taxonomy and Evaluation of Dense Two-Frame Stereo Correspondence Algorithms. International Journal of Computer Vision 47, 7–42 (2002)
5. http://vision.middlebury.edu/stereo/
6. von Neumann, J.: Theory of Self-Reproducing Automata. University of Illinois Press, Urbana (1966)
7. Scharstein, D., Szeliski, R.: High-accuracy stereo depth maps using structured light. In: IEEE Computer Society Conference on Computer Vision and Pattern Recognition, vol. 1, pp. 195–202 (2003)

MyCites: An Intelligent Information System for Maintaining Citations

George Papadakis[1,2] and Georgios Paliouras[1]

[1] Institute of Informatics and Telecommunications,
National Centre for Scientific Research "Demokritos", Athens, Greece
[2] Department of Electrical and Computer Engineering,
National Technical University of Athens, Greece
{gpapad,paliourg}@iit.demokritos.gr

Abstract. The evaluation of their research work and its effect has always been one of scholars' greatest concerns. The use of citations for that purpose, as proposed by Eugene Garfield, is nowadays widely accepted as the most reliable method. However, gathering a scholar's citations constitutes a particularly laborious task, even in the current Internet era, as one needs to correctly combine information from miscellaneous sources. There exists therefore a need for automating this process. Numerous academic search engines try to cover this need, but none of them addresses successfully all related problems. In this paper we present an approach that facilitates to a great extent citation analysis by taking advantage of new algorithms to deal with these problems.

Keywords: information extraction, citation matching, name disambiguation, mixed citation problem, split citation problem, string distance metrics.

1 Introduction

In the last decade there has been a strong interest and considerable effort in developing on-line services that provide access to academic databases. These attempts have culminated in the development of search engines that specialize in scholarly literature. The most notable of them are *Scopus[1]*, *Web of Science (WoS)[2]* and *Google Scholar(GS)[3,]* which are based on huge academic databases gathered from numerous sources. Some of these engines (e.g. Scopus and WoS) use structured sources, such as databases of publishers, in order to warrantee the precision of the provided information. Others (e.g. GS) emphasize on retrieving as much information as is available, automatically from unstructured data, such as Web sites. To the best of our knowledge there is currently no engine that addresses adequately both aspects of the problem.

One of the most valuable features of academic search engines is *citation analysis*, that is looking for papers that refer to a specific publication. In this way they automate

---

[1] http://www.scopus.com/scopus/home.url
[2] http://scientific.thomson.com/products/wos/
[3] http://scholar.google.com/

J. Darzentas et al. (Eds.): SETN 2008, LNAI 5138, pp. 371–376, 2008.

a laborious yet essential task of scholars, that of gathering citations in order to evaluate the influence of their research work. High recall engines, particularly GS, seem to gain in popularity, but there is still enough room for improvement, as they usually contain a relatively large portion of duplicate data and noise. The following problems are particularly important and hard to solve:

Definition 1. *Citation Matching* (CM) is the problem where, given two lists of publications, X and Y, the goal is to find for each x ($\in$ X) a set of $y_1, y_2, ..., y_n$ ($\in$ Y) such that both x and y_i ($1 \leq i \leq n$) in fact pertain to the same publication. Among the main causes of CM are the lack of a fixed format for citations, the various names that are attributed to a single author and errors in the parsing software.

Definition 2. *Mixed Citation* (MC) [1] is the problem where, given a collection of publications, C, by an author, a_i, the goal is to accurately identify publications by another author a_j in C, when a_i and a_j have <u>identical</u> name spellings.

Definition 3. *Split Citation* (SC) [2] is the problem where given two lists of author names and associated publications, X and Y, the goal is to find for each author name x ($\in$ X) a set of author names, $y_1, y_2, ..., y_n$ ($\in$ Y) such that both x and y_i ($1 \leq i \leq n$) are name variants of the same author.

We should point out that the MC and SC problems are so closely related to each other that are rarely succinctly distinguished. They are regarded as a single problem called *name disambiguation* or *name equivalence identification*. Along with the CM problem they belong to the broader *Identity Uncertainty Problem* or *Record Linkage*.

In this paper we propose new methods that deal with these problems and are embedded in a simple information system intended to automate the maintenance of citations for scholars and research groups through a user-friendly interface.

2   Related Work

Two teams have primarily worked on the SC and MC problems. The first one concentrated on the MC problem and tested supervised classification ([3]) and unsupervised clustering ([4],[5]) methods, concluding that the latter generally perform better, while not requiring processed datasets for training. Their clustering approaches presume though a predefined number of clusters, thus limiting their applicability. The second research group addresses both the MC ([1]) and SC ([1],[2]) problems, primarily concentrating on the scalability of their algorithms. The proposed matching methods are generally based on common features of publications: co-authors' names, paper and conference/journal title, with the first proving to be the most robust one.

As far as the CM problem is concerned, the term was initially coined in [6] and [7] by the creators of Citeseer, where they also presented four different methods based on simple string matching methods. In [8] another method is proposed, based on relational probability models (RPMs), while in [9] an innovative algorithm is presented based on conditional random fields (CRFs). It is worth noting that all of these methods were applied to the same dataset and are thus directly comparable, with the last one achieving the best performance.

Finally, [10] summarizes, categorizes and compares the most robust and efficient methods for string matching, that are at the heart of all the above-mentioned methods.

3 Application Use Cases

In this section we will analyze briefly the main functionality of the application that we have developed, based on the proposed approach. We do this by going through the steps that comprise a thorough search for a scholar.

1. *Fetch all publications that contain the given scholar in their author list.* This is done by issuing the appropriate query to the GS search engine, gathering and feeding the results to the wrapper we have developed for processing the returned HTML pages. The wrapper identifies the html tags that define the information of a single article and then the tags that encompass each attribute of that specific paper (title, authors, URL etc).

2. *Apply the Citation Matching algorithm for processing the gathered publications.* There is considerable noise in the form of duplicate articles in GS results and, therefore, a pre-processing stage that refines the data gathered by the initial query is indispensable. Otherwise the performance of the name disambiguation algorithm would be substantially degraded by the duplicates. Our method for solving this problem is presented in section 4.

3. *Present the user with the results of the CM algorithm for verification.* In this stage, the user is given the chance to amend potential mistakes or omissions of the CM algorithms. Specifically, the user is provided with all the necessary information (title, authors, URLs etc) so as to be able to judge whether two articles that were alleged to match are in fact different articles and thus have to be dissociated or whether two separate publications are duplicates and must be matched.

4. *Apply the Name Disambiguation algorithm.* This is the most critical step of each search as it entails the identification of all separate scholars that contribute to all papers maintained by our application, those already stored and those acquired during the current search. We address this problem by the algorithm presented in section 5.

5. *Present the user with the results of Name Disambiguation for verification.* The purpose of this step is to amend once again the results of the automatic processing so as to ensure that the data stored in the database is as accurate as possible. This is a critical step since potential errors that are not detected are perpetuated in subsequent runs of the name disambiguation algorithm, thus degrading its performance. By the end of this process, every piece of information concerning the separate scholars is stored.

6. *Search for citations.* Having completed the previous steps, the user can now search for the citations of a specific paper or start a new search for a different scholar. Every new search goes through the above steps before giving the user the chance to commence a new one.

4 Citation Matching

The most common forms of duplicate citations that appear in the results of GS and thus need to be addressed by our algorithm are the following:

1. *spelling mistakes*
2. *author's names concatenated with paper title (usually preceding it)*
3. *conference/journal title concatenated with paper title (usually following it)*
4. *title swapped with another information field*

In this context, our algorithm acts as follows:

1. It initially orders the retrieved papers by the number of citations, based on the assumption that among the multiple appearances of a single paper, the one with the most citations will probably contain the correct information, as it is highly unlikely that a paper is cited more frequently in a wrong way than in the right one. With this initial sorting we ensure that the more correct the information of a paper, the fewer times it is compared to another one.
2. It then checks the contents of the database before processing each paper, so as to avoid repeating the same process. We assume that the problem has been resolved for the stored papers, since the user has verified the stored data.
3. For each paper that is not stored in the database, its title is compared with that of its preceding ones, so as to find the most similar paper. The comparisons are done using the *SoftTFIDF string distance metric in combination with the Jaro metric* [10], which has proven the most suitable metric for the three first problems mentioned above, that is for matching strings that are sets of words (tokens).
4. If the best matching does not exceed a user-defined threshold, the algorithm checks whether the title has moved to another field of the paper's description (fourth case). This is done by forming a new string for each paper in the list, comprising the paper title, co-author names and conference/journal title. These strings are compared using again *SoftTFIDF with Jaro* and if their similarity exceeds another threshold, they are considered identical.

5 Name Disambiguation

In this section we introduce a new clustering method for solving simultaneously both the Mixed and the Split Citation problems. Our approach generally exploits the same article features as other methods proposed in the literature: co-authors' names, paper title, URLs of the papers and the name of the scholar. However, what differentiates our algorithm from the others is the range of its applicability. Our goal is to develop a method that given *any* dataset of citations identifies *all* separate scholars it contains *without any prior knowledge about them*. To achieve this goal the algorithm is based on the following basic principles:

1. Every author of a single paper corresponds to a separate scholar.
2. Every scholar can match with only one author of a single paper.

3. It is time and memory consuming to repeat the matching process for the authors of the already stored publications. After all, the data stored in the database has already been verified by the user and is thus assumed to be accurate.

Abiding by these principles our algorithm creates a graph of related authors, aiming to partition the graph in its connected components. In this context, a new node is initially added to the graph for each author of the **stored** papers. The nodes that correspond to the same scholar are then directly connected, in order to ensure that they are included in the same connected component after partitioning the graph.

Then, for each of the **new** papers, the most similar author name in the graph is found. The similarity is calculated by the *Jaro* string matching method. If this similarity exceeds the respective threshold, the value of the following formula is calculated for the data of the current two publications:

Tot_sim = β * co-author similarity + γ * title similarity + δ * URL similarity

These similarities are calculated using again the *combined SoftTFIDF with Jaro metric*. If the value of Tot_sim exceeds another threshold, the two authors are considered identical and their nodes are connected with a new edge.

With the completion of this process, the graph is partitioned in its connected components, each of which contains all information about a unique scholar: all variants of a scholar's name, along with all the papers the scholar has authored.

6 Experiments and Results

In order to measure the performance of our algorithms, we need to test them over a fairly large dataset, covering various influential factors, such as the scientific field of the papers, the nationality of the authors, etc. However due to time limitations, we performed a limited set of initial experiments, using the work of one of the authors of this paper as a seed, moving to the work of those who cite his papers, and so on. Despite its limited nature, the dataset included authors of various nationalities, some of which (e.g. Korean names) amplify the mixed citation problem. The algorithms were evaluated with the use of standard information retrieval criteria: *precision*[4], *recall*[5] and *f-measure* (their harmonic mean), and the results are presented in Table 1.

Table 1. Evaluation of our algorithms

Algorithm	Proposed Matches	False Matches	True Matches	Missed Matches	Precision	Recall	F-Measure
Citation Matching	70	22	56	8	68,57%	85,71%	76,19%
Name Disambiguation	732	40	719	27	94,54%	96,24%	95,38%

These initial results seem promising, but we need to acknowledge that the sample for the citation matching algorithm is too limited to draw safe conclusions. Furthermore, the performance of the name disambiguation algorithm is degraded by the large

---

[4] *Precision = (proposed matches – false matches) / proposed matches*
[5] *Recall = (true matches – missed matches) / true matches*

portion of citing authors that appear only once in the dataset. Therefore, a large-scale experiment may lead to different results.

7 Conclusions

We presented a new approach for addressing important problems in using academic search engines for citation analysis, namely the citation matching and mixed and split citation problems. The proposed methods were successfully embedded in an information system that aims to facilitate the maintenance of citations for scholars. The methods were evaluated, giving initial encouraging results.

There is undoubtedly great potential in evolving our system. First of all, we plan to add support for additional major academic search engines, such as Scopus and WoS. Combining the contents of these on-line sources will significantly increase the comprehensiveness of our system, Furthermore, based on the aforementioned performance of our methods there is evidently enough room for improvement. We primarily need to refine our citation matching algorithm and generalize its applicability to other academic search engines.

References

1. Lee, D., On, B.W., Kang, J., Park, S.: Effective and Scalable Solutions for Mixed and Split Citation Problems in Digital Libraries. In: Proceedings of the 2nd International Workshop on Information Quality in Information Systems, pp. 69–76 (2005)
2. On, B.W., Lee, D., Kang, J., Mitra, P.: Comparative Study of Name Disambiguation Problem using a Scalable Blocking-based Framework. In: Proceedings of the 5th ACM/IEEE-CS Joint Conference on Digital Libraries, pp. 344–353 (2005)
3. Han, H., Giles, L., Zha, H., Li, C., Tsioutsiouliklis, K.: Two Supervised Learning Approaches for Name Disambiguation in Author Citations. In: Proceedings of the 4th ACM/IEEE-CS Joint Conference on Digital Libraries, pp. 296–305 (2004)
4. Han, H., Xu, W., Zha, H., Giles, C.: A Hierarchical Naive Bayes Mixture Model for Name Disambiguation in Author Citations. In: Proceedings of the 2005 ACM Symposium on Applied Computing, pp. 1065–1069 (2005)
5. Han, H., Zha, H., Giles, C.: Name Disambiguation in Author Citations using a K-way Spectral Clustering Method. In: Proceedings of the 5th ACM/IEEE-CS Joint Conference on Digital Libraries, pp. 334–343 (2005)
6. Giles, C., Bollacker, K., Lawrence, S.: Citeseer: An Automatic Citation Indexing system. In: Proceedings of the Third ACM Conference on Digital Libraries, pp. 89–98 (1998)
7. Lawrence, S., Giles, C., Bollacker, K.: Digital Libraries and Autonomous Citation Indexing. IEEE Computer Society 32(6), 67–71 (1999)
8. Pasula, H., Marthi, B., Milch, B., Russel, S., Shpitser, I.: Identity uncertainty and citation matching. In: Advances in Neural Information Processing Systems (NIPS), vol. 15 (2003)
9. Wellner, B., McCallum, A., Peng, F., Hay, M.: An Integrated, Conditional Model of Information Extraction and Coreference with Application to Citation Matching. In: Proceedings of the 20th conference on Uncertainty in artificial intelligence, pp. 593–601 (2004)
10. Cohen, W., Ravikumar, P., Fienberg, S.: A Comparison of String Distance Metrics for Name-Matching Tasks. In: Proceedings of International Joint Conferences on Artificial Intelligence (IJCAI 2003) Workshop on Information Integration on the Web (2003)

Improving the Integration of Neuro-Symbolic Rules with Case-Based Reasoning

Jim Prentzas[1,2], Ioannis Hatzilygeroudis[1], and Othon Michail[1]

[1] University of Patras, School of Engineering
Dept of Computer Engin. & Informatics, 26500 Patras, Hellas (Greece)
[2] Dept of Informatics & Computer Technology, TEI of Lamia
35100 Lamia, Greece
dprentzas@teilam.gr, ihatz@ceid.upatras.gr

Abstract. In this paper, we present an improved approach integrating rules, neural networks and cases, compared to a previous one. The main approach integrates neurules and cases. Neurules are a kind of integrated rules that combine a symbolic (production rules) and a connectionist (adaline unit) representation. Each neurule is represented as an adaline unit. The main characteristics of neurules are that they improve the performance of symbolic rules and, in contrast to other hybrid neuro-symbolic approaches, they retain the modularity of production rules and their naturalness in a large degree. In the improved approach, various types of indices are assigned to cases according to different roles they play in neurule-based reasoning, instead of one. Thus, an enhanced knowledge representation scheme is derived resulting in accuracy improvement. Experimental results demonstrate its effectiveness.

1 Introduction

Approaches integrating rule-based and case-based reasoning have given effective knowledge representation schemes and are becoming increasingly popular in various fields [7], [8]. However, a more interesting approach is one integrating more than two reasoning methods towards the same objective. In [6], such an approach integrating three reasoning schemes, namely rules, neurocomputing and case-based reasoning in an effective way is introduced. To this end, neurules and cases are combined. Neurules are a type of hybrid rules integrating symbolic rules with neurocomputing in a seamless way. Their main characteristic is that they retain the modularity of production rules and also their naturalness in a large degree. In that approach, on the one hand, cases are used as exceptions to neurules, filling their gaps in representing domain knowledge and, on the other hand, neurules perform indexing of the cases facilitating their retrieval. Finally, it results in accuracy improvement.

In this paper, we enhance the above approach by employing different types of indices for the cases according to different roles they play in neurule-based reasoning. In this way, an improved knowledge representation scheme is derived as various types of neurules' gaps in representing domain knowledge are filled in by indexed cases. Experimental results demonstrate the effectiveness of the presented approach.

J. Darzentas et al. (Eds.): SETN 2008, LNAI 5138, pp. 377–382, 2008.

The rest of the paper is organized as follows. Section 2 presents neurules, whereas Section 3 presents methods for constructing the indexing scheme of the case library. Section 4 describes the hybrid inference mechanism. Section 5 presents experimental results regarding accuracy of the inference process.

2   Neurules

Neurules are a type of hybrid rules integrating symbolic rules with neurocomputing giving pre-eminence to the symbolic component. Neurocomputing is used within the symbolic framework to improve the performance of symbolic rules [3], [6]. In contrast to other hybrid approaches, the constructed knowledge base retains the modularity of production rules, since it consists of autonomous units (neurules), and also retains their naturalness in a large degree, since neurules look much like symbolic rules [3], [4]. Also, the inference mechanism is a tightly integrated process, which results in more efficient inferences than those of symbolic rules [3], [6]. Explanations in the form of if-then rules can be produced [5], [6].

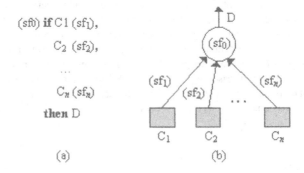

Fig. 1. (a) Form of a neurule (b) a neurule as an adaline unit

The form of a neurule is depicted in Fig.1a. Each condition C_i is assigned a number sf_i, called its *significance factor*. Moreover, each rule itself is assigned a number sf_0, called its *bias factor*. Internally, each neurule is considered as an adaline unit (Fig.1b). The *inputs* C_i ($i=1,...,n$) of the unit are the *conditions* of the rule. The weights of the unit are the significance factors of the neurule and its bias is the bias factor of the neurule. Each input takes a value from the following set of discrete values: [1 (true), 0 (false), 0.5 (unknown)]. This gives the opportunity to easily distinguish between the falsity and the absence of a condition in contrast to symbolic rules. The *output D*, which represents the *conclusion* (decision) of the rule, is calculated via the formulas:

$$D = f(\mathbf{a}), \quad \mathbf{a} = sf_0 + \sum_{i=1}^{n} sf_i \ C_i$$

$$f(\mathbf{a}) = \begin{cases} 1 & \text{if } \mathbf{a} \geq 0 \\ -1 & \text{otherwise} \end{cases}$$

where **a** is the *activation value* and *f(x)* the *activation function*, a threshold function. Hence, the output can take one of two values ('-1', '1') representing failure and success of the rule respectively. The general syntax of a condition C_i and the conclusion *D* is:

<condition>::= <variable> <l-predicate> <value>
<conclusion>::= <variable> <r-predicate> <value>

where <variable> denotes a *variable*, that is a symbol representing a concept in the domain, e.g. 'sex', 'pain' etc, in a medical domain. <l-predicate> denotes a symbolic or a numeric predicate. The symbolic predicates are {is, isnot} whereas the numeric predicates are {<, >, =}. <r-predicate> can only be a symbolic predicate. <value> denotes a value. It can be a *symbol* or a *number*. The significance factor of a condition represents the significance (weight) of the condition in drawing the conclusion(s).

The neurule-based inference engine gives pre-eminence to symbolic reasoning, based on a backward chaining strategy [3], [6]. As soon as the initial input data is given and put in the working memory, the output neurules are considered for evaluation. One of them is selected for evaluation. A neurule fires if the output of the corresponding adaline unit is computed to be '1' after evaluation of its conditions. A neurule is said to be 'blocked' if the output of the corresponding adaline unit is computed to be '-1' after evaluation of its conditions. Inference stops either when one or more output neurules are fired (success) or there is no further action (failure).

During inference, a conclusion is rejected (or not drawn) when none of the neurules containing it fires. This happens when: (i) all neurules containing the conclusion have been examined and are blocked or/and (ii) a neurule containing an alternative conclusion for the specific variable fires instead.

3 Indexing

Indexing concerns the organization of the available cases so that combined neurule-based and case-based reasoning can be performed. Indexed cases fill in gaps in the domain knowledge representation by neurules and during inference may assist in reaching the right conclusion. To be more specific, cases may enhance neurule-based reasoning to avoid reasoning errors by handling the following situations:

(a) Examining whether a neurule misfires. If sufficient conditions of the neurule are satisfied so that it can fire, it should be examined whether the neurule misfires for the specific facts, thus producing an incorrect conclusion.

(b) Examining whether a conclusion was erroneously rejected (or not drawn).

In the approach in [6], the neurules contained in the neurule base were used to index cases representing their exceptions. A case constitutes an exception to a neurule if its attribute values satisfy sufficient conditions of the neurule (so that it can fire) but the neurule's conclusion contradicts the corresponding attribute value of the case. In this approach, various types of indices are assigned to cases. More specifically, indices are assigned to cases according to different roles they play in neurule-based reasoning and assist in filling in different types of gaps in the knowledge representation by neurules. Assigning different types of indices to cases can produce an effective approach combining symbolic rule-based with case-based reasoning [1].

In this new approach, a case may be indexed by neurules and by neurule base conclusions as well. In particular, a case may be indexed as:

(a) *False positive (FP)*, by a neurule whose conclusion is contradicting. Such cases, as in our previous approach, represent exceptions to neurules and may assist in avoiding neurule misfirings.

(b) *True positive (TP)*, by a neurule whose conclusion is endorsing. The attribute values of such a case satisfy sufficient conditions of the neurule (so that it can fire) and the neurule's conclusion agrees with the corresponding attribute value of the case. Such cases may assist in endorsing correct neurule firings.

(c) *False negative (FN)*, by a conclusion erroneously rejected (or not drawn) by neurules. Such cases may assist in reaching conclusions that ought to have been drawn by neurules (and were not drawn). If neurules with alternative conclusions containing this variable were fired instead, it may also assist in avoiding neurule misfirings. 'False negative' indices are associated with conclusions and not with specific neurules because there may be more than one neurule with the same conclusion in the neurule base.

4 The Hybrid Inference Mechanism

The combined inference process mainly focuses on the neurules. The indexed cases are considered when: (a) sufficient conditions of a neurule are fulfilled so that it can fire, (b) all output or intermediate neurules with a specific conclusion variable are blocked and thus no final or intermediate conclusion containing this variable is drawn.

In case (a), firing of the neurule is suspended and case-based reasoning is performed for cases indexed as 'false positives' and 'true positives' by the neurule and cases indexed as 'false negatives' by alternative conclusions containing the neurule's conclusion variable. Cases indexed as 'true positives' by the neurule endorse its firing whereas the other two sets of cases considered (i.e., 'false positives' and 'false negatives') prevent its firing. The results produced by case-based reasoning are evaluated in order to assess whether the neurule will fire or whether an alternative conclusion proposed by the retrieved case will be considered valid instead.

In case (b), the case-based module will focus on cases indexed as 'false negatives' by conclusions containing the specific (intermediate or output) variable.

The basic steps of the inference process are the following:

1. Perform neurule-based reasoning for the neurules.
2. If sufficient conditions of a neurule are fulfilled so that it can fire, then
   2.1. Perform case-based reasoning for the 'false positive' and 'true positive' cases indexed by the neurule and the 'false negative' cases associated with alternative conclusions containing the neurule's conclusion variable.
   2.2. If none case is retrieved or the best matching case is indexed as 'true positive', the neurule fires and its conclusion is inserted into the working memory.
   2.3. If the best matching case is indexed as 'false positive' or 'false negative', insert the conclusion supported by the case into the working memory and mark the neurule as 'blocked'.

3. If all intermediate neurules with a specific conclusion variable are blocked, then
    3.1. Examine all cases indexed as 'false negatives' by the corresponding intermediate conclusions, retrieve the best matching one and insert the conclusion supported by the retrieved case into the working memory.
4. If all output neurules with a specific conclusion variable are blocked, then
    4.1. Examine all cases indexed as 'false negatives' by the corresponding final conclusions, retrieve the best matching one and insert the conclusion supported by the retrieved case into the working memory.

5 Experimental Results

In this section, we present experimental results using datasets acquired from [2]. The experimental results involve evaluation of the presented approach combining neurule-based and case-based reasoning and comparison with our previous approach [10]. 75% and 25% of each dataset were used as training and testing sets respectively. Each initial training set was used to create a combined neurule base and indexed case library. For this purpose, each initial training set was randomly split into two disjoint subsets, one used to create neurules and one used to create an indexed case library. More specifically, 2/3 of each initial training set was used to create neurules by employing the 'patterns to neurules' module [4] whereas the remaining 1/3 of each initial training set constituted non-indexed cases. Both types of knowledge (i.e., neurules and non-indexed cases) were given as input to the indexing construction module presented in this paper producing a combined neurule base and an indexed case library which will be referred to as NBRCBR. Neurules and non-indexed cases were also used to produce a combined neurule base and an indexed case library according to [6] which will be referred to as NBRCBR_PREV.

Inferences were run for both NBRCBR and NBRCBR_PREV using the testing sets. Inferences from NBRCBR_PREV were performed using the inference mechanism combining neurule-based and CBR as described in [6]. Inferences from NBRCBR were performed according to the inference mechanism described in this paper. Table 1 presents such experimental results regarding inferences from NBRCBR and NBRCBR_PREV. It presents results regarding classification accuracy of the integrated approaches and the percentage of test cases resulting in neurule-based reasoning errors that were successfully handled by case-based reasoning. Column '% FPs handled' refers to the percentage of test cases resulting in neurule misfirings (i.e., 'false positives') that were successfully handled by case-based reasoning. Column '% FNs handled' refers to the percentage of test cases resulting in having all output neurules blocked (i.e., 'false negatives') that were successfully handled by case-based reasoning. 'False negative' test cases are handled in NBRCBR_PREV by retrieving the best-matching case from the whole library of indexed cases. As can be seen from the table, the presented approach results in improved classification accuracy. Furthermore, in inferences from NBRCBR the percentages of both 'false positive' and 'false negative' test cases successfully handled are greater than the corresponding percentages in inferences from NBRCBR_PREV.

Table 1. Experimental results

Dataset	NBRCBR			NBRCBR_PREV		
	Classification Accuracy	% FPs Handled	% FNs Handled	Classification Accuracy	% FPs Handled	% FNs Handled
Car (1728 patterns)	96.04%	52.81%	64.07%	92.49%	15.51%	20,36%
Nursery (12960 patterns)	98.92%	58.68%	52.94%	97.68%	6.60%	18.82%

We also tested a nearest neighbor approach working alone in these two datasets (75% of the dataset used as case library and 25% of the dataset used as testing set). We used the similarity measure presented in Section 5. The approach classified the input case to the conclusion supported by the best-matching case retrieved from the case library. Classification accuracy for car and nursery dataset is 90.45% and 96.67% respectively. So, both integrated approaches perform better. This is due to the fact that the indexing schemes assist in focusing on specific parts of the case library.

References

1. Agre, G.: KBS Maintenance as Learning Two-Tiered Domain Representation. In: Aamodt, A., Veloso, M.M. (eds.) ICCBR 1995. LNCS, vol. 1010, pp. 108–120. Springer, Heidelberg (1995)
2. Asuncion, A., Newman, D.J.: UCI Repository of Machine Learning Databases, Irvine, CA. University of California, School of Information and Computer Science (2007), http://www.ics.uci.edu/~mlearn/MLRepository.html
3. Hatzilygeroudis, I., Prentzas, J.: Neurules: Improving the Performance of Symbolic Rules. International Journal on AI Tools 9, 113–130 (2000)
4. Hatzilygeroudis, I., Prentzas, J.: Constructing Modular Hybrid Rule Bases for Expert Systems. International Journal on AI Tools 10, 87–105 (2001)
5. Hatzilygeroudis, I., Prentzas, J.: An Efficient Hybrid Rule-Based Inference Engine with Explanation Capability. In: Proceedings of the 14th International FLAIRS Conference, pp. 227–231. AAAI Press, Menlo Park (2001)
6. Hatzilygeroudis, I., Prentzas, J.: Integrating (Rules, Neural Networks) and Cases for Knowledge Representation and Reasoning in Expert Systems. Expert Systems with Applications 27, 63–75 (2004)
7. Marling, C.R., Sqalli, M., Rissland, E., Munoz-Avila, H., Aha, D.: Case-Based Reasoning Integrations. AI Magazine 23, 69–86 (2002)
8. Prentzas, J., Hatzilygeroudis, I.: Categorizing Approaches Combining Rule-Based and Case-Based Reasoning. Expert Systems 24, 97–122 (2007)

Rule-Based Fuzzy Logic System for Diagnosing Migraine

Svetlana Simić[1], Dragan Simić[2], Petar Slankamenac[1], and Milana Simić-Ivkov[3]

[1] Institute of Neurology, Clinical Centre Vojvodina, Hajduk Veljkova 1-9,
21000 Novi Sad, Serbia
dsimic@Eunet.rs
[2] Novi Sad Fair, Hajduk Veljkova 11, 21000 Novi Sad, Serbia
dsimic@nsfair.co.rs
[3] Clinic of Dermatovenerology, Clinical Centre Vojvodina, Hajduk Veljkova 1-9,
21000 Novi Sad, Serbia
ssimic@uns.ns.ac.yu

Abstract. This research focussed on diagnosing migraine types in working people employing the rule-based fuzzy logic system. Migraine is not a disease which typically shortens one's life. However, it can be a serious social as well as a health problem. Approximately 27 billion euros per year are lost through reduced work productivity in the European Community. The diagnostic criteria developed by the International Headache Society (IHS) have been used in epidemiological researches, but there is no such tool which helps physicians make diagnoses. The rules were facilitated by the application of the IHS criteria for migraine types. Clinical experience was used to extend the established rules and improve the system. The proposed system is in the starting phase of the implementation at the Clinical Centre Vojvodina, Institute of Neurology in Novi Sad.

Keywords: migraine, diagnosing, rule-based fuzzy logic.

1 Introduction

Migraine is not a disease which typically shortens one's life. However, it can be a serious social as well as a health problem. Migraine is a common disorder amongst working people. A lot of people with headache are unaware of the type of headache they are suffering from. People do not always visit physicians for their complaints, and when they do, physicians do not always make a correct diagnosis.

Headache, and particularly migraine, create a burden for patients, their families and economy. Approximately 13 billion dollars per year are lost through reduced work productivity in the United States [1]. Migraine is the most expensive neurological problem in the European Community costing 27 billion euros annually [2].

While the diagnostic criteria developed by the International Headache Society (IHS) have been widely validated, there is no reliable tool that would help physicians make the diagnosis [3]. Our efforts are directed towards developing a software tool for daily clinical practice in order to distract physician's attention from migraine.

J. Darzentas et al. (Eds.): SETN 2008, LNAI 5138, pp. 383–388, 2008.
© Springer-Verlag Berlin Heidelberg 2008

Now, for the time being, the tool is developed for migraine - migraine without aura and migraine with aura - and other non-migraine headache types. The results of our research are that 20% of those suffering from migraine suffer from migraine with aura, while 80% suffer from migraine without aura. The results coincide with research presented in [4]. Our plan is to classify other non-migraine primary headaches as well as a tension-type headache, cluster headache and other primary headaches in future researches.

The rest of the paper is organised as follows. Section 2 presents basic headaches classification and designing the health questionnaire. Rule-based fuzzy logic (RBFL) as a successful technique for knowledge-based decision support is outlined in section 3. Section 4 describes results of the health questionnaire and some improvements to the RBFL system. Section 5 describes related work and section 6 concludes the paper.

2 Headaches and Designing a Health Questionnaire

The International Classification of Headache Disorders 2nd edition (ICHD) is perhaps the single most important document to read for physicians taking an interest in the diagnosis and management of headache patients. All headache disorders are classified into two major groups: 1) Primary headaches and 2) Secondary headaches. Each group is further subdivided into types, subtypes and sub-forms. Primary headaches are thus subdivided into: 1) Migraine, 2) Tension-type headache, 3) Cluster headache and other trigeminal autonomic cephalalgias and 4) Other primary headaches. There are only six types of migraine, yet these are further divided into 17 subtypes. Considering the wide range of headache types, our research focussed on: migraine without aura, migraine with aura and other non-migraine primary headaches. The diagnostic criteria for migraine without aura according to the ICHD are as follows:

A. At least 5 attacks fulfilling criteria B-D;   *From the second questionnaire*
B. Headache attacks lasting;   *Question 3, answer b) from 4 hours to 3 days*
C. Headache has at least two of the following characteristics:
    1. location;   *Question 6, answer a) unilateral*
    2. quality   *Question 4, answer a) pulsating*
    3. pain intensity   *Question 5, answer b) temperate or c) strong*
    4. aggravation of routine physical activity   *Questions 7 and 8, answer a) yes*
D. During headache at least one of the following:
    1. nausea and/or vomiting
        Question 9.1, answer a) yes; Question 9.2, answer a) yes, or b) no
    2. photophobia and phonophobia;   *Questions 9.3 and 9.4, answer a) yes*
E. Not attributable to another disorder;   *From the second questionnaire*

The diagnostic criteria for migraine with aura are as follows:
B. Aura consisting of at least one of the following, but no motor weakness:
    1. full reversible visual symptoms; *Question 15, a) numbing, or b) insensitivity*
    2. full reversible sensory symptoms; *Question 16, a) flickering light, or b) loss of eyesight*
    3. full reversible dysphasic speech disturbance; *Question 10, answer a) yes*

3 Rule-Based Fuzzy Logic System

Rule-based fuzzy logic (RBFL) is considered to be a successful technique for knowledge-based decision support in many domains, including medicine. In this research the rule-based fuzzy logic system was used as a tool for diagnosing migraine types. Rule-based fuzzy logic systems contain four components as shown in fig. 1.

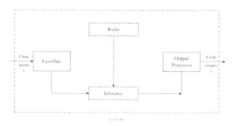

Fig. 1. Rule-Based Fuzzy Logic System - basic model

Rules are the heart of the fuzzy logic system, and may be provided by an expert or can be extracted from numerical data. In either case, the rules we are interested in can be expressed as a collection of IF - THEN statements and are provided as the collection of IF-THEN statements made by questioned people - patients.

The responses, given by the questioned patients suffering from headache are observed as a fuzzy process. While the same response applies to the different headache types, some of them apply to just one type. However, a patient may suffer from one or more headache types, but in one attack, the patient suffers from the specific one. The results of the questionnaire show the presence of both types, simultaneously. The proposed software tool could be coded as a flexible rule-based model with parameters tuning, according to [5], and presented system in [6].

4 First Results and Improvements

After inserting the responses of 80 questioned subjects, an analysis of headache types, received as inserted results of responses, was carried out. It was not expected to obtain a lot of non-migraine primary headaches diagnoses (table 1.). The high specificity of ICHD diagnostic criteria, while ideal for clinical trails, appears to lack the necessary sensitivity for migraine mandated in routine clinical practice. This lack of clinical

Table 1. First results of Migraine health questionnaire

Headache type	Number	%
Migraine without aura	7	8.75
Migraine with aura	0	0
Migraine	7	8.75
Non-migraine primary headaches	*73*	*91.25*

sensitivity may be a contribution factor to the continued underdiagnosis and under-treatment of migraine [7].

Working experience of physicians was used in order to extend the established rules and improve the system for suggesting diagnosis in patients suffering from headaches. This clearly shows that in order to establish the rules it is not enough to use defined standards or professional textbooks and manuals, but is necessary to include physicians with, ad hoc, their practical experience.

Based on our own clinical experience, we have made a decision to apply the modified established ICHD criteria. The improvement of the system has been carried out by adding, subtracting or changing the criteria but by increasing the significance of some, making it possible to establish the diagnosis in patients suffering from the migraine with more accuracy as presented in Table 2.

Table 2. Improving the rules and assessment headache type of Migraine health questionnaire

Headache type	Number	Surely assessed	%
	4	1	5
Migraine without aura	4	3	5
	8		10
Migraine with aura	2	1	2.5
Migraine	10		12.5
Non-migraine primary headaches	*70*		*87.5*

The improvement and the extension of the decision-making logic were conducted in a manner where ICHD criteria for migraine without aura B, C1, C2, and C3 were kept as obligatory conditions for migraine without aura determination. Other criteria, C4, D1, and D2, are optionally used with altered logic among them. Considering that C4, D1 and D2 represent optional decision-making criteria of migraine diagnoses, the term "surely assessed" is introduced. The term "surely assessed" shows the degree of assessed diagnoses of migraine types. Since there are 4 possible values observed, the value of "surely assessed" can vary from 0 to 3. It's obvious that 0 represents low level migraine type diagnosis assessment. The values 1 and 2 increase the value of accurate assessment in suggested diagnosis, while for "surely assessed" it could be said that the highest value is 3.

Criteria for migraine with aura B1, B2, and B3 were kept as obligatory conditions for migraine with aura determination.

Two types of results could be observed: 1) The prevalence of patients suffering from migraine in the general working population; 2) The prevalence of patients suffering from migraine with or without aura.

Epidemiological data indicates that migraine affects over 10% of the world population. The prevalence of migraine in Norway is 12%, similar to other western countries - Denmark 10%, France 12%the USA 12%, Canada 15%. In other cultures the migraine prevalence is lower: Japan 8%, Hong Kong 1% [8]. On the other hand it is shown that 20% of those suffering from migraine suffer from migraine with aura while 80% suffer from migraine without aura [4]. Some other researches show that only 30% of attacks are migraine with aura (without aura 70% / with aura 30%) [9].

Our results are in the expected range: 1) The prevalence of migraine in working people is 12.5% in our research, which represents a similar value to other western countries; 2)In the research of migraine types 20% of those suffering from migraine suffer from migraine with aura, while 80% suffer from migraine without aura.

5   Related Work

In general, in their original form, rule-based fuzzy logic systems were initially used in: 1) forecasting time-series, and 2) knowledge mining using surveys [10]. Studies involving medical patients: 1) A fuzzy logic system applied to umbilical Acid-Base assessment, which can provide vital information on the infant's health and guide requirements for neonatal care [11], 2) AIDA Cefalee is a diagnostic expert system able to suggests the correct ICDH-II diagnosis [12].

Headache research in (post) transitional countries: 1) The prevalence study comparing the lifestyles of female Belgrade students with migraine and non-migraine primary headache [13], The epidemiological survey of the prevalence of migraine, tension-type headache and chronic daily migraine in Georgia [14].

In [15], the discussion is carried out in this manner: "Because the questionnaire asked for the accompanying symptoms and features of all headaches, a classification of the headache subtype could be generated by the computer, based on the IHS criteria", but does not clearly show which method or technique is used.

6   Conclusion and Future Work

Migraine is common in working people and represents a significant health as well as a social and economic problem. The paper presents one of the approaches to diagnosing migraine types. Our research demonstrated that the rule-based fuzzy logic method is a promising support in diagnostic decision-making in medicine.

The rules in our research were facilitated by the application of the ICHS criteria for headache types. Physicians' clinical experience was used in order to extend the established rules and improve the system. After the general rules were improved, our results were in the expected range: 1) The prevalence of migraine in working people is 12.5%, which represents a similar value to other western countries; 2) The migraine types - 20% of those suffering from migraine suffer from migraine with aura while 80% suffer from migraine without aura according with similar research values.

Now, for the time being, the tool is developed for migraine – migraine without aura and migraine with aura, but our plan is to classify other non-migraine primary headaches: tension-type headache, cluster headache and other primary headache as well. This system for diagnosing primary headache types is the part of a project conducted at the Institute of Neurology of the Clinical Centre Vojvodina in Novi Sad [16], aimed at improving the primary headache diagnosis and headache severity assessment in patients suffering from headache.

References

1. Hu, X.H., Makson, L.E., Lipton, R.B., Stewart, W.F., Berger, M.L.: Burden of migraine in the United States: Disability and economic cost. Arch. Intern. Medicine 159, 813–818 (1999)
2. Andlin-Sobocki, P., Jonsson, B., Wittchen, H.U., Olsen, J.: Cost of disorder of the brain in Europe. Journal European Neurology 12(suppl. 1), 1–27 (2005)
3. El Hasnaoui, A., Vraly, M., Blin, P.: Assessment of migraine severity using the MIGSEV scale: relationship to migraine features and quality of life, vol. 24, pp. 262–270. Blackwell Publishing Ltd., Cephalalgia (2004)
4. Evans, R.W., Ninan, T.M.: Handbook of Headache. Lippincott Williams & Wilkins (2005)
5. Angelov, P.P.: Evolving Rule-based Models – A Tool for Design of Flexible Systems. Studies in Fuzziness and Soft Computing (Springer Berlin / Heidelberg), vol. 92. Physica-Verlag, Heidelberg (2002)
6. Lemke, F., Müller, J.A.: Carcinogenicity Prediction of Aromatic Compounds Based on Molecular Description (2000),
   http://www.knowledgeminer.net.pdf/carcino.pdf
7. Cady, R.K., Schreiber, C.P., Farmer, K.U.: Understanding the Patient With Migraine: The Evolution From Episodic Headache to Chronic Neurologic Disease. A Proposed Classification of Patients With Headache, Headache 44, 426–435 (2004)
8. Hagen, K., Zwart, J.-A., Vatten, L., Stovner, L., Bovin, G.: Prevalence of migraine and non-migranious headache – head HUNT, a large population-based study. Cephalalgia 20, 900–906 (2000)
9. Continuum (life long learning in neurology). Lippincott Williams & Wilkins (2005)
10. Mendel, J.M.: Uncertain Rule-Based Fuzzy Logic Systems: Introduction and New Direction. Prentice-Hall, Englewood Cliffs (2001)
11. Ozen, T., Gribaldi, J.M.: Investigating Adaptation in Type-2 Fuzzy Logic Systems Applied to Umbilical Acid-Based Assessment. In: Proceedings of 3rd International Conference of the European Network of Intelligent Technologies EUNITE (2003)
12. Simone, R., De., C.G., Ranieri, A., Bussone, G., Cortelli, P., D'Amico, D.: Validation of AIDA Cefalee, a computer-assisted diagnosis database for the management of headache patients. Journal Neurological Science 28 (suppl. 2), S213-S216 (2007)
13. Vlajinac, H., Špetić, S., Džoljić, E., Maksimović, J., Marinković, J., Kostić, V.: Some lifestyle habits of female Belgrade university students with migraine and non-migraine primary headache. The Journal of Headache and Pain 4(2), 67–71 (2003)
14. Katsarava, Z., Kukava, M., Mirvelashvili, E., Tavadze, A., Dzagnidze, A., Djibuti, M., Steiner, T.J.: A pilot methodological validation study for a population-based survey of the prevalences of migraine, tension-type headache and chronic daily headache in the country of Georgia. The Journal of Headache and Pain 8(2), 77–82 (2007)
15. Pop, P.H.M., Gierveld, C.M., Karis, H.A.M., Tiedink, H.G.M.: Epidemiological aspects of headache in a workplace setting and the impact on the economic loss. European Journal of Neurology 9, 171–174 (2002)
16. Simić, S.: Assessment of quality of life in patients with migraine and tension type headache, Master thesis, Faculty of Medicine, University Novi Sad (2006)

An Individualized Web-Based Algebra Tutor Based on Dynamic Deep Model Tracing

Dimitrios Sklavakis and Ioannis Refanidis

University of Macedonia, Department of Applied Informatics,
Egnatia 156, P.O. Box 1591, 540 06 Thessaloniki, Greece
{dsklavakis,yrefanid}@uom.gr

Abstract. This paper describes the motivations and goals of the MATHESIS project which concerns the development of an intelligent authoring environment for cognitive math tutors. It also describes the first implemented component of the project, the MATHESIS algebra tutor, a cognitive web-based tutor for algebraic expressions' expanding and factoring. The tutor uses cognitive model tracing by dynamically generating the plausible steps, checking them against student's solution steps and intervening when errors occur. Additionally, the tutor monitors the student's mastery of knowledge from problem to problem, i.e. the various cognitive skills. The tutor will be used as a prototype for the development of an ontology that will contain all of the tutor's knowledge. This ontology will eventually guide the creation of the authoring tools that will make faster and easier the creation of other cognitive tutors.

Keywords: intelligent tutoring systems, model tracing, cognitive tutors, web based, authoring systems, ontologies.

1 Introduction

One-to-one tutoring has proven to be the most effective way of teaching. Professor B. S. Bloom and his colleagues [1] found that the average student under tutoring was about two standard deviations above the average of the conventional class (30 students to one teacher). The successful implementation of the one-to-one tutoring model by Intelligent Tutoring Systems (ITS) today poses the problem of how to develop ITS that provide the same tutoring quality with a human tutor. Cognitive Tutors, a special kind of model-tracing tutors developed at Carnegie Mellon University based on the ACT-R [2] theory of cognition and learning, have shown significant success in domains like mathematics and computer programming. However, Cognitive Tutors are hard to author. The development of the problem solving as well as the teaching knowledge requires considerable amounts of time and the recruitment of Ph.D. level scientists in education, cognitive science and artificial intelligence programming.

This paper presents the MATHESIS project which aims at developing authoring tools for Cognitive Tutors in mathematics as well as an initial product of the project, the MATHESIS Algebra Tutor, a cognitive tutor for mathematics in the domain of expanding and factoring basic algebraic expressions.

J. Darzentas et al. (Eds.): SETN 2008, LNAI 5138, pp. 389–394, 2008.

The rest of the paper is structured as follows: Section 2 describes the motivation and goals of the MATHESIS project. Section 3 presents the MATHESIS algebra tutor. Finally, Section 4 presents related work whereas Section 5 concludes the paper and poses future directions of research.

2 The MATHESIS Project: Motivation and Goals

The motivation for the MATHESIS project is the principled design and successful implementation of Cognitive Tutors in U.S. secondary education schools [3]. In order to better understand the MATHESIS project goals and components, a brief description of the ACT-R theory and the design principles it entails is given below.

2.1 The MATHESIS Project Motivation: Cognitive Tutors

Central to the ACT-R theory is the concept of *cognitive skill* defined as a set of *production rules* that describe the problem-solving steps. These production rules are IF-THEN rules which match the problem's goal(s) and current state and produce new sub-goals. For example, a production rule for monomial multiplication could be: IF the goal is to *multiply-monomials* THEN *multiply-coefficients* AND *multiply-mainParts*. Production rules form the *procedural knowledge* of a Cognitive Tutor. They operate (match) on *facts* which describe the problem's states (initial, intermediate, goal). Facts are implemented as lists of property-value pairs and form the *declarative knowledge* of a Cognitive Tutor. For example a fact could be: (current-operation multiply-monomials).

The declarative and procedural knowledge form the *cognitive model* of a Cognitive Tutor which implements the problem-solving knowledge of the domain to be taught. The *tutoring model* of a Cognitive Tutor is based on *model tracing* and *knowledge tracing*.

The model tracing algorithm matches the student's problem-solving steps with the ones produced by the cognitive model. As far as the student's solution remains on a correct path the tutor remains silent. Otherwise it provides feedback as soon as an error occurs. The tutor can also provide help for the correct step(s) upon student request. Therefore model tracing keeps track of the cognitive skills' acquisition inside a problem.

The knowledge tracing algorithm keeps track of cognitive skills' acquisition from problem to problem. The model tracing algorithm provides a percentage of skill acquisition and the knowledge tracing algorithm adjusts the proposed problems according to that percentage. In this way, Cognitive Tutors allow for self-pacing of the student through the curriculum.

2.2 The MATHESIS Project Goals

Despite Cognitive Tutors' efficiency, it is currently estimated that 1 hour of tutoring takes 200-300 hours of development [4]. The main reason for this is the *knowledge acquisition bottleneck*: extracting the knowledge from the domain experts and encoding it into a program. Knowledge reuse appears as a necessity to overcome the knowledge acquisition bottleneck. Since expert knowledge and especially tutoring knowledge is so hard to create, re-using it is of paramount importance.

One widely used and quite promising technology for knowledge reuse is *ontological engineering* [5]. In the case of cognitive tutors, ontology engineering is the task of defining the cognitive model (facts, production rules) and tutoring model (user interface, model tracing and knowledge tracing) of the tutor and encode them in an ontology using specially designed environments for ontology management. This is the first research goal of the MATHESIS project. We believe that an efficient representation of a cognitive tutor's models in an ontology will provide a search space for the problem of cognitive tutor's authoring.

The second research goal is to develop the authoring tools that will help human authors search through this ontology space and therefore make their authoring faster and easier.

For the development and implementation of our research goals a bottom-up approach seems more appropriate. First, we need to implement a working prototype of a cognitive tutor. Then, the knowledge embedded in this tutor will be used to develop an ontology. Finally, based on the ontology we will develop the authoring tools whose purpose will be to guide the search through the ontology and help human authors.

3 The MATHESIS Algebra Tutor

The MATHESIS Algebra Tutor is a mathematics cognitive tutor for algebraic expressions' expanding and factoring. The domain of mathematics was chosen because it lends itself to bottom-up acquisition of cognitive skills and demands heavy reuse of them as well. In addition, adequate teaching expertise for developing the cognitive model of the tutor is available on behalf of one of the authors.

Three were the main issues that defined the overall architecture: a) the tutor interface should be web-based; we believe that the future of learning belongs to the world wide web and the tutor must be there, b) the model-tracing algorithm requires constant interaction between the cognitive model with the interface; therefore they should lie at the same side, that is the client side and c) the tutor should be able to be broken into pieces to produce the ontology and be reassembled back by the authoring tools.

The achievement of these requirements led us to implement the tutor using HTML for the interface and JavaScript for the cognitive and tutoring models. The primary

Fig. 1. The Algebraic Expression, Answering and Rough Space Input Controls

interface element is Design Science's WebEq [6] Input Control applet, an editor for displaying and editing mathematical expressions. It provides the same functionality as Equation Editor for Microsoft Word. There are three such input controls, the algebraic expression, answering space and rough space input controls (Figure 1).

3.1 The Tutor's Cognitive Model

The top-level cognitive skills that the tutor teaches are the following: monomial multiplication, division and power, monomial-polynomial and polynomial-polynomial multiplication, parentheses elimination, collect like terms, identities (square of sum and difference, product of sum-difference, cube of sum and difference), factoring (common factor, term grouping, identities, trinomial).

These cognitive skills are further decomposed in more simple ones. As an example we will consider the *multiply-monomials* skill. This is decomposed in two others, *multiply-coefficients* and *multiply-mainParts*. The *multiply-mainParts* is further decomposed in finding common variables and adding their exponents and finding non common variables and copying their exponents. This decomposition is implemented through JavaScript functions that correspond to the production rules and JavaScript data structures (simple variables, arrays, custom objects) that correspond to the facts. There are also relevant functions for common error checking like omitting variables, or not adding the exponents of common variables.

3.2 The Tutoring Model: Deep Cognitive Model Tracing

Equipped with such a detailed cognitive model, the MATHESIS tutor is able to exhibit expert human-like performance. The tutor makes all the cognitive tasks explicit to the student through the structure of the interface. First, the tutor parses the algebraic expression and creates all the relevant facts (kind of operations and priorities of them). The student must select a part of the algebraic expression and then select the operation he/she thinks corresponds to that part. Then the tutor, based on the parsed knowledge described above, checks the proposed operation against the selected expression. For example, the tutor checks if the student has selected only one operation or more, if the operation selected has the right priority to be performed, if the proposed operation matches with the part of the expression selected by the student. Only then the tutor proceeds to perform the operation.

In this stage, the tutor guides step by step the student with appropriate messages. Of course, in every step the tutor calculates the result, gets the student's answer and checks it for correctness. If any partial result is incorrect then the tutor displays the appropriate messages and asks again for that result in order to proceed. In the polynomial multiplication $(4 - 3x) * (5 - 2x)$, the tutor prompts the student for each one of the four monomial multiplications that must be performed, i.e. $4 * 5$, $4 * (-2x)$, $-3x * 5$ and $-3x * (-2x)$. For each one of the monomial multiplications the tutor behaves as if it was teaching the monomial multiplications as separate exercises, performing all the necessary cognitive tasks and checks. That's what we call *deep cognitive model tracing*. It must be pointed out that this deep cognitive model tracing is what makes one-to-one tutoring so effective and it is not an easily implemented feature even for a cognitive tutor. It is possible only with detailed cognitive task analysis which has knowledge reuse as a primary design parameter.

3.3 The Student Model: Knowledge Tracing

Based on such a detailed cognitive model and the deep model tracing feature, the MATHESIS algebra tutor keeps a detailed *student model*, that is which skills the student has mastered and to what extent from problem to problem. For each supported cognitive skill, e.g. monomial multiplication, the tutor keeps counters for the correct and incorrect answers of the student. With this simple mechanism, the tutor keeps track of the mastery level of each cognitive skill as a percentage calculated by the formula

$$mastery\ level = \frac{correct\ answers}{correct + incorrect\ answers} * 100\% \tag{1}$$

This percentage is time-stamped, i.e. the tutor keeps the date when a percentage changes, creating an accurate image of the mastery level change over time for every cognitive skill. It is important to stress out that the tutor updates the student model for all the cognitive skills that are present in a specific exercise. For example, when the student has to perform a *multiply-monomials* task, he/she must perform many *multiply-monomials* tasks. The tutor will update and time-stamp the mastery levels for this skill too. This behaviour is what we call *broad knowledge monitoring* and is a direct consequence of the *deep cognitive model tracing* feature of the tutor.

Although such a detailed, broad and dynamic student model gives the ability to the tutor to be highly adaptive as to what must be taught to every individual student, for the moment, the student model is just presented to the student and to the human tutor(s) that are responsible for assessing the students' knowledge. It is in our plans to design and implement a module that would use the student model to automate the selection of exercises to present to the student according to his/her mastery level of the various skills and the skills covered by each exercise.

4 Related Work

Despite their performance, Cognitive Tutors are proprietary, stand-alone applications that provide tutoring for a pre-programmed set of problems and of course they have the same high-cost demand in time and human resources [7].

To overcome these limitations, Carnegie Mellon University researchers have been developing the Cognitive Tutor Authoring Tools (CTAT) [8], a set of software tools that intend to make cognitive tutors' development easier and faster. The tools mainly support the authoring of *example-tracing tutors*. In these tutors, instead of writing production rules the author records the correct (or incorrect) answer for every step in the solution and the tools match these answers with the student's answers. Based on these tutors the tools provide debugging of the production rules that the author has to find and write by himself. In addition, the rules are stored in files from where they must manually be loaded and executed. Therefore the authoring knowledge remains isolated and en-coded. It is the ultimate goal of the MATHESIS project to re-code and open that knowledge through an ontology.

5 Discussion and Further Work

The MATHESIS algebra tutor has not been evaluated in a real school environment since it is still under development and testing. However, we got positive feedback when it was demonstrated to a few teachers of mathematics in Greek secondary education. Of course, being a research prototype, it needs more development and testing. Significant tutoring issues, like the granulation of the tutoring steps and pedagogical issues, like how to use the student model, are open.

What is important is the fact that the tutor's overall architecture and design will allow us to proceed to the second step of the MATHESIS project and develop an ontology that will contain all the knowledge now embedded to the HTML and JavaScript code. The ontology will make this authoring knowledge open and therefore reusable.

References

1. Bloom, B.S.: The 2 Sigma Problem: The Search of Methods for Group Instruction as Effective as One-to-One Tutoring. Educational Researcher 13(6), 4–16 (1984)
2. Anderson, J.R., Corbett, A.T., Koedinger, K.R., Pelletier, R.: Cognitive Tutors: Lessons Learned. The Journal of the Learning Sciences 4(2), 167–207 (1995)
3. Carnegie Learning, http://www.carnegielearning.com/products.cfm
4. Murray, T.: An overview of intelligent tutoring system authoring tools: Updated Analysis of the State of the Art. In: Murray, Ainsworth, Blessing (eds.) Authoring Tools for Advanced Technology Learning Environments, pp. 491–544. Kluwer Academic Publishers, Netherlands (2003)
5. Aitken, S.J., Sklavakis, D.: Integrating problem solving methods into CYC. In: Dean, T. (ed.) Proceedings of the International Joint Conference on Artificial Intelligence 1999, pp. 627–633. Morgan Kaufman Publishers, San Francisco (1999)
6. Design Science, http://www.dessci.com/en/products/webeq/
7. Aleven, V., McLaren, B., Sewall, J., Koedinger, K.R.: The Cognitive Tutor Authoring Tools (CTAT): Preliminary Evaluation of Efficiency Gains. In: Ikeda, M., Ashley, K.D., Chan, T.-W. (eds.) ITS 2006. LNCS, vol. 4053, pp. 61–70. Springer, Heidelberg (2006)
8. Cognitive Tutors Authoring Tools (CTAT), http://ctat.pact.cs.cmu.edu/

Design and Optimization of IIR Digital Filters with Non-standard Characteristics Using Continuous Ant Colony Optimization Algorithm

Adam Slowik and Michal Bialko

Department of Electronics and Computer Science, Koszalin University of Technology,
Sniadeckich 2 Street, 75-453 Koszalin, Poland
aslowik@ie.tu.koszalin.pl

Abstract. In this paper method of design and optimization of stable IIR digital filters with non-standard amplitude characteristics using continuous ant colony optimization algorithm ACO_R is presented. In proposed method (named ACO-IIRFD) dynamical changes of parameters in designed filters are introduced. Due to these dynamical changes of filter parameters, design of IIR digital filters with small deviations between designed filter characteristics and assumed characteristics is possible. Three IIR digital filters with amplitude characteristics: linearly-falling, linearly-growing, and non-linearly-growing, which can have application in amplitude equalizers, are designed using proposed method.

1 Introduction

Methods based on evolutionary algorithms [1] are used to design of digital filters since several years. As an example we can mention following papers [2-5]. However, only few papers describe applications of ant colony optimization algorithm to design digital filters; one of them is article [6]. Generally, a very few number of articles in this subject is connected with fact, that ant colony optimization algorithms [7] concern global optimization with discrete domains. However, in the case of many optimization tasks, the domains are continuous (as for example in design of digital filters). To use the ant colony optimization algorithms to the problem of continuous optimization some modifications of these algorithms have been created in last years. These modifications are as follows: (1995) CACO (*Continuous Ant Colony Optimization*) [8], (2000) API [9], (2002) CIAC (*Continuous Interacting Ant Colony*) [10], and (2008) ACO_R [11]; the last algorithm (in opposition to other continuous ant colony optimization algorithms) maintains a general idea of ant colony optimization algorithms depending on step-by-step building of solution, and shows to be a very effective tool in continuous functions optimization, including multi-modal functions [11].

Generally, the transfer function $H(z)$ of designed IIR (n-th order) digital filter in z domain is described as follows:

$$H(z) = \frac{b_0 + b_1 \cdot z^{-1} + b_2 \cdot z^{-2} + ... + b_{n-1} \cdot z^{-(n-1)} + b_n \cdot z^{-n}}{1 - \left(a_1 \cdot z^{-1} + a_2 \cdot z^{-2} + ... + a_{n-1} \cdot z^{-(n-1)} + a_n \cdot z^{-n}\right)} \quad (1)$$

J. Darzentas et al. (Eds.): SETN 2008, LNAI 5138, pp. 395–400, 2008.

Stability of IIR digital filter depends on poles location of transfer function (1) in z domain. If all poles are inside unitary circle (it means $|z| = 1$), then filter is stable. The main goal in digital filter design is searching a such set of values of a_i, and b_i coefficients for given characteristic, that obtained filter fulfills design assumptions and is stable.

2   ACO-IIRFD Method

Method ACO-IIRFD (based on ACO_R algorithm [11]) is operating according to following steps: **In the first step**, the initial set T consisting of $SizeT$ potential solutions of given optimization problem is randomly created. Each solution t_i from the set T consists of $2 \cdot n + 1$ filter coefficients which are coded in the following order:

$$t_i = [b_0 \quad b_1 \quad b_2 \quad ... \quad b_{n-1} \quad b_n \quad a_1 \quad a_2 \quad ... \quad a_{n-1} \quad a_n] \tag{2}$$

that is: $t_{i,1}=b_0$, $t_{i,2}=b_1$, $t_{i,3}=b_2$, ..., $t_{i,2n}=a_{n-1}$, $t_{i,2n+1}=a_n$. The value of each filter coefficient variable $t_{i,j}$ ($j \in [1; 2 \cdot n+1]$) is in the range $t_{i,j} \in [Low_j; High_j]$; in this paper, it is assumed, that the initial values are: $High_j = 1$, $Low_j = $ -1. For each j-th in the i-th solution, its value is selected using following formula:

$$t_{i,j} = R \cdot (High_j - Low_j) + Low_j \tag{3}$$

where: R-pseudo-random number with uniform distribution in the range $[0; 1)$. **In the second step**, each potential solution from the set T is evaluated using objective function $COST(.)$. The amplitude characteristics $H(f)$ is obtained:

$$H(f) = 20 \cdot log_{10} \left(\sqrt{H_{real}(f)^2 + H_{imag}(f)^2} \right) [dB] \tag{4}$$

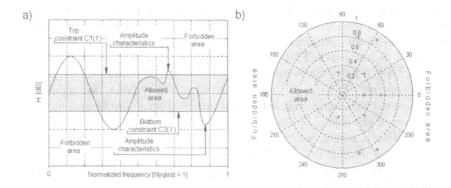

Fig. 1. Amplitude characteristics of designed digital filter with assumed constraints (a), poles of stable transfer function (b)

Accordingly to Figure 1, the objective function $COST(.)$ is defined as follows:

$$COST(.) = \sum_{i=1}^{k} Error\,(f_i) + \sum_{i=1}^{m} Stab_i \tag{5}$$

$$Error(f_i) = \begin{cases} |H(f_i) - C1(f_i)|, & \text{when } H(f_i) > C1(f_i) \\ |H(f_i) - C2(f_i)|, & \text{when } H(f_i) < C2(f_i) \\ 0, & \text{when } H(f_i) \in [C2(f_i); C1(f_i)] \end{cases} \tag{6}$$

$$Stab_i = \begin{cases} |z_i| \cdot w, & \text{when } (|z_i| \geq 1) \\ 0, & \text{when } (|z_i| < 1) \end{cases} \tag{7}$$

where: m - number of poles of transfer function ($m=n$), k - number of frequency samples ($k=256$), w - value of penalty (assumed $w=10^5$), f_i - value of i-th normalized frequency, $|z_i|$ - module value of i-th pole of transfer function in z domain. The result of evaluation of i-th solution using function $COST(.)$ is stored in variable $eval_i$ assigned to each solution t_i from the set T. The ACO-IIRFD algorithm, presented in this paper minimizes the value of objective function $COST(.)$. **In the third step**, the solutions t_i from the set T are sorted according to $eval_i$ values. After sorting process, the best solution (with the smallest value of $eval$) is located under index number 1 in the set T, and the worst solution (with the highest value of $eval$) is stored under index number $SizeT$ in the set T. **In the fourth step**, the value of ω_i is computed for each solution t_i. This value determines, the selection chance of the i-th solution from the set T to creation of new solution (new set of a_i, b_i coefficients), and is higher when i-th solution is located under lower index in the set T. The value of ω_i is defined as follows:

$$\omega_i = \frac{1}{q \cdot SizeT \cdot \sqrt{2 \cdot \pi}} \cdot exp\left(-\frac{(i-1)^2}{2 \cdot q^2 \cdot SizeT^2}\right) \tag{8}$$

The variable ω_i is defined by a function with normal distribution for argument i, mean equal to 1.0, and standard deviation $q \cdot SizeT$, where $q \in [0; 1]$ is the parameter of algorithm. In the case, when the value of q is small, then better solutions, that is, solutions having smaller values of indices after sorting process are stronger preferred (solutions located close to the best solution are selected more often), and in the case, when the value of q is large, then selection of solutions is taken from wider range with respect to the best solution. **In the fifth step**, each "ant" marked as an Ant_g ($g \in [1; NoA]$, NoA represents number of ants, and it is a parameter of the algorithm) generates one new solution marked as the S_g. This operation is executed according to following schema, which is repeated separately for each "artificial ant":

a) choose one solution from the set T using roulette selection method [1] scaled by values ω_i, and remember its index in variable h;
b) compute value of standard deviation marked as the σ_j for each j-th variable of new created solution S_g using following formula:

$$\sigma_j = \xi \cdot \sum_{e=1}^{SizeT} \frac{|t_{e,j} - t_{e,h}|}{SizeT - 1} \tag{9}$$

where: ξ-real number higher than zero, and having similar sense as a coefficient of pheromone evaporation in discrete ant colony optimization algorithms. When the value of coefficient ξ is higher - the convergence of the algorithm is slower; c) determine mean value marked as a μ_j for each j-th variable of new created solution S_g (the value of j-th variable in the solution S_g will be selected using normal distribution with mean value μ_j and standard deviation σ_j), using following formula: $\mu_j = t_{h,j}$; d) the value of the j-th variable for new created solution S_g is defined as follows: $S_{g,j} = randn\,(\mu_j, \sigma_j)$, where: $randn(\mu_j, \sigma_j)$-random value with normal distribution having mean value μ_j, and standard deviation σ_j. In the sixth step, each new solution S_g is evaluated using function $COST(.)$, and the obtained value is assigned to variable $eval_g$. In the seventh step, the new obtained solutions are included into the solution set T, which size is increased to the value $SizeT+NoA$. Next, all solutions are sorted with respect to the value of parameter $eval$ (the best solution with the smallest value $eval$ obtains index number 1). Then, NoA last solutions are deleted from set T, and the original size of this set equal to $SizeT$ is restored. In the eighth step, it checked whether $COST(t_1)=0$. If yes, then solution t_i is remembered as the *TheBest* solution, and the objective function $COST(.)$ is modified by decreasing the value of accepted deviations of amplitude characteristics, and then the new value $eval_i$ is computed for each solution t_i. If $COST(t_1)\neq0$, then ninth step of the proposed algorithm is executed. In the ninth step, it is checked whether the result stored in solution t_1 has not been improved by d iterations of the algorithm. If the result stored in solution t_1 has not been improved, then the result stored in *TheBest* is returned, and the operation of algorithm is stopped. If the result stored in solution t_1 has been improved before reaching of d iterations of algorithm, then the third step of the ACO-IIRFD algorithm is executed.

3   Description of Experiments

Proposed method has been tested by the design of the 10 order ($n=10$) three IIR digital filters with amplitude characteristics: linearly growing (for normalized frequency $f=0$ the value of gain was equal to -40 [dB], for normalized frequency $f=1$ the value of gain was equal to 0 [dB]), linearly falling (for normalized frequency $f=0$ the value of gain was equal to 0 [dB], for normalized frequency $f=1$ the value of gain was equal to -40 [dB]), and non-linearly growing (the attenuation is represented by quadratic characteristics; for normalized frequency $f=0$ the value of gain was equal to -40 [dB], and for normalized frequency $f=1$ the value of gain was equal to 0 [dB]. The values of parameters in the ACO-IIRFD method were as follows: filter order $n=10$, size of solutions set $SizeT=800$, number of ants $NoA=100$, $q=0.01$, $\xi=0.085$, $d=200$. At the start was assumed that the deviation values of the attenuation characteristics from ideal case can not be higher than 10 [dB] at any frequency point. Then, during the algorithm operation the values of assumed attenuation deviations were decreased by 1 [dB] step, and after achieving the value of deviation equal to 1 [dB], the values of attenuation deviations were decreased by 0.1 [dB] step. The main purpose of the experiment was a design of 10-order digital filters having possibly lowest deviations of their amplitude characteristics from ideal characteristics.

In Figures 2a - 2c, the differences (deviations) between ideal characteristics H, and characteristics H' obtained using ACO-IIRFD method for designed filters are presented and in Figures 2d - 2f, the number of generations required to obtain designed digital filters with prescribed deviations of attenuation values between obtained characteristics, and ideal characteristics is shown.

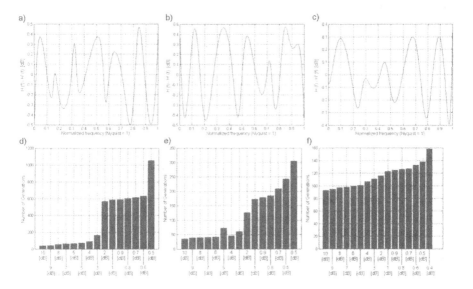

Fig. 2. Deviations between ideal characteristics, and characteristics obtained using ACO-IIRFD method for 10-order digital filters having characteristics: linearly growing (a), linearly falling (b), and non-linearly growing (c); Number of generations required to obtain designed digital filters with prescribed deviations of attenuation values of amplitude characteristics: linearly growing (d), linearly falling (e), and non-linearly growing (f)

It can be seen that in the case of filters with linearly growing (Figure 2a), and linearly falling (Figure 2b) characteristics, the deviations of attenuation values between obtained characteristics and ideal characteristics do not exceed +/- 0.5 [dB], however, in the case of filter with non-linearly growing characteristics (Figure 2c), these deviations of attenuation values do not exceed +/- 0.4 [dB]. All designed filters are stable (all poles are located inside unitary circle in the z plane). In performed experiments the ACO-IIRFD algorithm has been stopped after 1267 generations for linearly growing characteristics, after 728 generations for linearly falling characteristics, and after 452 generations for non-linearly growing characteristics. To obtain smaller deviations of attenuation values higher order filter is required.

4 Conclusion

It has been shown that it is possible to design and optimize digital filters with non-standard amplitude characteristics using the continuous ant colony optimization algorithm. Three digital filters designed using described ACO-IIRFD method fulfill all design assumptions and are stable. The full automation of the design and optimization process of digital filters with non-standard amplitude characteristics is possible with the use of proposed ACO-IIRFD method, and the expert knowledge concerning the filter design, and digital signal processing is not required. The ACO_R algorithm is newly developed technique (in year 2008) for continuous optimization based on ants colony. Because of that, this paper is probably the first application of this algorithm to design and optimization of stable IIR digital filters with non-standard amplitude characteristics.

References

1. Michalewicz, Z.: Genetic Algorithms + Data Structures = Evolution Programs. Springer, Heidelberg (1992)
2. Erba, M., Rossi, R., Liberali, V., Tettamanzi, A.G.B.: Digital Filter Design Through Simulated Evolution. In: Proceedings of ECCTD 2001, Espoo, Finland, August 2001, vol. 2, pp. 137–140 (2001)
3. Slowik, A., Bialko, M.: Evolutionary Design of IIR Digital Filters with Non-Standard Amplitude Characteristics. In: 3rd National Conference on Electronics, Kolobrzeg, June 2004, pp. 345–350 (2004)
4. Nurhan, K.: Digital IIR filter design using differential evolution algorithm. EURASIP Journal on Applied Signal Processing 8, 1269–1276 (2005)
5. Nurhan, K., Bahadir, C., Tatyana, Y.: Performance comparison of genetic and differential Evolution algorithms for digital FIR filter design. In: Yakhno, T. (ed.) ADVIS 2004. LNCS, vol. 3261, pp. 482–488. Springer, Heidelberg (2004)
6. Karaboga, N., Kalinli, A., Karaboga, D.: Designing digital IIR filters using ant colony optimisation algorithm. Engineering Applications of Artificial Intelligence 17(3), 301–309 (2004)
7. Dorigo, M., Maniezzo, V., Colorni, A.: Ant System: Optimization by a colony of cooperating agents. IEEE Transactions on SMC-B 26(1), 29–41 (1996)
8. Bilchev, G., Parmee, I.C.: The ant colony metaphor for searching continous design spaces. In: Fogarty, T.C. (ed.) AISB-WS 1995. LNCS, vol. 993, pp. 25–39. Springer, Heidelberg (1995)
9. Monmarche, N., Venturini, G., Slimane, M.: On how Pachycondyla apicalis ants suggest a new search algorithm. Future Generation Computer Systems 16, 937–946 (2000)
10. Dreo, J., Siarry, P.: A new ant colony algorithm using the heterarchical concept aimed at optimization of multiminima continous functions. In: Doringo, M., Di Caro, G., Samples, M. (eds.) Ant Algorithms 2002. LNCS, vol. 2463, pp. 216–221. Springer, Heidelberg (2002)
11. Socha, K., Doringo, M.: Ant colony optimization for continous domains. European Journal of Operational Research 185(3), 1155–1173 (2008)

An Empirical Study of Lazy Multilabel Classification Algorithms

E. Spyromitros, G. Tsoumakas, and I. Vlahavas

Department of Informatics,
Aristotle University of Thessaloniki,
54124 Thessaloniki, Greece
{espyromi,greg,vlahavas}@csd.auth.gr

Abstract. Multilabel classification is a rapidly developing field of machine learning. Despite its short life, various methods for solving the task of multilabel classification have been proposed. In this paper we focus on a subset of these methods that adopt a lazy learning approach and are based on the traditional k-nearest neighbor (kNN) algorithm. Two are our main contributions. Firstly, we implement BRkNN, an adaptation of the kNN algorithm for multilabel classification that is conceptually equivalent to using the popular Binary Relevance problem transformation method in conjunction with the kNN algorithm, but much faster. We also identify two useful extensions of BRkNN that improve its overall predictive performance. Secondly, we compare this method against two other lazy multilabel classification methods, in order to determine the overall best performer. Experiments on different real-world multilabel datasets, using a variety of evaluation metrics, expose the advantages and limitations of each method with respect to specific dataset characteristics.

1 Introduction

Traditional *single-label* classification is concerned with learning from a set of examples that are associated with a single label λ from a set of disjoint labels L, $|L| > 1$. If $|L| = 2$, then the learning task is called *binary* classification, while if $|L| > 2$, then it is called *multi-class* classification. In *multilabel* classification, each example is associated with a set of labels $Y \subseteq L$.

Multilabel classification methods can be categorized into two different groups [1]: i) *problem transformation* methods, and ii) *algorithm adaptation* methods. The first group of methods are algorithm independent. They transform the multilabel classification task into one or more single-label classification, regression or label ranking tasks. The second group of methods extend specific learning algorithms in order to handle multilabel data directly.

In this paper we focus on lazy multilabel classification methods of both categories that are based on the k Nearest Neighbor (kNN) algorithm. Among the strong points of these methods is that their time complexity scales linearly with respect to $|L|$. Furthermore, their main computationally intensive operation is the calculation of nearest neighbors, which is actually independent of $|L|$.

J. Darzentas et al. (Eds.): SETN 2008, LNAI 5138, pp. 401–406, 2008.

Two are our main contributions in this work. Firstly, we implement BRkNN, an adaptation of the kNN algorithm for multilabel classification that is conceptually equivalent to using the popular Binary Relevance problem transformation method in conjunction with the kNN algorithm, but $|L|$ times faster. We also identify two useful extensions of BRkNN that improve its overall predictive performance. Secondly, we compare this method against two other lazy multilabel classification methods, in order to determine the overall best performer.

The rest of this paper is structured as follows. Section 2 presents the BRkNN method and its extensions. Section 3 presents the setup of the experimental work and Section 4 discusses the results. Finally, Section 5 concludes this work.

2   BRkNN and Extensions

Binary Relevance (BR) is the most widely-used problem transformation method for multilabel classification. It learns one binary classifier $h_\lambda : X \rightarrow \{\neg\lambda, \lambda\}$ for each different label $\lambda \in L$. BR transforms the original data set into $|L|$ data sets D_λ that contain all examples of the original data set, labeled as λ if the labels of the original example contained λ and as $\neg\lambda$ otherwise. It is the same solution used in order to deal with a multi-class problem using a binary classifier, commonly referred to as one-against-all or one-versus-rest.

BRkNN is an adaptation of the kNN algorithm that is conceptually equivalent to using BR in conjunction with the kNN algorithm. Therefore, instead of implementing BRkNN, we could have utilized existing implementations of BR [2] and kNN [3]. However, the problem in pairing BR with kNN is that it will perform $|L|$ times the same process of calculating the k nearest neighbors. To avoid these redundant time-intensive computations, BRkNN extends the kNN algorithm so that independent predictions are made for each label, following a single search of the k nearest neighbors. This way BRkNN is $|L|$ times faster than BR plus kNN during testing, a fact that could be crucial in domains with a large set of labels and requirements for low response times. BRkNN was implemented within the MULAN multilabel classification software [2].

We propose two extensions to the basic BRkNN algorithm. Both are based on the calculation of *confidence* scores for each label $\lambda \in L$ from BRkNN. The confidence for a label can be easily obtained by considering the percentage of the k nearest neighbors that include it. Formally, let $Y_j, j = 1 \ldots k$, be the label sets of the k nearest neighbors of a new instance x. The confidence c_λ of a label $\lambda \in L$ is equal to:

$$c_\lambda = \frac{1}{k} \sum_{j=1}^{k} I_{Y_j}(\lambda)$$

where $I_{Y_j} : L \rightarrow \{0, 1\}$ is a function that outputs 1 if its input label λ belongs to set Y_j and 0 otherwise, called *indicator function* in set theory.

The first extension of BRkNN, called BRkNN-a, checks whether BRkNN outputs the empty set, due to none of the labels $\lambda \in L$ being included in at least half of the k nearest neighbors. If this condition holds, then it outputs the label with the highest confidence. It so deals with a general disadvantage of BR, that has not been raised in the past: as each label is independently predicted in BR, there exists a possibility that the empty set is given as the overall output. We hypothesize that better results will be obtained through the proposed extension that outputs the most probable label when this phenomenon arises.

The second extension of BRkNN, called BRkNN-b calculates the average size s of the label sets of the k nearest neighbors at a first step, $s = \frac{1}{k} \sum_{j=1}^{k} |Y_j|$, and then outputs the $[s]$ (nearest integer of s) labels with the highest confidence.

3 Experimental Setup

3.1 Datasets

We experiment with 3 datasets from 3 different application domains: The biological dataset *yeast* [4] is concerned with protein function classification. The image dataset *scene* [5] is concerned with semantic indexing of still scenes. The music dataset *emotions* [6] is concerned with the classification of songs according to the emotions they evoke.

Table 1 shows certain standard statistics of these datasets, such as the number of examples in the train and test sets, the number of numeric and discrete attributes and the number of labels, along with multilabel data statistics, such as the number of distinct label subsets, the label cardinality and the label density [1]. Label cardinality is the average number of labels per example, while label density is the same number divided by $|L|$.

Table 1. Standard and multilabel statistics for the data sets used in the experiments

Dataset	Examples	Attributes		Labels	Distinct Subsets	Label Cardinality	Label Density
		Numeric	Discrete				
scene	2712	294	0	6	15	1.074	0.179
emotions	593	72	0	6	27	1.868	0.311
yeast	2417	103	0	14	198	4.327	0.302

3.2 Evaluation Methodology

We perform two sets of experiments. In the first one, we compare BRkNN to its extensions. In the second one, we compare the best version of BRkNN in each dataset to two other lazy multilabel classification methods, LPkNN and MLkNN, in order to make a final recommendation.

LPkNN is simply the pairing of the Label Powerset (LP) problem transformation method [2] with the kNN algorithm. LP considers each different subset of L that appears in the training set as a different label of a single-label classification task. LPkNN has not been discussed in the related literature to the best of our knowledge.

MLkNN [7] is another adaptation of the kNN algorithm for multilabel data. What mainly differentiates this method from BRkNN is the use of prior and posterior probabilities which are directly estimated from the training set based on frequency counting. We implemented MLkNN in Java within the MULAN multilabel classification software [2] for the purposes of this study.

Each method was executed with a varying number of nearest neighbors. Specifically, the parameter k ranged from 1 to 30. The performance of each method for each k was evaluated using 10-fold cross-validation, in order to obtain an accurate performance estimate. In each fold, the following metrics were calculated [2], and eventually averaged over all folds:

- *Example-based.* Hamming loss, accuracy, F-measure and subset accuracy
- *Label-based.* Micro and macro version of F-measure

4   Experimental Results

4.1   Do the Proposed Extensions Improve BRkNN?

In this subsection we investigate whether BRkNN-a and BRkNN-b improve the performance of BRkNN. Table 2 reports the average performance of the three algorithms across all 30 values of the k parameter for each dataset. It presents results for all evaluation metrics mentioned in Section 3.2. The best result on each metric and dataset is shown with bold typeface. The last line contains for each algorithm the number of metrics for which it achieves the best result, while within parentheses there is the number of metrics for which BRkNN-a and BRkNN-b are better than the base BRkNN algorithm.

The results show that both extensions outperform the base BRkNN method in more than half of the 6 metrics on all datasets. BRkNN-a outperforms BRkNN in 6, 5 and 5 out of the 6 metrics in the scene, emotions and yeast datasets respectively. BRkNN-b outperforms BRkNN in 6, 4 and 4 out of the 6 metrics in the scene, emotions and yeast datasets respectively. These two pieces of evidence strongly support that both BRkNN-a and BRkNN-b are beneficial extensions.

Studying the performance of the algorithms at each individual dataset, we notice that BRkNN-a dominates in scene and emotions, while BRkNN-b dominates

Table 2. Experimental results of BRkNN, BRkNN-a and BRkNN-b on all datasets, averaged for all k

metric	scene			emotions			yeast		
	base	ext-a	ext-b	base	ext-a	ext-b	base	ext-a	ext-b
Hamming loss	0.0950	**0.0938**	0.0941	**0.1976**	0.1982	0.2175	**0.1974**	0.1975	0.2082
accuracy	0.6256	**0.7226**	0.7218	0.5215	**0.5441**	0.5430	0.5062	0.5080	**0.5346**
F-measure	0.6386	**0.7392**	0.7381	0.6275	0.6576	**0.6590**	0.5777	0.5795	**0.6652**
subset accuracy	0.5993	**0.6889**	0.6886	0.2895	**0.2971**	0.2759	0.1958	**0.1959**	0.1766
micro F-measure	0.6964	**0.7296**	0.7284	0.6499	**0.6577**	0.6509	0.6374	0.6380	**0.6567**
macro F-measure	0.6955	**0.7363**	0.7349	0.6224	**0.6303**	0.6294	0.3926	0.3931	**0.4261**
#wins (#better)	0	6 (6)	0 (6)	1	4 (5)	1 (4)	1	1 (5)	4 (4)

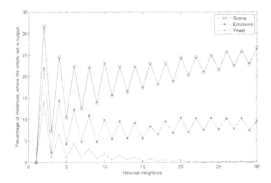

Fig. 1. Percentage of new instances, where BRkNN outputs the empty set (y axis), with respect to the number of nearest neighbors (k) (x axis) for all datasets

in yeast. This performance pattern correlates with the cardinality of the datasets, which is 1.074, 1.868 and 4.327 for the scene, emotions and yeast dataset respectively (see Table 3.1). Actually, it is natural for datasets of low cardinality, such as scene and emotions, to favor BRkNN-a over BRkNN, because the probability that the latter outputs the empty set increases in such datasets. This is clearly shown in Figure 1, which plots the percentage of the instances, where BRkNN outputs the empty set, for various values of the k parameter. BRkNN-a deals with exactly this problem of BRkNN. On the other hand, BRkNN-b works better in datasets with larger cardinality, as it includes a mechanism to predict the number of true labels associated with a new instance.

4.2 Comparison of BRkNN, LPkNN and MLkNN

Table 3 reports the average performance of the three algorithms across all 30 values of the k parameter for each dataset. It presents results for all evaluation metrics mentioned in Section 3.2. The best result on each metric and dataset is shown with bold typeface. The last line contains for each algorithm the number of metrics for which it achieves the best result.

Table 3. Experimental results of best version of BRkNN, LPkNN and MLkNN with normalization on all datasets, averaged for all k

	scene			emotions			yeast		
metric	BR-a	LP	ML	BR-a	LP	ML	BR-b	LP	ML
Hamming loss	0.0938	0.0955	**0.0884**	0.1982	0.2094	0.2003	0.2082	0.2143	**0.1950**
accuracy	**0.7226**	0.7181	0.6720	0.5441	**0.5600**	0.5233	**0.5346**	0.5280	0.5105
F-measure	**0.7392**	0.7343	0.6944	0.6576	**0.6662**	0.6352	**0.6652**	0.6375	0.5823
subset accuracy	**0.6889**	0.6854	0.6272	0.2971	**0.3287**	0.2780	0.1766	**0.2452**	0.1780
micro F-measure	0.7296	0.7249	**0.7316**	0.6577	**0.6649**	0.6509	**0.6567**	0.6415	0.6422
macro F-measure	**0.7363**	0.7323	0.7341	0.6303	**0.6505**	0.6110	0.4261	**0.4322**	0.3701
#wins	4	0	2	1	5	0	3	2	1

We notice that BRkNN-a and LPkNN dominate in the scene and emotions datasets respectively, while in the yeast dataset there is no clear winner. However BRkNN-b performs better in most measures, followed by LPkNN and finally MLkNN. There is no apparent explanation on why LPkNN performs better in the emotions dataset. We notice in Table 3.1 that this dataset has the highest label density, while the scene dataset where LPkNN has the worst performance has the lowest label density. However we cannot safely argue that high density datasets lead to improved performance of the LPkNN algorithm.

5   Conclusions

This paper has studied how the k Nearest Neighbor (kNN) algorithm is used for the classification of multilabel data. It presented BRkNN, an efficient implementation of the pairing of BR with kNN, along with two interesting extensions. Experimental results indicated that the proposed extensions are in the right direction. In addition, the paper compared experimentally BRkNN with two other methods (LPkNN and MLkNN) and reached to some interesting conclusions as to what kind of evaluation metrics and what kind of datasets are well-suited to the different methods.

References

1. Tsoumakas, G., Katakis, I.: Multi-label classification: An overview. International Journal of Data Warehousing and Mining 3, 1–13 (2007)
2. Tsoumakas, G., Vlahavas, I.: Random k-labelsets: An ensemble method for multi-label classification. In: Kok, J.N., Koronacki, J., Lopez de Mantaras, R., Matwin, S., Mladenič, D., Skowron, A. (eds.) ECML 2007. LNCS (LNAI), vol. 4701, pp. 406–417. Springer, Heidelberg (2007)
3. Witten, I., Frank, E.: Data Mining: Practical Machine Learning Tools and Techniques, 2nd edn. Morgan Kaufmann, San Francisco (2005)
4. Elisseeff, A., Weston, J.: A kernel method for multi-labelled classification. In: Advances in Neural Information Processing Systems 14 (2002)
5. Boutell, M., Luo, J., Shen, X., Brown, C.: Learning multi-label scene classification. Pattern Recognition 37, 1757–1771 (2004)
6. Trohidis, K., Tsoumakas, G., Kalliris, G., Vlahavas, I.: Multilabel classification of music into emotions. In: Proc. 9th International Conference on Music Information Retrieval (ISMIR 2008), Philadelphia, PA, USA (2008)
7. Zhang, M.L., Zhou, Z.H.: A k-nearest neighbor based algorithm for multi-label classification. In: Proceedings of the 1st IEEE International Conference on Granular Computing, pp. 718–721 (2005)

An Algorithm of Decentralized Artificial Immune Network and Its Implementation

Mariusz Święcicki

Cracow University of Technology, Institute of Computer Modelling, Cracow, Poland
mswiecic@gmail.com

Abstract. In this article we have presented a mechanism of antigen transformation as it occurs in the natural system. Using this mechanism as a base the article has defined paradigms which were used to determine the architecture of the decentralized artificial immune network. The achievements results have been shown that the proposed learning mechanisms and the decentralized architecture consisting of computational nodes is calibrated and it can be successfully used in the distributed computational environments such as Grid and Clusters systems.

Keywords: immune algorithms, decentralized immune network, data mining.

1 Introduction

In our decade we could notice a great applying artificial intelligence systems, which are based on artificial immune algorithms. The qualities of these algorithms appeared to be so interesting that they were quickly applied to numerous problems connected with discrete optimization, pattern recognition and data mining technique.

However, in case of application of immune algorithms in data mining technique certain complication may occur. These problems are caused by the fact that in the aforementioned fields we are using large sets of data about learning process. So far, all system learning algorithms have been based on interactions which are taking place among lymphocytes in data processing, which are sequential and centralized.

This article presents artificial immune network algorithm which is a decentralized algorithm both in learning process, and during the process of answer generating. The level of algorithm decentralization can be subject to change, and then the whole system will be a calibrating system.

2 Assumptions of Decentralized Artificial Immune Network

Sequentiality of artificial immune network algorithm implementations causes that even with contemporary computational possibilities of modern computers; they cannot always be implemented to solve the tasks which belong to great scale problems. An example of such issues can be a classification of multidimensional data. In these matters large sets of data during learning are used. Therefore, sequentiality and

J. Darzentas et al. (Eds.): SETN 2008, LNAI 5138, pp. 407–412, 2008.

centralization of artificial immune network learning algorithm both create a significant computation problem. In this paragraph, we go to the definition of introductory
assumptions, which decentralized immune network algorithm is based on. While
defining this decentralized immune network algorithm three vital aspects have been
taken into consideration.

2.1 Paradigm of Decentralized Artificial Immune Network

Firstly, as the presented description implies, the process of introductory immune response has a decentralized character when it comes to identification of antigen or
antigens. Because in the beginning phase, B lymphocytes do not recognize the whole
antigen but only its individual proteins, which have been appropriately transformed
by the presenting cell, T-type auxiliary cells are engaged in this process [1, 2, 3].
They inform about the identification of alien proteins by means of release of chemical
substances which cause stimulation of B lymphocytes.

Secondly, the process of identification of antigen takes place in several independent places. It means that the same antigen can transformed by the antigen presenting
cells (APC) and identified by auxiliary T-type cells and B cells concurrently in different places inside the organism.

Thirdly, the immune system can form - prepare immune response for a number of
different antigens at same moment.

Bearing in mind the aforementioned assumption, we will obtain an decentralized
algorithm of an artificial immune network.

2.2 Level of Decentralization of Artificial Immune Network

The parameter will define the level of decentralization of artificial immune network
algorithm. This parameter will refer to the learning phase and the response phase.
Implementation of this parameter results from the second and the third clause of the
paradigm. In the natural immune system each antigen is transformed together with
other antigens. This process takes place in a lymph gland. Following this example in
case of artificial immune system, a lymph gland of the natural system can be
compared to computational node, i.e. from a computer.

$$\overline{\overline{Decent}} \ \% \ = \ \left(1 - \frac{\frac{1}{N}\sum_{i=1}^{N}\overline{\overline{ag}}_{i}}{\overline{\overline{AG}}} \right) \cdot 100 \tag{1}$$

Where:

ag_i – i^{th} subset of the learning set

AG – a set of antigens, the learning set and $AG = ag_1 \cup \ldots \cup ag_N$

N –number of divisions AG learning set

$\overline{\overline{AG}}$ - number of elements of the learning set (set of antigens)

$\overline{\overline{ag}}_{i}$ - number of elements of ith learning subset

The function of the computational node is similar to the function of a lymph gland. A
lymph gland provides the environment for the process of hypermutation, i.e. the process of learning. Inside a lymph gland antigen identification occurs and the response of

the natural immune system is prepared. In the computational node of the decentralized immune network the learning process is conducted for a given set of antigens. It means that it is responsible for the creation of an appropriate set of antibodies for a given subset of antigens. In the computational node the response of the network is generated for a given set of antigens based on the set of memory cells stored in the computational node. Formula (1) presents a parameter has been introduced to the description of the level of decentralization.

3 Algorithm of Decentralized Immune Network

Algorithm of decentralized immune network consists of two parts. The first part of the algorithm is connected with the learning process of decentralized immune network. The second part is connected with the computation of response by means of decentralized immune network.

3.1 Learning Algorithm

Learning algorithm of immune network consists of two phases. In the first phase of this algorithm a division of a learning set – antigen set – occurs. The second phase of the algorithm covers a learning procedure with the application of a training set. The process of network training takes place in the computational node which, similarly to a lymph gland, provides the environment in the learning process. The computational node provides computational capacity in order to conduct learning process. The architecture of the decentralized immune network in a general case is made of N computational nodes. The computational nodes are connected with each other by means of local area network or wide area network. The number of computational nodes decides directly on the level of decentralization of the artificial immune network. Each of the nodes, which are included in the architecture of the artificial immune network fulfills the same functions.

3.1.1 Decentralization of the Antigen Set
This type of division is connected with the division of antigen set into subdivisions and it results from the second and the third paradigm introduced in the previous paragraph of this article.

---

*N – on the basis of the parameter the decentralization degree **Decent** and set number **AG** a number of computational nodes is calculated*
For i=1 to N
  *1 Create from **AG** an ag$_i$ subset of the learning set using **CRITERION** function as a criterion, and the number of N computational nodes*
  2 Send ag$_i$ to ith computational node
  3 Initiate a network training process in ith computational node

---

Fig. 1. Algorithm – Decentralization of antigen set

This type of division is used in case of numerous antigen sets and it aims at speeding up the learning process by its decentralization. In this way the architecture we are discussing here, provides calibration in case of the learning process of the immune network.

3.1.2 Course of the Learning Process in the Computational Nodes

The actual process of learning of the decentralized immune network takes place in the computational nodes which are equivalent of lymph glands. After sending a subset of antigens the learning process is initiated. In case of this model of the decentralized network, a learning algorithm has been used, which is well described in literature [1, 2, 3]. Therefore, we are not going to describe in details the mechanism of this algorithm operation, except form enumerating the most important stages of this algorithm.

For each $Ag_j \notin AG$

    1   Determining affinity between jth antigen and the set of antibodies **AB**
    2   creating A set consisting of N antibodies of the highest level of affinity
    3   N antibodies undergo the process of proportionally to their affinity towards antigens;
    4   created set of C clones undergoes mutation; probability of mutation is reversely proportional to affinity of parent antibodies
    5   From C set , choosing ζ% antibodies which have the highest affinity ratio and creating M memory cells from them
    6   Determining affinity between **Ag**, *and the created M memory cells*
    *7 * **Cells death**: *elimination of all memory cells from M set for which affinity value fulfills the* $D > \sigma_d$
    8   Determining affinity between antibodies, which are being cloned.
*9 * **Suppression**: *elimination of these clones from M set, for which the formula is true* $D < \sigma_s$:

Fig. 2. Algorithm – Sub-network learning algorithm executed in the computational node

Learning Algorithm of the decentralized immune network consists of nine basic steps. In the first step affinity between j^{th} antigen and a set of antibodies is determined. On the basis of this operation, a subset of the best stimulated antibodies is created. These antibodies are subject to the process of cloning according to the mechanism of cloning selection. In further steps of the described algorithm newly created clones undergo hypermutation and then some newly created clones are removed. It happens when an antibody, which is to function as a memory cell in the process of hypermutation, has been deeply changed. It can be expected that affinity of this memory cell for a given antigen may have insignificant value. In the last point the set of memory cells undergoes the process of suppression. It means that some memory cells which similar to one another will be removed. This process prevents excessive redundancy of memory cells set.

3.1.3 Generating Responses by the Decentralized Immune Network

When the parameter *Decent%*, which defines the level of decentralization of the set of antibodies, is bigger than zero, responses by the immune network is a bit more complicated, because the set of antibodies has a scattered character. It means that each computational node included in the decentralized architecture of the immune network has a unique set of antibodies at its disposal. In this case, the antigen is presented to each set of memory cells found in individual computational nodes. As a result of this action, an antibody of the highest affinity to the presented antigen in a given computational node is sent to the central node. In this central node among other antibodies obtained from individual computational nodes an antibody of the highest affinity to a given antigen is selected.

4 Result of Experiments

In this subchapter we will present the results of the tests of the decentralized immune network. The conducted tests aimed at confirmation of the computational effectiveness of the decentralized artificial immune network mainly in two aspects. First of all, they were to prove that learning of the decentralized immune network using large set of data is far more effective than in the case of commonly used centralized algorithms. The confirmation of the effectiveness and calibration of the discussed architecture of the artificial immune network can decide on practical application of this algorithm in data mining technique. The second aspect is response generation. In both mentioned cases the fundamental criterion, by means of which we will assess the presented algorithm, is the time of learning and the time of response.

4.1 A Phase of Learning of Decentralized Artificial Immune Network

Table 1 shows a total value of learning time for the immune network depending on the number of computational nodes. According to the expectations for the network which consists of only one computational node and the level of decentralization *Decent%* which equals zero, the time of learning takes the highest value. But with 10 computational nodes and the level of decentralization of the learning set occupying 90 percent of the learning time for a given learning set, it takes the lowest value in this testing environment

Table 1. Relation between the learning time and the number of nodes, decentralization level

No computational nodes	1	2	3	4	5	6	7	8	9	10
Decent% [%]	0	50	66	75	80	83	85	87	89	90
learning time per node[s]	963	439	256	172	116	90	73	60	53	49
Total learning time [s]	963	878	770	687	580	542	508	479	481	490

In presented table we can observe a certain anomaly connected with the excessive drop of learning time value than it would result from the number of network computational nodes only. As it could be expected a summary value of the learning time should be kept more or less on the same level. However as the presented table imply, the value of the summary time decreases by implementation of new computational nodes. The reason for such anomalies is time complexity of the learning algorithm. Time complexity of this algorithm is not a linear complexity. As it results from the analysis of the structure of this algorithm, the time complexity of the network learning algorithm is at least $O(n^2)$. Additionally, we should take into consideration the time complexity of the algorithm, which calculates the value of affinity of antibodies, which comes down to time complexity of matrix multiplication algorithm.

$$T_C(n) = c \cdot n^2 + O_{AN}(n^k) \geq T_D(n) = \sum_{i=1}^{N}(c_i \cdot (\tfrac{n}{N})^2 + O_{AN}((\tfrac{n}{N})^k)) = \max_{1 \leq i \leq N}(c_i \cdot \tfrac{n^2}{N_i} + O_{AN}((\tfrac{n}{N})^k)) \quad (2)$$

Where:

n – size of the learning set

T_C – function defining time of learning time computation without set decentralization

T_D – function defining time of learning time computation with set decentralization

c_i– proportion constant

N – number of computational nodes

$O_{AN}(n^k)$– time complexity of numerical algorithms in a given computational environment k≥ 2, e.g. matrix multiplication algorithm

Formula (2) provides estimation of computation time complexity of the learning algorithm in case when it is realized by N computational nodes. The left side of the presented inequality refers to the realization of the learning process in centralized version. The right side of the formula is connected with the realization of the learning algorithm in decentralized version. It can be seen that the a paradigm of independency of the learning processes made it possible to divide the learning set into independent subsets. The division of the learning set caused the decrease of the duration time of the learning process, firstly, through parallel transformation of individual learning sets. Secondly, through the decrease of the size of the learning set, on which time complexity of the learning algorithm depends on. The presented formula implies that the division of the learning set into subsets with the same number of elements provides the largest decrease of the learning time of the decentralized network.

5 Summary

Introducing the notion of a computational node, which is the equivalent of a lymph gland in the natural system and basing on the defined paradigms, the architecture of the artificial immune network has been presented which is calibrated. Here, we have suggested learning mechanism of the artificial immune network which have an influence on decentralization of computation executed by the network, namely, learning through decentralization of antigen set. The tests show that the proposed learning mechanism and the decentralized architecture consisting of computational nodes is calibrated and it can be successfully used in the distributed computational environments such as grid and clusters.

References

1. De Castro, L.N., Von Zuben, F.J.: An Evolutionary Immune Network for Data Clustering. In: Proc. of the IEEE SBRN, pp. 84–89 (2000a)
2. De Castro, L.N., Von Zuben, F.J.: The Clonal Selection Algorithm with Engineering Applications. In: GECCO 2000 - Workshop Proceedings, pp. 36–37 (2000b)
3. Święcicki, M., Wajs, W., Wais, P.: An artificial immune algorithms apply to preprocessing signals. In: Bubak, M., van Albada, G.D., Sloot, P.M.A., Dongarra, J. (eds.) ICCS 2004. LNCS, vol. 3037, pp. 703–707. Springer, Heidelberg (2004)

Predicting Defects in Software Using Grammar-Guided Genetic Programming

Athanasios Tsakonas and Georgios Dounias

University of the Aegean, Department of Finance and Management Engineering,
Fostini 31 str., Chios, Greece
tsakonas@stt.aegean.gr, g.dounias@aegean.gr

Abstract. The knowledge of the software quality can allow an organization to allocate the needed resources for the code maintenance. Maintaining the software is considered as a high cost factor for most organizations. Consequently, there is need to assess software modules in respect of defects that will arise. Addressing the prediction of software defects by means of computational intelligence has only recently become evident. In this paper, we investigate the capability of the genetic programming approach for producing solution composed of decision rules. We applied the model into four software engineering databases of NASA. The overall performance of this system denotes its competitiveness as compared with past methodologies, and is shown capable of producing simple, highly accurate, tangible rules.

Keywords: Software engineering, defect prediction, genetic programming.

1 Introduction

Addressing software quality can ensure cost reduction and efficient resource allocation. A major factor for the assessment of software code is whether the code module is prone to defects in the future. To estimate the software quality several metrics have been developed in the past. Static code metrics [9],[6] are inexpensive, easy to calculate, and they are widely used. However, these measurements have been criticized [4][5][16] on their effectiveness and efficiency, as standalone instruments. Later work [11], has shown that applying data mining techniques can dramatically increase the power of the aforementioned metrics. The main target of such a data-mining task is to effectively predict whether modules will present code defects in the future, so as the management could efficiently allocate resources for monitoring them. Genetic programming (GP) [7] is a computational intelligence methodology which carries expedient attributes such as variable length solution representation, and functional solution nodes. It has been applied in numerous problems, and its domains of applications are constantly increasing. This work inherits recent advances on genetic operators' adaptive rates [18]. The data mining task in this work is two-fold. Firstly, we aim to produce simple and comprehensible rules that can be used without the assistance of software. Secondly, we seek for high classification rates, if possible better to those found in literature. The paper is organized as follows. Next section

J. Darzentas et al. (Eds.): SETN 2008, LNAI 5138, pp. 413–418, 2008.

describes the background, presenting the software defects prediction domain and the grammar guided genetic programming. Following this section, we deal with the design and the implementation of the system. Next, the results and a following discussion are presented. The paper ends with our conclusion and a description of future work.

2 Background

2.1 Software Defects Prediction

Among the principal tasks during software project management is the assessment of the software cost. Additionally, extensive assessment is required for high assurance software. This software cost is affected directly by the software quality. To address this need, there have been developed various techniques of software code assessment, such as the static code metrics. The available metrics for the code derive from the work of [9] and [6].

2.2 Grammar-Guided Genetic Programming

Genetic programming [7] is an extension to the genetic algorithms concept. The main advance is the ability to maintain a population consisted of variable-length, tree-structured individuals, in which each node can have functional ability. By applying grammars, a genotype - a point in the search space- corresponds always to a phenotype - a point in the solution space, an approach known as legal search space handling method [9]. We apply legal search in this work using a context-free grammar [2][3][8][13][17][18][19].

3 Design and Implementation

3.1 Data Pre-processing

We have tested the methodology in four software engineering data sets: CM1, KC1, KC2 and PC1. These datasets have been addressed in [11] and [12]. All software modules come from NASA and their metrics have become recently available by the PROMISE repository of public domain software engineering data sets. Table 1 summarizes the features of this data. Further details for each feature can be found in [9] and [6].

3.2 Genetic Programming Setup

To improve the search process and control the solution size, an adaptive scheme for the operation rates was followed. These parameters were adapted from past work of the authors [18], and they do not necessarily represent the best values for these datasets.

Table 1. NASA software metrics data examined

Name	Data set	Total instances	Defects	No defects	Language
CM1	Spacecraft instrument	498	49	449	C
KC1	Storage management for ground data	2109	326	1783	C++
KC2	Science data processing	552	105	415	C++
PC1	Flight software for earth orbiting satellite	1109	1032	77	C

During the run, the actual training data set is used to evaluate candidate solutions. However, in order to promote a candidate as the solution of the run, in our approach it is required that at least one of the following conditions applies:

- this candidate achieves higher fitness score in the validation set too,
- the absolute difference between validation fitness and training fitness score is smaller.

The first rule is the common approach used in all validation models; the second rule is introduced in this work, and it was experimentally observed to produce solutions that carried significantly higher generalization ability in the problems encountered. In other words, this approach promotes solutions that demonstrate no overfitting to one of the sets (either the actual training set or the validation set), but it rather requires the fitness improvement in one set to be in step with the other [14].

3.3 Fitness Function

In order to validate this software engineering data, various measures have been proposed in literature. In [1], the following measures have been used, in a genetic programming model for a number of generic problems encountered:

$$\text{Recall} = pd = \frac{tp}{tp + fn} \tag{1}$$

$$\text{TNRate} = \frac{tn}{tn + fp} \tag{2}$$

$$\text{fitness} = \text{support} = \text{Recall} \cdot \text{TNRate} \tag{3}$$

In [10] the fitness measure that involves the *accuracy*, is proposed based on results that show that this metric presented the smaller deviation in classification success between the training and the test set, for a number of experiments. On the other hand, in [1], when using the *Recall* and the *TNRate*, there is an equivalent treatment for both classes as far as the classification reward is concerned, irrespectively of the relative size for each class. Hence we adopted the latter measure for our fitness function. Additionally, another metric is calculated in our experiments, *precision*, to allow future comparisons:

$$prec = \frac{tp}{fp + tp} \qquad (4)$$

This *precision* measure is analogous to the *support* measure we have used in our system, as it can be seen in the equation (5). Hence, using the *support* measure as a fitness measure is also in concordance to literature that requires a system scoring also high *precision* values (i.e. aiming for high *support* values can assist in qualifying high *precision* rates).

$$\frac{1}{support} = \left(\frac{1}{prec} - 1\right) \cdot \frac{fn}{tn} + \frac{fn}{tp} + \frac{fp}{tn} + 1 \qquad (5)$$

Having discussed the system design, in the following session we describe the results of the application of our methodology in the four software engineering databases.

4 Results and Discussion

We performed 10-fold cross validation. Table 2 summarizes our best results found for each measure during the 10-fold run, in the test set, and includes the mean and the standard deviation of these results. The solution for the CM1 problem is as follows:

```
       If count_of_lines_of _comments > -0.94 then true else false
```
The promoted solution for the KC1 data is:
```
       If  essential_complexity < 0.76 then
               (if total_operands < -0.95 then false else true)
       else false
```
For the KC2 problem, the system derived the following rule:
```
       If design_complexity = 0.46 then
               (if line_count_of_code < 0.95 then false else true )else

               (if unique_operands > -0.90 then true else false)
```
Finally, the following rule was found for the PC1 data:
```
       If program_length < -0.53 then
               (if difficulty > -0.21 then
               (if total_operators > -0.68 then true else false)
               else
       (if software_size_lines_of_code < -0.93 then true
       else false))
       else false
```

In all problems, our model succeeded in producing small, easily comprehensible results that need not any further software to be applied in practice. In Table 3, we compare the best-promoted solutions of our system during the 10-fold validation, to the best models found in literature.

5 Conclusion and Further Research

This paper described an effort to address the software quality domain, by using computational intelligence for effective decision-making. Our approach makes use of the genetic programming paradigm, in its grammar-guided advance, in order to produce decision rules. Further tuning is enforced to the genetic operators, and special use of the validation set fitness is applied. The model is applied on four databases that are consisted of software metrics of NASA's developed software. In two of the databases, our model is proved superior to the existing literature in both comparison variables, and in the rest two databases, the system is shown better in one of the two variables.

Table 2. Grammar-guided GP, 10-Fold Cross Validation Results

	CM1			KC1		
	Best	Mean	Std.Dev	Best	Mean	Std.Dev
Support	0.7085	0.5982	0.0538	0.5731	0.5579	0.0107
PD	1.0000	0.5967	0.2344	0.8750	0.7544	0.0935
PF	0.2000	0.2719	0.0724	0.2569	0.3135	0.0399
PREC	0.3077	0.1905	0.0586	0.3553	0.3062	0.0312
Accuracy	0.8750	0.7295	0.0768	0.7393	0.6967	0.0309
Generation		38	60		55	77
Size		191	224		259	196
	KC2			PC1		
	Best	Mean	Std.Dev	Best	Mean	Std.Dev
Support	0.7127	0.6697	0.0304	0.7508	0.6442	0.0548
PD	0.8182	0.7482	0.0594	0.8750	0.7441	0.0615
PF	0.1428	0.1929	0.0400	0.0000	0.2911	0.2301
PREC	0.5714	0.5039	0.0488	1.0000	0.9728	0.0207
Accuracy	0.8302	0.7830	0.0336	0.8559	0.7414	0.0547
Generation		30	36		82	62
Size		260	203		296	201

Table 3. Results comparison

Model		PD	PF		PD	PF
		CM1			PC1	
Menzies et al. [11]		0.350	0.100		0.240	0.240
Menzies et al. [12]		0.710	0.270		0.480	0.170
This paper	(#8)	1.000	0.311	(#3)	0.757	0.125
		KC1			KC2	
Menzies et al. [11]		0.500	0.150		0.450	0.150
This paper	(#6)	0.818	0.275	(#7)	0.800	0.142

Moreover, the system managed to produce small and comprehensible solutions that do not require a computing environment to apply. The application of our system to such data is a straightforward process, and adds little complexity to the classification task of the modules. Hence we believe that software engineers can easily adapt such a data mining system, which can then be used in an inexpensive way, combined with the static metrics calculation. Further investigation involves the application of our methodology into more software quality problems, involving other databases, in an attempt to provide a transparent view on its effectiveness for this class of problems.

References

1. Berlanga, F.J., del Jesus, M.J., Herrera, F.: Learning compact fuzzy rule-based classification systems with genetic programming. In: 4th Conference of the European Society for Fuzzy Logic and Technology (EUSFLAT 2005), Barcelona, pp. 1027–1032 (2005)
2. Blickle, T., Theile, L.: A mathematical analysis of tournament selection. In: Eshelman, L.J. (ed.) Proc. of the 6thInternational.Conference on Genetic Algorithms, pp. 9–16. Lawrence Erlbaum Associates, Hillsdale (1995)
3. Eads, D., Hill, D., Davis, S., Perkins, S., Ma, J., Porter, R., Theiler, J.: Genetic Algorithms and Support Vector Machines for Time Series Classification. In: Proc. SPIE, vol. 4787, pp. 74–85 (2002)
4. Fenton, N., Ohlsson, N.: Quantitative Analysis of Faults and Failures in a Complex Software System. IEEE Trans. Software Eng., 797–814 (2000)
5. Fenton, N.E., Pfleeger, S.: Software Metrics: A Rigorous and Practical Approach. Int'l Thompson Press (1997)
6. Halstead, M.: Elements of Software Science. Elsevier, Amsterdam (1977)
7. Koza, J.R.: Genetic Programming: On the Programming of Computers by Means of Natural Selection. MIT Press, Cambridge (1992)
8. Koza, J., Bennett, F., Andre, D., Keane, M.: Genetic Programming III: Automatic Programming and Automatic Circuit Synthesis. Morgan Kaufmann, San Francisco (2003)
9. McCabe, T.: A Complexity Measure. IEEE Trans. Software Eng. 4, 308–320 (1976)
10. Menzies, T., Dekhtyar, A., Distefano, J., Greenwald, J.: Problems with Precision: A Response to Comments on; Data Mining Static Code Attributes to Learn Defect Predictors. IEEE Trans. on Soft. Eng. 33(9), 637–640 (2007)
11. Menzies, T., DiStefano, J., Orrego, A., Chapman, R.: Assessing Predictors of Software Defects. In: Proc. Workshop Predictive Software Models (2004)
12. Menzies, T., Greenwald, J., Frank, A.: Data Mining Static Code Attributes to Learn Defect Predictors. IEEE Trans. on Soft. Eng. 32(11) (January 2007)
13. Montana, D.J.: Strongly Typed Genetic Programming. Evolutionary Computation 3(2) (1995)
14. Quinlan, J.R.: Bagging, boosting, and C4.5. In: Proc. 13th Nat. Conf. Art. Intell., pp. 725–730 (1996)
15. Singleton, A.: Genetic Programming with C++. BYTE Magazine (February 1994)
16. Shepperd, M., Ince, D.: A Critique of Three Metrics. J. Systems and Software 26(3), 197–210 (1994)
17. Tsakonas, A., Dounias, G.: Hierarchical Classification Trees Using Type-Constrained Genetic Programming. In: Proc. of 1st Intl. IEEE Symposium in Intelligent Systems, Varna, Bulgaria (2002)
18. Tsakonas, A., Dounias, G.: Evolving Neural-Symbolic Systems Guided by Adaptive Training Schemes: Applications in Finance. App. Art. Intell. 21(7), 681–706 (2007)
19. Yu, T., Bentley, P.: Methods to Evolve Legal Phenotypes. In: Eiben, A.E., Bäck, T., Schoenauer, M., Schwefel, H.-P. (eds.) PPSN 1998. LNCS, vol. 1498, pp. 280–291. Springer, Heidelberg (1998)

A Clustering Framework to Build Focused Web Crawlers for Automatic Extraction of Cultural Information

George E. Tsekouras[*], Damianos Gavalas, Stefanos Filios, Antonios D. Niros, and George Bafaloukas

Department of Cultural Technology and Communication, University of the Aegean, 81100, Mytilene, Lesvos, Greece
gtsek@ct.aegean.gr, dgavalas@aegean.gr

Abstract. We present a novel focused crawling method for extracting and processing cultural data from the web in a fully automated fashion. After downloading the pages, we extract from each document a number of words for each thematic cultural area. We then create multidimensional document vectors comprising the most frequent word occurrences. The dissimilarity between these vectors is measured by the Hamming distance. In the last stage, we employ cluster analysis to partition the document vectors into a number of clusters. Finally, our approach is illustrated via a proof-of-concept application which scrutinizes hundreds of web pages spanning different cultural thematic areas.

Keywords: web crawling, HTML parser, document vector, cluster analysis, Hamming distance, similarity measure, filtering.

1 Introduction

Web crawlers typically perform a simulated browsing of the web by extracting links from pages, downloading all of them and repeating the process ad infinitum. This process requires enormous amounts of hardware and network resources, ending up with a large fraction of the visible web on the crawler's storage array. When information about a predefined topic is desired though, a specialization of the aforementioned process called "focused crawling" is used [1]. When searching for further relevant web pages, the focused crawler starts from the given pages and recursively explores the linked web pages [1, 2]. While the simple crawlers perform a breadth-first search of the whole web; focused crawlers explore only a small portion of the web using a best-first search guided by the user interest and based on similarity estimations [1, 2]. To maintain a fast information retrieval process, a focused web crawler has to perform web document classification under certain similar characteristics. One of the most efficient approaches to maintain this issue is to use cluster analysis.

This article introduces a novel clustering-based focused crawler that involves: (a) creation a multidimensional document vector comprising the most frequent word occurrences; (b) calculation of the Hamming distances between the cultural-related documents; (c) partitioning of the documents into a number of clusters.

---

[*] Corresponding author.

J. Darzentas et al. (Eds.): SETN 2008, LNAI 5138, pp. 419–424, 2008.

The remainder of the article is organized as follows: Section 2 describes the structure of the focused web-crawler. Section 3 discusses the experimental results and Section 4 presents the conclusions of the present work.

2 The Proposed Method

Prior to performing clustering of web documents, our algorithm involves the documents retrieval (crawling) and parsing and also the calculation of their distance vector and distance matrix. The high-level process of crawling, parsing, filtering and clustering of the downloaded web pages is illustrated in Figure 1. In detail, our algorithm is described within the following subsections.

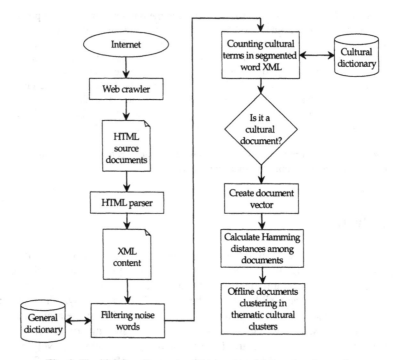

Fig. 1. The high-level process of the proposed focused web-crawler

2.1 Crawling Procedure

For the retrieval of web pages we utilize a simple recursive procedure which enables breadth-first searches through the links in web pages across the Internet. The application downloads the first document, retrieves the web links included within the page and then recursively downloads the pages where these links point to, until the requested number of documents has been downloaded. The respective pseudo-code implementation is given in Figure 2.

```
Initialize:
UrlsDone = ; UrlsTodo = {'firstsite.seed.htm', 'secondsite.seed.htm'..}
Repeat:
    url = UrlsTodo.getNext()
    ip = DNSlookup( url.getHostname() )
    html = DownloadPage( ip , url.getPath() )
    UrlsDone.insert( url )
    newUrls = parseForLinks( html )
    For each newUrl
    If not UrlsDone.contains( newUrl )
    then UrlsTodo.insert( newUrl )
```

Fig. 2. The web crawling algorithm

2.2 Parsing of HTML Documents

The parser maintains the title and the 'clear' text content of the document and the URL addresses where the document links point to. The title and the document's body content are then translated to XML format. The noisy words such as articles and punctuation marks are filtered. The documents not including sufficient cultural content are deleted. The documents based on their URL addresses are retrieved and parsed on the next algorithm's execution round, unless they have been already appended in the 'UrlsDone' list.

2.3 Calculation of the Document Vectors

For each parsed document we calculate the respective document vector denoted as DV. The DV indicates the descriptive and most useful words included within the document and their frequencies of appearance. For instance, if a document D_i includes the words a, b, c and d with frequency 3, 2, 8 and 6 respectively, then its document vector will be: $DV_i = [3a, 2b, 8c, 6d]$. The dimension of DV_i equals the number of the words it includes (for the previous example it is $|DV_i| = 4$) and varies for each document. Next, we reorder each DV_i in descending order of words frequencies so the vector of the previous example becomes: $DV_i = [8c, 6d, 3a, 2b]$. Finally, we filter the DV_i maintaining only a specific number of T words, so that all DV_is are of equal dimension. Thus, for $T=2$, the vector of the previous example becomes: $DV_i = [8c, 6d]$. The filtering excludes some information, since we have no knowledge of which words with small frequencies are included in each document. If no filtering is applied, then the worse case scenario for a dictionary of W words is a W-dimensions DV_i, where each word of the dictionary appears only once in a document.

2.4 Calculation of the Document Vectors Distance Matrix

We calculate the distance matrix DM of the parsed web documents. Each scalar element $DM_{i,j}$ of this matrix equals the distance between DV_i and DV_j document vectors: $DM_{i,j} = d(DV_i, DV_j)$, which represents the dissimilarity of the words included within the corresponding documents, i.e. the more different the words (and their frequencies) of documents D_i and D_j, the higher the $DM_{i,j}$ value. The distance matrix values $DM_{i,j}$ are calculated based on the Hamming distance calculation method: if a word w with frequency f appears in D_i and not in D_j or vice-versa, $DM_{i,j}$ increases by f. Otherwise, w appears on D_i with frequency f_i and on D_j with frequency f_j and therefore, $DM_{i,j}$ increases by the absolute value of the difference between f_i and f_j ($|f_i - f_j|$). The distance matrix provides the input of our clustering algorithm presented in the following section.

2.5 Clustering Process

Let $X = \{DV_1, DV_2, ..., DV_N\}$ be a set of N document vectors of equal dimension. The potential of the i-th document vector is defined as follows,

$$Z_i = \sum_{j=1}^{N} S_{ji} , \qquad (1 \le i \le N) \tag{1}$$

where S_{ji} is the similarity measure between DV_j and DV_i given as follows,

$$S_{ji} = \exp\{-\alpha \cdot d(DV_j, DV_i)\}, \qquad \alpha \in (0,1) \tag{2}$$

A document vector that appears a high potential value is surrounded by many neighboring document vectors. Therefore, a document vector with a high potential value is a good nominee to be a cluster center. Based on this remark the potential-based clustering algorithm is described as follows,

Step 1). Select values for the design parameters $\alpha \in (0,1)$ and $\beta \in (0,1)$. Initialy, set the number of clusters equal to $n=0$.

Step 2). Using eqs (1) and (2) and the distance matrix calculate the potential values for all document vectors DV_i $(1 \le i \le N)$.

Step 3). Set $n=n+1$.

Step 4). Calculate the maximum potential value: $Z_{max} = \max_{1 \le i \le N} \{Z_i\}$ $\qquad (3)$

Select the document vector $DV_{i_{max}}$ that corresponds to Z_{max} as the center element of the n-th cluster: $C_n = DV_{i_{max}}$.

Step 5). Remove from the set X all the document vectors having similarity (given in eq. (2)) with $DV_{i_{max}}$ greater than β and assign them to the n-th cluster.

Step 6). If X is empty stop. Else turn the algorithm to step 2. ∎

The implementation of the above algorithm requires a priori knowledge of the values for α and β. To simplify the approach, we set $\alpha = 0.5$ and we calculate the optimal value of β by using the following cluster validity index,

$$V = \frac{COMP}{SEP} \tag{7}$$

Where *COMP* is cluster compactness measure and *SEP* the cluster separation measure, which are respectively given as,

$$COMP = \sum_{k=1}^{n} Z_k^C \tag{4}$$

and

$$SEP = g\left(\min_{i,j}\{d(C_i, C_j)\} \right) \tag{5}$$

where in (4) Z_k^C is the potential value of the k-th cluster center and in (5) the function g is a strictly increasing function. Here, we choose the following form,

$$g(x) = x^q, \qquad q \in (1, \infty) \tag{6}$$

Table 1. Comparative classification results with respect to the Harvest Ratio

Category	Best-First Search	Accelerated Focused Crawler [4]	First-Order Crawler [5]	Proposed Method
Cultural conservation	48.7%	65.3%	68.9%	69.6%
Cultural heritage	52.1%	72.4%	73.7%	77.2%
Painting	70.6%	72.5%	76.4%	77.1%
Sculpture	52.0%	67.2%	71.6%	72.1%
Dancing	66.8%	73.8%	88.8%	90.6%
Cinematograph	67.4%	84.7%	90.2%	85.4%
Architecture Museum	55.4%	50.5%	78.3%	72.1%
Archaeology	59.7%	59.8%	80.0%	84.9%
Folklore	60.8%	64.9%	82.5%	84.0%
Music	65.2%	85.4%	93.0%	93.4%
Theatre	71.5%	87.2%	91.5%	92.0%
Cultural Events	58.8%	74.0%	90.6%	87.1%
Audiovisual Arts	63.3%	68.4%	82.8%	84.7%
Graphics Design	68.8%	69.0%	78.6%	82.9%
Art History	48.7%	59.6%	60.9%	59.0%

In the above equation, the parameter q is used to normalize the separation measure in order to cancel undesired effects related to the number of document vectors and the number of clusters, as well. To this end, the main objective is to select the value of the parameter β, which minimizes the validity index V.

3 Experimental Evaluation

Target topics were defined and the page samples were obtained through meta-searching the Yahoo search engine. We choose 15 categories related to cultural information, which are depicted in Table 1. For each category, we downloaded 1000 web pages to train the algorithm. After generating the dictionary, for each category, we selected the 200 most frequently reported words, using the inverse document frequency (IDF) for each word [3]. In the next step, we defined the dimension of each document vector as $T=30$. Note that, we can keep the feature space dimension large, but in this case the computational cost will increase.

To test the method, we downloaded another 1000 pages for each category and we utilized the well-known Harvest Ratio [4, 5] to compare the method with other algorithms. The results are presented in Table 1, where we can easily verify that, except of few cases, our method outperformed the rest of the methods.

4 Conclusions

We have shown how cluster analysis can be efficiently incorporated into a focused web crawler. The basic idea of the approach is to create multidimensional document vectors each of which corresponds to a specific web page. The dissimilarity between two distinct document vectors is measured using the Hamming distance. Then, we use a clustering algorithm to classify the set of all document vectors into a number of clusters, where the respective cluster centers are objects from the original data set that satisfy specific conditions. The classification of unknown web pages is accomplished by using the minimum Hamming distance. Several experimental simulations took place, which verified the efficiency of the proposed method.

References

[1] Huang, Y., Ye, Y.-M.: wHunter: A Focused Web Crawler – A Tool for Digital Library. In: Chen, Z., Chen, H., Miao, Q., Fu, Y., Fox, E., Lim, E.-p. (eds.) ICADL 2004. LNCS, vol. 3334, pp. 519–522. Springer, Heidelberg (2004)
[2] Zhu, Q.: An algorithm for the focused web crawler. In: The Proceedings of the 6th International Conference on Machine Learning and Cybernetics, Hong Kong, (2007)
[3] Tsekouras, G.E., Anagnostopoulos, C.N., Gavalas, D., Economou, D.: Classification of Web Documents using Fuzzy Logic Categorical Data Clustering. In: Boukis, C., Pnevmatikakis, A., Polymenakos, L. (eds.) Artificial Intelligence and Innovations: From Therory to Applications, pp. 93–100. Springer, Heidelberg (2007)
[4] Chakrabarti, S., Punera, K., Subramanyam, M.: Accelerated focused crawling through online relevance feedback. In: WWW, pp. 148–159 (2002)
[5] Xu, Q., Zuo, W.: First-order Focused Crawling. In: The Proceedings of the International Conference on WWW 2007, Banff, Alberta, Canada (2007)

Non-negative Matrix Factorization for Endoscopic Video Summarization

Spyros Tsevas[1], Dimitris Iakovidis[1], Dimitris Maroulis[1], Emmanuel Pavlakis[2], and Andreas Polydorou[2]

[1] Dept. of Informatics and Telecommunications, University of Athens
Panepistimiopolis, GR-15784, Athens, Greece
{s.tsevas,dimitris.iakovidis}@ieee.org, dmarou@di.uoa.gr
[2] Department of Surgery, Aretaieion Hospital, V. Sofias 76 avenue, 115 27 Athens, Greece
egp@otenet.gr, apolyd@med.uoa.gr

Abstract. Wireless Capsule Endoscopy (WCE) has been introduced as a non-invasive colour imaging technique for the inspection of the small intestin along with the rest of the gastrointestinal tract. Each WCE examination results in a 50,000-frames video that has to be visually inspected frame-by-frame by the doctor and this may be a highly time-consuming task even for the experienced gastroenterologist. In this paper we propose a novel approach that leads to a summarized version of the original video enabling significant reduction in the video assessment time without losing any critical information. It is based on symmetric non-negative matrix factorisation initialized by the fuzzy c-means algorithm and it is supported by non-negative Lagrangian relaxation to extract a subset of video frames containing the most representative scenes from an entire examination. The experimental evaluation of the proposed approach was performed using previously annotated endoscopic videos from various sites of the small intestine.

Keywords: Non-negative matrix factorisation, wireless capsule endoscopy, video, summarisation.

1 Introduction

Wireless Capsule Endoscopy (WCE) [1] represents a major departure from conventional endoscopy which is inefficient for the examination of the small intestine and is usually uncomfortable for the patient. By using the WCE technique, the physician can efficiently diagnose a range of gastrointestinal disorders, including ulcer, unexplained bleeding, and polyps. One of the major challenges that WCE imposes is the size of the resulting video which results in a more than an hour of intense labour for the physician in order to examine the whole frame sequence [2] while this manual examination process does not guarantee that some abnormal regions are missed.

Several computational approaches coping with the analysis of the WCE video have been proposed [3-14]; however, to the best of our knowledge, no major contribution has been made to the reduction of the time required for visual inspection of the WCE video. To cope with this issue, we propose an effective computational approach that

J. Darzentas et al. (Eds.): SETN 2008, LNAI 5138, pp. 425–430, 2008.
© Springer-Verlag Berlin Heidelberg 2008

drastically reduces the video frames to be inspected enabling this way faster inspection of the video sequence. The proposed approach [17] applies a methodology based on non-negative matrix factorisation (NMF) [18, 19] to summarize the WCE video by keeping the most representative scenes from the whole examination.

The rest of this paper consists of three sections. Section II provides a description of the proposed methodology. Section III, presents the results of its experimental application on WCE video data, and Section IV summarises the conclusions that can be derived from this study.

2 Methodology

The proposed approach for WCE video summarisation is based on the data reduction methodology described in [17] and it takes place in three steps. In the first step Fuzzy C-Means (FCM) is applied on the input video stream to group its frames into a predefined number of clusters, whereas in the second and in the third step two NMF algorithms are subsequently applied on the clustered frames so that they extract only some representative video frames from the whole video.

Given a non-negative $m \times n$ matrix $\mathbf{V}$, the NMF algorithms seeks to find non-negative factors $\mathbf{W}$ and $\mathbf{H}$ of $\overline{\mathbf{V}}$ such that:

$$\mathbf{V} \approx \overline{\mathbf{V}} = \mathbf{W} \times \mathbf{H} \qquad (1)$$

where $\mathbf{W} \in \mathfrak{R}^{m \times k}$ and $\mathbf{H} \in \mathfrak{R}^{k \times n}$.

The dimensionality and the initial values of $\mathbf{W}$ and $\mathbf{H}$ (or just $\mathbf{H}$ in certain algorithms) are determined by means of the FCM algorithm. FCM performs soft clustering of the video frames so that they belong to more than a single cluster. The memberships of each frame to the different clusters are stored in a $k \times n$ matrix $\mathbf{U}_{FCM}$.

The neighbouring frames in the original m-dimensional vector space, are determined by calculating the $n \times n$ matrix of the Euclidean distances which is used for the calculation of the geodesic distance matrix $\mathbf{D}_G$ that contains the geodesic distances (shortest paths) between the vectorial representations of the frames. Next, $\mathbf{D}_G$ is transformed into a pairwise similarity matrix according to the exponential weighting scheme in Eq. (2),

$$V = e^{\frac{D_G}{r}} \qquad (2)$$

V is going to be used as an input to the FCM.

The dimension k of $\mathbf{U}_{FCM}$ is set equal to the predefined number of clusters c, whereas $\mathbf{W}$ and $\mathbf{H}$ are initialized with the m-dimensional cluster centroids and the values of the membership matrix of the converged FCM, respectively.

The symmetric NMF (SymNMF) which for a square matrix is:

$$\mathbf{V} \approx \mathbf{H} \times \mathbf{H}^T \qquad (3)$$

SymNMF is applied on $\mathbf{V}$ so that it "unfolds" the clusters and makes them more transparent. For the calculation of $\mathbf{H}$ we followed the iterative approach described in [17, 22] initializing $\mathbf{H}$ with $\mathbf{U_{FCM}}$. Iteration takes place until the objective function of the SymNMF converges to a small positive value close to zero.

The final step of the methodology imposes orthogonality constraints on the output of the SymNMF so as to extract the most representative members of a given cluster. It is implemented by means of an NMF multiplicative update iterative algorithm known as Non-negative Lagrangian Relaxation (NLR) [17, 23].

In NLR the entries are viewed as cluster indicators and as a result the interpretation of the results at convergence is straightforward allowing this way a relatively easy interpretation of the cluster structure.

3 Results

In order to illustrate the performance of the proposed summarization approach, a number of experiments were conducted on a controlled dataset comprising of annotated video frames with ground truth information provided by expert endoscopists who visually inspected and annotated each frame. A total of 281,000 WCE frames were obtained with identical imaging settings from different patients and two kinds of abnormal findings were identified; ulcers and bleeding. As each finding was visible in more than a single frame, neighbourhoods of frames were extracted for each finding. This process led to the extraction of a total of eight neighbourhoods of frames with abnormal findings which were further balanced at 40 frames per category by random sampling of the larger set to avoid bias. The final dataset dataset consists of 4 ulcer neighbourhoods that contain 11, 8, 12 and 9 frames and 4 bleeding neighbourhoods that consists of 12, 19, 5 and 4 frames; summing up to 40 and 58 frames of ulcers and bleedings respectively. Aiming to investigate the discrimination between abnormal and normal tissues, a total of 40 frames of normal tissues, was appended. The normal frames were extracted from randomly sampled sites of normal tissues over the whole dataset leading to the formation of a 7 neighbourhoods of normal frames. The resulting dataset consists of 120 frames ($n=120$).

In order to reduce the computational cost and the detail of each frame the video frames were rescaled from 260×260 pixels to 91×91 pixels ($m=8281$). Experimentation showed that the use of smaller frames was not beneficial for the overall results.

The images were converted to greyscale and used to form the initial dataset matrix of $m×n$ dimensions. By following the process described in the previous section we calculated the similarity matrix V according Eq. (2), with $r=100$ [17], so as to proceed with the FCM calculations. FCM was executed for 3 clusters.

In the following, we subsequently applied SymNMF and NLR to V. It can be observed that still after SymNMF the cluster structure is not clear. After the application of NLR the clusters are not really separated, though NLR enforces orthogonality. This lack of strict orthogonality is due to the fact that the number of iterations of the SymNMF and NLR accordingly are finite. Actually, only a part of the examples are strictly 'orthogonal' to the members of other clusters. These members form the Most Representative Examples (MREs) of the cluster.

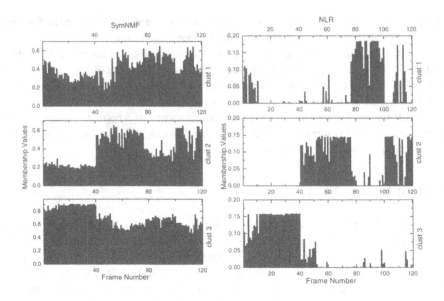

Fig. 1. Results of the SymNMF (left) and NLR (right)

In order to extract the MREs of each cluster, we apply the orthogonality condition with a mild deviation from the strict orthogonality according to [17]. Thus, we apply a threshold T to the entries of **X**. The value of T controls the degree of summarization of the WCE video. Large values of T lead to more examples (frames) in the resulting set of MREs. Figure 2 illustrates how the total number of frames in the resulting video varies with T, as well as the percentage reduction in the total number of frames of the original video. From these figures it is obvious that for threshold values close to 1E-5 the total number of frames per cluster is substantially reduced. Moreover, the total number of frames may be reduced down to 10% of the number of frames of the original video, and since the number of frames is proportional to the visual inspection time, a 90% reduction in this time is feasible.

Such a frame reduction would actually be worthless if the remaining frames would not contain representatives from all the possible abnormal findings, since missing even a sing abnormality could be critical for the patient. A thorough examination of the frames comprising the summarised video validated that the proposed approach did not miss any abnormal finding. The summarized video was containing at least one representative frame from each neighbourhood of frames of abnormal findings. For certain values of T the proposed approach missed some neighbourhoods of normal frames, but this is insignificant considering that it does not have any implications for the patient. For T below 1E-5 the number of frames became too small and many neighbourhoods didn't have any representative frames in the final video.

By integrating a time stamp to each representative frame we can offer the physician the ability to return to the corresponding frame of initial video so as to further examine the area of interest.

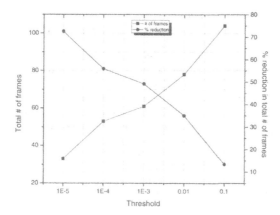

Fig. 2. Total number of frames and percentage reduction in the total number of frames for different values of threshold T

4 Conclusions

The novel approach to WCE video summarisation that we presented is based on the application of the NMF on the video frames according to the methodology proposed in [17]. The results of its experimental evaluation on annotated WCE videos with multiple findings showed that a significant reduction in the total number of frames of the original video without any loss of patient-critical information is feasible, leading to a significant reduction of the visual inspection time required per endoscopic examination.

Our future work includes utilization of various image features for the discrimination of other types of abnormal findings such as polyps and cancer, as well as further experimentation with many annotated WCE videos and investigation of memory-efficient techniques to perform NMF on large WCE video streams.

References

1. Iddan, G., Meron, G., Glukhovsky, A., Swain, P.: Wireless capsule endoscopy. Nature 405 (6785), 417–418 (2000)
2. Maieron, A., Hubner, D., Blaha, B., Deutsch, C., Schickmair, T., Ziachehabi, A., Kerstan, E., Knoflach, P., Schoefl, R.: Multicenter retrospective evaluation of capsule endoscopy in clinical routine. Endoscopy 36(10), 864–868 (2004)
3. Coimbra, M.T., Cunha, J.P.S.: MPEG-7 visual descriptors - contributions for automated feature extraction in capsule endoscopy. IEEE Transactions on Circuits and Systems for Video Technology 16(5), 628–636 (2006)
4. Li, B., Meng, M.Q.-H.: Analysis of the gastrointestinal status from wireless capsule endo-scopy images using local color feature. In: ICIA 2007 International Conference on Infor-mation Acquisition, July 8-11, 2007, pp. 553–557 (2007)
5. Mackiewicz, M., Berens, J., Fisher, M., Bell, D.: Colour and texture based gastrointestinal tissue discrimination. In: ICASSP, IEEE International Conference on Acoustics, Speech and Signal Processing - Proceedings, vol. 2, pp. II597-II600 (2006)

6. Berens, J., Mackiewicz, M., Bell, D.: Stomach, intestine and colon tissue discriminators for wireless capsule endoscopy images. In: Progress in Biomedical Optics and Imaging, Proceedings of SPIE, vol. 5747, pp. (I): 283–290 (2005)

7. Lee, J., Oh, J., Shah, S.K., Yuan, X., Tang, S.J.: Automatic classification of digestive organs in wireless capsule endoscopy videos. In: Proceedings of the ACM, Symposium on Applied Computing, pp. 1041–1045 (2007)

8. Bourbakis, N.: Detecting abnormal patterns in WCE images. In: Proceedings - BIBE 2005, 5th IEEE Symposium on Bioinformatics and Bioengineering, pp. 232–238 (2005)

9. Hwang, S., Oh, J., Cox, J., Tang, S.J., Tibbals, H.F.: Blood detection in wireless capsule endoscopy using expectation maximization clustering. In: Progress in Biomedical Optics and Imaging, Proceedings of SPIE, vol. 6144 I (2006)

10. Kodogiannis, V.S., Boulougoura, M.: Neural network-based approach for the classification of wireless-capsule endoscopic images. In: Proceedings of the International Joint Conference on Neural Networks, vol. 4, pp. 2423–2428 (2005)

11. Kodogiannis, V.S., Boulougoura, M., Lygouras, J.N., Petrounias, I.: A neuro-fuzzy-based system for detecting abnormal patterns in wireless-capsule endoscopic images. Neurocomputing 70(4-6), 704–717 (2007)

12. Li, B., Meng, M.Q.-H.: Wireless capsule endoscopy images enhancement by tensor based diffusion. In: Annual International Conference of the IEEE Engineering in Medicine and Biology Proceedings, pp. 4861–4864 (2006)

13. Vilariño, F., Kuncheva, L.I., Radeva, P.: ROC curves and video analysis optimization in intestinal capsule endoscopy. Pattern Recognition Letters 27(8), 875–881 (2006)

14. Vilariño, F., Spyridonos, P., Pujol, O., Vitrià, J., Radeva, P., De Iorio, F.: Automatic detection of intestinal juices in wireless capsule video endoscopy. In: Proceedings - International Conference on Pattern Recognition, vol. 4, pp. 719–722 (2006)

15. Vilariño, F., Spyridonos, P., Vitrià, J., Malagelada, C., Radeva, P.: A machine learning framework using SOMs: Applications in the intestinal motility assessment. In: Martínez-Trinidad, J.F., Carrasco Ochoa, J.A., Kittler, J. (eds.) CIARP 2006. LNCS(LNAI), vol. 4225, pp. 188–197. Springer, Heidelberg (2006)

16. Wadge, E., Boulougoura, M., Kodogiannis, V.: Computer-assisted diagnosis of wireless-capsule endoscopic images using neural network based techniques. In: Proceedings of the 2005 IEEE International Conference on Computational Intelligence for Measurement Systems and Applications, CIMSA 2005, pp. 328–333 (2005)

17. Okun, O., Priisalu, H.: Unsupervised data reduction. Signal Processing 87(9), 2260–2267 (2007)

18. Lee, D.D., Seung, H.S.: Unsupervised learning by convex and conic coding. Adv. Neural Inf. Process. Systems 9, 515–521 (1997)

19. Lee, D.D., Seung, H.S.: Learning the parts of objects by non-negative matrix factorization. Nature 401, 788–791 (1999)

20. Lee, D.D., Seung, H.S.: Algorithms for non-negative matrix factorization. Adv. Neural Inf. Process. Systems 13, 556–562 (2000)

21. Saul, L.K., Lee, D.D.: Multiplicative updates for classification by mixture models. Adv. Neural Inf. Process. Systems 14, 897–904 (2002)

22. Ding, C., He, X., Simon, H.D.: On the equivalence of nonnegative matrix factorization and spectral clustering. In: Proceedings of the SIAM International Conference on Data Mining, Newport Beach, CA, April 2005, pp. 606–610 (2005)

23. Ding, C., He, X., Simon, H.D.: Nonnegative Lagrangian relaxation of K-means and spectral clustering. In: Proceedings of the Sixteenth European Conference on Machine Learning, Porto, Portugal, October 3–7, 2005, pp. 530–538 (2005)

Nature Inspired Intelligence for the Constrained Portfolio Optimization Problem

Vassilios Vassiliadis and Georgios Dounias

Management and Decision Engineering Laboratory, Department of Financial & Management Engineering, School of Business Studies, University of the Aegean, 31 Fostini Str. GR-821 00, Greece
v.vassiliadis@fme.aegean.gr, g.dounias@aegean.gr

Abstract. In this paper, we apply a basic Bee Colony Optimization algorithm in order to find a high-quality solution for the constrained portfolio optimization problem. Moreover, we use a basic Ant Colony Optimization algorithm and a Tabu Search metaheuristic approach as a benchmark. Our findings indicate that nature-inspired methodologies are able to find feasible solutions for dynamic optimization problems in a reasonable amount of time in contrast with the simple tabu search.

Keywords: Bee Colony Optimization, Nature-Inspired Intelligence, Portfolio Optimization.

1 Introduction

Dynamic optimization problems, such as the constrained portfolio optimization problem, cannot be efficiently solved using traditional approaches such as quadratic programming [1]. In the traditional portfolio optimization problem, an investor constructs a portfolio of assets with the aim of minimizing its risk for a standard level of return [2,16]. However, in order to make this formulation more real we should impose certain constraints such as cardinality constraints, floor and ceiling constraints.

Nature-inspired intelligence introduces methodologies (ACO, PSO etc.) that are able to find high-quality solutions in a reasonable amount of time. These methodologies are based on the way natural systems work. NI techniques have some differences with Genetic Algorithms. Firstly, GAs and NI algorithms share different mechanisms in finding feasible solutions (evolution vs real-life systems). Also, GAs maintain a pool of solutions rather just one, as ACO and BCO do.

In this paper, we propose an algorithm based on the way a real bee colony behaves, namely Bee Colony Optimization (BCO) algorithm, in order to solve the constrained portfolio optimization problem. The main characteristic of this algorithm lies on the way real bees interact with each other in order to find a high-quality food source [3]. Moreover, we apply an ant colony optimization algorithm to the same problem, for comparison reasons. ACO is based on the foraging behavior of a real ant colony [4]. Finally, we apply a simple Tabu search as a benchmark. In tabu search, the fundamental concept is that of a move, a systematic operator that, given a single starting

J. Darzentas et al. (Eds.): SETN 2008, LNAI 5138, pp. 431–436, 2008.

solution, generates a number of other possible solutions, without revisiting previously encountered solutions [1].

The main aim of this paper is to show that nature-inspired schemes provide high-quality solutions in dynamic optimization problems, such as the constrained portfolio optimization problem. The paper is organized as follows. Section 1 is the introduction. Section 2 contains the basic literature review on the portfolio optimization problem. Section 3 presents the main aspects of the portfolio optimization problem. Section 4 briefly describes the intelligent approaches used. In section 5 we present our findings. Finally, section 6 concludes and proposes further research on this direction.

2 Literature Review

The portfolio optimization problem has been widely analyzed in the literature. For reasons of convenience, we are going to present only a number of selected works on this field. Genetic algorithms have been used for the portfolio optimization problem [1,7]. In some cases, the performance of GA is compared with heuristic algorithms such as simulated annealing and tabu search and found to be superior. Heuristic algorithm is a method which employs heuristics (domain specific search strategies) in order to facilitate efficient search of the search space. In some cases, GAs have been hybridized with other AI methods such as neural networks [4]. Also, evolutionary strategies have been used widely [3,5,8]. In addition, NI algorithms such as ACO, PSO have been used to tackle with the same problem [2,6,7,9]. Their results are superior to other traditional methodologies.

To summarize this literature review, we could note that AI methodologies such as neural networks, genetic algorithms and evolutionary strategies outperform traditional ones. Nature-inspired algorithms such as PSO and ACO have proven to be very effective, as well. Some of the main characteristics of NI techniques are their speed of convergence to an acceptable solution and their ability to explore the solution space in a simple and at the same time intelligent way. However, little attention has been given to other nature-inspired methodologies such as BCO, regarding the portfolio optimization problem. So, our aim in this current work is to assess the performance of other nature-inspired algorithms, as well.

3 Problem Formulation

The classical portfolio optimization problem, as introduced by Markowitz [14] in the modern portfolio theory (MPT) can be described as follows: minimize σ^2_p over x_i's, s.t. $r_p = r^*$, $\Sigma_i x_i = 1$, $x_i > 0$, for all x_i's, where, σ^2_p is the portfolio's variance, σ_{ij} is the covariance of assets i, j, x_i is the weight of asset i, r_p is the portfolio's expected return, r_i is the individual asset's expected return, r^* is the lower threshold constraint for the portfolio's expected return.

The main objective is to minimize the risk of the portfolio, which is defined by the variance, for a given level of expected return.

In order to make the problem more realistic, we introduce several realistic constraints. Firstly, the capital allocated in each asset in the portfolio (asset's weight) cannot exceed defined upper and lower thresholds. These kinds of constraints are known as floor and ceiling constraints [3]: $x_i \geq x_l \wedge x_i \leq x_u$.

The second condition allows short-selling. This means, that assets can take negative weights, as well.

Finally, in order for our formulation to be functional, the target return (r^*) must be larger than the return of the minimum variance portfolio. In other case, a rational investor would choose the minimum variance portfolio. Due to the fact that the target return is selected exogenously, and might therefore be below the expected return of the minimum variance portfolio, we reformulate the original objective function by maximizing the expected return, diminished by the incurred risk: maximize ($a*r_p -$ ($1-a)*\sigma^2_p$), where $a \in [0,1]$.

4 Methodological Issues

In this paper, we apply a Bee Colony Optimization algorithm (BCO) for the constrained portfolio optimization problem. BCO is based on the interaction between real bees when searching for food [12]. Particularly, in real bee colonies, there are three kinds of bees: employed bees, onlooker bees and scouts. Employed bees carry with them information about a particular food source, its distance and direction from the nest and the profitability of the source and share this information with a certain probability. Communication among bees through a process called "waggle dance" [10,12]. Onlooker bees wait to the nest and usually find a food source through the information shared by employed bees. When a bee abandons its food source, it becomes a scout and searches the environment surrounding the nest for new food sources [12].

In artificial bee colonies each solution corresponds to a portfolio of assets. BCO searches for the best possible combination of assets and gradient descent search is used for the calculation of the assets' weights. Firstly, the employed bees start searching for solutions and then return to the hive and share the information, regarding the solutions found, through a probability rule (artificial "waggle dancing"). Next, the onlooker bees select a solution, using the "roulette wheeling" mechanism, based on the calculated probabilities, and explore its neighborhood. The process described above is repeated for a number of generations. The Bee Colony Optimization algorithm can be found in the following url: http://www.fme.aegean.gr/nicpob.asp .

The second nature-inspired algorithm is the Ant Colony Optimization algorithm (ACO). ACO is an intelligent methodology which exploits the foraging behavior of real ant colonies [11]. The ACO algorithm we used in this paper can be found in the following url: http://www.fme.aegean.gr/nicpob.asp .

Finally, the benchmark algorithm combines tabu search and gradient descent search. The tabu search mechanism is responsible for the selection of assets and the gradient descent search tries to find the optimal weights by minimizing the objective function.

5 Results

We conducted a set of simulations in order to demonstrate the efficiency of the nature-inspired algorithms. Our database comprises of daily returns of stocks from the S and P's 500 stock index[1] for the year 2007. Returns for stocks were calculated based on historical prices using the following formula: $r_t = \log(P_t/P_{t-1})$ [7]. The calculation of variance was based on historical stock prices, as well. In our approach, we split the whole database in an estimation period (03/01/2007-03/07/2007) and an out-of-sample period (05/07/2007-31/12/2007). The portfolios have been constructed based on the estimation period.

We have excluded from our database these stocks appearing to have missing values. So, our database comprises of 312 stocks.

Before running the set of simulations, the open parameters of the algorithms and the problem were adjusted[2]. The values of the parameters can be found at the following url: http://www.fme.aegean.gr/nicpob.asp .

The following table (table 5) summarizes the results of the simulations[3].

Table 5. Experimental results

	ACO	BCO	TABU-GDS[4]
Portfolio	130,177,8,53,97,148, 155,150,190,136	66,58,49,114,145,294, 53,212,104,41	51,74,53,179,48,211, 280,212,205,130
Weights	0.19,0.35,-0.01,0.45, -0.04,0.04,-0.10,0.01, -0.05,0.16	0.07,0.27,0.14,0.14,0.09, 0.18, 0.33,-0.04,0.02,-0.19	0.01,0.05,0.58,-0.19, 0.12,0.19,0.10,-0.01, 0.01,0.13
In-sample			
Expected Return	0.0012	0.0012	0.0012
Variance	$2.2056*10^{-5}$	$1.7912*10^{-5}$	$2.2867*10^{-5}$
Sharpe Ratio	0.2552	0.2833	0.2507
Sortino Ratio	0.3773	0.4551	0.3630
Out-of-sample			
Expected Return	$-4.5598*10^{-4}$	$5.9724*10^{-4}$	$-2.1550*10^{-4}$
Variance	$9.5717*10^{-5}$	$9.9711*10^{-5}$	$9.8766*10^{-5}$
Sharpe Ratio	-0.0467	0.0597	-0.0218
Sortino Ratio	-0.0655	0.0936	-0.0308

The first column of the table shows the results for the ant colony optimization algorithm, the second column refers to the bee colony optimization algorithm and the third one to the benchmark method. The first row of each method shows the assets that comprise the portfolios. The second row contains the capital allocations of these

---

[1] The database was provided by Kepler Asset Management LLC.

[2] The open parameters of the algorithm were selected based on a set of simulations. In future work, we intend to continue the experimentation on the configuration settings, in order to have a better insight on the validity of the presented results.

[3] Some of the results are presented in more than two decimals. This is due to the fact that if we limit these numbers to two decimals some of the results would appear to have zero values. Also, Table 5 is presented in a condensed way due to lack of space.

assets in each portfolio. As we can see, every weight is smaller than 0.60 and higher than -0.60, which means that the floor and ceiling constraints are successfully met. In addition, short selling is observed in some assets (negative weights). Then, we present the in-sample and out-of-sample results: expected return for portfolio, variance of portfolio, Sharpe ratio and Sortino ratio. Sharpe ratio measures the risk premium per unit of risk. Finally, Sortino ratio is a similar measure to Sharpe ratio. However, it takes into consideration only positive returns.

Based on the above results, we could comment on the following findings:

- The portfolios constructed by the proposed approaches are different, apart from some assets. This indicates that each method converged to a different solution.
- In terms of the value of the objective function (variance), the BCO achieved the smallest variance among the three methods. As we can see the simplified tabu metaheuristic yields the worst result. This could be indicative of the potential of nature-inspired approaches to approximate high-quality solutions. Moreover, BCO achieves better results compared to ACO.
- BCO yields the best results in terms of Sharpe and Sortino ratio, as well.
- BCO yields the best results in the out-of-sample period, as well. However, we cannot ensure the validity and the accuracy of our proposed method in this period, due to the lack of a forecasting mechanism for the returns. This explains the unsatisfactory results.

Based on the above set of configuration parameters, BCO needed almost twice as much time as ACO, in order to reach a feasible high-quality solution. However, as we can observe by the simulation results, BCO found a better portfolio than ACO. Nevertheless, this fact could not be used as a comparison factor between the two intelligent approaches.

6 Conclusion and Further Research

In this study, we proposed a nature-inspired approach, namely BCO, for the constrained portfolio optimization problem. Also, we applied another intelligent algorithm, ACO. Finally, we used a simplified metaheuristic, tabu search, as a benchmark methodology. The present work shows that NI techniques such as BCO and ACO are able to converge faster to near-optimal solutions. Also, they are simpler to implement and sometimes they prove to be more effective in hard optimization tasks such as the portfolio optimization problem.

As future work, we intend to perform additional simulation experiments in order to define the optimal configuration settings for the nature-inspired algorithms. By doing so, we should achieve better results. Furthermore, we intend to apply and compare BCO or ACO to other formulations of the portfolio optimization problem and to other databases, as well. Finally, we intend to apply competitive approaches such as Genetic Algorithms and Genetic Programming to this kind of problem.

References

1. Chang, T.J., Mead, N., Beasley, E.J., Sharaiha, Y.M.: Heuristics for cardinality constrained portfolio optimization. Comp. & Op. Res. 27, 1271–1302 (2000)
2. Maringer, D.: Small is beautiful. Diversification with a limited number of assets (2006), http://www.essex.ac.uk/ccfea
3. Maringer, D.: Portfolio Management with Heuristic Optimization. In: Advances in Computational management Science, vol. 8. Springer, Heidelberg (2005)
4. Lazo, J.G.L., Vellasco, M.M.R., Pacheco, M.A.C.: A hybrid genetic-neural system for portfolio selection and management. In: 6th Int. Conference on Engineering Applications of Neural Networks, EANN 2000, Kingston Upon Thames (2000)
5. Streichert, F., Ulmer, H., Zell, A.: Evolutionary algorithms and the cardinality constrained optimization problem. In: Operations Research Proceedings, Int. Conference of Operations Research, pp. 253–260 (2003)
6. Xu, F., Chen, W., Yang, L.: Improved Particle Swarm Optimization for realistic portfolio selection. In: 8th ACIS Int. Conference on Software Engineering, Artificial Intelligence, Networking, and Parallel/Distributed Computing, China (2007)
7. Oh, K.J., Kim, T.Y., Min, S.: Using genetic algorithm to support portfolio optimization for index fund management. Expert Sys. with Appl. 28, 371–379 (2005)
8. Korczak, J.J., Lipinski, P., Roger, P.: Evolution strategy in portfolio optimization. In: Artificial Evolution, 5th Int. Conf., Le Creusot, France, pp. 156–167 (2002)
9. Kendall, G., Su, Y.: A particle swarm optimization approach in the construction of optimal risky portfolios. In: Proceedings of the 23rd IASTED Int. Multi-Conference: Artificial Intelligence and Applications, Innsbruck, Austria (2005)
10. Karaboga, D., Basturk, B.: On the performance of artificial bee colony (ABC) algorithm. Applied Soft Computing 8, 687–697 (2008)
11. Dorigo, M., Stutzle, T.: Ant Colony Optimization. MIT, Cambridge (2004)
12. Passino, K.M., Seeley, T.D., Vissher, P.K.: Swarm cognition in honey bees. Behavioural Ecology and Sociobiology 62, 401–411 (2008)
13. Yang, X.S.: Engineering Optimizations via nature-inspired virtual bee algorithms. In: Mira, J., Álvarez, J.R. (eds.) IWINAC 2005. LNCS, vol. 3562, pp. 317–323. Springer, Heidelberg (2005)
14. Markowitz, H.: Portfolio Selection: efficient diversification of investments, 2nd edn. B. Blackwell, Cambridge (1991)

CLIVE – An Artificially Intelligent Chat Robot for Conversational Language Practice

John Zakos[1,2] and Liesl Capper[1]

[1] MyCyberTwin, Sydney, Australia
{j.zakos,l.capper}@mycybertwin.com
[2] Griffith University, Gold Coast, Australia
j.zakos@griffith.edu.au

Abstract. This paper presents an artificially intelligent chat robot called CLIVE. The aim of CLIVE is to provide to a useful and engaging method for people learning a foreign language, to practice their conversational skills. Unlike other systems that focus on providing a limited or structured tutoring experience for language learning, CLIVE has the ability of holding open, natural human-like conversations with people on a wide range of topics. This provides users with a life-like experience that is a more natural way of learning a new language. Experiments were conducted between CLIVE and real human users and an analysis of the conversations shows that CLIVE performs with accuracy and is an accepted method of language practice amongst users.

Keywords: Artificial intelligence, chat robot, language, conversation, intelligent interaction, humanized character.

1 Introduction

Human machine conversation is still a relatively developing area of artificial intelligence, even since the invention of ELIZA [1] by Weizenbaum a number of decades ago. ELIZA was presented as a computer program that could perform natural language communication with a human. Real human users could talk to ELIZA, which was initially created to respond like a psychotherapist, and have an engaging and meaningful conversation with her. Since then, artificial intelligent conversational systems, called chat robots or 'chatbots', have been used for a range of applications and topic areas including customer service [2], entertainment [3], religion [4] and intelligent tutoring [5]. Education is an important area in which chat robots have and can make significant contributions to learning [6]. In particular, people learning a new language are faced with challenges in improving and developing their conversational skills. Not only are language learners required to gain an understanding and familiarity with vocabulary, grammar, pronunciation and comprehension, but they are also required to develop their social and conversational skills. Applications of chat robot technology, such as the Let's chat system [7], can help people in developing their conversational language skills. Let's Chat is a prototype chat robot system recently

J. Darzentas et al. (Eds.): SETN 2008, LNAI 5138, pp. 437–442, 2008.

proposed that simulates a social and conversational environment with a person. The premise is that by conversationally interacting with Let's Chat in a natural and humanized way, people can acquire good new language skills that they can then utilize as confident participants in real life conversation.

Following this notion, CLIVE is proposed as an intelligent chatbot that can hold conversations with people for the purpose of effective, natural language learning. Unlike Let's Talk, that is a prototype with limited interaction capabilities and knowledge areas, CLIVE is presented as a very capable, interesting and engaging chat robot personality that people can naturally interact with for effective for language learning. A key aspect of CLIVE is that ability to chat with foreign speaker who have no or very limited knowledge of the language being learned. This paper presents CLIVE as a capable and new method of language learning.

The remainder of this paper is organized into 4 further sections. The next section describes CLIVE and the methodology used to create him. Section 3 presents the experimental setup. Section 4 presents the results of the experiments and an analytical discussion. In section 5 a conclusion is drawn and an outline for future research is described.

2   CLIVE

CLIVE is an artificially intelligent chatbot that has the ability of holding real conversations with people, via an instant messaging user interface, for the main purpose of conversational language practice.

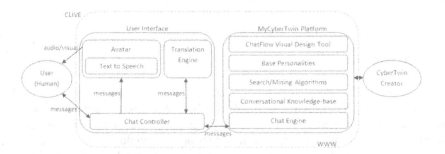

Fig. 1. Overview of the technology architecture for CLIVE

He was created using the artificial intelligence chatbot technology platform MyCyberTwin [8]. MyCyberTwin provides a toolbox of AI features that are essentially responsible for providing CLIVE perceived intelligence and ability to chat. Figure 1 presents an overview of CLIVE's web architecture.

2.1   User Interaction

Users can chat to CLIVE via an instant messaging interface, sending him textual input messages and receiving text and voice audio responses. CLIVE has the ability to

Fig. 2. The user interface to CLIVE for native Greek speakers learning English

understand any of the world's major languages, making his application to international audiences possible. A key feature of CLIVE is his ability to accept an input message in a given language *A* and respond in another language *B*. This makes it possible for native speakers of language *A* to learn language *B*, with limited or no prior knowledge of language *B*.

While CLIVE can be configured to handle any language pairs, the remainder of this section describes the user interface for CLIVE from the perspective of a Greek native chatting to an English speaking CLIVE to practice their English conversational skills. Figure 2 shows how the user interface to CLIVE looks to a native speaker of Greek who is learning English. At the top of the interface, CLIVE is presented as a life-like 3D avatar that blinks, talks and moves his head in a natural way, imitating human form. Below the avatar is the chat history box, that shows the running history of the conversation between CLIVE and the user. Below this is the user input message box, where a user enters the input message to send to CLIVE.

In Figure 2, CLIVE is setup to respond in English to native Greek speakers. This means that a Greek user can send CLIVE a message in either Greek of English during the conversation. If the user's level of English is good, then they may choose to enter English when they are feeling confident and familiar in how to express themselves. Alternatively, they can enter their message to CLIVE in their native language of Greek. Regardless of whether the message is submitted in English or Greek, CLIVE will respond to the user in English but provide a translation of the response in Greek. This Greek translation can be viewed by the user when they clicking on the orange translate button on the right hand side of the line in the chat history. When a user enters an input message in their native language Greek, the chat controller automatically detects this and makes a request to the translation engine for it to be translated into English, since this is the language that CLIVE is configured in. The English translation is returned to the chat controller and it is then sent to the MyCyberTwin platform that in turn responds with a response in English and it is this message that is shown back to the user as CLIVE's response. CLIVE also provides an audio speech utterance of the response, so the user can hear the pronunciation of the sentence and each of the words within. The combination of this audio utterance along with the English response and its respective Greek translation, provide useful information to the user to learn the English language. Users can have a conversation with CLIVE about anything and everything. If at any stage CLIVE is not confident in responding with a specific relevant response, he will respond with a general comment and try to keep the user engaged and prolong the conversation.

2.2 MyCyberTwin

MyCyberTwin is a commercial artifical intelligence technology platform that gives its users the ability to create compelling chatbots, called CyberTwins, that have the ability to imitate a real human in holding a conversation with particularly good accuracy. The key advantage in using the MyCyberTwin platform over other technology or techniques is that it provides a powerful toolbox approach to creating an artificially intelligent chatbot. MyCyberTwin offers its users easy to use visual tools that allows chat robots to be created easily and effectivly. Upon using the MyCyberTwin platform, a creator of a CyberTwin has the core components available to use for building their CyberTwin. These include a ChatFlow visual design tool, base personalities, search/mining algorithms, a conversational knowledge-base dictionary and a chat engine. The combinational use of these core components make MyCyberTwin an ideal platform to use for the creation of an artificial intelligent character such as CLIVE. With MyCyberTwin, CLIVE could be created in a timely manner as the MyCyberTwin toolset made the chat flow design and content submission easy tasks to accomplish. But most importantly, this was because MyCyberTwin's in-built artificial intelligence capabilities removed the need for in-depth, technical design of the chatbot by the creator.

3 Experimental Setup

MyCyberTwin was used to create an English speaking CLIVE character. His base personality was set to be warm hearted and intellectual in nature. Content was added to his content profile in the form of questions and answers, statements and utterances on a range of topics. The chat flow design was created to make CLIVE be more responsive in nature rather than controlling. The main aim of experimentation was to: 1) Ascertain whether CLIVE could hold a decent conversation when input messages are given in English. 2) Ascertain whether CLIVE could hold a decent conversation when the input messages are given in Greek. 3) To gain an initial insight from users as to whether CLIVE would make useful language learning tool.

CLIVE was made available to up to 10 Greek and English speaking users via a URL that pointed to a web page that could be accessed with an Internet browser. Users were told that CLIVE is a virtual character created for conversational English practice and asked to engage in natural conversation with him, just as if they were chatting to a real human. The conversations were recorded and analyzed.

4 Results and Analysis

In most English only speaking conversation, CLIVE performed exceptionally well, holding a fairly intelligent, thoughtful and natural conversation. These conversations were considered to be the baseline to assess the quality of the Greek conversations to. Figure 3 shows a resulting conversation between CLIVE and a Greek speaker.

CLIVE: Hi! I have been waiting to chat with you. What have you been doing?
Human: ήμουνα στην θάλασσά   (*I was at the beach*)
CLIVE: You were? Why is that?
Human: μου αρέσει να κολυμπώ   (*I like to swim*)
CLIVE: It's good to have a clear idea of what you like.
Human: πες μου κάτι άλω.. για την οικογένειά σού
            (*Tell me something else... about your family*)
CLIVE: I consider myself family to all the MyCyberTwins alive. Tell me
            more about your family.

Fig. 3. Sample of a conversation between CLIVE and a Greek speaking native

In this conversation CLIVE also performed quite well, understanding the human and responding relevantly in the majority of cases. Overall, CLIVE performed exceedingly well, responding relevantly and accurately throughout the majority of conversations. In the English conversations CLIVE responded accurately over 95% of the time. Sometimes, obviously, users input obscure and complicated messages during a conversation and CLIVE is unable to comprehend the meaning.

Analysis of the Greek conversations showed that, although CLIVE performed well, he did not respond with the same level of accuracy. In nearly all of the cases where CLIVE was not accurate in responding, it was because of the lack of quality of the Greek to English translation that is performed internally by the translation engine. Since it is responsibility of the translation engine to accurately convert a Greek input message into a English message before it is sent to the MyCyberTwin chat engine for processing, it plays an important role in ensuring the quality of the chat. The limitations of the quality of translations could be overcome by developing a conversational Greek knowledge-base dictionary in MyCyberTwin. This way the CLIVE user interface can send original Greek input messages to MyCyberTwin for processing rather than relying on sending a translated English form that could be inaccurate. All of the human users commented on how useful they found CLIVE to be as a tool to practice their conversational skills. Greek users would like to see the translations improved.

Although they could understand the English to Greek translations that were not grammatically correct, they felt that accurate translations would be required to make CLIVE a more appreciated and useful system as a reliable tool for language learning.

5 Conclusion

In this paper, an artificially intelligent chatbot called CLIVE has been presented. CLIVE can talk to people and ultimately be provided as an alternative application for conversational language practice. Experimental results show that CLIVE has the ability to hold conversations to a good level of accuracy and prove to be useful and accepted tool for language practice. Future research will be focused on more formal user testing to test CLIVE's applicability and suitability across a wider user demographic and across additional language pairs. More MyCyberTwin knowledge dictionaries will be developed so that the system relies less on the translation engine to convert a message from one language to another. This includes the creation of a Greek conversational knowledge dictionary using the MyCyberTwin platform. This should improve the quality of the conversations and the overall language learning experience.

References

1. Weizenbaum, J.: ELIZA - A Computer Program For the Study of Natural Language Communication Between Man and Machine. Communications of the ACM 9(1), 36–45 (1966)
2. Stockburger, S., Fernandez, T.: Virtual Onsite Support: Using Internet Chat and Remote Control to Improve Customer Service. In: Proceedings of the 30th Annual ACM SIGUCCS Conference on User services, pp. 143–147 (2002)
3. Brooks, A., Jesse, G., Hoffman, G., Lockerd, A., Lee, H., Breazeal, C.: Robot's Play: Interactive Games with Sociable Machines. Computers in Entertainment 2(3), 10 (2004)
4. Abu Shawar, B., Atwell, E.: Accessing an Information System by Chatting. In: Proceedings of 9th International Conference on the Application of Natural Language to Information Systems, pp. 396–401 (2004)
5. Spierling, U.: Beyond Virtual Tutors: Semi-autonomous Characters as Learning Companions. In: International Conference on Computer Graphics and Interactive Techniques (2005)
6. Kerly, A., Hall, P., Bull, S.: Bringing Chatbots into Education: Towards Natural Language Negotiation of Open Learner Models. Knowledge-Based Systems 20(2), 177–185 (2007)
7. Stewart, I., File, P.: Let's Chat: A Conversational Dialogue System for Second Language Practice. Journal Computer Assisted Language Learning 20(2), 97–116 (2007)
8. MyCyberTwin, http://mycybertwin.com

Author Index